Ecology: Evolution, Biodiversity and Conservation

Ecology: Evolution, Biodiversity and Conservation

Editor: Kelsey Malone

Callisto Reference
www.callistoreference.com

Callisto Reference,
118-35 Queens Blvd., Suite 400,
Forest Hills, NY 11375, USA

Visit us on the World Wide Web at:
www.callistoreference.com

ISBN: 978-1-64116-268-5 (Hardback)

The publisher's policy is to use permanent paper from mills that operate a sustainable forestry policy. Furthermore, the publisher ensures that the text paper and cover boards used have met acceptable environmental accreditation standards.

Printed in China.

Cataloging-in-Publication Data

Ecology : evolution, biodiversity and conservation / edited by Kelsey Malone.
p. cm.
Includes bibliographical references and index.
ISBN 978-1-64116-268-5
1. Ecology. 2. Biology. 3. Evolution (Biology). 4. Biodiversity. 5. Biodiversity conservation.
6. Ecosystem management. I. Malone Kelsey.
QH541 .E26 2020
577--dc23

Table of Contents

Preface

This book has been an outcome of determined endeavour from a group of educationists in the field. The primary objective was to involve a broad spectrum of professionals from diverse cultural background involved in the field for developing new researches. The book not only targets students but also scholars pursuing higher research for further enhancement of the theoretical and practical applications of the subject.

The field of study that focuses on the interactions among organisms and their environment is called ecology. It is a branch of biology that overlaps with the disciplines of genetics, ethology and evolutionary biology. The applications of ecology are prevalent in the fields of conservation biology, city planning, wetland management, community health, natural resource management and human social interaction. The understanding of evolution, biodiversity and the need for conservation is vital for a complete understanding of ecology. Adaptation, natural selection, inheritance, populations and life history are some of the concepts intrinsic to both evolutionary biology and ecology. Biodiversity is the variety of life and life's processes. It encompasses the genetic differences among living organisms, the ecosystems and communities they inhabit and the evolutionary and ecological processes required for their subsistence. The conservation and management of biodiversity requires innovative approaches to address the ecological scope of biodiversity. This book contains some path-breaking studies in the field of ecology. It is a valuable compilation of topics, ranging from the basic to the most complex theories and insights in evolution, biodiversity and conservation. It is an essential guide for both academicians and those who wish to pursue this discipline further.

It was an honour to edit such a profound book and also a challenging task to compile and examine all the relevant data for accuracy and originality. I wish to acknowledge the efforts of the contributors for submitting such brilliant and diverse chapters in the field and for endlessly working for the completion of the book. Last, but not the least; I thank my family for being a constant source of support in all my research endeavours.

Editor

Forest Structure and the Species Composition of the Parque Estadual Mata Atlântica, Located in Goiás State, Brazil

Ismael Martins Pereira,[1] Jovan Martins Rios,[1] Vagner Santiago do Vale,[1] Vera Lúcia Gomes-Klein,[2] Ademilson Coneglian,[1] and Ednaldo Cândido Rocha[1]

[1] *Universidade Estadual de Goiás (UEG), Câmpus Ipameri, Ipameri, GO, Brazil*
[2] *Department de Botânica, Universidade Federal de Goiás (UFG), Avenida Esperança, S/N, Câmpus Samambaia (Campus II), Goiânia, GO, Brazil*

Correspondence should be addressed to Ismael Martins Pereira; ismaelmpufg@gmail.com

Academic Editor: L. M. Chu

The Parque Estadual Mata Atlântica (PEMA), Goiás state, Brazil, is an Atlantic Forest reserve which has not been studied yet. The aims were to investigate plant biodiversity and characterize the flora to support the management plan. We adopt the point-centered quarter method to inventory the tree-shrub and lianas with the diameter at breast height more than or equal to 5 cm. The study was performed through 1.86 km of transect with 187 sampled points. We sampled 748 individuals and identified 109 species, 82 genera, and 51 families. The Shannon-Wiener diversity index (H') was 3.97, and Pielou's evenness (J) was 0.846, both considered high. The Fabaceae family was the most diverse with 20 species. *Aspidosperma polyneuron*, an endangered tree species, had greater relative dominance and importance value. The floristic similarity analysis indicated that the PEMA vegetation belongs to the seasonal forest, currently very fragmented and degraded. Due to several anthropogenic threats, we suggest the following recommendations: prevent the exploitation of species, avoid hunting, control invasive grasses, prevent forest fires, and promote environmental education. This study demonstrates the importance of PEMA for species conservation. Therefore, this study is essential for the development of management plan and the conservation of PEMA biodiversity.

1. Introduction

The Atlantic Forest Domain (AFD) is the second largest American rainforest [1], recognized as a global biodiversity hotspot [2]. This forest originally covered ca. 148 million hectares; approximately 92% is distributed in 17 Brazilian states, and the remainder belongs to Argentina and Paraguay [3, 4].

The classification of Brazilian vegetation recognizes two groups in the Atlantic Forest Domain: rainforests and seasonal forests [5]. The rainforests occur near the coast. The Seasonal Dry Neotropical Forests (SDNF) are inland in the country as enclaves in the Cerrado and Caatinga biomes [5, 6]. These forests occur in disjunct patches scattered throughout the Neotropic, and the major biomes where SDNF is in contact are the savanna woodland (e.g., the Cerrado), lowland tropical rainforest, and montane forest [7]. An estimate indicates that the extent of SDNF makes up about 15% of the Cerrado area (ca. 300,000 km^2) in small areas or up to tens of thousands of square kilometers, for example, "Mato Grosso Goiano" in central region of Goiás state and "Triângulo Mineiro" in Minas Gerais state [8]. Other disjunctions of SDNF in the valleys of São Francisco and Paraná rivers in Goiás and Tocantins states are not measured [9].

Despite the importance of the Atlantic Forest, it is very fragmented and degraded, and therefore the remnants are important for conservation [10]. Thus, the conservation of AFD becomes a challenge due to various factors (fires, deforestations, and fragmentation), especially because of the high level of human disturbance [11]. It is necessary to conduct floristic studies and to create new reserves, especially in higher diversity areas [12–15].

The aims of the present study were to inventory the flora of tree-shrubs and lianas of Parque Estadual Mata Atlântica (PEMA), located in the Água Limpa municipality, Goiás state, Brazil, to improve the knowledge of local flora, and to support the management plan of PEMA.

2. Materials and Methods

2.1. Characterization of the Study Area. PEMA was established by Decree 6.442 of April of 2006. The PEMA is managed by "Secretaria do Meio Ambiente, Recursos Hídricos, Infraestrutura, Cidades, e Assuntos Metropolitanos (SECIMA)" of Goiás State, Brazil. It covers about 940 hectares and is located in the municipality of Água Limpa, southern region of Goiás state (18°21′03″ S; 48°76′85″ W) (Figure 1).

The average elevation of PEMA is 611 m, bordered by Brisas Reservoir that belongs to the Itumbiara hydroelectric dam, Goiás, Brazil (Figure 1). The soils are Dystrophic Litholic Neosols-Eutrophic Haplic Cambisols to red Eutrophic Argisols [16]. The climate is characterized by two distinct seasons, a hot and rainy season from November to March and a dry season from April to October [17]. The climate is classified as Aw (Köppen) (tropical with summer rains). The average annual temperature is 23.5°C, and average annual rainfall is 1.314 mm for Água Limpa city, Goiás [17].

2.2. Vegetation Inventory and Data Analysis. We used the point-centered quarter method (PCQM) for phytosociological survey. The distance between sampling points was 10 m, totaling 187 sampling points in 1.86 km of nine transects. We registered the distance from the point, height, and diameter from all trees, shrubs, and lianas, live or dead, whose diameter at breast height (DBH) was equal to or larger than 5 cm. We analyzed the following phytosociological parameters: Shannon-Wiener diversity index (H'), Pielou's evenness (J) [18], relative density (RD), relative dominance (RDo), relative frequency (RF), and importance value (IV) [19].

The species accumulation curve was constructed to evaluate sampling sufficiency. The estimation of species richness was performed using the Jackknife 1 estimator [20] in the EstimateS program version 9.1 [21]. These results were used for construction of observed (Mao Tao) and estimated species accumulation curves.

The presence/absence matrix for all species founded in PEMA was compared with other domains: Atlantic Forest Domain (AFD), including Seasonal Dry Neotropical Forests (SDNF), Cerrado, and Amazonian Rainforest (ARF). The categorization of the main phytophysiognomy in which the species occurs was based on the literature records [8, 22–24]. This data matrix was used to perform a cluster analysis using the Jaccard Similarity Index as a measure of similarity between PEMA and the other biomes by Unweighted Pair Group Method with Arithmetic Mean (UPGMA). Statistical analyses were done in R program (R Core Team, 2017) (https://www.R-project.org/), using the "vegan" package (Oksanen et al., 2017) (https://CRAN.R-project.org/package=vegan) (last access in 04/09/2017).

The individuals were identified in situ or by botanical samples in laboratory analysis. The fertile botanical material will be deposited in the Universidade Estadual de Goiás (UEG) herbarium. Species nomenclature was standardized according to the Plant List (http://www.theplantlist.org/), the International Plant Name Index (http://www.ipni.org), and Jardim Botânico do Rio de Janeiro (http://floradobrasil.jbrj.gov.br/reflora/floradobrasil/FB90947). The APG IV was used for the botanic classification of species, genera, and families [25].

3. Results

The study was performed through 1.86 km of transect with 187 sampled points. We sampled 748 individuals and identified 109 species, 82 genera, and 51 families. The Shannon-Wiener diversity index (H') was 3.97, and Pielou's evenness (J) was 0.846. The estimated richness was higher than the observed richness (Figure 2).

The most representative families in species number were Fabaceae (20), Malvaceae (7), Myrtaceae (6), Apocynaceae (5), Bignoniaceae, Euphorbiaceae, Rubiaceae, and Sapindaceae, with four species each. Chrysobalanaceae, Meliaceae, Moraceae, Rutaceae, and Salicaceae were represented by three species each, totaling 63% of all identified species (Figure 3).

The species with the highest relative density (RD) were *Casearia gossypiosperma* (7.63%), *Myrcia splendens* (5.62%), *Metrodorea stipularis* (4.42%), *Myracrodruon urundeuva* (3.75%), *Syagrus oleracea* (3.75%), *Anadenanthera colubrine/Myrcia rostrata* (3.61%), *Trichilia claussenii* (2.75%), *Apuleia leiocarpa* (2.68%), and *Casearia rupestris* (2.41%). These species contributed with approximately 40.43% of the living individuals (Table 1). Another 39 species contributed with only a single individual. The four species with highest relative density (RD) are the same with regard to relative frequency (RF): *C. gossypiosperma* (6.12%), *M. splendens* (5.65%), *M. stipularis,* and *S. oleracea* (3.93%). The species with highest values of the relative dominance (RDo) were *Aspidosperma polyneuron* (15.04%), followed by *A. colubrina* (10.56%), *Hymenaea courbaril* (7.99%), *Ceiba speciosa* (6.68%), *Enterolobium contortisiliquum* (5.36%), *Cariniana estrellensis* (5.06%), *Attalea phalerata* (4.24%), *Apuleia leiocarpa* (3.28%), *Aspidosperma discolor* (3.27%), and *M. urundeuva* (2.38%). The sum of top 10 species in terms of IV corresponded to 43.21% from the total. Dead individuals presented the highest IV and the second largest relative density (RD = 7.23).

Six endangered plants were identified in the PEMA* (*A. leiocarpa*, garapa; *Astronium fraxinifolium*, gonçalo alves; *A. polyneuron*, peroba rosa; *Bowdichia virgilioides* Kunth, sucupira preta; *Cedrela fissilis*, cedro; and *M. urundeuva*, aroeira) (Table 1).

Cluster analysis of floristic similarities showed that about 80% of the PEMA species were shared with the SDNF. The PEMA species compared to other biomes showed two well-defined groups (PEMA-Dry Neotropical Seasonal Forest and AFD), indicating >75% of species similarity from PEMA with AFD (Figure 4). Additionally, 68% of the PEMA species also occur in Cerrado (Figure 4). The group including only

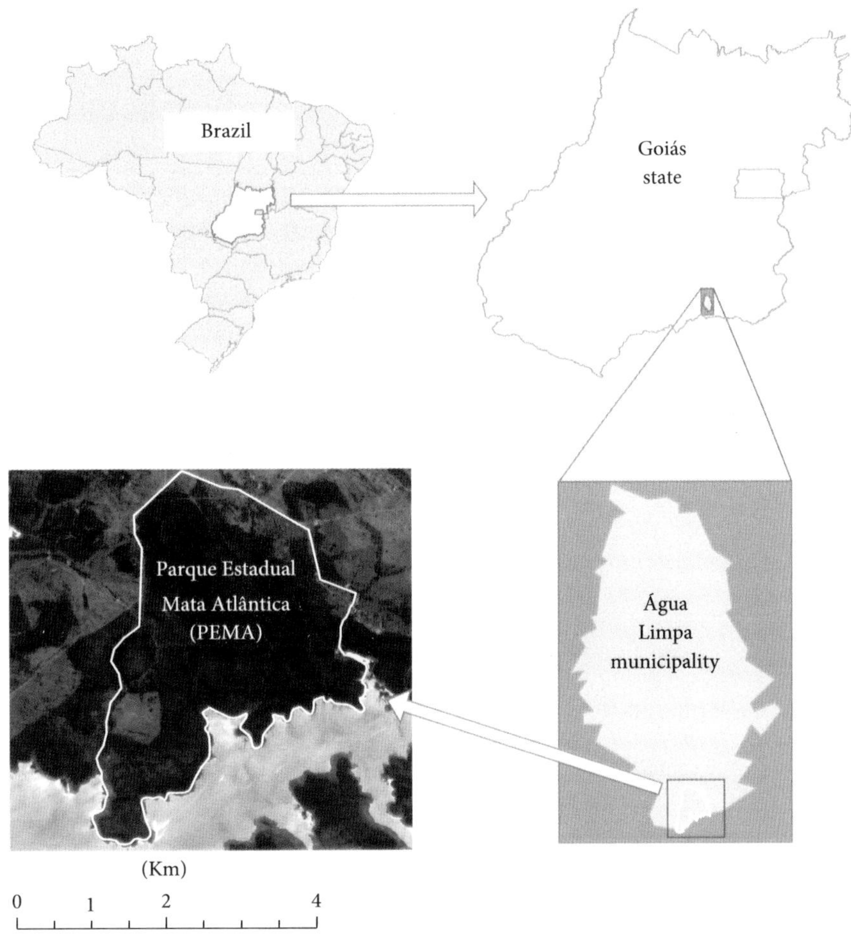

FIGURE 1: Parque Estadual Mata Atlântica (PEMA) in Água Limpa municipality, Goiás state, Brazil. Satellite image: Google Earth, 2016.

the Amazon Rainforest suggests that only about 27% of the PEMA species also occur in the Amazon (Figure 4).

4. Discussion

Previous phytosociological studies in the Seasonal Dry Neotropical Forests indicated index rarely exceeding 3.5 [26]. However, the Shannon-Wiener diversity index obtained in this study ($H' = 3.97$), as well the evenness ($J = 0.846$), was higher. These values are equivalent to other studies in well-preserved semideciduous seasonal forest remnants [26]. For example, Pereira et al. [27] obtained a diversity index of $H' = 3.58$ in a seasonal forest located in the same river basin of the PEMA. Thus, PEMA has a great biodiversity in the shrub-tree strata [28]. Moreover, the diversity estimator proposed by Heltshe and Forrester [20] indicated that the estimated richness was higher than observed richness (Figure 2). Besides, the sample sufficiency measured by the species accumulation curve indicates that it has not reached an asymptote (see Figure 2). Schilling and Ferreira [29] reported that community boundaries in tropical forests are difficult to define, especially due to high species richness, so the curve does not stabilize even with large samples.

The Fabaceae family presented the greater diversity, since it is a characteristic family from seasonal forests, which is well represented in most studies in Atlantic Forest Domain [29, 30]. In PEMA, the common pattern for forests was observed, where few species are abundant and many species are rare [30]. For example, the 10 species with the highest relative density contributed approximately 40.43% of the living individuals. Another 39 species contributed to only one individual. However, many of these species can be considered numerically rare only for a given area at a given time but not necessarily from a biological viewpoint [31]. For example, the dominant species *Aspidosperma polyneuron* is common in semideciduous seasonal forests (SSF) [32, 33] but is listed as endangered species because of its high economic value and declining population [34, 35].

A total of 54 dead individuals represented about 7% of the sampled plants and the highest IV (18.87) and RF (7.85) and second largest RD. All these values are considered relatively high when compared to those found in riparian forest (IV = 9.92; RD = 3.88) [27]. The high rate of dead individuals resulted from disturbances like fires, since their thin bark does not tolerate high temperatures; therefore, tree mortality influences the forest dynamics [36].

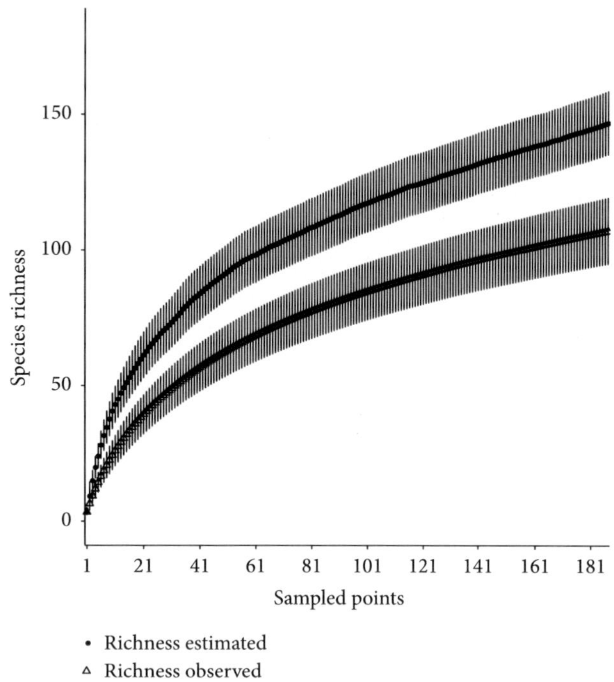

FIGURE 2: Accumulation curves of observed and estimated species from Parque Estadual Mata Atlântica (PEMA). Dots and triangles indicate the species richness; bars indicate confidence intervals ($p < 0.05$).

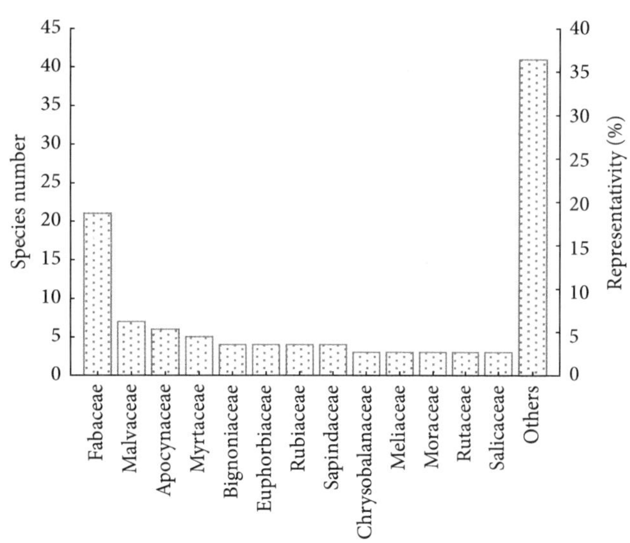

FIGURE 3: The most representative families in species number of the Parque Estadual Mata Atlântica (PEMA), Goiás, Brazil.

For the lianas, the relative density (RD = 4.15%) and importance value (IV 8.95) were relatively low, totaling only 31 individuals. However, the lianas can be represented by high percentages (15–25%) of the individuals in tropical forests [37]. In the Amazon, the lianas percentages may reach up to 44% of the forest species, caused mainly by changes in environmental gradients [38]. Thus, like the dead individuals, the lianas influence the dynamics, structure, and the diversity of tropical forests [39]. The increase in density of lianas, mainly due to forest disturbances, becomes an impact indicator [38, 40]. Although the high densities of lianas can be considered a negative factor to forest [41, 42], lianas are important component of forest ecosystems, originally recognized by Darwin.

The results of cluster analysis showed that about 80% of the PEMA species were common to the SDNF, and a similarity >75% of the PEMA species also occurred in AFD (Figure 4). PEMA is a transitional area between Cerrado and AFD. Thus, PEMA is clearly SDNF belonging to the inland Atlantic Forest Domain [26]. This forest is conditioned to dual climate seasonality: a season with intense summer rains, followed by a dry period [26, 43]. A total of 68% of the PEMA species also occurred in Cerrado. The similarity of SDNF to Cerrado was already reported by Françoso et al. [30] who found that the Atlantic Forest Domain has the greatest number of flora shared with the Cerrado, and it is the most important neighboring biome, influencing the Cerrado floristic composition.

The Atlantic Forest Domain is one of the most threatened tropical biomes, with much of the forest in small and isolated remnants (less than 50 ha) [44]. In general, the AFD is extremely fragmented, with only 11.7% from its original cover [4], mainly the Seasonal Dry Neotropical Forests [12]. This forest destruction is caused mainly by agricultural expansion [45], occurring in Brazilian central region and ecotones areas of Cerrado and Atlantic Forest Domain [28], resulting in biodiversity loss [11]. This study emphasizes the importance of floristic inventories and biodiversity assessments, necessary to understanding the current richness and conservation of forest biodiversity [43]. Therefore, the biological knowledge is important for guidance of conservation polices [30], and a holistic approach is necessary to solve the conservation problems [46]. Thus, PEMA is extremely important for biological conservation due to attributes such as high biodiversity. For example, in the fauna group, PEMA houses at least 23 species of medium and large mammals, seven of which are listed as endangered in Brazil [47]. For the plant group, this reserve has at least six endangered species [48]. Therefore, this protected area shelters 13 endangered species of plants and animals, showing their importance for biodiversity maintenance.

5. Conclusions

PEMA is an important reserve for the conservation of the biodiversity of Seasonal Dry Neotropical Forests, currently threatened by human activities. The reserve contains at least 13 threatened species of plants and animals. In addition, it has a high biodiversity and, because this forest is well conserved, is a priority in conservation and management actions. Thus, this study provides data for the development of a management plan and the conservation of PEMA biodiversity.

Table 1: Phytosociological features of Parque Estadual Mata Atlântica (PEMA), Goiás.

Species	Biome	Family	IN	RD%	RDo%	RF%	IV
Dead individuals	—	—	54	7, 23	3.64	7.85	18.72
**Aspidosperma polyneuron* Müll.Arg.	1, 2	Apocynaceae	10	1.34	15.04	1.41	17.79
Anadenanthera colubrina (Vell.) Brenan	1, 2, 3	Fabaceae	27	3.62	10.56	3.14	17.31
Casearia gossypiosperma Briq.	1, 2, 3	Salicaceae	57	7.63	2.35	6.12	16.10
Myrcia splendens (Sw.) DC.	1, 2	Myrtaceae	42	5.62	1.03	5.65	12.31
Metrodorea stipularis Mart.	1, 2, 3	Rutaceae	33	4.42	1.96	3.93	10.31
Hymenaea courbaril L.	1, 2, 3	Fabaceae	7	0.94	7.99	0.94	9.87
Cariniana estrellensis (Raddi) Kuntze	1, 2, 3, 4	Lecythidaceae	16	2.14	5.06	2.51	9.72
Ceiba speciosa (A. St.-Hil.) Ravenna	1, 2, 3	Malvaceae	8	1.07	6.68	1.26	9.00
Cipós or lianas	—	—	31	4.15	0.41	4.40	8.95
**Apuleia leiocarpa* (Vogel) J.F.Macbr.	1, 2, 3	Fabaceae	20	2.68	3.28	2.98	8.94
**Myracrodruon urundeuva* Allemão	1, 2, 3	Anacardiaceae	28	3.75	2.38	2.67	8.80
Syagrus oleracea (Mart.) Becc.	2	Arecaceae	28	3.75	0.88	3.93	8.55
Myrcia rostrata DC.	—	Myrtaceae	27	3.62	1.14	3.14	7,90
Attalea phalerata Mart. ex Spreng.	1, 2, 3, 4	Arecaceae	10	1.34	4.24	1.41	6.99
Enterolobium contortisiliquum (Vell.)	1, 2, 3	Fabaceae	5	0.67	5.36	0.79	6.82
Trichilia claussenii C. DC.	1, 2, 3	Meliaceae	22	2.95	0.84	2.67	6.46
Terminalia argentea Mart.	2, 3, 4	Combretaceae	9	1.21	3.20	1.41	5.82
Aspidosperma discolor A. DC.	1, 2, 3, 4	Apocynaceae	7	0.94	3.27	1.10	5.30
Croton floribundus Spreng.	1, 2	Euphorbiaceae	14	1.87	1.40	1.88	5.16
Casearia rupestris Eichler	1, 2, 3	Salicaceae	18	2.41	0.36	2.20	4.97
Trichilia catigua A. Juss.	1, 2, 3, 4	Meliaceae	17	2.28	0.42	2.04	4.74
Piptadenia gonoacantha (Mart.)	1, 2, 3	Fabaceae	13	1.74	0.97	1.88	4.60
Jacaratia spinosa (Aubl.) A.DC.	1, 2, 3	Caricaceae	8	1.07	1.54	1.26	3.87
Aspidosperma subincannum Mart.	2, 3	Apocynaceae	13	1.74	0.33	1.57	3.64
Aspidosperma eburneum Allemão ex Sal.	1, 4	Apocynaceae	6	0.80	1.57	0.94	3.31
Celtis iguanaea (Jacq.) Sarg.	1, 2, 3	Cannabaceae	10	1.34	0.35	1.26	2.94
Dilodendron bipinnatum Radlk.	2, 3	Sapindaceae	8	1.07	0.57	1.26	2.90
Tabebuia roseoalba (Ridl.) Sandwith	1, 2	Bignoniaceae	10	1.34	0.33	1.10	2.77
Unidentified species 5	—	—	8	1.07	0.65	0.63	2.35
Licania alba (Bernoulli) Cuatrec.	—	Chrysobalanaceae	8	1.07	0.28	0.94	2.29
Apeiba tibourbou Aubl.	2, 3, 4	Malvaceae	6	0.80	0.53	0.94	2.28
Cordiera sessilis (Vell.) Kuntze	2, 3	Rubiaceae	6	0.80	0.21	0.94	1.95
Copaifera langsdorffii Desf.	1, 2, 3	Fabaceae	4	0.54	0.68	0.63	1.84
Guazuma ulmifolia Lam.	1, 2, 3, 4	Sapotaceae	5	0.67	0.38	0.79	1.84
**Cedrela fissilis* Vell.	1, 2, 3	Meliaceae	5	0.67	0.38	0.79	1.83
Schefflera morototoni (Aubl.) Maguire	1, 2, 3, 4	Araliaceae	4	0.54	0.62	0.63	1.78
Protium heptaphyllum (Aubl.) Marchand	1, 2, 3, 4	Burseraceae	4	0.54	0.60	0.63	1.77
Rhamnidium elaeocarpum Reissek	1, 2, 3	Rhamnaceae	5	0.67	0.24	0.79	1.69
Bauhinia forficata Link	1, 2, 3	Fabaceae	5	0.67	0.14	0.79	1.59
Inga edulis Mart.	1, 2	Fabaceae	5	0.67	0.29	0.63	1.59
Qualea parviflora Mart.	3	Vochysiaceae	4	0.54	0.39	0.63	1.55
Alibertia edulis (Rich.) A. Rich. ex DC.	1, 2, 3, 4	Rubiaceae	5	0.67	0.08	0.79	1.53
Pouteria torta (Mart.) Radlk.	3, 4	Sapindaceae	4	0.54	0.24	0.63	1.40
Unonopsis guatterioides (A. DC.) R.E. Fr.	1, 2, 3, 4	Annonaceae	4	0.54	0.20	0.63	1.36
Cardiopetalum calophyllum Schltdl.	1, 2, 3, 4	Annonaceae	4	0.54	0.19	0.63	1.35
Erythrina verna Vell.	1	Fabaceae	4	0.54	0.13	0.63	1.30
Eugenia leitonii D. Legrand	1	Myrtaceae	2	0.27	0.71	0.31	1.30
Ouratea castaneifolia (DC.) Engl.	1, 2, 3	Ochnaceae	4	0.54	0.10	0.63	1.26
**Astronium fraxinifolium* Schott	2, 3	Anacardiaceae	4	0.54	0.06	0.63	1.22
Matayba guianensis Aubl.	1, 2, 3, 4	Sapindaceae	3	0.40	0.48	0.31	1.20
Inga vera Willd.	1, 2, 3	Fabaceae	4	0.54	0.11	0.47	1.12
Cecropia pachystachya Trécul	1, 2, 3, 4	Urticaceae	3	0.40	0.19	0.47	1.06

TABLE 1: Continued.

Species	Biome	Family	IN	RD%	RDo%	RF%	IV
Trema micrantha (L.) Blume	1, 2, 3	Cannabaceae	4	0.54	0.05	0.47	1.06
Diospyros inconstans Jacq.	1, 2, 3, 4	Ebenaceae	3	0.40	0.16	0.47	1.03
Handroanthus impetiginosus (Mart. ex DC.)	1, 2, 3	Bignoniaceae	2	0.27	0.43	0.31	1.02
Myrocarpus peruiferum L.F.	1, 2, 4	—	3	0.40	0.08	0.47	0.95
Luehea divaricata Mart. & Zucc.	1, 2, 3	Malvaceae	1	0.13	0.63	0.16	0.92
Albizia polycephala (Benth.) Killip	1, 2	Fabaceae	2	0.27	0.25	0.31	0.83
Sterculia striata A. St.-Hil. & Naudin	2, 3, 4	Malvaceae	2	0.27	0.23	0.31	0.81
Cupania vernalis Cambess.	1, 2, 3	Sapindaceae	3	0.40	0.09	0.31	0.81
Drimys brasiliensis Miers	1, 2, 4	Winteraceae	3	0.40	0.06	0.31	0.77
Sapium glandulosum (L.) Morong	1, 2, 3	Euphorbiaceae	2	0.27	0.19	0.31	0.77
Styrax camporum Pohl.	1, 2	Sterculiaceae	2	0.27	0.16	0.31	0.75
Matayba guianensis Aubl.	1, 4	Sapindaceae	2	0.27	0.14	0.31	0.72
Piper aduncum L.	1, 2, 3, 4	Piperaceae	2	0.27	0.09	0.31	0.67
Albizia niopoides (Spruce ex Benth.)	1, 2, 3	Fabaceae	2	0.27	0.08	0.31	0.67
Zanthoxylum rhoifolium Lam.	1, 2, 3	Rutaceae	2	0.27	0.06	0.31	0.65
Ficus citrifolia Mill.	1, 2, 3	Moraceae	2	0.27	0.06	0.31	0.64
Psidium cattleianum Afzel. ex Sabine	1, 2	Myrtaceae	3	0.40	0.03	0.16	0.59
Styrax pohlii A. DC.	3	Styracaceae	1	0.13	0.30	0.16	0.59
Unidentified species 3	—	—	1	0.13	0.26	0.16	0.55
Luehea grandiflora Mart.	2, 3, 4	Malvaceae	1	0.13	0.20	0.16	0.49
Sloanea guianensis (Aubl.) Benth.	1, 2	Elaeocarpaceae	1	0.13	0.19	0.16	0.48
Maclura tinctoria (L.) D.Don ex Steud.	1, 2, 3, 4	Moraceae	1	0.13	0.18	0.16	0.47
Hymenolobium petraeum Ducke	—	Fabaceae	1	0.13	0.17	0.16	0.46
Zanthoxylum riedelianum Engl.	1, 2, 3	Rutaceae	2	0.27	0.02	0.16	0.44
Nectandra lanceolata Nees & Mart.	1, 2, 3	Lauraceae	1	0.13	0.11	0.16	0.40
Andira anthelmia (Vell.) J.F.Macbr.	1	Fabaceae	1	0.13	0.10	0.16	0.39
Hirtella glandulosa Spreng.	1, 2, 3, 4	Chrysobalanaceae	1	0.13	0.08	0.16	0.37
Sorocea bonplandii (Baill.) W.C.Burger	1, 2, 3	Moraceae	1	0.13	0.06	0.16	0.35
Platypodium elegans Vogel	2, 3	Fabaceae	1	0.13	0.06	0.16	0.35
Inga cylindrica (Vell.) Mart.	1	Fabaceae	1	0.13	0.06	0.16	0.35
Aralia warmingiana (Marchal) J. Wen	1, 2, 3, 4	Araliaceae	1	0.13	0.05	0.16	0.34
Cnidoscolus urens (L.) Arthur	1, 2	Euphorbiaceae	1	0.13	0.05	0.16	0.34
Agonandra brasiliensis Miers ex Benth.	1, 2	Opiliaceae	1	0.13	0.04	0.16	0.33
Handroanthus ochraceus (Cham.) Mattos	3	Bignoniaceae	1	0.13	0.03	0.16	0.33
Roupala montana (Huber) K.S. Edwards	3	Proteaceae	1	0.13	0.03	0.16	0.33
Aloysia virgata (Ruiz & Pav.) Juss.	1, 2	Verbenaceae	1	0.13	0.03	0.16	0.32
Croton urucurana Baill.	1, 2	Euphorbiaceae	1	0.13	0.02	0.16	0.32
Unidentified species 1	—	—	1	0.13	0.02	0.16	0.31
Himatanthus sucuuba (Spruce ex Müll.)	4	Apocynaceae	1	0.13	0.01	0.16	0.30
Handroanthus vellosoi (Toledo) Mattos	1, 2, 3	Bignoniaceae	1	0.13	0.01	0.16	0.30
Hirtella glabrata Pilg.	2	Chrysobalanaceae	1	0.13	0.01	0.16	0.30
Terminalia glabrescens Mart.	1, 2, 3	Combretaceae	1	0.13	0.01	0.16	0.30
Inga sessilis (Vell.) Mart.	2, 3	Fabaceae	1	0.13	0.01	0.16	0.30
Cordiera macrophylla (K.Schum.) Kuntze	2, 3	Rubiaceae	1	0.13	0.01	0.16	0.30
Unidentified species 2	—	—	1	0.13	0.01	0.16	0.30
Aspidosperma sp.	—	Apocynaceae	1	0.13	0.01	0.16	0.30
Erythrina falcata Benth.	1, 2	Fabaceae	1	0.13	0.01	0.16	0.30
Luehea paniculata Mart.	1, 2, 3	Malvaceae	1	0.13	0.01	0.16	0.30
Piper tuberculatum Jacq.	2, 3, 4	Piperaceae	1	0.13	0.01	0.16	0.30
Tocoyena formosa (Cham. & Schltdl.)	2, 3	Rubiaceae	1	0.13	0.01	0.16	0.30
Unidentified species 4	—	—	1	0.13	0.01	0.16	0.30
Croton floribundus Spreng.	1, 2	Euphorbiaceae	1	0.13	0.01	0.16	0.30
Pseudobombax tomentosum (Mart. & Zucc.)	2, 3	Malvaceae	1	0.13	0.01	0.16	0.30

Table 1: Continued.

Species	Biome	Family	IN	RD%	RDo%	RF%	IV
Myrcia undulata O.Berg	1	Myrtaceae	1	0.13	0.01	0.16	0.30
Casearia decandra Jacq.	1, 2	Salicaceae	1	0.13	0.01	0.16	0.30
Pterodon emarginatus Vogel	1	Fabaceae	1	0.13	0.01	0.16	0.30
Dipteryx alata Vogel	1	Fabaceae	1	0.13	0.01	0.16	0.13
* *Bowdichia virgilioides* Kunt	1	Fabaceae	1	0.13	0.01	0.16	0.13

1: Atlantic Forest (AF); 2: Seasonal Dry Neotropical Forest (SDNF); 3: Cerrado; 4: Amazonian Rainforest. NI: number of individuals; RD: relative density (%); RDo: relative dominance (%); RF: relative frequency (%); IV: importance value. *Endangered species in the PEMA. The categorization of the main phytophysiognomy in which the species occurs followed Oliveira-Filho and Fontes (2000), Ratter et al. (2003), Oliveira-Filho et al. (2005), and Lorenzi (2014, 2016ab).

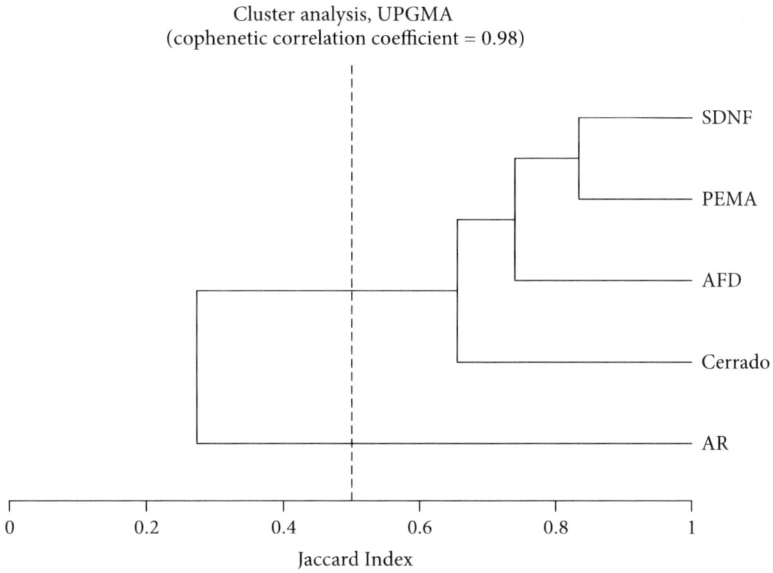

Figure 4: Cluster analysis for correlation of PEMA vegetation and Atlantic Forest Domain (AFD), Amazonian Rainforest (AR), Cerrado, and Seasonal Dry Neotropical Forest (SDNF).

Conflicts of Interest

The authors declare that there are no conflicts of interest regarding the publication of this manuscript.

Acknowledgments

The authors acknowledge "Secretaria do Meio Ambiente, Recursos Hídricos, Infraestrutura, Cidades, e Assuntos Metropolitanos (SECIMA)" for granting the license for the study of PEMA. The authors also acknowledge "Pró-Reitoria de Pesquisa e Pós-Graduação (PrP) da Universidade Estadual de Goiás (UEG)" for the scientific initiation scholarship to Jovan Martins Rios.

References

[1] M. Tabarelli, C. A. Peres, and F. P. L. Melo, ""few winners and many losers" paradigm revisited: emerging prospects for tropical forest biodiversity," *Biological Conservation*, vol. 155, pp. 136–140, 2012.

[2] N. Myers, R. A. Mittermeler, C. G. Mittermeler, G. A. B. da Fonseca, and J. Kent, "Biodiversity hotspots for conservation priorities," *Nature*, vol. 403, no. 6772, pp. 853–858, 2000.

[3] R. C. Forzza, J. F. A. Baumgratz, C. E. M. Bicudo et al., *Catálogo de Plantas e Fungos do Brasil*, Rio de Janeiro: Jardim Botânico do Rio e Janeiroe, 1st edition, 2010.

[4] M. C. Ribeiro, J. P. Metzger, A. C. Martensen, F. J. Ponzoni, and M. M. Hirota, "The Brazilian Atlantic forest: how much is left, and how is the remaining forest distributed? Implications for conservation," *Biological Conservation*, vol. 142, no. 6, pp. 1141–1153, 2009.

[5] IBGE, *Manual Técnico da Vegetação Brasileira*, Instituto Brasileiro de Geografia e Estatística, 2012.

[6] J. R. Stehmann, R. C. Forzza, A. Salino, M. Sobral, D. Pinheiro, and L. H. Y. Kamino, Eds., *Plantas da Floresta Atlântica*, Rio de Janeiro: Jardim Botânico do Rio de Janeiro, 2009.

[7] R. T. Pennington, M. Lavin, and A. Oliveira-Filho, "Woody plant diversity, evolution, and ecology in the tropics: perspectives from seasonally dry tropical forests," *Annual Review of Ecology, Evolution and Systematics*, vol. 40, no. 7th, pp. 437–457, 2009.

[8] A. T. Oliveira-Filho, J. A. Jarenkow, and M. J. N. Rodal, "Floristic relationships of seasonally dry forests of eastern South America based on tree distribution patterns," in *Neotropical Savannas and Seasonally Dry Forests: Plant Diversity, Biogeography, and Conservation*, J. A. Pennington, R. T. Lewis, and G. P. Ratter, Eds., pp. 159–192, CRS Press, Edinburg, UK, 2006.

[9] B. A. Pereira, F. Venturoli, and F. A. Carvalho, "Florestas estacionais no cerrado: uma visão geral," *Pesquisa Agropecuária Tropical*, vol. 41, no. 3, pp. 446–455, 2011.

[10] T. W. Gillespie, B. Lipkin, L. Sullivan, D. R. Benowitz, S. Pau, and G. Keppel, "The rarest and least protected forests in biodiversity hotspots," *Biodiversity and Conservation*, vol. 21, no. 14, pp. 3597–3611, 2012.

[11] B. Doug, E. Pipa, L. Katherine, M. May-Tobin, S. Roquemore, and E. Saxon, *The Root of the Problem*, UCS Publications, Cambridge, UK, 2011.

[12] R. F. Haidar, J. M. Fagg, J. R. Pinto et al., "Florestas estacionais e áreas de ecótono no estado do Tocantins, Brasil: parâmetros estruturais, classificação das fitofisionomias florestais e subsídios para conservação," *Acta Amazonica*, vol. 43, no. 3, pp. 261–290, 2013.

[13] A. M. Z. Martini, P. A. Fiaschi, A. M. Amorin, and J. L. Paixão, "A hot-point within a hot-spot: a high diversity site in Brazil's Atlantic Forest," *Biodiversity and Conservation*, vol. 16, no. 11, pp. 3111–3128, 2007.

[14] I. M. Pereira, V. L. Gomes-Klein, and M. Groppo, "Distribution and conservation of davilla (dilleniaceae) in Brazilian Atlantic Forest using ecological niche modeling," *International Journal of Ecology*, vol. 2014, Article ID 819739, 2014.

[15] M. Q. Matos and J. M. Felfili, "Florística, fitossociologia e diversidade da vegetação arbórea nas matas de galeria do Parque Nacional de Sete Cidades (PNSC), Piauí, Brasil," *Acta Botanica Brasilica*, vol. 24, no. 2, pp. 483–496, 2010.

[16] IBGE, Instituto Brasileiro de Geografia e Estatística. 2006. Mapas de Solo do Brasil. Mapas de Unidade de Relevo do Brasil. Disponível em: <ftp://geoftp.ibge.gov.br/mapas_tematicos/mapas_murais>.

[17] C. A. Alvares, J. L. Stape, P. C. Sentelhas, J. L. De Moraes Gonçalves, and G. Sparovek, "Köppen's climate classification map for Brazil," *Meteorologische Zeitschrift*, vol. 22, no. 6, pp. 711–728, 2013.

[18] A. E. Magurran, *Ecological Diversity and its Measurement*, Princeton University Press, London, UK, 1988.

[19] D. Muller-Dombois and H. Ellenberg, *Aims and Methods of Vegetation Ecology*, John Wiley & Sons, New York, NY, USA, 1974.

[20] J. F. Heltshe and N. E. Forrester, "Estimating species richness using the jackknife procedure.," *Biometrics*, vol. 39, no. 1, pp. 1–11, 1983.

[21] R. K. Colwell, A. Chao, N. J. Gotelli et al., "Models and estimators linking individual-based and sample-based rarefaction, extrapolation and comparison of assemblages," *Journal of Plant Ecology*, vol. 5, no. 1, pp. 3–21, 2012.

[22] D. A. Carvalho, A. T. Oliveira-Filho, E. d. Vilela, and N. Curi, "Florística e estrutura da vegetação arbórea de um fragmento de floresta semidecedual às margens do reservatório da usina hidrelétrica Dona Rita (Itambé do Mato Dentro, MG)," *Acta Botanica Brasilica*, vol. 14, no. 1, pp. 37–55, 2000.

[23] H. Lorenzi, *Árvores Brasileiras*, Instituto Plantarum, Nova Odessa, Brazil, 6 edition, 2014.

[24] H. Lorenzi, *Árvores Brasileiras*, Instituto Plantarum, Nova Odessa, Brazil, 7 edition, 2016.

[25] M. W. Chase, M. J. M. Christenhusz, M. F. Fay et al., "An update of the Angiosperm Phylogeny Group classification for the orders and families of flowering plants: APG IV," *Botanical Journal of the Linnean Society*, vol. 181, no. 1, pp. 1–20, 2016.

[26] S. d. Lopes, I. Schiavini, A. P. Oliveira, and V. S. Vale, "An ecological comparison of floristic composition in seasonal semideciduous forest in Southeast Brazil: implications for conservation," *International Jounal of Forest Research*, vol. 2012, 2012.

[27] I. Pereira, G. Jerônimo, K. Fonseca, and A. Coneglian, "Floristic and phytosociological study of gallery forest in Cerrado," *Revista Agrotecnologia - Agrotec*, vol. 6, no. 2, pp. 1–19, 2015.

[28] N. S. Soares, C. A. Gonçalves, G. M. Araújo, and C. Lomônaco, "Floristic composition and abundance in forest fragments: a case study from Southern Goiás, Brazil," *Bioscience Journal*, vol. 31, no. 4, pp. 1238–1252, 2015.

[29] A. C. Schilling and L. Ferreira, "Curva de acumulação de espécies e suficiência amostral em florestas tropicais," *Revista Brasileira de Botânica*, vol. 31, pp. 179–187, 2008.

[30] R. D. Françoso, R. F. Haidar, and R. B. Machado, "Tree species of South America central savanna: endemism, marginal areas and the relationship with other biomes," *Acta Botanica Brasilica*, vol. 30, no. 1, pp. 78–86, 2016.

[31] A. N. Oliveira and I. L. Amaral, "Florística e fitossociologia de uma floresta de vertente na Amazônia Central, Amazonas, Brasil," *Acta Amazonica*, vol. 34, no. 1, pp. 21–34, 2004.

[32] Y. R. Boloventa, *Populações Arbóreas Nativas Estão Regenerando no Maior Fragmento de Floresta Estacional Semidecidual do Norte do Paraná*, Universidade Estadual de Londrina, Londrina, Brazil, 2009.

[33] L. A. B. Jorge, T. M. Millani, R. C. B. Fonseca, and A. A. Arruda, "Diameter structure and spatial arrangement of the most abundant species in a seasonal semideciduous forest fragment in Botucatu, Southeastern Brazil," *Floresta e Ambiente*, vol. 22, no. 3, pp. 355–367, 2015.

[34] G. Martinelli, T. Messina, and T. M. L. S. Filho, *Livro vermelho da flora do Brasil - Plantas raras do Cerrado*, vol. 1, Rio de Janeiro: Andrea Jakobsson : Instituto de Pesquisas Jardim Botânico do Rio de Janeiro : CNCFlora, 1st edition, 2014.

[35] G. Martinelli and M. A. Morais, Livro Vermelho da Flora do Brasil,.

[36] W. A. Hoffmann, R. Adasme, M. Haridasan et al., "Tree topkill, not mortality, governs the dynamics of savanna-forest boundaries under frequent fire in central Brazil," *Ecology*, vol. 90, no. 5, pp. 1326–1337, 2009.

[37] C. B. Murray-Smith, N. A. Brummitt, A. T. Oliveira-Filho et al., "Plant diversity hotspots in the Atlantic coastal forests of Brazil," *Conservation Biology*, vol. 23, no. 1, pp. 151–163, 2009.

[38] E. A. Oliveira, B. S. Marimon, T. R. Feldpausch et al., "Diversity, abundance and distribution of lianas of the CerradoAmazonian forest transition," *Plant Ecology Diversity*, vol. 7, no. 1-2, pp. 231–240, 2017.

[39] L. G. Lohmann, C. D. Bell, M. F. Calió, and R. C. Winkworth, "Pattern and timing of biogeographical history in the Neotropical tribe Bignonieae (Bignoniaceae)," *Botanical Journal of the Linnean Society*, vol. 171, no. 1, pp. 154–170, 2013.

[40] B. L. P. Villagra, E. P. C. Gomes, R. J. Burnham, and S. R. Neto, "Diversity and abundance of climbers from the Atlantic Forest, Southeastern Brazil," *Biodiversity and Conservation*, vol. 22, no. 11, pp. 2505–2517, 2013.

[41] A. N. Oliveira, I. L. Amaral, M. B. Ramos, and K. M. Formiga, "Aspectos florísticos e ecológicos de grandes lianas em três ambientes florestais de terra firme na Amazônia Central," *Acta Amazonica*, vol. 38, no. 3, pp. 421–430, 2008.

[42] S. A. Schnitzer and F. Bongers, "Increasing liana abundance and biomass in tropical forests: emerging patterns and putative mechanisms," *Ecology Letters*, vol. 14, no. 4, pp. 397–406, 2011.

[43] S. Jayakumar, S. S. Kim, and J. Heo, "Floristic inventory and diversity assessment - a critical review," *Proceedings of the National Academy of Ecology and Environmental Sciences*, vol. 1, pp. 151–168, 2011.

[44] P. K. Lira, L. R. Tambosi, R. M. Ewers, and J. P. Metzger, "Land-use and land-cover change in Atlantic Forest landscapes," *Forest Ecology and Management*, vol. 278, pp. 80–89, 2012.

[45] B. Phalan, M. Bertzky, S. H. M. Butchart et al., "Crop expansion and conservation priorities in tropical countries," *PLoS ONE*, vol. 8, no. 1, Article ID e51759, 2013.

[46] D. Lindenmayer and M. Hunter, "Some guiding concepts for conservation biology," *Conservation Biology*, vol. 24, no. 6, pp. 1459–1468, 2010.

[47] E. C. Rocha, K. L. Soares, and I. M. Pereira, "Medium-and large-sized mammals in Mata Atlântica State Park, southeastern Goiás, Brazil," *Check List*, vol. 11, no. 6, article no. 1801, 2015.

[48] G. Martinelli and M. A. Morais, *Livro Vermelho da Flora do Brasil*, Instituto de Pesquisas do Jardim Botânico do Rio de Janeiro, 2013.

2

Benthic Macroinvertebrates Diversity as Bioindicator of Water Quality of some Rivers in East Kalimantan, Indonesia

Fatmawati Patang, Agoes Soegianto, and Sucipto Hariyanto

Department of Biology, Faculty of Sciences and Technology, Universitas Airlangga, Campus C Unair., Jl. Dr. Ir. Soekarno, Surabaya, Indonesia

Correspondence should be addressed to Agoes Soegianto; agoes_soegianto@unair.ac.id

Academic Editor: Panos V. Petrakis

The objectives of this study were to clarify and evaluate the water quality of a number of rivers in East Kalimantan province of Indonesia. For this purpose, our study successfully utilized the benthic macroinvertebrates diversity as well as physical-chemical parameters of river's water. For instance, based on the values of Average Score per Taxon (ASPT) and the National Sanitation Foundation-Water Quality Index (NSF-WQI), Karang Mumus River was categorized as polluted with *Chironomus* sp. and *Melanoides tuberculata* as codominant taxa. In addition, Jembayan River exhibited doubtful or moderate quality containing *M. tuberculata* and *A. parvula* as codominant taxa. However, Pampang River was found to be the cleanest river with Odonata and Baetidae families as codominant taxa.

1. Introduction

As open water ecosystems, rivers get strongly influenced by the surrounding environment. Water quality of a river is influenced by several parameters like land use, settlement patterns, farming, and industrial activities around that river [1]. For example, rivers in East Kalimantan province of Indonesia are also facing the problems with recent activities of residents like coal mines construction and oil-palm plantations exactly along the river banks. These activities destroy the water quality of rivers and consequently lead to a certain change in the benthic macroinvertebrates community structure. The study of biodiversity, species abundance, dominance, and distribution of macroinvertebrate fauna to determine the extent of changes in their structure and composition associated with water quality changes therefore should be conducted [1, 2].

Studying macroinvertebrate diversity is one of the most effective and inexpensive ways to estimate the ecological quality of the waters [2, 3]. For instance, measurement of the physical and chemical properties of water can also be utilized to estimate its quality but such measurements cannot exactly represent the actual state of the waters. Therefore it is necessary to combine physical, chemical, and biological evaluation along with other monitoring methods to provide a comprehensive picture of environmental water quality [4, 5]. Biological monitoring using macroinvertebrates has been found accurate and advantageous compared with using other organisms because macroinvertebrates are extremely sensitive to organic pollutants, widely distributed, and easy and economical to sample [2, 6].

Research reports on the use of macroinvertebrates to assess the water quality in aquatic ecosystems have been extensively published by several researchers [2, 3, 6–14]. For example, the use of the ecological index and the macroinvertebrate biotic index has been widely developed in America and Europe [3, 14–16]. On the other hand, to the best of our knowledge limited studies were found in literature utilizing the application of the biotic indexes to evaluate the river water quality in Indonesia and particularly in Kalimantan. Therefore, we envisioned first ever use of macroinvertebrates to assess the water quality in Indonesia to clarify and confirm the presence of harmful pollutant in Indonesian river water. For this purpose, we conducted this study to analyze and observe the changes in water quality of three rivers in East Kalimantan, Indonesia, and successfully utilized macroinvertebrates as a bioindicator of water quality. Our approach was

TABLE 1: Detailed information of sampling stations.

River's name	Sampling station	Geographical position	Surrounding environment
Karang Mumus	K-1	00°30.491′S and 117°09.426′E	Populated area, population density: 12,785 people/km^2, harbor with motor ship activities, usage of river for bathing, washing, and latrines
	K-2	00°29.073′S and 117°09.064′E	
	K-3	00°28.238′S and 117°09.442′E	
Jembayan	J-1	00°33.138′S and 117°01.100′E	Activity of coal-mining and oil-palm plantation, population density: 66 people/km^2
	J-2	00°33.057′S and 117°00.790′E	
	J-3	00°33.085′S and 117°00.574′E	
Pampang	P-1	00°19.927′S and 117°11.555′E	Natural habitat, activity of oil-palm plantation, population density: 397 people/km^2
	P-2	00°20.076′S and 117°11.874′E	
	P-3	00°19.850′S and 117°11.132′E	

FIGURE 1: Sampling locations of benthic macroinvertebrate and water quality parameters (source: Google Earth Pro 7.3.1.4507, Build Date February 6, 2018).

to determine current status of these rivers using the Shannon-Weaver diversity index [17], Average Score per Taxon (ASPT) [18], and the National Sanitation Foundation-Water Quality Index (NSF-WQI) [19].

2. Materials and Methods

2.1. Study Area. The study was conducted mainly on three rivers of East Kalimantan with the help of three sampling stations of each river (Figure 1). Selection of the sampling stations was based on the possible pollutant loads and the magnitude of human activities along the rivers. Detailed location information of these sampling sites, and the latitude and longitude of all stations, are presented in Table 1.

2.2. Sampling and Identification of Benthic Macroinvertebrates. Minimum of 100 individuals of benthic macroinvertebrates were collected, stored, and transported by each sampling station [20, 21]. Furthermore, sampling of macroinvertebrates was conducted in December 2015 (rainy season) and June 2016 (dry season). Samples of benthic macroinvertebrate were collected using Surber net (30 × 30 cm^2) for rocky substrate, Ekman-grab (25 × 25 cm^2) for muddy substrate, and kick net for habitat containing dense aquatic plants. Organisms collected were rinsed with water, separated from debris and sediment using forceps, and finally preserved in 70% ethanol. All macroinvertebrates were identified to the family level using appropriate references [22–35].

2.3. Water Quality Measurement. A wide range of water quality parameters were measured at all sampling locations, with key parameters being dissolved oxygen (DO), pH, biological oxygen demand (BOD), temperature, total phosphate, nitrate, turbidity, and total dissolved solid (TDS). The analytical methods of water quality parameters were followed by the Standard Method for the Examination of Water and Wastewater [35].

2.4. Data Analysis. After identification and enumeration of macroinvertebrates, we then calculated the diversity index (H), the dominance index (C), the Evenness index (E), the important value index (IVI) of each species, and the Average Score per Taxon (ASPT).

The Shannon and Weaver [17] diversity index was estimated by the following equation:

$$H = -\sum\left(\frac{n_i}{N}\right)\ln\left(\frac{n_i}{N}\right), \tag{1}$$

where n_i is number of individuals of ith species, N is total number of individuals, ln is the natural log, and $\sum$ is the sum of the calculations.

The Evenness index (E) is computed from Pielou's index [36]:

$$E = \frac{H}{\ln S}, \tag{2}$$

where H is Shannon–Weaner diversity index and ln S is natural log of the total number of species recorded.

In addition, Simpson's dominance index [37] was calculated by the following equation:

$$C = \sum\left(\frac{n_i}{N}\right)^2 \tag{3}$$

Importance value index (IVI) was calculated by [38]

$$\begin{aligned} IVI &= \text{Relative density} + \text{relative frequency} \\ &\quad + \text{relative dominance}, \end{aligned} \tag{4}$$

Table 2: New weight score (W_i) for 8 parameters on NSF – WQI [39].

No.	Parameter	Original weight score	Modified weight score
1	DO	0.17	0.20
2	pH	0.11	0.13
3	BOD	0.11	0.13
4	Temperature	0.10	0.12
5	Total phosphate	0.10	0.12
6	Nitrate	0.10	0.12
7	Turbidity	0.08	0.10
8	Total solid	0.07	0.08
9	Fecal coliform	0.16	-
	Total	1	1

where relative density = (density of a specie/total density of all species) × 100, relative frequency = (frequency of a specie/total frequency of all species) × 100, and relative dominance = (dominance of a specie/total dominance of all species) × 100. The value of IVI may range from 0 to 3.00 (or 300%). This value is referred to the importance percentage. The importance value or the importance percentage gives an overall estimation of the influence of importance of a species in the community.

Furthermore, the Average Score per Taxon (ASPT) represents the average tolerance score of all taxa within the community and ASPT value can be calculated by dividing the Biological Monitoring Working Party (BMWP) over the number of families represented in the sample. The BMWP system considers the sensitivity of invertebrates to pollution; families are assigned a score which is the sum of the values for all families present in the sample. Values greater than 100 are associated with clean streams [40, 41]. On the other hand, scores of heavily polluted streams are less than 10 [41]. The ASPT value equals the average of the tolerance scores of all macroinvertebrate families found and ranges from 0 to 10. The index values for ASPT are classified into four categories as follows (>6: clean water; 5-6: doubtful quality; 4-5: probable moderate pollution; <4: probable severe pollution) [16].

Also, National Sanitation Foundation-Water Quality Index (NSF-WQI) is often used to determine the level of water quality based on nine parameters such as BOD, DO, nitrate, total phosphate, temperature, turbidity, total solids, pH, and fecal coliform. In this study, eight parameters were applied without fecal coliform; hence, there was a modification of weight as shown at Table 2. The modification was allowed if the water quality parameter number was reduced and modified total weight score remained 1. Weight score modification of each parameter was proportional to its original weight score [39]. Furthermore, the weight score of each parameter (W_i) was multiplied by the subindex value of each parameter (L_i). For obtaining score of subindex, we used the online NSF-WQI Calculator at www.water-research.net/watrqualindex/waterqualityindex.htm. Finally, scores from all of parameters are summed up using the following formula:

$$NSF\text{-}WQI = \sum_{i=0}^{n} W_i L_i, \quad (5)$$

where NSF-WQI is Water Quality Index Score, W_i is the weight score of ith parameter, and L_i is the subindex value of ith parameter.

The index of water quality (NSF-WQI) values is classified into five categories as follows: 0-25: very bad; 26-50: bad; 51-70: medium or moderate; 71-90: good; >91-100: excellent [19]. A Principal Component Analysis (PCA) with a focus on sampling site was performed to examine the relationships between biotic measures used (H, C, E, and ASPT) and physical-chemical variables (BOD, DO, nitrate, total phosphate, temperature, turbidity, total solids, and pH) [42, 43]. The PCA was carried out using the open source software (PAST program Version 3 b7).

3. Results and Discussion

The diversity index (H), the dominance index (C), and the Evenness index (E) are frequent tools to predict the conditions of an aquatic environment based on the biological components. Figure 2 shows that the highest H value of benthic macroinvertebrates in December 2015 was recorded at P-3 (2.602) and the lowest H at K-2 (0.383). The highest value of C was noted at K-2 (0.776) and lowest at P-3 (0.09). The highest E was found at J-2 (0.982) and the lowest E at K-2 (0.552). In June 2016, the highest H value was observed at J-2 (1.941) and the lowest H at K-2 (0.2). The highest C was recorded at P-1 and the lowest C at K-2. The highest E was noted at J-1 and lowest E at K-2 (Figure 3).

In December 2015, the highest number of taxa (taxa richness) of benthic macroinvertebrates was found in the Pampang River (21 taxa) and the lowest in the Karang Mumus River (4 taxa) (Figure 4). In June 2016, the highest macroinvertebrate taxa was found in the Pampang River (14 Taxa) and lowest in the Karang Mumus River (6 Taxa) (Figure 4).

The important value index (IVI) of each species is presented in Table 3. During sampling in December 2015, *Chironomus* sp. and *Melanoides tuberculata* were codominant taxa in Karang Mumus River. Four taxa: *Melanoides tuberculata*, *Macrobrachium* sp., *Acentrella parvula*, and *Chironomus* sp., were codominant in Jembayan River. In Pampang River, the codominant taxon was *Coenagrion* sp. During sampling period of June 2016, *Melanoides tuberculata* and *Chironomus* sp. were codominant in Karang Mumus River.

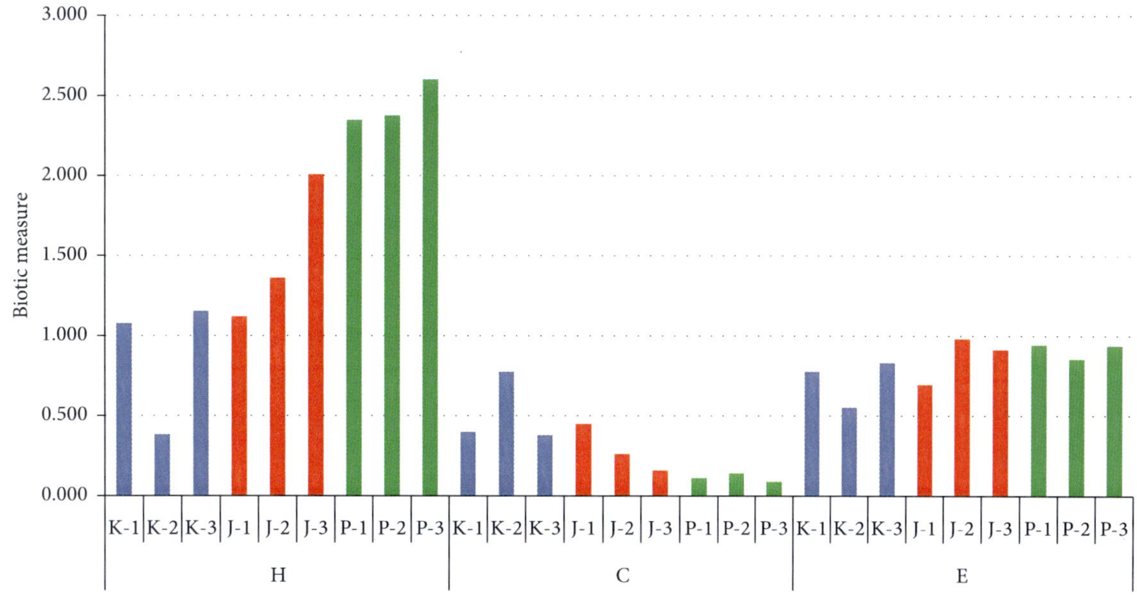

FIGURE 2: Diversity Index (H), Dominance Index (C), and Evenness Index (E) in December 2015.

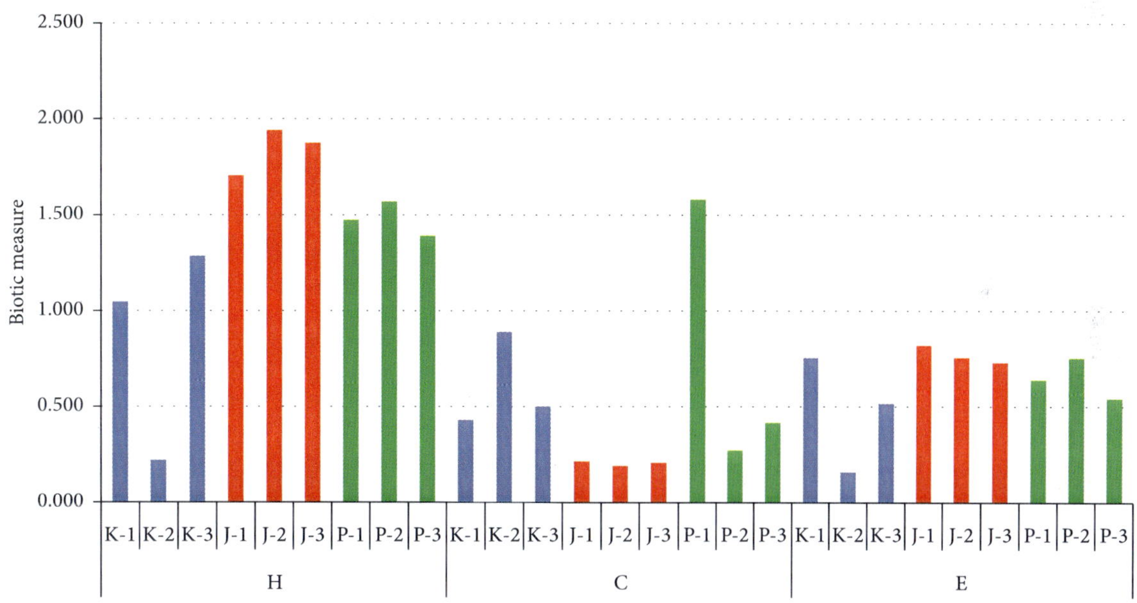

FIGURE 3: Diversity Index (H), Dominance Index (C), and Evenness Index (E) in June 2016.

In the Jembayan River, we noted that *Melanoides tuberculata*, *Acentrella parvula*, and *Baetis flavistriga* were codominant taxa. *Acentrella parvula* and *Baetis flavistriga* were also codominant taxa in Pampang River.

Water quality measurements based on biotic index (ASPT) are shown in Table 4. In December 2015, according to ASPT criteria, Karang Mumus River was in moderate to severe polluted category, Jembayan River in doubtful quality, and Pampang River in doubtful to clean quality. In June 2016, Karang Mumus River was in doubtful to severe polluted category, Jembayan River was in doubtful quality, and Pampang River was in clean quality.

Water quality parameters in December 2015 and June 2016 of each station are presented in Tables 5 and 6. Based on the WQI values, in December 2015, the water quality of Karang Mumus River was in bad condition, and Jembayan and Pampang Rivers were in moderate condition, respectively (Table 5). In June 2016, Karang Mumus, Jembayan, and Pampang Rivers were in moderate condition (Table 6).

According to the PCA, during the rainy season (October 2015), sampling sites J-1, J-2, and J-3 (Jembayan) show a strong affinity to high ASPT, pH, DO, and E values. Sampling sites P-1, P-2, and P-3 (Pampang) demonstrate a strong affinity to high H and nitrate values, while sampling sites K-1, K-2,

TABLE 3: Important value index (IVI) of benthic macroinvertebrate.

Species	Important Value Index (%)					
	December 15			June 16		
	Karang Mumus	Jembayan	Pampang	Karang Mumus	Jembayan	Pampang
Chironomus sp.	97.33	26.49	13.58	48.62	13.91	17.25
Melanoides tuberculata	44.07	50.06	13.39	68.32	42.26	4.57
Acentrella parvula	17.57	27.19	8.01	7.30	25.31	57.38
Thiara scabra	14.50	0.00	0.00	7.77	0.00	0.00
Gomphus sp.	13.88	0.00	0.00	6.59	0.00	12.01
Brotia testudinaria	12.65	0.00	8.20	7.77	11.57	0.00
Macrobrachium sp.	0.00	33.86	0.00	6.59	12.06	0.00
Aeshna sp.	0.00	14.99	0.00	0.00	0.00	0.00
Lestes sp.	0.00	13.34	8.39	0.00	14.83	12.84
Ceriagrion tenellum	0.00	7.61	3.62	0.00	0.00	0.00
Coenagrion sp.	0.00	7.38	29.13	0.00	0.00	13.86
Parathelphusa pantherina	0.00	6.67	0.00	6.59	3.25	0.00
Culex sp.	0.00	5.73	0.00	0.00	0.00	0.00
Calopteryx damselfly	0.00	0.00	9.35	0.00	0.00	8.52
Melanoides sp.*1*	0.00	0.00	9.16	0.00	0.00	0.00
Melanoides sp.*3*	0.00	0.00	8.01	0.00	0.00	0.00
Libellula sp.	0.00	0.00	8.01	0.00	3.25	3.80
Tanypus sp.	0.00	0.00	7.81	0.00	6.65	0.00
Plexippus sp.	0.00	0.00	7.62	0.00	0.00	0.00
Corbicula sp.	0.00	0.00	6.85	0.00	0.00	0.00
Macrodiplax balteata	0.00	0.00	4.39	0.00	3.10	0.00
Melanoides sp.*2*	0.00	0.00	4.39	0.00	0.00	0.00
Trichoceridae	0.00	0.00	4.20	0.00	0.00	0.00
Oxycerini sp.	0.00	0.00	3.81	0.00	0.00	0.00
Caenis sp.	0.00	0.00	3.81	0.00	0.00	0.00
Baetis flavistriga	0.00	0.00	3.62	7.77	21.92	26.36
Aphylla williamsoni	0.00	0.00	3.43	0.00	0.00	0.00
Micrasema sp.	0.00	0.00	3.43	0.00	0.00	0.00
Leucrocuta sp.	0.00	0.00	3.43	0.00	0.00	0.00
Isoperla sp.	0.00	0.00	3.43	0.00	0.00	0.00
Oligochaeta	0.00	0.00	3.43	0.00	0.00	0.00
Hydrometridae	0.00	0.00	3.23	0.00	0.00	0.00
Amphypsyche sp.	0.00	0.00	3.23	0.00	0.00	0.00
Pila ampullacea	0.00	0.00	0.00	6.83	0.00	0.00
Pila globosa	0.00	0.00	0.00	6.36	3.40	0.00
Dryops	0.00	0.00	0.00	6.36	0.00	3.64
Hemiptera (Cicadas)	0.00	0.00	0.00	6.36	0.00	3.64
Eristalis tenax (Syrphidae)	0.00	0.00	0.00	6.12	0.00	0.00
Tiara sp.*1*	0.00	0.00	0.00	0.00	11.28	0.00
Tiara scabra	0.00	0.00	0.00	0.00	10.50	0.00
Nerita sp.	0.00	0.00	0.00	0.00	3.56	0.00
(Undetermined Ephemeroptera sp.1)	0.00	0.00	0.00	0.00	3.10	0.00
Baetisca sp.	0.00	0.00	0.00	0.00	3.10	0.00
Trichoptera (Hydropsychidae)	0.00	0.00	0.00	0.00	0.00	9.20
(Undetermined Ephemeroptera sp.2)	0.00	0.00	0.00	0.00	0.00	7.35
Maladeraholosaricea	0.00	0.00	0.00	0.00	0.00	3.80
(Undetermined Ephemeroptera sp.3)	0.00	0.00	0.00	0.00	0.00	3.49

TABLE 4: Biotic index (ASPT) of benthic macroinvertebrates in all sampling locations.

	Biotic Index (ASPT)	
Sampling station	December 2015	June 2016
K-1	5.00	3.00
K-2	4.00	4.00
K-3	3.50	5.80
J-1	5.67	5.25
J-2	6.00	6.00
J-3	5.70	5.38
P-1	6.69	6.10
P-2	5.87	6.00
P-3	5.93	6.20

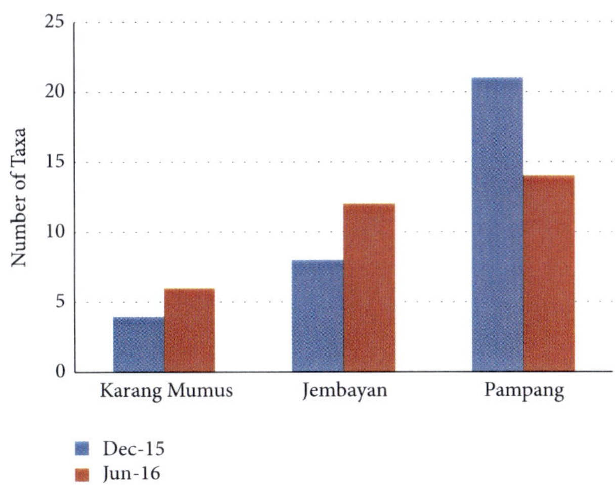

FIGURE 4: Taxa Richness of benthic macroinvertebrate.

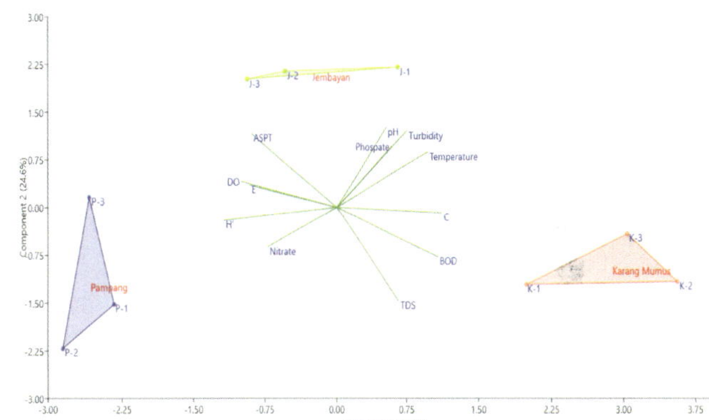

FIGURE 5: Diagram of sampling sites by Principal Component Analysis (PCA), based on the biotic measures and physical-chemical variables in East Kalimantan of Indonesia in December 2015.

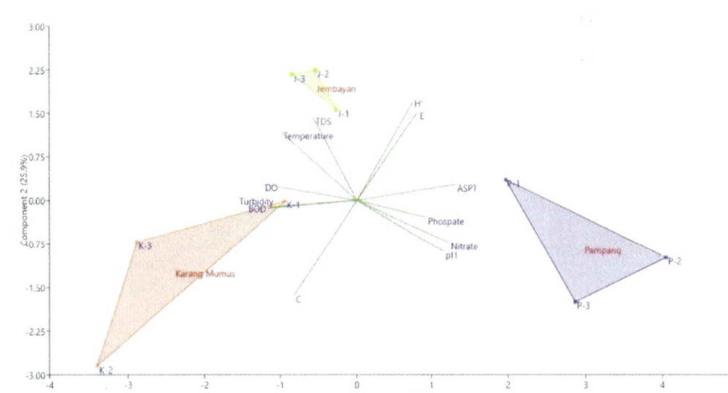

FIGURE 6: Diagram of sampling sites by Principal Component Analysis (PCA), based on the biotic measures and physical-chemical variables in East Kalimantan of Indonesia in June 2016.

and K-3 (Karang Mumus) present a strong affinity to high values of C, BOD, and TDS (Figure 5). In June 2016 (dry season), Jembayan River (J-1, J-2, and J-3) shows a strong affinity to DO, temperature, TDS, H, and E; Pampang River (P-1, P-2, and P-3) presents a strong affinity to pH, phosphate, nitrate, and ASPT; and Karang Mumus (K-1, K-2, and K-3) demonstrates a strong affinity to turbidity, BOD, and C (Figure 6).

In December 2015 (rainy season), 16 taxa of macroinvertebrates were noted in P-3. The high diversity of macroinvertebrates at P-3 was supported by moderate water quality (highest WQI) (Table 5) during this period. According to PCA, a strong affinity to high H (diversity index) and nitrate was also shown by Pampang River (Figure 5). Dense riparian vegetation found on the banks of the river may provide high nutrient sources for macroinvertebrates. These plants also play a prominent role in the remediation of contaminated water by pesticides and detergent active ingredients before entering the rivers [44]. In contrast, station K-2 had very low diversity of macroinvertebrate. The intense human activities (for examples the uses of this river for bathing, washing, and latrines) which produced the high BOD and TDS values (Figure 5) and lack of riparian vegetation near this station were probably major causes of the very low diversity of macroinvertebrates in this station. We noted during our survey in December 2015 that water hyacinth (*Eichhornia crassipes*) was more dominant in this station.

In both December 2015 and June 2016, *M. tuberculata* and *Chironomus* sp. were codominant taxa in Karang Mumus River. *Chironomus* sp. is a species indicator of waters contaminated by high load of organic waste. In fact, it can live in waters even at low oxygen levels [45, 46]. The high dominance

TABLE 5: Water quality parameters in December 2015 in each station.

		December 15											
		Karang Mumus				Jembayan				Pampang			
No.	Parameter	K-1	K-2	K-3	Mean ± SD	J-1	J-2	J-3	Mean ± SD	P-1	P-2	P-3	Mean ± SD
1	Dissolved Oxygen	5.97	6.05	3.97	5.33±1.18	7.23	7.15	7.17	7.18±0.04	7.35	7.33	7.4	7.36±0.04
2	pH	5.6	5.55	5.19	5.45±0.22	5.86	5.98	5.65	5.83±0.17	3.53	3.7	5.98	4.40±1.37
3	BOD	79.81	66.71	61.27	69.27±9.53	1.27	0.1	0.96	0.78±0.61	3.35	2.57	0.1	2.01±1.70
4	Temperature Change	30	29.81	29.97	29.93±0.10	29.7	29.92	29.38	29.67±0.27	27.13	25.54	25.61	26.09±0.90
5	Total Phosphate	0.28	0.16	0.68	0.37±0.27	0.58	0.12	0.66	0.45±0.29	0.05	0	0.11	0.05±0.06
6	Nitrates	0.1	0.04	0	0.05±0.05	0.38	0.43	0.15	0.32±0.15	0.38	1.8	0.14	0.78±0.89
7	Turbidity	15.2	20.3	32.6	22.70±8.94	33.6	39.9	18.6	30.70±10.94	2.46	1.19	2.23	1.96±0.68
8	TDS	245	258	270	257.67±12.50	104	103	96	101.00±4.36	216	214	102	177.33±65.25
	$WQI = \sum(Wi \times Ii)$	42.23	42.53	34.71	39.82±4.43	50.99	56.6	51.62	53.07±3.07	50.19	51.47	62.14	54.60±6.56
	Criteria	B	B	B	B	M	M	M	M	B	M	M	M

Note: B: bad; M: medium or moderate.

Table 6: Water quality parameters in June 2016 in each station.

No.	Parameter	June 2016 Karang Mumus K-1	K-2	K-3	Mean ± SD	Jembayan J-1	J-2	J-3	Mean ± SD	Pampang P-1	P-2	P-3	Mean ± SD
1	Dissolved Oxygen	8.09	7.84	8.23	8.05±0.20	7.56	7.82	7.66	7.68±0.13	6.66	7	6.78	6.81±0.17
2	pH	7	7.01	7	7.00±0.01	6.9	6.9	7.01	6.94±0.06	7.39	7.4	7.4	7.40±0.01
3	BOD	1.11	1.98	1.54	1.54±0.43	1.26	1.49	1.68	1.47±0.21	1.03	1.05	1.04	1.04±0.01
4	Temperature Change	30	30	30	30.00±0.00	30	30	30	30.00±0.00	29	27	26.6	27.53±1.29
5	Total Phosphate	0	0	0	0.00±0.00	0.06	0.04	0.09	0.06±0.02	0	0.72	0.08	0.26±0.39
6	Nitrates	0.26	0.14	0.01	0.14±0.13	0.13	0.10	0.14	0.12±0.02	0.26	0.69	0.45	0.47±0.22
7	Turbidity	15.2	20.3	32.6	22.70±8.94	12.3	8.2	16.8	12.43±4.30	17.3	2.24	2.2	7.25±8.71
8	TDS	16	17.8	16.7	16.83±0.91	23.9	58.3	85.5	55.90±30.87	12	27	12	17.00±8.66
	$WQI = \sum(Wi \times Ii)$	63.04	60.07	60.79	61.30±1.55	62.76	63.64	61.73	62.71±0.96	63.36	60.39	66.16	63.30±2.89
	Criteria	M	M	M	M	M	M	M	M	M	M	M	M

Note: M: medium or moderate.

index indicates that the waters were instable causing imbalance of ecosystem. Under this condition, usually only certain types of organism like *Chironomus* sp. can survive because of their ability to tolerate the high organic contamination [45]. Some of Chironomidae larvae are efficient indicators of mesotrophic waters, and these are usually found at location having high decomposed organic matter. Thus, presence of the Chironomidae family (indicated by high percentage of IVI) reflected that the Karang Mumus River was in polluted category. Likewise presence of *Melanoides tuberculata* of the Thiaridae family is well known as a species that can tolerate the presence of low dissolved oxygen and high suspended particulate matter in river's water [46]. Gastropod especially *Melanoides* is very abundant in waters affected by agricultural waste even at low level of dissolved oxygen [47]. *M. tuberculata* is tolerant not only in oligotrophic ecosystems [45] but also in ecosystem contaminated by low level of organic matter [47]. *M. tuberculata* remains active at night, and it likes the temperature in the range of 18–32°C. *M. tuberculata* has an operculum that can protect itself from drought so that it can survive on dry land and high salinity [48, 49]. In addition, operculum also serves to increase their tolerance to toxic chemicals in the environment. So, these taxa are recommendable to be used as a bioindicator of polluted ecosystems [48, 49].

In both October 2015 and June 2016, *M. tuberculata* and *Acentrella parvula* were codominant taxa in Jembayan River. *A. parvula* (family of Baetidae) is also known as one of benthic macroinvertebrates which is intolerant to the contaminants. Therefore, Baetidae family can also be used as a bioindicator of low levels of organic matter contamination [50].

In October 2015, *Coenagrion* sp. (order: Odonata) and *A. parvula* were codominant organism in Pampang River. Odonata is a facultative or intermediate organism that can survive in moderate level of the environmental change. This group can survive in waters containing organic matter. However, they are quite sensitive to water quality degradation [46]. In June 2016, Pampang River is dominated by *Acentrella parvula* and *Baetis flavistriga*. These two taxa are insects from the Baetidae family which can serve as low organic pollutant bioindicators [50, 51]. During dry season (June 2016), it is most likely that Pampang River was contaminated by low level of organic matter due to the decreased water debit and deceased input of allochthonous materials into the water body.

Water Quality Index (WQI) serves as single index that describes water quality of certain location at certain time. In December 2015, according to the WQI values, Karang Mumus River was in poor water quality but in June 2016 it was improved to moderate quality (Tables 5 and 6). Meanwhile Jembayan and Pampang Rivers were found in moderate quality in both December 2015 and June 2016. Regarding the low value of WQI in October 2015 (rainy season) compared to June 2016 (dry season), we concluded that this can be flood water and run-off during the rainy season bringing allochthonous materials into the water body or from the resuspension of the sediment (autochthonous) materials. Similar effects were observed in Shiroro Lake [52] and Gbako River, Nigeria [53]. The accumulation of these materials in the water body led to decreased water quality. Moreover, Karang Mumus is located near to populated area and harbor with intense loading-unloading and motor ship activities, and the low WQI in Karang Mumus River was also most probably due to all those activities. For Jembayan River, coal-mining activities and oil-palm plantations around this river are the potential activities which contribute to worsening the river water quality and benthic macroinvertebrate diversity. On the other hand, the potential sources affecting the water quality of Pampang River were oil-palm and pecan plantation activities along this river.

4. Conclusion

Based on the ASPT and WQI values, our study clarified and estimated that Karang Mumus River recently received certain pollutants and can be categorized as dangerously polluted river. In fact, macroinvertebrates in the river were dominated by *Chironomus* sp., and *Melanoides tuberculata* which are clearly indicated. On the other hand, Jembayan River was found to be of doubtful or moderate quality according to ASPT and WQI values with *M. tuberculata* and *A. parvula* as codominant taxa. Furthermore, Pampang River was the cleanest river based on ASPT and WQI values, and Odonata and Baetidae families were codominant in the river. These two families are quite sensitive to water quality degradation and only tolerant to low organic pollution. However, if organic pollution continues at this rate, the water of these rivers will become seriously harmful.

Disclosure

The Permanent Address of Fatmawati Patang is Department of Biology, Faculty of Mathematics and Natural Science, Universitas Mulawarman, Kampus Gunung Kelua, Jl. Barong Tongkok, Samarinda, Indonesia.

Conflicts of Interest

The authors declare no conflicts of interest regarding the use of research contents and publication of this paper.

Acknowledgments

The first author would like to thank the Indonesian Ministry of Research, Technology and Higher Education for providing generous financial support for scholarship and research equipment. In addition, technical assistance of Mr. Setiyanto is gratefully acknowledged.

References

[1] U. Suriawiria, *Water in a Healthy Life and Environment*, Alumni, Bandung, Indonesia, 2003.

[2] D. Setiawan, "The study of macrozoobenthos community at downstream waters of Lematang river surrounding in Pasar Bawah, Lahar Regency," *Science Research Journal*, vol. 9, pp. 12–14, 2009.

[3] A. E. Ogbeibu and B. J. Oribhabor, "Ecological impact of river impoundment using benthic macro-invertebrates as indicators," *Water Research*, vol. 36, no. 10, pp. 2427–2436, 2002.

[4] J. A. Sciortino and R. Ravikumar, Fishery Harbour Manual on the Prevention of Pollution – Bay of Bengal Programme, BOBP for Fisheries Management, BOBP/MAG/22, Madras, India, 1999.

[5] United States Environmental Protection Agency, *Basic Information on Water Quality Criteria*, 2018, https://www.epa.gov/wqc/basic-information-water-quality-criteria#aquatic.

[6] H. Kalyoncu and H. Gülboy, "Benthic macroinvertebrates from dariören and isparta streams (Isparta/Turkey)-biotic indices and multivariate analysis," *Journal of Applied Biological Sciences*, vol. 3, no. 1, pp. 79–86, 2009.

[7] A. Oliveira and M. Callisto, "Benthic macroinvertebrates as bioindicators of water quality in an atlantic forest fragment," *Iheringia - Serie Zoologia*, vol. 100, no. 4, pp. 291–300, 2010.

[8] C. B. Uherek and F. B. Pinto Gouveia, "Biological monitoring using macroinvertebrates as bioindicators of water quality of maroaga stream in the maroaga cave system, presidente figueiredo, amazon, Brazil," *International Journal of Ecology*, vol. 2014, 7 pages, 2014.

[9] D. Kartikasari, C. Retnaningdyah, and E. Arisoesilaningsih, "Application of water quality and ecology indices of benthic macroinvertebrateon evaluate water quality of tertiary irrigation in Malang district," *The Journal of Tropical Life Science*, vol. 1, pp. 193–201, 2013.

[10] D. M. Rosenberg and V. H. Resh, "Introduction to freshwater biomonitoring and benthic macroinvertebrates," in *Freshwater Biomonitoring and Benthic Macroinvertebrates*, Chapman and Hall, New York, NY, USA, 1993.

[11] I. Czerniawska-Kusza, "Comparing modified biological monitoring working party score system and several biological indices based on macroinvertebrates for water-quality assessment," *Limnologica*, vol. 35, no. 3, pp. 169–176, 2005.

[12] J. D. Elias, J. N. Ijumba, Y. D. Mgaya, and F. A. Mamboya, "Study on freshwater macroinvertebrates of some Tanzanian rivers as a basis for developing biomonitoring index for assessing pollution in tropical African regions," *Journal of Ecosystems*, vol. 2014, 8 pages, 2014.

[13] K. Suleiman and I. Abdullahi, "Biological assessment of water quality: a study of Challawa river water Kano, Nigeria," *Bayero Journal of Pure and Applied Sciences*, vol. 4, no. 2, pp. 121–127, 2012.

[14] M. M. Roozbahani, S. M. B. Nabavi, P. Farshchi, and A. Rasekh, "Studies on the benthic macroinvertebrates diversity species as bio-indicators of environmental health in Bahrekan Bay (Northwest of Persian Gulf)," *African Journal of Biotechnology*, vol. 9, no. 51, pp. 8763–8771, 2010.

[15] M. N. Varnosfaderany, E. Ebrahimi, N. Mirghaffary, and A. Safyanian, "Biological assessment of the Zayandeh Rud River, Iran, using benthic macroinvertebrates," *Limnologica*, vol. 40, no. 3, pp. 226–232, 2010.

[16] S. M. Mandaville, "Bioassessment of Freshwaters Using Benthic Macroinvertebrates-A Primer, First Ed., Project E-1, Soil & Water Conservation Society of Metro Halifax. VIII, Chapters I-XXVII, Appendices A-D. 244p," 2002.

[17] C. E. Shannon, *The Mathematical Theory of Communication*, The University of Illinois Press, Urbana, Ill, USA, 1949.

[18] P. D. Armitage, D. Moss, J. F. Wright, and M. T. Furse, "The performance of a new biological water quality score system based on macroinvertebrates over a wide range of unpolluted running-water sites," *Water Research*, vol. 17, no. 3, pp. 333–347, 1983.

[19] B. Oram, *Calculating NSF Water Quality Index*, 2010, http://www.water-research.net/Watershed/temperature.htm.

[20] W. M. H. W. A. Ghani, C. S. Md Rawi, S. A. Hamid, and S. A. Al-Shami, "Efficiency of different sampling tools for aquatic macroinvertebrate collections in malaysian streams," *Tropical Life Sciences Research*, vol. 27, no. 1, pp. 115–134, 2016.

[21] J. Soedarso and Y. Wardiatno, *Assessment of the Quality Status of the River with Macrozoobenthos as Indicator*, Pena Nusantara, Bogor, Indonesia, 2015.

[22] C. W. Heckman, *Encyclopedia of South American Aquatic Insects Ephemeroptera Illustrated Keys to Known Families, Genera, and Species in South America*, Springer Science+Business Media Dordrecht, New York, NY, USA, 2002.

[23] C. W. Heckman, *Encyclopedia of South American Aquatic Insects Plecoptera Illustrated Keys to Known Families, Genera, and Species in South America*, Springer Science+Business Media Dordrecht, New York, NY, USA, 2003.

[24] C. W. Heckman, *Encyclopedia of South American Aquatic Insects Odonata Anisoptera Illustrated Keys to Known Families, Genera, and Species in South America*, Springer Science+Business Media Dordrecht, New York, NY, USA, 2006.

[25] C. W. Heckman, *Encyclopedia of South American Aquatic Insects: Odonata - Zygoptera*, Springer Science+Business Media, Dordrecht, Netherlands, 2008.

[26] J. Gooderham, *The Waterbug Book*, CSIRO Publishing, Victoria, Australia, 2002.

[27] J. Oscoz, D. Galicia, and R. Miranda, *Identification Guide of Freshwater Macroinvertebrates of Spain*, Springer Netherlands, Dordrecht, 2011.

[28] I. D. Hodkinson, R. W. Merritt, and K. W. Cummins, "An Introduction to the Aquatic Insects of North America," *Journal of Animal Ecology*, vol. 50, no. 1, p. 330, 1981.

[29] R. W. Pennak, *Freshwater Invertebrates of United States*, John Wiley and Sons, New York, NY, USA, 1978.

[30] J. H. Epler, *Identification Manual for the Larval Chironomidae (Diptera) of North and South Carolina*, North Carolina Department of Environment and Natural Resources Division of Water Quality, 2001.

[31] B. Dharma, "Indonesian Shell," Edisi I dan II. PT. Sarana Graha, Jakarta, 1988.

[32] W. T. Edmonson, *Fresh Water Biology*, John Wiley and Sons, New York, NY, USA, 1963.

[33] J. G. Needham and P. R. Needham, *A Guide to the Study of Fresh-Water Biology*, Holden Day, Inc., San Francisco, Calif, USA, 1962.

[34] P. Barnard, *Tropical Asian Streams Macrozoobenthos, Ecology and Conservation*, Hong Kong University Press, Hong Kong, China, 1999.

[35] E. W. Rice, R. B. Baird, A. D. Eaton, and L. S. Clesceri, *Standard Methods for the Examination of Water and Wastewater*, American Public Health Association, American Water Works Association, Water Environment Federation, 22nd edition, 2012.

[36] A. J. Kohn, *Ecological Diversity*, Wiley-Interscience Publication, London, UK, 1975.

[37] E. P. Odum, *Fundamentals of Ecology*, W.B. Sounders Company Ltd., Philadelphia, Pa, USA, 1971.

[38] J. Brower, J. H. Zar, and C. N. von Ende, *Field and Laboratory Methods for General Ecology*, McGraw-Hill Education, Boston, Mass, USA, 4th edition, 1997.

[39] H. Effendi, Romanto, and Y. Wardiatno, "Water Quality Status of Ciambulawung River, Banten Province, Based on Pollution Index and NSF-WQI," *Procedia Environmental Sciences*, vol. 24, pp. 228–237, 2015.

[40] C. F. Mason, *Biology of Freshwater Pollution*, Prentice Hall, New York, NY, USA, 4th edition, 2002.

[41] N. Arslan, A. Salur, H. Kalyoncu, D. Mercan, B. Barisik, and D. A. Odabasi, "The use of BMWP and ASPT indices for evaluation of water quality according to macroinvertebrates in Kucuk Menderes River (Turkey)," *Biologia (Poland)*, vol. 71, no. 1, pp. 49–57, 2016.

[42] P. S. Kumar and A. B. Khan, "The distribution and diversity of benthic macroinvertebrate fauna in Pondicherry mangroves, India," *Aquatic Biosystems*, vol. 9, no. 1, article 15, 2013.

[43] M. T. Suriano and A. A. Fonseca-Gessner, "Structure of benthic macroinvertebrate assemblages on a gradient of environmental integrity in Neotropical streams," *Acta Limnologica Brasiliensia*, vol. 25, no. 4, pp. 418–428, 2013.

[44] P. Legendre and L. Legendre, *Numerical Ecology*, Elsevier, Amsterdam, The Netherlands, 1998.

[45] L. Mariantika and C. Retnaningdyah, "The change of benthic macroinvertebrate community structure due to human activity in the spring channel of the source of clouds of Singosari sub-district, Malang Regency," *Jurnal Biotropika*, vol. 2, pp. 254–259, 2014.

[46] A. T. Sastrawijaya, *Environmental Pollution*, Rineka Cipta, Jakarta, Indonesia, 1991.

[47] A. C. Rehn, "Benthic macroinvertebrates as indicators of biological condition below hydropower dams on west slope Sierra Nevada streams, California, USA," *River Research and Applications*, vol. 25, no. 2, pp. 208–228, 2009.

[48] A. J. Mitchell, M. S. Hobbs, and T. M. Brandt, "The effect of chemical treatments on red-rim melania Melanoides tuberculata, an exotic aquatic snail that serves as a vector of trematodes to fish and other species in the USA," *North American Journal of Fisheries Management*, vol. 27, no. 4, pp. 1287–1293, 2007.

[49] Y. Wang, H. Wei, J. Wang et al., "Electropolymerized polyaniline/manganese iron oxide hybrids with an enhanced color switching response and electrochemical energy storage," *Journal of Materials Chemistry A*, vol. 3, no. 41, pp. 20778–20790, 2015.

[50] I. Alhejoj, E. Salameh, and K. Bandel, "Mayflies (Order Ephemeroptera): An Effective Indicator of Water Bodies Conditions in Jordan," *International Journal of Scientific Research in Environmental Sciences*, vol. 2, no. 10, pp. 361–370, 2014.

[51] V. M. Junqueiraand and S. C. M. Campos, "Adaptation of the BMWP method for water quality evaluation to Rio das Velhas watershed (Minas Gerais, Brazil)," *Acta Limnologica Brasiliensia*, vol. 10, pp. 125–135, 1998.

[52] R. Koloanda and A. Oladimeji, "Water quality and some nutrient levels in Shiroro Lake, Niger State, Nigeria," *International Journal of Aquatic Science*, vol. 19, no. 2, 2004.

[53] U. N. Keke, F. O. Arimoro, Y. I. Auta, and A. V. Ayanwale, "Temporal and spatial variability in macroinvertebrate community structure in relation to environmental variables in Gbako River, Niger State, Nigeria," *Tropical Ecology*, vol. 58, no. 2, pp. 229–240, 2017.

Analysis of the Spatial Relationship between Cattle and Wild Ungulates across Different Land-Use Systems in a Tropical Savanna Landscape

Geoffrey W. Kinga[1] **, John Mironga,**[2] **and Wilfred O. Odadi**[1]

[1] *Department of Natural Resources, Egerton University, P.O. Box 536–20115, Egerton, Kenya*
[2] *Department of Geography, Egerton University, P.O. Box 536–20115, Egerton, Kenya*

Correspondence should be addressed to Geoffrey W. Kinga; jeffwaki@yahoo.com

Academic Editor: Béla Tóthmérész

In many African savanna landscapes, domestic and wild herbivores cooccur across different land-use systems, but the role of land-use in shaping their spatial relationship is poorly understood. We evaluated the spatial relationship between cattle and wild herbivores categorized by body sizes and feeding habits across different land-use types, namely, private ranches (PR), transitional lands (TRL), and pastoral grazing areas (PGA), in Laikipia County, Kenya. Cattle and wild herbivores spatial distribution data were obtained from Kenya's Department of Resources Survey and Remote Sensing (DRSRS). Spatial relationships between cattle and different wild herbivore guilds were analyzed using Ripley's bivariate K_{12} function. In PR, wild herbivore guilds showed significant attraction to cattle at short distances. In TRL, wild grazers, mixed feeders, megaherbivores, and medium-sized ungulates exhibited significant attraction to cattle. Additionally, repulsion was observed between cattle and browsers at short distances under this land-use system. In PGA, wild grazers, mixed feeders, and megaherbivores repelled strongly with cattle at short distances while browsers and medium-sized ungulates were significantly attracted to cattle. Cattle and wild herbivores were more randomly and independently distributed in PR than in TRL and PGA. These spatial relationships imply better coexistence between cattle and wild herbivores in PR than in TRL and PGA.

1. Introduction

Savannas are the most widespread ecosystems in the tropics and subtropics. The tropical and subtropical savannas cover nearly a third of the world's land surface and over 50% of Africa [1]. Sub-Saharan Africa supports approximately 162 million poor livestock keepers [2, 3], majority of which are found in savanna rangelands. Cattle population in the Sub-Saharan Africa is estimated at 191 million heads, more than 50% of these cattle are in East Africa [4–6]. In addition to supporting people and livestock production, African savanna rangelands are also important biodiversity reservoirs. These landscapes not only support large population of hoofed herbivores (ungulates) but also very diverse community of indigenous large mammals [7, 8]. Approximately, 46 species of extant ungulates are endemic to the African savanna biome, exceeding that of any other continent [7]. Notably, however, livestock (cattle, goats, and sheep) usually occur in savanna ecosystems in higher numbers than wild herbivores given the advantage of animal husbandry practices such as veterinary care, predator control, and supplementation of feed and water provision, thus putting even higher demands to the ecosystem [9].

Conversion of grassland ecosystems into croplands, afforestation, urbanization, and other detrimental human activities is a global problem [10]. Increasing human population has led to increased need for food production and this has resulted in expansion of croplands; close to 20% of savanna grasslands in Africa has been converted to cropland and urban areas [9]. Furthermore, livestock production has also increased owing to the increasing demand for protein from the increasing human population, a trend projected to continue [11, 12]. These changes continue to increase pressure on rangeland resources, reducing their sustainability and capacity to provide critical ecosystem services and support biodiversity conservation and livestock production. While numerous protected areas have been established across the savanna ecosystems for the purpose of wildlife conservation, large populations of wildlife occur outside protected areas [13] where they share habitats with livestock.

Habitat sharing between domestic and wild ungulates may result in different kinds of ecological interactions between these two herbivore guilds, including competition [14–16], facilitation [15, 17], and transmission of diseases and parasites [18, 19]. However, the intensity of these interactions may vary depending on various factors, for instance, the digestive system and mouth morphology [20, 21], body size and feeding habits of the wild ungulates involved, animal type and intensity [22], season (dry or wet), and land-use type where domestic and wild herbivores cooccur [15, 23, 24]. The kind of ecological interactions and the various factors that affect their strength may determine the nature and magnitude of spatial relationship between domestic and wild ungulates. Previous studies in African savanna ecosystems have shown that livestock has potential or actual adverse effects on native wild ungulates [25–27]. While wild herbivores adapt to presence of competing cattle by using refuge habitat or by modifying their diet, spatial partitioning has also been shown as a strategy employed by wild ungulates [28]. Despite the fact that livestock and wild herbivores cooccur across different land-use systems, the influence of land-use systems on spatial relationship between these ungulate guilds remains poorly understood. Understanding the effect of land-use systems on the spatial relationship between domestic (livestock) and wild herbivore guilds is necessary for their improved management on human-dominated savanna landscapes.

We investigated the influence of land-use system on spatial relationship between cattle and wild mammalian herbivores in Laikipia County, Kenya. The study area is a typical example of a tropical savanna which is of profound socioeconomic and ecological significance. Laikipia was ideal for the study because it hosts a mixture of livestock and a diverse assemblage of wild herbivores. Domestic and wild herbivores commonly share habitats across much of the Laikipia landscape because only approximately 2% of the landscape is formally protected exclusively for wildlife conservation [23, 29].

In this study, we compared the spatial relationship between herds of cattle and mammalian herbivores among three land-use types, namely, private ranches (PR), transitional lands (TRL), and pastoral grazing areas (PGA). Considering the varying densities of cattle and wild herbivores, management regimes, and forage resources availability in the three land-use types, we expected that the spatial relationship between these herbivore guilds would vary among land-use types. Additionally, we expected that the strength of these effects would vary depending on wild herbivore body size (megaherbivores and medium-sized herbivores) and feeding habits (grazers, browsers, and mixed feeder).

2. Materials and Methods

2.1. Study Area. The research was conducted in Laikipia County, Kenya. The county covers approximately 9666 km^2 [23, 30] and lies between latitude $0°17'S–0°52'N$ and longitude $36°13'E–37°23'E$ at an elevation of between 1700 and 2000 meters above sea-level (Figure 1).

The study area is located in the central part of Kenya, where it largely falls on the northern side of the equator (0°). To the southwest are the Aberdare highlands and to the southeast is the Mt. Kenya. A climatic gradient occurs longitudinally in Laikipia due to the presence of Mt. Kenya (5199 m) and Aberdare ranges (3999 m). Laikipia experiences a mean annual rainfall range of between 300 mm and 750 mm with a weak trimodal distribution. Long rains fall between March and June, continental rains occur between August and September, and short rains fall between October and December [31], but with marked spatial-temporal variability. The study area has numerous rivers and streams primarily originating from the two major water towers: Aberdares and Mt. Kenya. Various streams join to form two perennial rivers (Ewaso Ng'iro and Ewaso Narok) which flow through Laikipia plains to Samburu [23, 30]. Streams which originate within Laikipia conduct water seasonally only during the rainy season, otherwise remaining dry for the other part of the year. There are two swamps and several springs, dams, and depressions which fill with water during the rainy season [32].

Laikipia is a species-rich savanna rangeland in both animal and plant life-forms. It harbors over 95 mammal species of which 25 are ungulate species, 540 bird species, and close to 1000 species of invertebrates [29]. Common ungulates include plains zebra *(Equus burchelli)*, impala *(Aepyceros melampus)*, dik-dik *(Madoqua kirkii)*, Grant's gazelle *(Gazella granti)*, African elephant *(Loxodonta africana)*, hartebeest *(Alcelaphus buselaphus)*, oryx *(Oryx beisa)*, eland *(Taurotragus oryx)*, giraffe *(Giraffa camelopardalis)*, buffalo *(Syncerus caffer)*, gerenuk *(Litocranius walleri)*, black rhino *(Diceros bicornis)*, white rhino *(Ceratotherium simum)*, Grevy's zebra *(Equus grevyi)*, and waterbuck *(Kobus ellipsiprymnus)* [33]. Common large carnivores include lion *(Panthera leo)*, cheetah *(Acinonyx jubatus)*, leopard *(Panthera pardus)*, African wild dog *(Lycaon pictus)*, and spotted hyena *(Crocuta crocuta)*.

Habitat types in Laikipia are largely characterized by the dominant plant community or plant species. In the grassland and open woodlands, several species of *Acacia* and *Commiphora* dominate the woody vegetation layer. Various graminoids and forb species dominate the understory vegetation. *Acacia* and *Commiphora* woodland is predominant in the dry central and the northern part of Laikipia; *Acacia mellifera* is the dominant species in this habitat type. In the dry upland forest habitat type, African olive *(Olea africana)* and cedar *(Juniperus procera)* dominate. In the evergreen bushland forest, the dominant vegetation type is *Euclea divinorum* with *Acokanthera schimperi* and *Carissa spinosa* also being present. In the west Laikipia and along the escarpments *Leleshwa* bushes *(Tarchonanthus camphoratus)* and the sand olive *(Dodonaea angustifolia)* exist especially in the overgrazed areas. The yellow fever

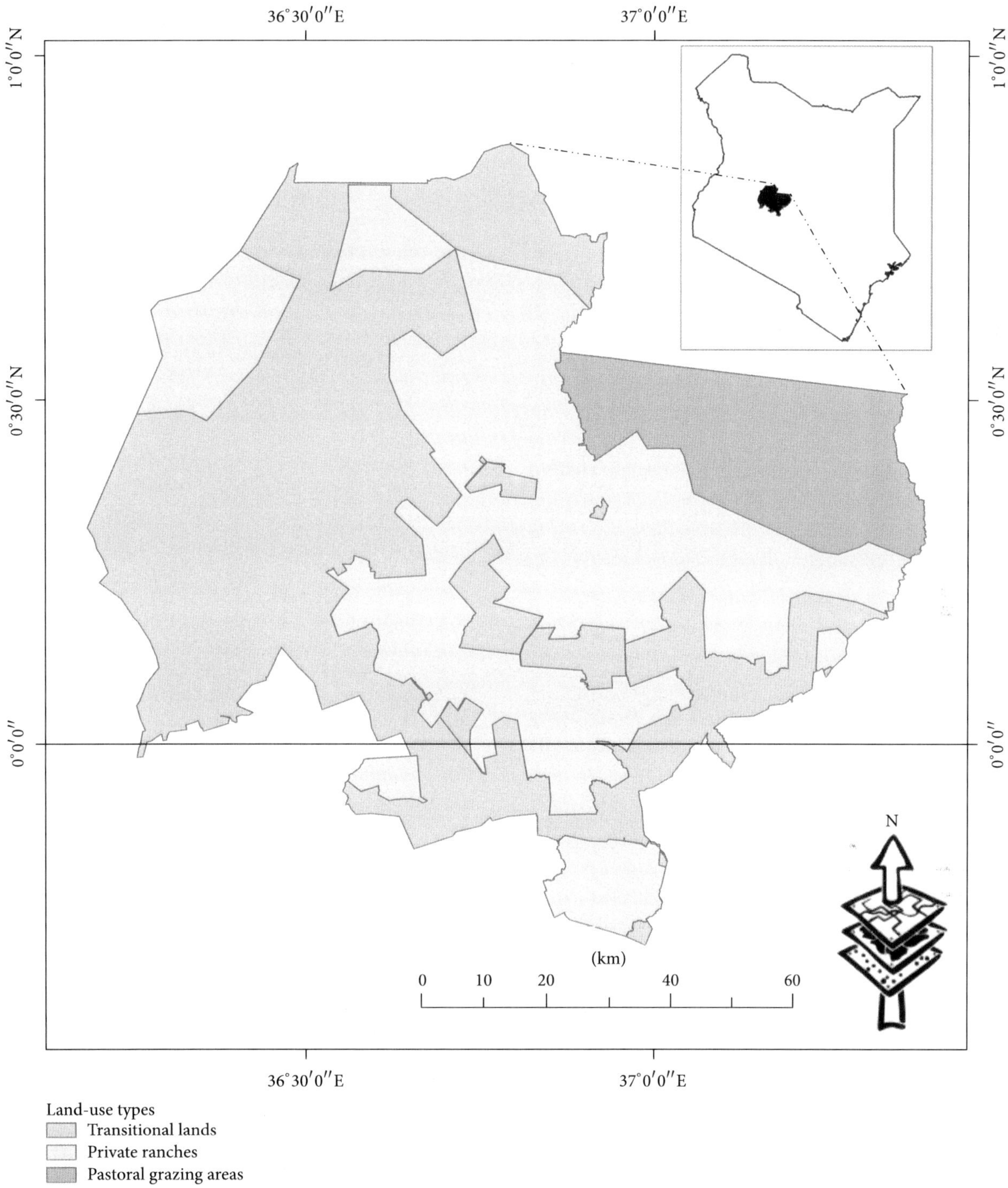

FIGURE 1: Laikipia county broad land-use types (equator: 0°0′0″). Source: LWF, 2012.

tree *(Acacia xanthophloea)* dominates along the riverine and papyrus dominates in the wetlands [29]. Dominant grasses include *Pennisetum stramineum, Lintonia nutans, Themeda triandra, P. mezianum, Cynodon dactylon, Cynodon plectostachyus, Tragus berteronianus, Cymbopogon pospischilii, Digitaria milanjiana, Bothriochloa insculpta,* and *Brachiaria lachnantha* [34, 35]. Forbs and herbs also constitute the plant community in Laikipia.

Laikipia County communities engage in various socioeconomic activities; agriculture in various forms dominates. In the northern part of the county where rainfall is scarce, pastoralism is the dominant economic activity being practiced on communally owned land. Agropastoralism and small scale mixed farming dominate the rest of the county especially in areas where land ownership is small-holding and rainfall permits arable farming. Agropastoralism refers to land-use

TABLE 1: Categorization of Laikipia wild herbivores based on body size and feeding habits.

Wild herbivore species	Body size		Feeding habits		
	Megaherbivore	Medium-sized	Grazers	Browsers	Mixed-feeders
African elephant	Yes				Yes
Rhino	Yes			Yes	
Giraffe	Yes			Yes	
Burchell's zebra		Yes	Yes		
Grevy's zebra		Yes	Yes		
Thomson's gazelle		Yes	Yes		
Grant's gazelle		Yes	Yes		
Hartebeest		Yes	Yes		
Impala		Yes			Yes
Cape buffalo		Yes	Yes		
Eland		Yes			Yes
Warthog		Yes	Yes		
Beisa oryx		Yes			Yes
Greater kudu		Yes		Yes	
Lesser kudu		Yes		Yes	
Waterbuck		Yes	Yes		
Reedbuck		Yes	Yes		
Gerenuk		Yes		Yes	

Source: author, 2017.

where a parcel of land is either on crop farming (during wet season) or on livestock (during dry periods). Where land owners or government entities have large tracts of land, cattle ranching, game ranching, and tourism are practiced [36]; moreover, large scale intensive horticultural farming is done in the uplands. Livestock in Laikipia include cattle, donkeys, sheep and goats, and camels. An aerial survey conducted in 2012 estimated that Laikipia hosts approximately 149,910 heads of cattle and 1,454 donkeys. Sheep and goats were estimated at 380,312 while camels were estimated at 4,150 [37].

2.2. Study Design. In this study, we assessed the spatial relationship between cattle and wild herbivores of different feeding habits and body sizes (Table 1) across three broad land-use categories discernible in the study area. Based on feeding habits, wild herbivores were grouped as grazers, browser, and mixed feeders. Grazers comprise herbivores that consume herbaceous plants including grass, grass-like plants, and forbs; browsers predominantly consume leaves, twigs or reproductive parts of shrubs, woody vines, and tree while mixed feeders graze or browse depending on habitat and/or season [38, 39]. Based on the body size, the wild herbivores were classified as megaherbivores or medium-sized ones. Megaherbivores refer to herbivores whose individuals may weigh over 1000 kg while medium-sized herbivores refer to herbivores that weigh above 20 kg [15, 40].

Land-use was classified based on (1) the dominant economic activity, (2) presence of cattle and wildlife, and (3) land ownership regime. This categorization is similar to the previous categorizations in the study area [23, 33]. The land-use categories used were private ranches (PR; also known as prowildlife properties or ranches), pastoral grazing areas (PGA; also known as group ranches), and transitional lands (TRL). These categories are briefly described below.

2.2.1. Private Ranches. This land-use type comprises large scale land holdings which are acquired or leased to individuals or private entities by the government. The private proprietors manage these properties which can either be fenced or unfenced. The "owners" of these properties utilize the land primarily for wildlife conservation while at the same time keeping livestock as a secondary utility for varied reasons; therefore wildlife density is usually higher compared to livestock density [23, 33]. Cattle are usually accompanied by a herdsman. Private ranches tend to provide sufficient pasture to both wild and domestic ungulates due to various land-use management interventions; however, range utilization by cattle is seasonal depending on forage availability.

2.2.2. Pastoral Grazing Areas. These are properties which are registered under a limited number of families from the local pastoral community who communally own and manage their land. Unlike the private ranches where property management is centralized, some group ranches have a much decentralized land management system where each group ranch elects a committee which manages grazing, tourism, and other land utilization activities [33]. The group ranch members may practice mixed farming but they are largely pastoralists; livestock density is usually higher than wildlife and to some extent exceeding the recommended stocking density thus displacing wildlife [23, 33]. Livestock herds are normally accompanied by herdsmen. Due to the "open-access" nature of grazing and unregulated stocking rate in this land-use type, forage resources get exhausted quickly especially during the dry seasons forcing herders to move their livestock to other areas in search of adequate pasture.

2.2.3. Transitional Lands. These are lands which have been subdivided into small plots (1–10 ha) and are owned by the small scale holders by way of having free-hold title deed. Land owners manage their individual properties. Occupants practice mixed farming or agropastoralism when rain permits; in the unoccupied plots, the pastoral communities dominate with their livestock thus displacing wildlife in those areas. There are also some large scale farms and ranches which may or may not tolerate wildlife but do not actively favor wildlife conservation; as a result, wildlife densities are varied in these transitional properties [23]. Cattle herds are usually accompanied by herders. In this land-use type, livestock utilize the range seasonally depending on forage availability. During the dry seasons, herders move with their livestock to other areas with abundant pasture.

2.3. Data Acquisition and Processing. Livestock and wild ungulates spatial data collected during November 2012 aerial census was obtained from the Department of Resources Survey and Remote Sensing (DRSRS), a Kenyan government department mandated with collecting, storing, analyzing, updating, and disseminating geospatial information on natural resources. DRSRS has over the years conducted aerial surveys for both wildlife and livestock in Laikipia where each detected group of either livestock or wild ungulates is mapped using the Global Positioning System (GPS) and the individuals counted. The spatial data was processed based on species and the land-use type in which the species occurred; individual species vector data (points) were then grouped according to the respective wild ungulate category based on feeding habit or body size. Spatial analysis was done using *splancs* package [41] in the R environment [42].

2.4. Data Analysis. Ripley's bivariate K_{12} function [43] was used to evaluate the spatial relationship between cattle herds and wild herbivore herds at different scales of distance at an interval of 250 meters up to a maximum distance (s) of 5000 meters. This function assesses whether the events (cattle herds and wild herbivore herds) being studied are aggregated, independently distributed, or segregated. It measures the average amount of event 2 (wild herbivore herds) located within a distance (s) of randomly chosen event 1 (cattle herd), divided by the overall density of event 2 [44]. Equation (1) defines Ripley's bivariate K_{12} function:

$$\widehat{K}_{12}(s) = \frac{1}{\lambda_2} E(N_{2s}), \quad (1)$$

$$\lambda_2 = \frac{N_2}{A}, \quad (2)$$

where

N_{2s} is number of type 2 events within a distance s of an arbitrary type 1 event,

λ_2 is the intensity of the type 2 events,

N_2 is number of groups of type 2 event,

A is unit area occupied by type 1 and type 2 events,

s is distance of type 2 events from type 1 events (radius).

TABLE 2: Cattle and wild herbivore events in different land-use systems.

Ungulate guild	Land-use type PR	TRL	PGA
Cattle	212	399	22
Wild grazers	1025	379	23
Browsers	271	28	6
Mixed feeders	564	57	24
Megaherbivores	460	45	22
Medium sized ungulates	1431	421	31

Source: author, 2017.

The corresponding linearized $\widehat{L}_{12}$ function was calculated from the resultant $\widehat{K}_{12}$ function to enable the graphical interpretation of the relationship between cattle and wild ungulate guilds. The linearized $\widehat{L}_{12}$ was given by

$$\widehat{L}_{12} = \sqrt{\frac{\widehat{K}_{12}(s)}{\pi}} - s. \quad (3)$$

Assuming two events are independently distributed, the location of type 2 events (wild ungulate groups) should be randomly distributed with respect to the location of type 1 events (cattle herds) [44, 45]. From equation 2, the expected number of type 2 events within a distance s of a randomly chosen type 1 event when distribution is uniform is $\lambda_2 \pi s^2$. Therefore, if two types of events are independent, the estimate of bivariate $\widehat{K}_{12}$ should equal the area occupied by type 2 events; $\widehat{K}_{12}(s) = \pi s^2$ and $\widehat{L}_{12}(s) = 0$ [44, 46]. If there are more events of type 2 within a distance s of an arbitrary type 1 event than expected under the assumption of independence, then $\widehat{K}_{12}(s) > \pi s^2$ and $\widehat{L}_{12}(s) > 0$, an indication of positive dependence between the two types of events, and this is interpreted as an attraction or aggregation between the two events. If $\widehat{K}_{12}(s) < \pi s^2$ and $\widehat{L}_{12}(s) < 0$, it indicates negative dependence which is interpreted as a repulsion or segregation between the two events [46, 47]. Cattle herds' location remained unchanged, while wild ungulates locations were randomly shifted 100 times within the respective land-use type. The land-use types (study area subunits) were assumed to be torus to enable the shifting of event 2 locations [44, 48]. The 95% upper and the lower confidence bands (envelopes) for the $\widehat{L}_{12}$ function were derived to show where the observed data had greater variation from the random simulations. Table 2 shows the number of cattle and wild herbivore events in each land-use type used in the analysis.

3. Results

3.1. Cattle versus Wild Herbivores of Different Feeding Habits. Cattle and wild grazers showed a significant attraction

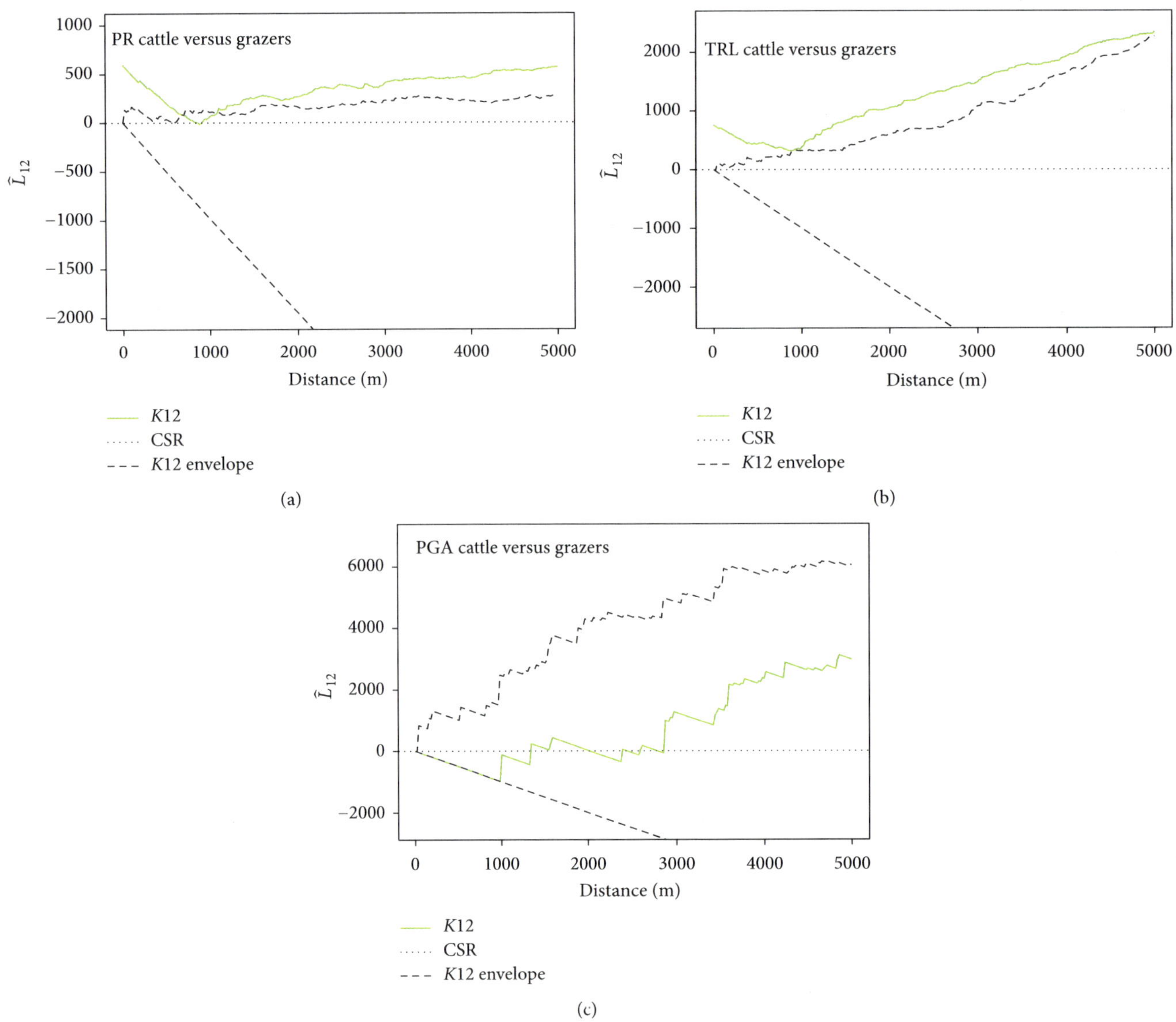

FIGURE 2: Spatial relationship between cattle and wild grazers across land-use types: (a) private ranches (PR), (b) transitional lands (TRL), and (c) pastoral grazing areas (PGA). Green solid line K_{12} is Ripley's bivariate function estimate; dashed lines (K_{12} envelope) are the 95% upper and lower confidence envelopes while the dotted line is the reference for complete spatial randomness (CSR). Source: author, 2017.

towards each other up to approximately 700 m and from approximately 1400 m to 5000 m in PR (Figure 2(a)). In TRL, the two guilds exhibited significant attraction at all distances up to 5000 m (Figure 2(b)). In PGA, cattle and wild grazers exhibited two different departures from independence. At distances up to approximately 1400 m, between 2000 m and 2600 m, and at around 2900 m, linearized K_{12} function (L_{12}) was less than 0, indicating repulsion. However, below 1000 m, it was not clear whether repulsion was significant or not because L_{12} overlapped with the lower confidence envelope. Attraction was observed between approximately 1400 m and 2000 m, between 2600 m and 2800 m, and from approximately 2900 m to 5000 m. The observed L_{12} curve did not go beyond the upper envelope and thus the attraction between cattle and wild grazers was not significant in PGA unlike in PR and TRL (Figure 2(c)).

Cattle and browsers exhibited both attraction and repulsion in the PR; at short distances of up to approximately 700 m, there was significant attraction between the two guilds. From 700 m up to 5000 m, the L_{12} curve was less than 0 indicating repulsion, notably; the repulsion was weak and near complete spatial randomness ($L_{12} = 0$) (Figure 3(a)). In TRL, cattle and browsers exhibited repulsion. At short distances up to approximately 600 m, the repulsion appeared strong though it could not be established whether it was significant or not because the L_{12} curve overlapped with the lower confidence envelope; beyond this distance up to 5000 m, the repulsion was moderate (Figure 3(b)). In PGA, significant attraction was exhibited at short distances up to approximately 500 m; beyond 500 m up to 5000 m, cattle and browsers tended to show strong attraction which was not significant (Figure 3(c)).

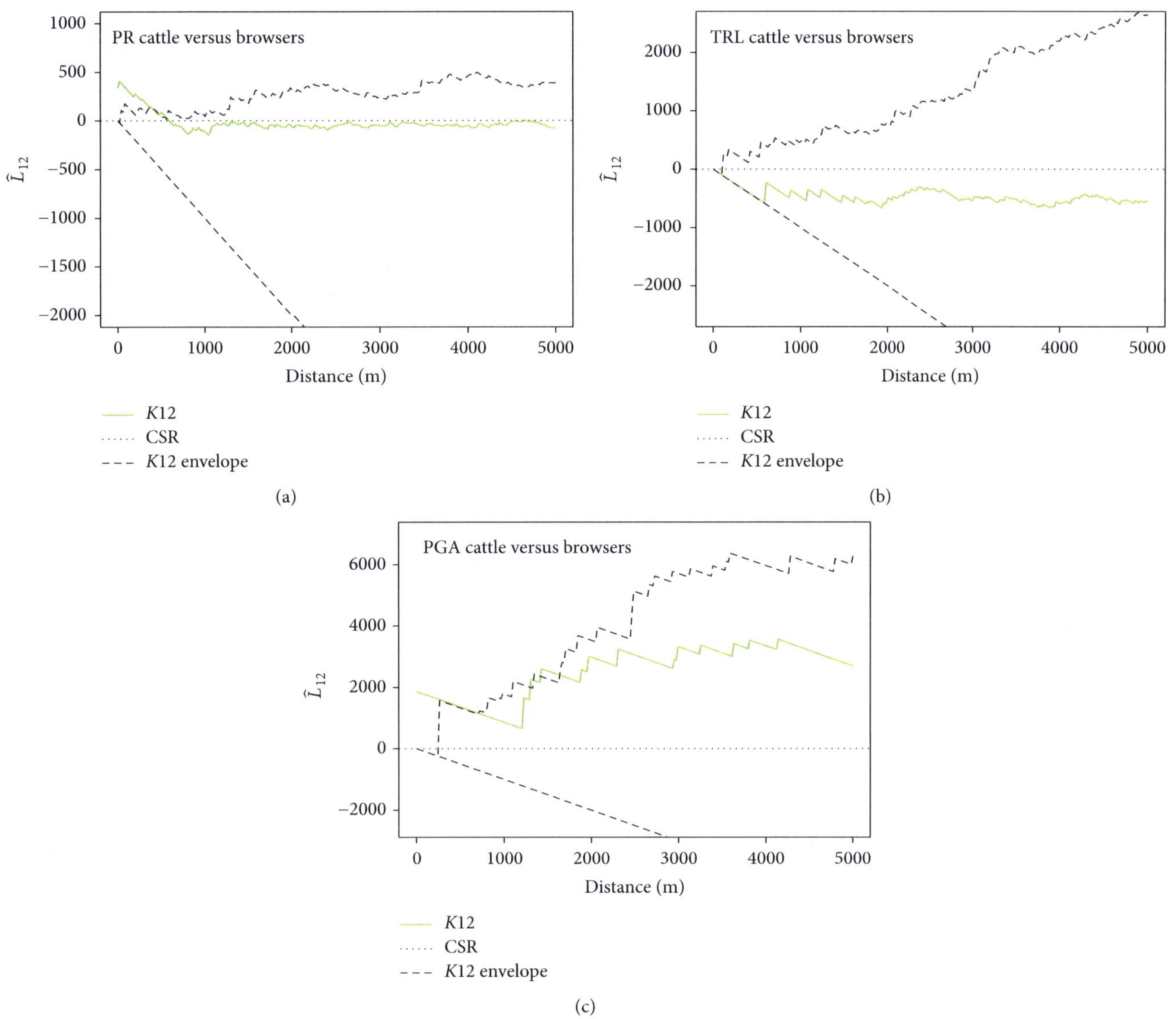

FIGURE 3: Spatial relationship between cattle and wild browsers across land-use types: (a) private ranches (PR), (b) transitional lands (TRL), and (c) pastoral grazing areas (PGA). Green solid line K_{12} is Ripley's bivariate function estimate; dashed lines (K_{12} envelope) are the 95% upper and lower confidence envelopes while the dotted line is the reference for complete spatial randomness (CSR). Source: author, 2017.

Cattle and mixed feeders exhibited two different departures from independence across all the three land-use systems unlike in the cases of cattle and grazers and cattle and browsers. In the PR, significant attraction was observed up to approximately 700 m followed by a weak repulsion up to around 1200 m. A weak attraction was observed from 1200 m to 5000 m, the observed spatial relationships between cattle and mixed feeders in PR conspicuously oscillated very close to complete spatial randomness (Figure 4(a)). In TRL, there was a weak but statistically significant attraction between cattle and mixed feeders up to approximately 400 m followed by repulsion between 500 m and approximately 3300 m. A short stint of weak attraction was observed followed by weak repulsion up to around 4400 m and then an attraction up to 5000 m (Figure 4(b)). In the PGA, there was strong repulsion at short distances (<300 m) but it could not be established whether it was significant or not because L_{12} overlapped with the lower confidence envelope. This was followed by moderate attraction at all distances up to 5000 m (Figure 4(c)).

3.2. Cattle versus Wild Herbivores of Different Body Sizes. A weak significant attraction existed between cattle and megaherbivores in PR at distances less than approximately 700 m. Weak repulsion was exhibited between 700 m and 3300 m, followed by weak oscillations of attraction and repulsion up to around 3900 m. Further, a weak attraction was observed up to 5000 m (Figure 5(a)). In TRL, a weak significant attraction was observed at short distances (<300 m). From approximately 600 m up to 5000 m, repulsion between cattle and megaherbivores was observed (Figure 5(b)). In PGA, a very strong repulsion was exhibited at short distances (<400 m) followed by moderate attraction up to 5000 m (Figure 5(c)).

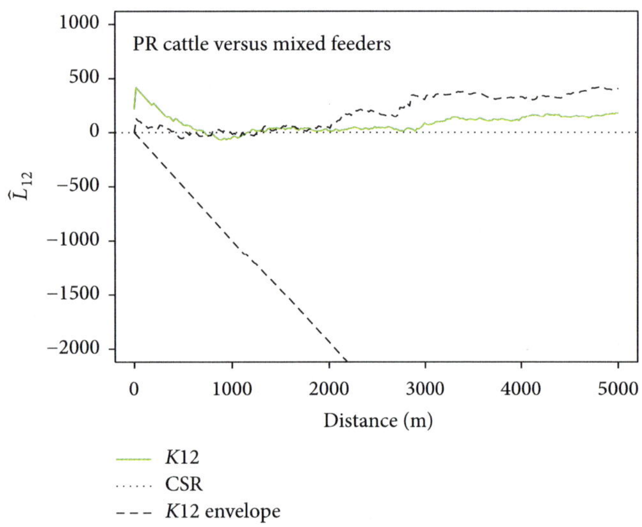

(a)

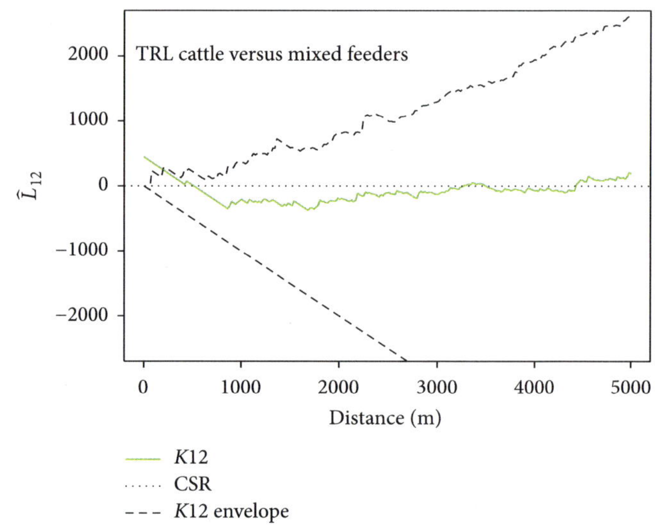

(b)

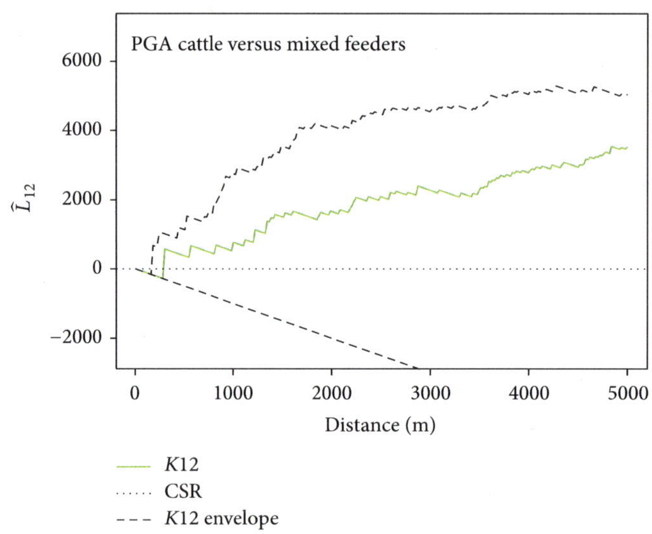

(c)

FIGURE 4: Spatial relationship between cattle and mixed feeders across land-use types: (a) private ranches (PR), (b) transitional lands (TRL), and (c) pastoral grazing areas (PGA). Green solid line K_{12} is Ripley's bivariate function estimate; dashed lines (K_{12} envelope) are the 95% upper and lower confidence envelopes while the dotted line is the reference for complete spatial randomness (CSR). Source: author, 2017.

A significant positive dependence between cattle and medium-sized ungulates was observed throughout all the scales of distances in the PR; however, the strength of the relationship weakened towards the 5000 m mark (Figure 6(a)). In the TRL, the relationship between the two ungulate guilds indicated a significant attraction up to 5000 m (Figure 6(b)). In PGA, cattle and medium-sized ungulates showed significant attraction at short distances (<400 m); a short stint of repulsion was exhibited between approximately 700 m and 1000 m followed by attraction behavior up to 5000 m (Figure 6(c)).

4. Discussion

The wild grazers largely exhibited significant attraction to cattle in PR and TRL unlike in PGA where they showed both repulsion and attraction. In a parallel study, mean NDVI in PR and TRL (excluding forested areas) was higher than in PGA (excluding forested areas) indicating higher forage availability (G.W. Kinga, unpublished data). The observed attraction between cattle and wild grazers in PR and TRL is consistent with previous studies where niche overlap between cattle and some wild grazers has been reported especially when forage resources are abundant [26, 49]. The observed spatial attraction in PR and TRL could be as a result of both habitat and dietary niche overlap between cattle and wild grazers. In PGA, the spatial repulsion at short distances (≤1400 m) could be as a result of differential use of space and/or forage resources which is associated with resource scarcity [26]. At moderate to larger scale distances (≥2900 m) cattle and wild grazers exhibited a positive relationship (attraction). The deviation of the observed spatial

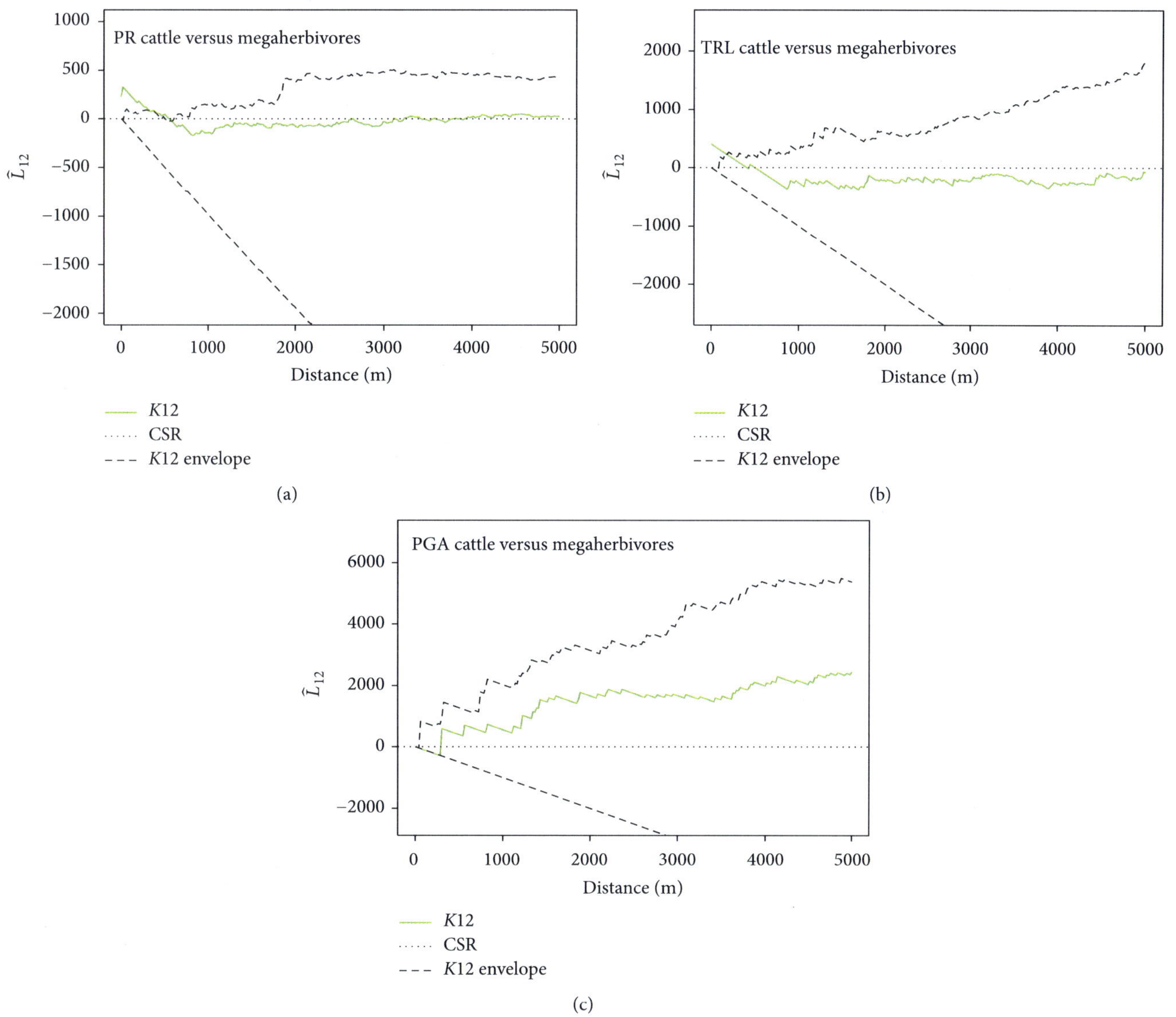

FIGURE 5: Spatial relationship between cattle and megaherbivores across land-use types: (a) private ranches (PR), (b) transitional lands (TRL), and (c) pastoral grazing areas (PGA). Green solid line K_{12} is Ripley's bivariate function estimate; dashed lines (K_{12} envelope) are the 95% upper and lower confidence envelopes while the dotted line is the reference for complete spatial randomness (CSR). Source: author, 2017.

relationship (L_{12} curve) from complete spatial randomness (CSR) was greatest in PGA followed by TRL and least in PR, suggesting that "randomness" was relatively higher in PR followed by TRL and finally the PGA, a phenomenon which is thought to correlate well with relative forage resources abundance across the different land-use types.

Browsers had a significant attraction to cattle at distances (≤700 m) in PR; however, they largely exhibited negative dependence in PR and TRL. Conversely, they generally exhibited attraction in PGA which was significant at distances (approximately ≤500 m). The observed repulsion between cattle and browsers in both PR and TRL could be due to lack of dietary niche overlap as the two guilds have exclusively different feeding styles as was observed in some browser species [50]. The attraction between cattle and browsers in PGA could possibly be due to sharing of same resource patches even though exploiting different forage materials. Although there was general repulsion in PR, its strength was very weak (slightly below CSR) unlike in TRL.

Mixed feeders exhibited both departures from CSR in all the three land-use types considered in this study. This is presumed to be due to their dynamic feeding style allowing them to conveniently adjust their spatial interaction with cattle. Attraction, independence, and repulsion at different scales of distance were evident (Figures 4(a)–4(c)), an observation closely consistent with [44] on three different mixed feeder species. Just like in the case of grazers and browsers, the departures from CSR were lowest in PR, moderate in TRL, and highest in PGA.

Megaherbivores exhibited attraction towards cattle at short distances (≤700 m and ≤500 m) in PR and TRL, respectively, possibly due to habitat niche overlap as compared

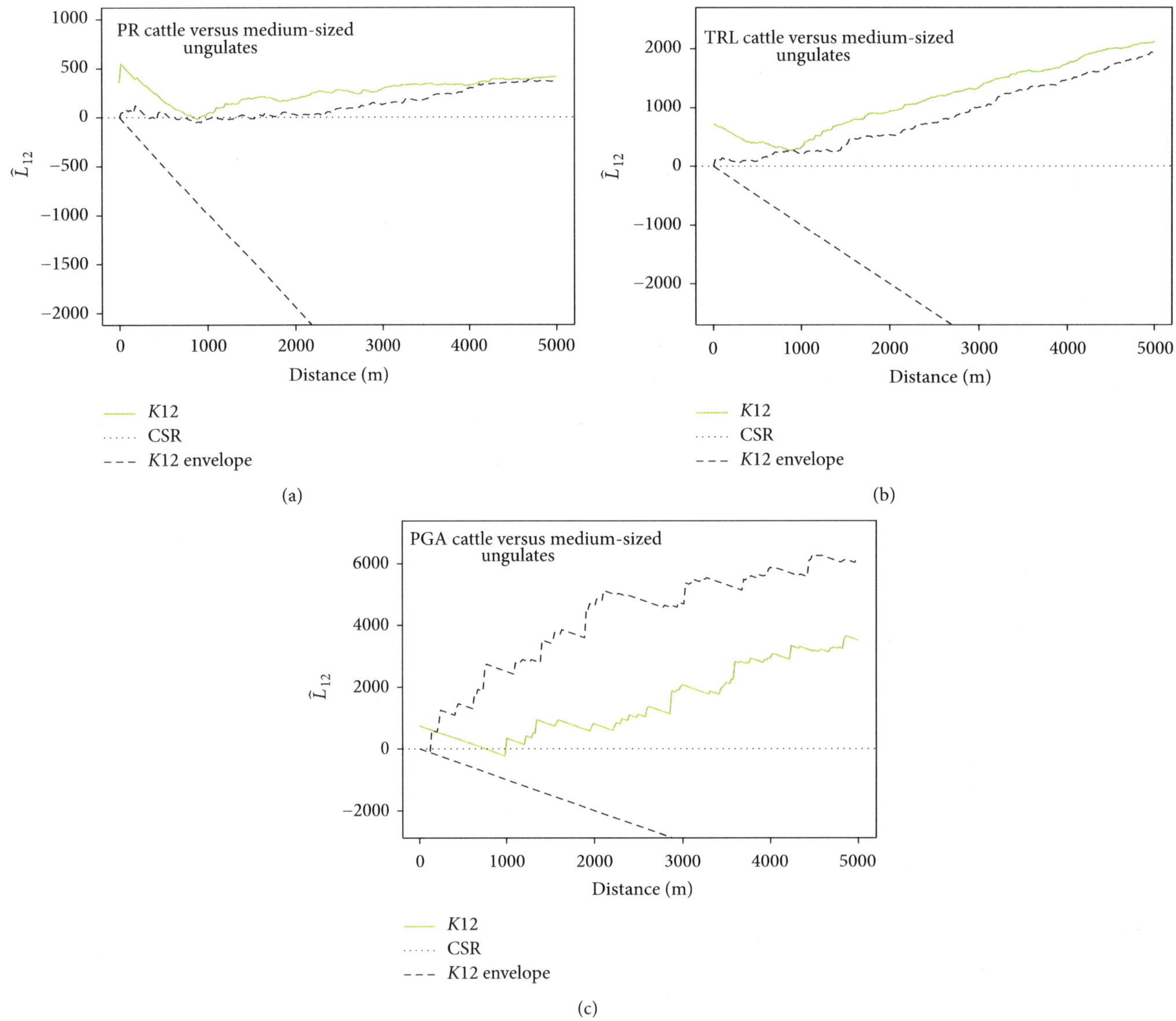

FIGURE 6: Spatial relationship between cattle and medium-sized ungulates across land-use types: (a) private ranches (PR), (b) transitional lands (TRL), and (c) pastoral grazing areas (PGA). Green solid line K_{12} is Ripley's bivariate function estimate; dashed lines (K_{12} envelope) are the 95% upper and lower confidence envelopes while the dotted line is the reference for complete spatial randomness (CSR). Source: author, 2017.

to dietary niche overlap. This is more so because no pure grazer was in this ungulate guild (megaherbivore). The spatial repulsion observed at fairly short to moderate distances in PR (≈700 m to ≈3300 m) and short to larger distances in TRL (≈600 m to 5000 m) broadly conforms with previous studies indicating that particular species of megaherbivores (elephants) segregate with cattle even at larger scales of distance [44]. This could be as a result of the two ungulate groups exploiting different habitat types within the respective land-use type. The variation of L_{12} curve from CSR was minimal in PR than in TRL, possibly indicating some degree of independence in PR than in TRL. The higher attraction in PGA especially at larger scales of distance could be as a result of the two guilds using same habitat patches within the land-use type.

The medium-sized ungulates broadly indicated attraction to cattle at nearly all scales of distance in all the three land-use types (Figures 6(a)–6(c)); this was attributed to the fact that some wild ungulates in this guild are pure grazers and mixed feeders that not only shared the same habitat but also used same dietary niche with cattle. This ungulate group showed greater deviation from CSR in the PGA than in TRL and PR, implying that more independence was exhibited in PR.

The observed spatial relationships (attraction, independence, and repulsion) between cattle and different wild herbivore guilds in the three different land-use types can largely be attributed to pasture resources availability and the density of cattle and wild herbivores in the specific land-use types. Pasture resource availability is a requisite conditional requirement for interspecific competition to occur between sympatric populations [14]; wild ungulates have previously employed spatial partitioning to counter the competitive effects from cattle [28]. And cattle therefore are a possible cause of the observed spatial relationships. On the other hand, the density of either cattle or wild herbivore guilds in the respective land-use type may potentially affect the spatial relationship between them. Cattle density has previously been highest in TRL followed by PR and PGA while wild ungulates density has been highest in PR followed by TRL and PGA having the lowest density [23]. The observed spatial relationships in this study cannot therefore be singly attributed to competition arising from pasture resource use or varying densities of cattle and specific wild ungulate guilds across the land-use types. However, it is also important to consider the departure of the observed spatial relationship (L_{12} curve) from CSR and how it changes across land-use systems and correlate it with other community attributes in future studies.

5. Conclusions

The five wild ungulate guilds considered in this study exhibited varying spatial relationships with cattle at different scales of distance, a phenomenon which is possibly attributed to habitat and/or dietary niche overlap amidst other possible biotic and abiotic factors. Stronger attractions were exhibited in PGA than in TRL and PR. Moreover, most important is the consistent observation that the variation of the observed spatial relationship (L_{12}) from CSR increased from PR through TRL to PGA for all the wild ungulate guilds analyzed. This is perceived as an indicator of randomness and therefore implying that both domestic and wild ungulate guilds were more independently distributed in PR followed by TRL and finally the PGA. This order is to some extent associated with forage resources available and accessible to wild and domestic ungulates in these three land-use types. The near independent spatial distribution between cattle and different wild herbivores in PR could possibly be an indicator of better coexistence between these two guilds especially in ranches which are better managed ecologically compared to the other land-use systems. The spatial attraction in PGA could be an indicator of complete niche overlap between cattle and wild herbivores in exploiting forage resources in this land-use type which are fairly scarce due to the uncontrolled stocking rate by the community and the "open-access" nature of resources in PGA, potentially leading to competition. The extent of departures from CSR in the three land-use systems shows clearly that cattle have less effect on wild herbivores in PR than in TRL and PGA; this is a possible indicator of better coexistence between the two herbivore guilds in PR than in the other two land-use systems. However, more studies on this subject area need to be done and specifically focus on individual species, effects of seasonality, predator presence, forage availability, habitat type, and patchiness and other anthropogenic factors that may influence the spatial relationship between cattle and wild ungulates.

Conflicts of Interest

The authors declare no conflicts of interest regarding publishing this paper.

Acknowledgments

The authors acknowledge the Department of Resources Survey and Remote Sensing (DRSRS) for making the data used in this study available besides allowing access to the library resources.

References

[1] I. Silberbauer-Gottsberger and G. Gottsberger, "Tropical savannas-introduction," *Tropical Biology and Conservation Management*, vol. X, http://www.eolss.net/sample-chapters/c20/e6-142-se-18.pdf.

[2] R. L. Kruska, R. S. Reid, P. K. Thornton, N. Henninger, and P. M. Kristjanson, "Mapping livestock-oriented agricultural production systems for the developing world," *Agricultural Systems*, vol. 77, no. 1, pp. 39–63, 2003.

[3] P. K. Thornton, R. L. Kruska, N. Henninger, P. M. Kristjanson, R. S. Reid, and T. P. Robinson, "Locating poor livestock keepers at the global level for research and development targeting," *Land Use Policy*, vol. 20, no. 4, pp. 311–322, 2003.

[4] FAO. (1999, 12/6/2017). *Cattle and small ruminant systems in sub-saharan africa*. Available: http://www.fao.org/docrep/005/Y4176E/y4176e0f.htm#bm15.

[5] T. P. Robinson, G. R. William Wint, G. Conchedda et al., "Mapping the global distribution of livestock," *PLoS ONE*, vol. 9, no. 5, Article ID e96084, 2014.

[6] "Atlas of african agriculture research and development: Revealing agriculture's place in africa," in *International Food Policy Research Institute*, K. Sebastian, Ed., WA, USA, 2014.

[7] J. T. Du Toit and D. H. M. Cumming, "Functional significance of ungulate diversity in African savannas and the ecological implications of the spread of pastoralism," *Biodiversity and Conservation*, vol. 8, no. 12, pp. 1643–1661, 1999.

[8] D. A. Frank, S. J. McNaughton, and B. F. Tracy, "The ecology of the earth's grazing ecosystems: Profound functional similarities exist between the Serengeti and Yellowstone," *Bioscience*, vol. 48, no. 7, pp. 513–521, 1998.

[9] D. Coventry, "World Resources 2000–2001: People and Ecosystems: The Fraying Web of Life," *Agriculture, Ecosystems & Environment*, vol. 86, no. 1, pp. 109-110, 2001.

[10] B. Deák, B. Tóthmérész, O. Valkó et al., "Cultural monuments and nature conservation: a review of the role of kurgans in the conservation and restoration of steppe vegetation," *Biodiversity and Conservation*, vol. 25, no. 12, pp. 2473–2490, 2016.

[11] C. L. Delgado, "Rising demand for meat and milk in developing countries: Implications for grasslands-based livestock production," *Grassland: A Global Resource*, pp. 29–39, 2005.

[12] P. K. Thornton, "Livestock production: Recent trends, future prospects," *Philosophical Transactions of the Royal Society B: Biological Sciences*, vol. 365, no. 1554, pp. 2853–2867, 2010.

[13] D. Western, S. Russell, and I. Cuthil, "The status of wildlife in protected areas compared to non-protected areas of Kenya," *PLoS ONE*, vol. 4, no. 7, Article ID e6140, 2009.

[14] B. Butt and M. D. Turner, "Clarifying competition: the case of wildlife and pastoral livestock in East Africa," *Pastoralism*, vol. 2, no. 1, article no. 9, 2012.

[15] W. O. Odadi, M. K. Karachi, S. A. Abdulrazak, and T. P. Young, "African wild ungulates compete with or facilitate cattle depending on season," *Science*, vol. 333, no. 6050, pp. 1753–1755, 2011.

[16] H. H. Prins, J. G. Grootenhuis, and T. T. Dolan, *Wildlife Conservation by Sustainable Use*, Springer Netherlands, Dordrecht, 2000.

[17] I. J. Gordon, "Facilitation of red deer grazing by cattle and its impact on red deer performance," *Journal of Applied Ecology*, vol. 25, no. 1, pp. 1–10, 1988.

[18] D. D. Maleko, G. N. Mbassa, W. F. Maanga, and E. S. Sisya, "Impacts of wildlife-livestock interactions in and around arusha national park, tanzania," *Current Research Journal of Biological Sciences*, vol. 4, pp. 471–476, 2012.

[19] H. Z. Dohna, D. E. Peck, B. K. Johnson, A. Reeves, and B. A. Schumaker, "Wildlife-livestock interactions in a western rangeland setting: Quantifying disease-relevant contacts," *Preventive Veterinary Medicine*, vol. 113, no. 4, pp. 447–456, 2014.

[20] I. J. Gordon and A. W. Illius, "Incisor Arcade Structure and Diet Selection in Ruminants," *Functional Ecology*, vol. 2, no. 1, p. 15, 1988.

[21] I. J. Gordon and A. W. Illius, "The functional significance of the browser-grazer dichotomy in African ruminants," *Oecologia*, vol. 98, no. 2, pp. 167–175, 1994.

[22] E. Tóth, B. Deák, O. Valkó et al., "Livestock Type is More Crucial Than Grazing Intensity: Traditional Cattle and Sheep Grazing in Short-Grass Steppes," *Land Degradation & Development*, 2016.

[23] N. J. Georgiadis, J. G. N. Olwero, G. Ojwang', and S. S. Romañach, "Savanna herbivore dynamics in a livestock-dominated landscape: I. Dependence on land use, rainfall, density, and time," *Biological Conservation*, vol. 137, no. 3, pp. 461–472, 2007.

[24] C. T. Robbins, D. E. Spalinger, and W. van Hoven, "Adaptation of ruminants to browse and grass diets: are anatomical-based browser-grazer interpretations valid?" *Oecologia*, vol. 103, no. 2, pp. 208–213, 1995.

[25] P. A. Stephens, C. A. D'Sa, C. Sillero-Zubiri, and N. Leader-Williams, "Impact of livestock and settlement on the large mammalian wildlife of Bale Mountains National Park, southern Ethiopia," *Biological Conservation*, vol. 100, no. 3, pp. 307–322, 2001.

[26] M. M. Voeten and H. H. T. Prins, "Resource partitioning between sympatric wild and domestic herbivores in the Tarangire region of Tanzania," *Oecologia*, vol. 120, no. 2, pp. 287–294, 1999.

[27] T. P. Young, T. M. Palmer, and M. E. Gadd, "Competition and compensation among cattle, zebras, and elephants in a semi-arid savanna in Laikipia, Kenya," *Biological Conservation*, vol. 122, no. 2, pp. 351–359, 2005.

[28] R. Bergström and C. Skarpe, "The abundance of large wild herbivores in a semi-arid savanna in relation to seasons, pans and livestock," *African Journal of Ecology*, vol. 37, no. 1, pp. 12–26, 1999.

[29] LWF, *A wildlife conservation strategy for laikipia county Laikipia Wildlife Forum*, Nanyuki, Kenya, 1st edition, 2012.

[30] M. D. Graham, I. Douglas-Hamilton, W. M. Adams, and P. C. Lee, "The movement of African elephants in a human-dominated land-use mosaic," *Animal Conservation*, vol. 12, no. 5, pp. 445–455, 2009.

[31] A. Ulrich, C. Ifejika Speranza, P. Roden, B. Kiteme, U. Wiesmann, and M. Nüsser, "Small-scale farming in semi-arid areas: Livelihood dynamics between 1997 and 2010 in Laikipia, Kenya," *Journal of Rural Studies*, vol. 28, no. 3, pp. 241–251, 2012.

[32] G. O. Ojwang, *On sampling variability of wildlife cropping-quota allocated on the basis of airborne sample surveys, MSc Thesis*, ITC, Enschede, 2000.

[33] S. R. Sundaresan and C. Riginos, "Lessons learned from biodiversity conservation in the private lands of laikipia, kenya," *A Journal of Natural and Social Sciences*, vol. 1081, 2010.

[34] T. P. Young, T. P. Young, B. D. Okello, D. Kinyua, and T. M. Palmer, "KLEE: A long-term multi-species herbivore exclusion experiment in Laikipia, Kenya," *African Journal of Range & Forage Science*, vol. 14, no. 3, pp. 94–102, 1997.

[35] T. P. Young, N. Patridge, and A. Macrae, "Long-term glades in acacia bushland and their edge effects in Laikipia, Kenya," *Ecological Applications*, vol. 5, no. 1, pp. 97–108, 1995.

[36] LWF, The contribution of the rural economy of laikipia as the basis of a model county.

[37] M. Kinnaird, OT. Brien, and G. Ojwang, *Sample count aerial surveys as a monitoring tool for wildlife and livestock: A case study from laikipia county*, 2012.

[38] J. B. Kauffman and D. A. Pyke, "Range ecology, global livestock influences," *Encyclopedia of biodiversity*, vol. 5, pp. 33–52, 2001.

[39] A. C. Treydte, S. Baumgartner, I. M. A. Heitkönig, C. C. Grant, and W. M. Getz, "Herbaceous forage and selection patterns by ungulates across varying herbivore assemblages in a South African savanna," *PLoS ONE*, vol. 8, no. 12, Article ID e82831, 2013.

[40] H. Fritz, P. Duncan, I. J. Gordon, and A. W. Illius, "Megaherbivores influence trophic guilds structure in African ungulate communities," *Oecologia*, vol. 131, no. 4, pp. 620–625, 2002.

[41] B. Rowlingson, P. Diggle, R. Bivand, G. Petris, and S. Eglen, Splancs, Spatial and space-time point pattern analysis. R package version,.

[42] R.Development.Core.Team, *R: A language and environment for statistical computing*, R Foundation for Statistical Computing, Vienna, Austria, 2008, http://www.R-project.org.

[43] B. S. Rowlingson and P. J. Diggle, "Splancs, Spatial point pattern analysis code in s-plus," *Computers Geosciences*, vol. 19, pp. 627–655, 1993.

[44] F. Hibert, C. Calenge, H. Fritz et al., "Spatial avoidance of invading pastoral cattle by wild ungulates: Insights from using point process statistics," *Biodiversity and Conservation*, vol. 19, no. 7, pp. 2003–2024, 2010.

[45] T. C. Bailey and A. C. Gatrell, *Interactive spatial data analysis, . Longman Scientific Technical*, vol. 413, Longman Scientific Technical, Essex, 1995.

[46] F. Goreaud and R. Pélissier, "Avoiding misinterpretation of biotic interactions with the intertype K 12-function: population independence vs. random labelling hypotheses," *Journal of Vegetation Science*, vol. 14, no. 5, pp. 681–692, 2003.

[47] S. N. Martens, D. D. Breshears, C. W. Meyer, and F. J. Barnes, "Scales of above-ground and below-ground competition in a semi-arid woodland detected from spatial pattern," *Journal of Vegetation Science*, vol. 8, no. 5, pp. 655–664, 1997.

[48] H. W. Lotwick and B. W. Silverman, "Methods for analysing spatial processes of several types of points," *Journal of the Royal Statistical Society. Series B (Methodological)*, vol. 44, no. 3, pp. 406–413, 1982.

[49] W. K. Ego, D. M. Mbuvi, and P. F. K. Kibet, "Dietary composition of wildebeest (Connochaetes taurinus) kongoni (Alcephalus buselaphus) and cattle (Bos indicus), grazing on a common ranch in south-central Kenya," *African Journal of Ecology*, vol. 41, no. 1, pp. 83–92, 2003.

[50] H. Fritz, M. De Garine-Wichatitsky, and G. Letessier, "Habitat use by sympatric wild and domestic herbivores in an African savanna woodland: The influence of cattle spatial behaviour," *Journal of Applied Ecology*, vol. 33, no. 3, pp. 589–598, 1996.

4

Experimental Study of Environmental Effects: Leaf Traits of Juvenile *Fagus sylvatica*, *Acer pseudoplatanus*, and *Carpinus betulus* are Comparableto Leaves of Mature Trees in Upper Canopies

Stephanie Stiegel and Jasmin Mantilla-Contreras

Institute of Biology and Chemistry, University of Hildesheim, 31141 Hildesheim, Germany

Correspondence should be addressed to Stephanie Stiegel; stiegel@uni-hildesheim.de

Academic Editor: Béla Tóthmérész

Morphological and functional leaf traits like leaf toughness and nutrient content are essentially influenced by the environment, especially through light and climatic conditions. Varying light conditions have been identified as a significant predictor for the variation of many leaf traits. However, the leaf acclimation to light is suggested to be of secondary importance. The aim of the experimental study was to analyse environmental effects (microclimate and soil moisture), which are present in upper canopies of forest stands, on leaf traits of juvenile *Fagus sylvatica* L. (European beech; Fagaceae), *Acer pseudoplatanus* L. (sycamore maple; Sapindaceae), and *Carpinus betulus* L. (hornbeam; Betulaceae). The experimental design managed to imitate two distinct microclimates causing different temperature and air humidity conditions. Furthermore, the irrigation treatment with different levels of applied water caused distinct soil moisture conditions in the trial pots. As a result of the treatments, leaves of *C. betulus* showed a tendency of decreased specific leaf area (SLA) caused by the treatment with warmer and drier microclimate. The environmental effect on SLA was even stronger with lower soil moisture conditions. Chlorophyll content showed lower values in treatments with higher soil moisture conditions in both greenhouses for *F. sylvatica* and *A. pseudoplatanus*. The trends are in accordance with combined effects of temperature, air humidity, and soil moisture on SLA, and increased leaf chlorophyll content caused by slight drought stress. Plants in the greenhouses were exposed to full sunlight indicating a microclimatic environment comparable to upper canopies in forest stands. The comparable SLA and chlorophyll content between leaves of mature *F. sylvatica* trees in upper canopies and juvenile trees of the greenhouses suggest similar environmental conditions instead of ontogenetic effects that are responsible for the formation of leaf trait characteristics.

1. Introduction

Morphological and functional leaf traits are essentially influenced by the environment, especially through light and climatic conditions. Changes in climate (temperature and air humidity) and light can affect leaf toughness and leaf nutrient content like nitrogen (N) and carbon (C) concentrations. While leaves of tree seedlings and saplings grow in a similar environment of the understorey, large trees need to produce leaves with a distinct development of traits that are acclimated to different environmental conditions in the canopy.

The formation of softer leaves with a thinner and larger leaf lamina, represented by a high specific leaf area (SLA), is a common response to humid environments [1–3]. In addition, high temperatures can also lead to an increase in SLA, but it strongly depends on sufficient soil moisture content [4]. Low soil moisture content rather leads to water stress situations for plants. In return, water stress can cause sclerophylly [5], resulting in thickened, hardened foliage that resists loss of moisture. Sclerophylly is based in the accumulation of phenolic compounds and lignification of leaf tissues [6, 7].

Leaf N concentrations are especially influenced by light conditions [8, 9]. Sun-exposed leaves usually show increased leaf N concentrations compared to shade leaves [10, 11]. Nonetheless, patterns of leaf N content also depend on the shade tolerance of tree species. In lower light environments, increasing the leaf N concentration is a strategy of N partitioning for more efficient light harvesting [12]. Furthermore, levels of humidity can also affect leaf nutrient concentrations. Low soil moisture conditions of dry environments, causing water stress to host plants, increase the N content in plant tissues [13].

Varying light conditions have been identified as a significant predictor for the variation of many leaf traits within forest canopies (e.g., [14, 15]). Leaf trait trends identified by Thomas [16] are likely to be influenced by the leaf acclimation both to environmental conditions (light) and to plant ontogeny (tree size). However, the leaf acclimation to light is suggested to be of secondary importance. Studies with a controlled light effect (e.g., comparison between leaves of open-grown saplings and upper canopy trees) indicate that ontogenetic changes in leaf toughness and herbivory cannot be fully accounted by environmental acclimation responses to sun and shade [17, 18]. Strong effects of tree size on leaf toughness are found independently of crown exposure [16]. Decreases in SLA and related leaf features (leaf tissue density and lignifications) at the end of tree ontogeny are also noted to be independent of sun and shade acclimation [19–22]. Furthermore, the magnitude of ontogenetic changes in traits is larger than what is documented for studies of light acclimation responses [18, 23–26].

The aim of the experimental study was to analyse environmental effects, which are present in upper canopies of forest stands, on leaf traits of juvenile *Fagus sylvatica* L. (European beech; Fagaceae), *Acer pseudoplatanus* L. (sycamore maple; Sapindaceae), and *Carpinus betulus* L. (hornbeam; Betulaceae). The two following hypotheses were tested: (1) higher temperatures (and lower air humidity) and lower soil moisture conditions increase leaf toughness, and (2) lower soil moisture conditions increase the leaf N content.

2. Materials and Methods

2.1. Set-Up. Plant individuals of *Fagus sylvatica* L. (European beech; Fagaceae), *Acer pseudoplatanus* L. (sycamore maple; Sapindaceae), and *Carpinus betulus* L. (hornbeam; Betulaceae) with an age of 2 years and a height about 50–80 cm (Müller Münchehof, Seesen, Germany) were planted in trial pots in August 2013 (see Figure 1(a)). A mixture of 50% soil (Fruhstorfer Erde Typ T, HAWITA, Vechta, Germany) and 50% sand (Estrich sand, grain size = 0–2 mm, Tönsmeier, Hildesheim, Germany) was used as substrate. Two greenhouses (size: 6x28 m) were installed with different UV permeable greenhouse films (FVG EURO 4 and FVG Sun 5 Clear ST, FVG Professional Gardening, Dernbach, Germany) in March 2014, creating distinct climatic conditions (see Figure 1(b)). A plant protection product (Micula, Biofa AG, Münsingen, Germany) was applied to the trees in April 2014 against eggs and individuals of sap-sucking insects that potentially occurred on the tree individuals avoiding herbivory on the experimental foliage material. All trial pots were protected against insects with mosquito nets.

Table 1: Treatment codes representing the combination of temperature and irrigation levels that were used in the experimental study.

Treatment	Lower irrigation	Higher irrigation
Lower temperature	LTLW	LTHW
Higher temperature	HTLW	HTHW

The experimental design consisted of 10 trial pots for each tree species in greenhouses 1 and 2. Manipulations of microclimate and soil moisture were used as treatments for the trial pots. Trial pots in greenhouses 1 and 2 were labelled with the codes LT (lower temperature) and HT (higher temperature) for the microclimate treatment, respectively. Irrigation codes LW (lower water amount) and HW (higher water amount) were added to the trial pots in each greenhouse according to the soil moisture treatment. Overall, four different treatments with five replicates existed for each tree species (see Table 1).

Trial pots were irrigated two or three times a week. The different amounts of irrigation levels were orientated to maximum and minimum values of the precipitation gradient of a field study (Artern: 59 l/m^2; Wahlsburg: 75 l/m^2) by Stiegel *et al.* [27]. The area of a trial pot was 0.03 m^2. Therefore, the amount of water was adapted to the size of the trial pot resulting in 1.77 l and 2.25 l per month for lower and higher water irrigation treatments, respectively. Based on tougher conditions in the trial pots compared to forest sites, trees were always irrigated when soil was dry to avoid withering of individuals. A total amount of 4.96 l and 7.08 l water per month (about three times higher than at the forest sites) was used for irrigation of each trial pot with lower and higher water irrigation, respectively.

2.2. Measurements. Greenhouse measurements took place in June 2014. Microclimate was assessed as air temperature and relative air humidity. Microclimatic data were measured every hour with data loggers (iButton, Model DS1923, Maxim Integrated, California, USA). Data loggers were installed in the centers at about 1.5 m height in both greenhouses. Soil moisture was measured as volumetric water content (% v v^{-1}) with a soil moisture sensor (FieldScout TDR 100 Soil Moisture Meter, Aurora, Illinois, USA) using 7.5 cm long rods. Five measurements of soil moisture were taken for every trial pot during the experimental period to calculate mean values for statistical analyses.

Specific leaf area (cm^2 g^{-1}) was assessed as an indicator for leaf toughness. It relates the area of a fresh leaf to its dry mass, and low SLA values are linked to structural defences [28]. Fully developed leaves were collected for all tree individuals in June 2014 (four leaves of *F. sylvatica* and *C. betulus* and three leaves of *A. pseudoplatanus*). All fresh leaves were scanned with a flat-bed scanner (CanoScan LiDE 110, Canon, Krefeld, Germany), analysing their areas with the computer image analysis system WinFOLIA (Régent Instruments Inc., Ville de Québec, QC, Canada). Then, each foliage sample was dried (70°C, 48 h) and weighed for calculation of SLA that

TABLE 2: Climatic conditions represented by temperature and relative air humidity during day (5 am–9 pm; n = 1020) and midday (11 am–2 pm; n = 60) in greenhouses 1 and 2 in June 2014. Values represent the median and interquartile ranges (IQR = first quartile, third quartile). Uppercase letters indicate significant differences of temperature and relative air humidity between the greenhouses using Mann-Whitney U test ($p < 0.05$).

Climate	**Temperature** (°C)		**Relative air humidity** (%)	
	Greenhouse 1	Greenhouse 2	Greenhouse 1	Greenhouse 2
Day	21.4 (16.7,27.9)^{A}	22.7 (17.2,30.0)^{B}	58 (40,80)^{A}	55 (37,79)^{A}
Midday	26.5 (22.3,31.5)^{A}	29.3 (24.1,35.9)^{B}	41 (33,52)^{A}	37 (28,49)^{B}

(a) (b)

FIGURE 1: Experimental study with the (a) trail pots harbouring *Fagus sylvatica*, *Acer pseudoplatanus*, and *Carpinus betulus* tree individuals (b) in greenhouses 1 (left) and 2 (right) at the installation site at the Samelsonplatz, Hildesheim.

was used as the mean per individual for further analyses of leaf toughness.

The chlorophyll content of leaves correlates with leaf N content [29], because up to 75% of N content is located in chloroplasts [30]. Chlorophyll content, as an indicator for leaf N content, was measured with a CCM-200 plus Chlorophyll Content Meter (Opti-Sciences Inc., Hudson, NH, USA). From each tree individual, four chlorophyll values measured as chlorophyll content index (CCI) were taken in June 2014. Average values were calculated for each treatment per tree species (n = 20). Since *A. pseudoplatanus* did not survive the treatment with higher temperature and lower water irrigation, neither SLA nor chlorophyll content could be assessed.

2.3. Statistics. For significant comparisons of measured parameters (microclimate, soil moisture conditions, and leaf traits) in the greenhouses, statistical analyses were performed with R Version 3.4.1 [31]. Statistical distributions of the parameters were assessed with the Shapiro-Wilk test. Depending on the statistical distributions, further tests for microclimate and soil moisture were performed with Mann-Whitney U test and t-test, respectively. Comparisons of SLA and chlorophyll content between the treatments were performed with ANOVA or Kruskal-Wallis and suitable post-hoc test using the R package *pgrimess* [32].

3. Results

Climatic conditions represented by temperature and relative air humidity differed between the two greenhouses (see Table 2). On average, temperature was increased about 1.3°C and 2.8°C, and relative air humidity decreased about 3% and 4% in greenhouse 2 compared to greenhouse 1 during day and midday, respectively. Humidity was only significantly distinct between the two greenhouses considering midday values.

TABLE 3: Comparisons of soil moisture conditions (volumetric water content) in the trial pots between lower and higher irrigation treatments for all three tree species (n = 10, per species). Presented are mean values with standard deviation. Uppercase letters indicate significant differences using t-test (*F. sylvatica*: $p = 0.003$, df = 13; *A. pseudoplatanus*: $p = 0.028$, df = 18; *C. betulus*: $p = 0.079$, df = 18).

Trial pot species	**Soil moisture** (% v v^{-1})	
	Lower irrigation	Higher irrigation
Fagus sylvatica	18.9 (± 2.2)^{A}	24.6 (± 4.3)^{B}
Acer pseudoplatanus	21.6 (± 3.9)^{A}	25.3 (± 3.0)^{B}
Carpinus betulus	19.6 (± 2.4)^{A}	22.4 (± 4.3)^{A}

Soil moisture conditions were increased through the treatment of higher irrigation for all tree species in both greenhouses. Average values of volumetric water content were increased about 6%, 4%, and 3% in the higher irrigated trial pots of *F. sylvatica*, *A. pseudoplatanus*, and *C. betulus*, respectively. Statistical analyses revealed significant differences for trial pots containing *F. sylvatica* and *A. pseudoplatanus* trees (see Table 3).

Average SLA values were highest for *F. sylvatica* (173 cm^2 g^{-1} ± 25.9 SD), intermediate for *A. pseudoplatanus* (167 cm^2 g^{-1} ± 20.3 SD), and lowest for *C. betulus* (155 cm^2 g^{-1} ± 20.8 SD). Differences of SLA between the treatments (microclimate and soil moisture) were only present for *C. betulus* but not for *F. sylvatica* and *A. pseudoplatanus* (see Figure 2). Specific leaf area of *C. betulus* was decreased in warmer temperatures of greenhouse 2 compared to greenhouse 1 but

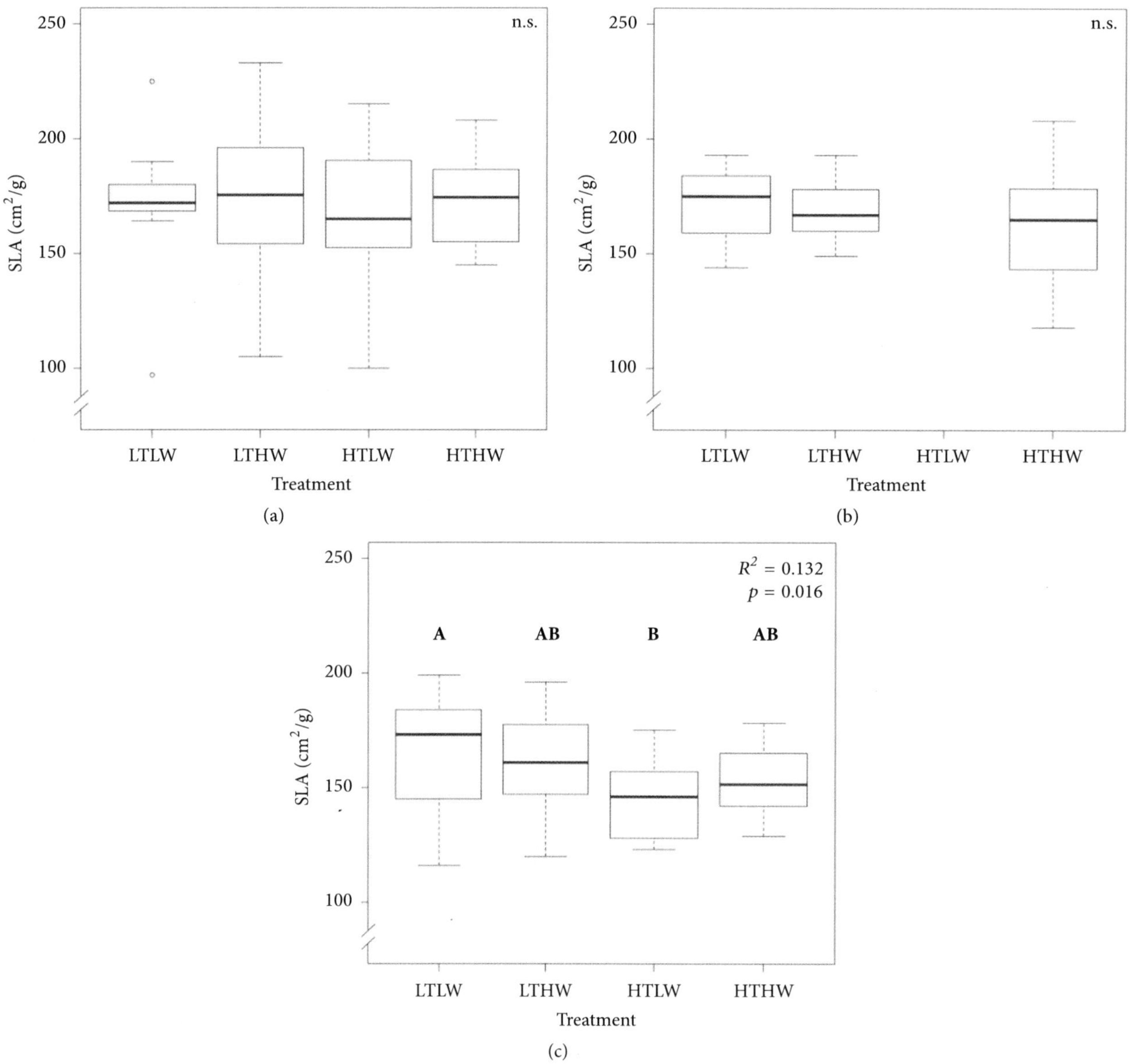

FIGURE 2: Specific leaf area (SLA) of (a) *Fagus sylvatica* (n = 20), (b) *Acer pseudoplatanus* (n = 15), and (c) *Carpinus betulus* (n = 18) for the different treatments in the greenhouse experiment. Boxplots are marked with uppercase letters indicating significant differences or with "n.s." for nonsignificant differences using (a) Kruskal-Wallis and post hoc test ($p < 0.05$; df = 3) and (b-c) ANOVA and Tukey's HSD ($p < 0.05$, df = 3). Treatments: LTLW, lower temperature and lower irrigation; LTHW, lower temperature and higher irrigation; HTLW, higher temperature and lower irrigation; HTHW, higher temperature and higher irrigation.

only differed significantly with the lower irrigation treatment compared to the other three treatment combinations.

Chlorophyll contents were on average similar for all tree species with values ranging between 7.1 CCI ± 2.7 SD (*A. pseudoplatanus*), 7.4 CCI ± 2.7 SD (*C. betulus*), and 8.2 CCI ± 2.0 SD (*F. sylvatica*). Depending on the experimental irrigation treatment, leaf chlorophyll content differed significantly for *F. sylvatica* and *A. pseudoplatanus* but not for *C. betulus* (see Figure 3). *Carpinus betulus* leaves showed lower chlorophyll content in treatments with higher soil moisture conditions in both greenhouses. Chlorophyll content of trees with the same irrigation treatment did not differ significantly between the two greenhouses.

4. Discussion

The experimental design managed to imitate two distinct microclimates causing different temperature and air humidity conditions. Furthermore, the irrigation treatment with different levels of applied water caused distinct soil moisture conditions in the trial pots. Different soil moisture conditions implicated varying water availability for the tree individuals.

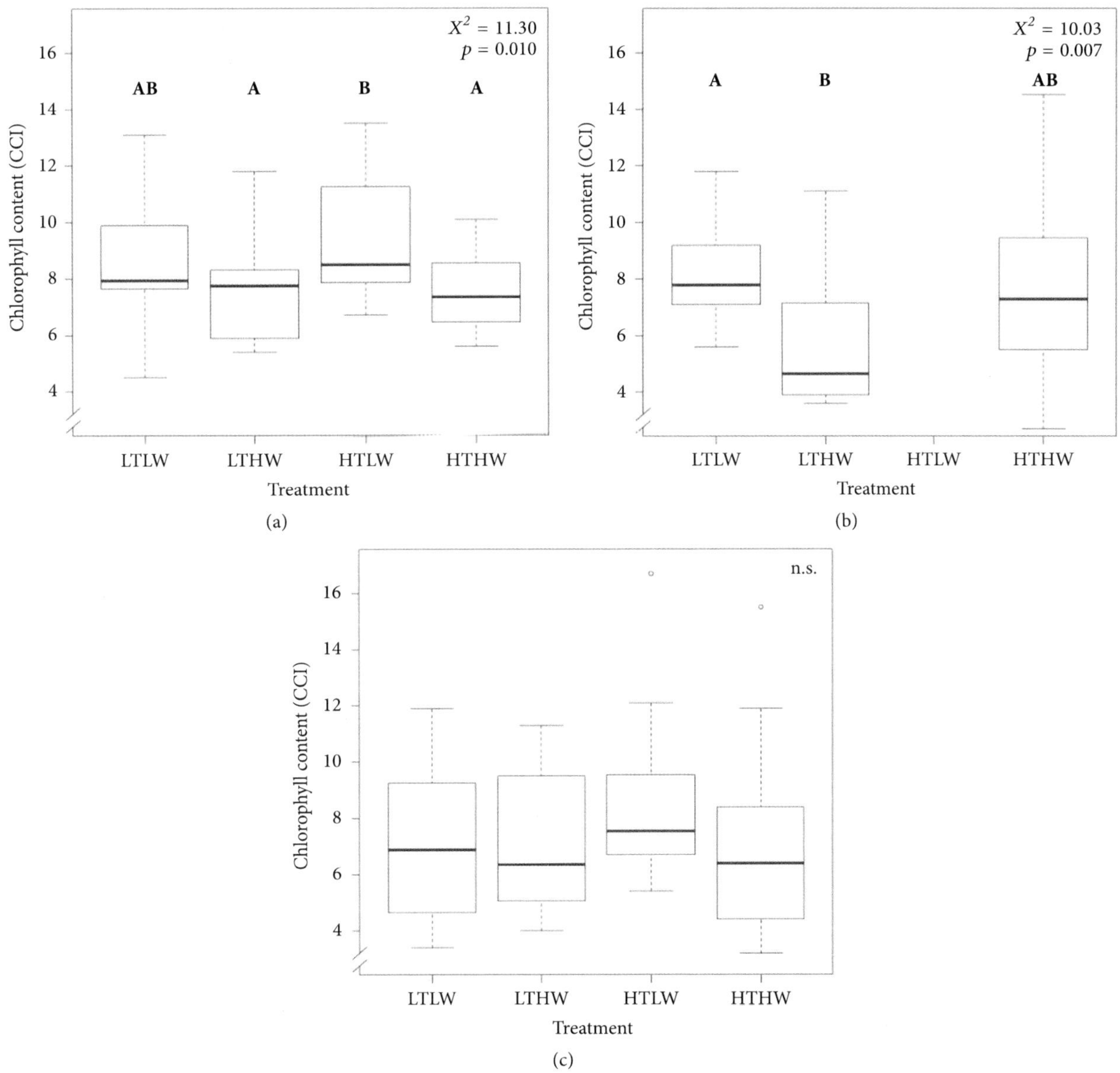

FIGURE 3: Chlorophyll content of (a) *Fagus sylvatica* (n = 20), (b) *Acer pseudoplatanus* (n= 15), and (c) *Carpinus betulus* (n = 20) for the different treatments in the greenhouse experiment. Boxplots are marked with uppercase letters indicating significant differences using Kruskal-Wallis and post hoc test ($p < 0.05$; df = 3) or with "n.s." for nonsignificant differences. Treatments: LTLW, lower temperature and lower irrigation; LTHW, lower temperature and higher irrigation; HTLW, higher temperature and lower irrigation; HTHW, higher temperature and higher irrigation.

As a result of the treatments, *C. betulus* showed varying leaf toughness (indicated by SLA), and *F. sylvatica* and *A. pseudoplatanus* differed in chlorophyll content (indicating leaf N content).

Plants in the greenhouses were exposed to full sunlight indicating a microclimatic environment comparable to upper canopies of mixed deciduous forest stands in Central Germany [27]. Average day temperatures in upper canopies of the field study are 1.5 and 2.8°C lower and relative air humidity conditions are 11 and 14% higher compared to greenhouses 1 and 2 of this experimental study, respectively. The magnitude of differences in temperature and relative air humidity between lower and upper canopies (1.1°C and 5%, respectively) of the forest stands [27] is comparable to the difference between the two greenhouses of this study (1.3°C and 3%, respectively).

The SLA of all three juvenile tree species (*F. sylvatica*, *A. pseudoplatanus*, and *C. betulus*) in this experimental greenhouse study is similar to SLA values of mature *F. sylvatica* in upper canopies, representing less than half of the SLA values that were found for the tree species in the understorey of forest stands [27]. The comparable leaf toughness between upper canopies (mature trees) and the greenhouses (juvenile trees) suggests similar environmental conditions instead of ontogenetic effects that are responsible for the formation of leaf trait characteristics. Upper canopies and the greenhouses were both exposed to full sunlight, and leaves of juvenile trees in the greenhouses with low SLA values exhibit the typical pattern of sun leaves. Generally, a decline of SLA can be induced by light conditions because light increases the leaf thickness [33]. Thicker sun leaves of *F. sylvatica* are characterized by a lower SLA compared to the much thinner blades of shade leaves [34].

Leaf chlorophyll content was decreased about 50% in the experimental study compared to values that were measured in the field study [27]. As part of the photosynthesis, chlorophyll content is dependent on the regulating influence of light conditions. Studies present contrasting results concerning patterns of chlorophyll content based on light conditions. Sun leaves of *F. sylvatica* show higher chlorophyll content than shade leaves [35], in contrast to shade leaves of other tree species that contain more chlorophyll than sun leaves [36]. High light conditions in the greenhouses might have caused the low chlorophyll content in leaves of the three tree species. In addition, this effect of decreasing chlorophyll content can be enhanced through impacts of water supply.

Leaves of *C. betulus* showed a tendency of decreased SLA caused by the treatment with warmer and drier microclimate. The environmental effect on SLA was even stronger with lower soil moisture conditions, thus supporting the first hypothesis. This trend is in accordance with combined effects of temperature and air humidity on SLA [1–3], considering also the important factor of soil moisture [4]. According to basic plant physiology, drought experiments with potted tree seedlings or saplings show a reduction in SLA with decreasing water supply [37, 38]. In contrast, *F. sylvatica* and *A. pseudoplatanus* did not demonstrate significant changes in SLA caused by the experimental treatments. While *F. sylvatica* and *A. pseudoplatanus* are most abundant in forest communities where soil drought is rare, *C. betulus* grows also in regions with regular or episodic summer drought [39]. *Carpinus betulus* is known to have smaller water flux levels per tree in contrast to greater water flux levels encountered in *F. sylvatica* and *A. pseudoplatanus* [40].

Generally, drought sensitivity varies between different tree species. Indeed, *C. betulus* reveals a lower drought sensitivity compared to *F. sylvatica* and *A. pseudoplatanus* [40]. Potentially, *C. betulus* reacted as an adaptation strategy with lower SLA, leading to an increase in leaf toughness, for the survival in a warm and dry environment, which was manipulated by the experimental treatments. This would also be in accordance with the range for SLA, which increases in the sequence *A. pseudoplatanus* < *F. sylvatica* < *C. betulus* [41]. Regarding *A. pseudoplatanus*, the fact that tree individuals grown under the treatment with lower water supply in the warmer and drier greenhouse did not survive has to be taken into account. Compared to the other two tree species, leaves of *A. pseudoplatanus* had the largest size, which potentially resulted in a higher water demand leading to the desiccation of the tree individuals.

In the experimental study, differences of chlorophyll content did not exist between the two greenhouses but regarding the irrigation treatment, leaves showed lower chlorophyll content in treatments with higher soil moisture conditions in both greenhouses. Significant differences were identified for the chlorophyll content of *F. sylvatica* and *A. pseudoplatanus*. The physiological state of plants, which influences photosynthetic processes, is strongly affected by the water supply. Impacts of drought stress on the variation of chlorophyll content have been well studied. Generally, drought stress can limit plant growth through variations of chlorophyll content, respiration, and nutrient metabolism [42]. Chlorophyll content decreases significantly through drought stress situations [43–45]. However, a slight drought stress can increase leaf chlorophyll content [46]. Potentially, the lower irrigation treatment caused minor drought stress conditions for the tree individuals of *F. sylvatica* and *A. pseudoplatanus* reacting with increased chlorophyll content. This is also in accordance with a higher drought sensitivity of *F. sylvatica* and *A. pseudoplatanus* compared to *C. betulus* [40]. In conclusion, the patterns of chlorophyll content, indicating leaf N concentrations, would support the second hypothesis for *F. sylvatica* and *A. pseudoplatanus* that lower soil conditions lead to an increase in leaf N content.

5. Conclusions

The comparable specific leaf area and chlorophyll content between leaves of mature *F. sylvatica* trees in upper canopies of forest stands and juvenile trees of the experimental greenhouse study suggest similar environmental conditions instead of ontogenetic effects that are responsible for the formation of leaf trait characteristics.

Conflicts of Interest

The authors declare that there are no conflicts of interest regarding the publication of this paper.

Acknowledgments

Special thanks are due to Tim Drissen, Pina Lammers, Ricarda Pätsch, Tsiry Rakatoariosa, Carina Rauf, Julia Treitler, Nastia Trenkamp, Robin Stadtmann, Marc Wätzold, and Rebecca Winter for their support along the experimental study. The authors acknowledge financial support for covering the costs to publish in open access by Stiftung Universität Hildesheim.

References

[1] S. A. Cunningham, B. Summerhayes, and M. Westoby, "Evolutionary divergences in leaf structure and chemistry, comparing rainfall and soil nutrient gradients," *Ecological Monographs*, vol. 69, no. 4, pp. 569–588, 1999.

[2] Ü. Niinemets, "Global-scale climatic controls of leaf dry mass per area, density, and thickness in trees and shrubs," *Ecology*, vol. 82, no. 2, pp. 453–469, 2001.

[3] A. Sellin, A. Tullus, A. Niglas, E. Õunapuu, A. Karusion, and K. Lõhmus, "Humidity-driven changes in growth rate, photosynthetic capacity, hydraulic properties and other functional traits in silver birch (Betula pendula)," *Ecological Research*, vol. 28, no. 3, pp. 523–535, 2013.

[4] Z. Z. Xu and G. S. Zhou, "Combined effects of water stress and high temperature on photosynthesis, nitrogen metabolism and lipid peroxidation of a perennial grass Leymus chinensis," *Planta*, vol. 224, no. 5, pp. 1080–1090, 2006.

[5] F. Bussotti, A. Bottacci, A. Bartolesi, P. Grossoni, and C. Tani, "Morpho-anatomical alterations in leaves collected from beech trees (Fagus sylvatica L.) in conditions of natural water stress," *Environmental and Experimental Botany*, vol. 35, no. 2, pp. 201–213, 1995.

[6] F. Bussotti, P. Grossoni, and A. Bottacci, "Sclerophylly in beech (Fagus sylvatica L.) trees: Its relationship with crown transparency, nutritional status and summer drought," *Forestry*, vol. 70, no. 3, pp. 267–271, 1997.

[7] P. Grossoni, F. Bussotti, C. Tani, E. Gravano, S. Santarelli, and A. Bottacci, "Morpho-anatomical alterations in leaves of Fagus Syl Vatica L. And Quercus Ilex L. in different environmental stress condition," *Chemosphere*, vol. 36, no. 4-5, pp. 919–924, 1998.

[8] P. B. Reich, M. B. Walters, and D. S. Ellsworth, "From tropics to tundra: global convergence in plant functioning," *Proceedings of the National Academy of Sciences of the United States of America*, vol. 94, no. 25, pp. 13730–13734, 1997.

[9] I. J. Wright, P. B. Reich, M. Westoby et al., "The worldwide leaf economics spectrum," *Nature*, vol. 428, no. 6985, pp. 821–827, 2004.

[10] M. Fortin and Y. Mauffette, "The suitability of leaves from different canopy layers for a generalist herbivore (Lepidoptera: Lasiocampidae) foraging on sugar maple," *Canadian Journal of Forest Research*, vol. 32, no. 3, pp. 379–389, 2002.

[11] K. R. Levesque, M. Fortin, and Y. Mauffette, "Temperature and food quality effects on growth, consumption and post-ingestive utilization efficiencies of the forest tent caterpillar Malacosoma disstria (Lepidoptera: Lasiocampidae)," *Bulletin of Entomological Research*, vol. 92, no. 2, pp. 127–136, 2002.

[12] U. Niinemets, "Distribution of foliar carbon and nitrogen across the canopy of Fagus sylvatica: Adaptation to a vertical light gradient," *Acta Oecologica*, vol. 16, no. 5, pp. 525–541, 1995.

[13] G. Rouault, J.-N. Candau, F. Lieutier, L.-M. Nageleisen, J.-C. Martin, and N. Warzée, "Effects of drought and heat on forest insect populations in relation to the 2003 drought in Western Europe," *Annals of Forest Science*, vol. 63, no. 6, pp. 613–624, 2006.

[14] T. Rijkers, T. L. Pons, and F. Bongers, "The effect of tree height and light availability on photosynthetic leaf traits of four neotropical species differing in shade tolerance," *Functional Ecology*, vol. 14, no. 1, pp. 77–86, 2000.

[15] D. B. Leal and S. C. Thomas, "Vertical gradients and tree-to-tree variation in shoot morphology and foliar nitrogen in an old-growth Pinus strobus stand," *Canadian Journal of Forest Research*, vol. 33, no. 7, pp. 1304–1314, 2003.

[16] S. C. Thomas, "Photosynthetic capacity peaks at intermediate size in temperate deciduous trees," *Tree Physiology*, vol. 30, no. 5, pp. 555–573, 2010.

[17] S. C. Thomas and W. E. Winner, "Photosynthetic differences between saplings and adult trees: An integration of field results by meta-analysis," *Tree Physiology*, vol. 22, no. 2-3, pp. 117–127, 2002.

[18] S. C. Thomas, A. J. Sztaba, and S. M. Smith, "Herbivory patterns in mature sugar maple: Variation with vertical canopy strata and tree ontogeny," *Ecological Entomology*, vol. 35, no. 1, pp. 1–8, 2010.

[19] M. E. Day, M. S. Greenwood, and A. S. White, "Age-related changes in foliar morphology and physiology in red spruce and their influence on declining photosynthetic rates and productivity with tree age," *Tree Physiology*, vol. 21, no. 16, pp. 1195–1204, 2001.

[20] Ü. Niinemets, "Stomatal conductance alone does not explain the decline in foliar photosynthetic rates with increasing tree age and size in Picea abies and Pinus sylvestris," *Tree Physiology*, vol. 22, no. 8, pp. 515–535, 2002.

[21] E. Nabeshima and T. Hiura, "Size dependency of photosynthetic water- and nitrogen-use efficiency and hydraulic limitation in Acer mono," *Tree Physiology*, vol. 24, no. 7, pp. 745–752, 2004.

[22] J. R. England and P. M. Attiwill, "Changes in leaf morphology and anatomy with tree age and height in the broadleaved evergreen species, Eucalyptus regnans F. Muell," *Trees - Structure and Function*, vol. 20, no. 1, pp. 79–90, 2006.

[23] D. S. Ellsworth and P. B. Reich, "Leaf mass per area, nitrogen content and photosynthetic carbon gain in Acer saccharum seedlings in contrasting forest light environments," *Functional Ecology*, vol. 6, no. 4, pp. 423–435, 1992.

[24] D. S. Ellsworth and P. B. Reich, "Water relations and gas exchange of Acer saccharum seedlings in contrasting natural light and water regimes," *Tree Physiology*, vol. 10, no. 1, pp. 1–20, 1992.

[25] T. W. Sipe and F. A. Bazzaz, "Gap partitioning among maples (Acer) in central New England: Shoot architecture and photosynthesis," *Ecology*, vol. 75, no. 8, pp. 2318–2332, 1994.

[26] M. Beaudet, C. Messier, D. W. Hilbert, E. Lo, Z. M. Wang, and M. J. Lechowicz, "Leaf- and plant-level carbon gain in yellow birch, sugar maple, and beech seedlings from contrasting forest light environments," *Canadian Journal of Forest Research*, vol. 30, no. 3, pp. 390–404, 2000.

[27] S. Stiegel, M. H. Entling, and J. Mantilla-Contreras, "Reading the leaves' palm: Leaf traits and herbivory along the microclimatic gradient of forest layers," *PLoS ONE*, vol. 12, no. 1, 2017.

[28] J. H. C. Cornelissen, S. Lavorel, E. Garnier et al., "A handbook of protocols for standardised and easy measurement of plant functional traits worldwide," *Australian Journal of Botany*, vol. 51, no. 4, pp. 335–380, 2003.

[29] A. K. Van den Berg and T. D. Perkins, "Evaluation of a portable chlorophyll meter to estimate chlorophyll and nitrogen contents in sugar maple (Acer saccharum Marsh.) leaves," *Forest Ecology and Management*, vol. 200, no. 1–3, pp. 113–117, 2004.

[30] M. B. Peoples and M. J. Dalling, "The interplay between proteolysis and amino acid metabolism during senescence and nitrogen reallocation," in *Senescence and aging in plants*, L. D. Noodén and A. C. Leopold, Eds., pp. 181–217, San Diego, USA, 1988, http://agris.fao.org/agris-search/search.do?recordID=US8866793.

[31] R Core Team, *A Language and Environment for Statistical Computing [Internet]*, R Foundation for Statistical Computing, Vienna, Austria, 2017, https://www.R-project.org, 2017.

[32] P. Giraudoux, "Miscellaneous functions for data analysis in ecology, with special emphasis on spatial data," R package version 1.6.7. https://CRAN.R-project.org/package=pgirmess, 2017.

[33] J. I. Yun and S. E. Taylor, "Adaptive implications of leaf thickness for sun- and shade-grown Abutilon theophrasti," *Ecology*, vol. 67, no. 5, pp. 1314–1318, 1986.

[34] H. K. Lichtenthaler, A. Ač, M. V. Marek, J. Kalina, and O. Urban, "Differences in pigment composition, photosynthetic rates and chlorophyll fluorescence images of sun and shade leaves of four tree species," *Plant Physiology and Biochemistry*, vol. 45, no. 8, pp. 577–588, 2007.

[35] H. K. Lichtenthaler, F. Babani, G. Langsdorf, and C. Buschmann, "Measurement of differences in red chlorophyll fluorescence and photosynthetic activity between sun and shade leaves by fluorescence imaging," *Photosynthetica*, vol. 38, no. 4, pp. 521–529, 2000.

[36] R. Matyssek, J. Fromm, H. Rennenberg, and A. Roloff, *Biologie der Bäume [Internet]*, Ulmer, Stuttgart, Germany, 2010, http://www.utb-shop.de/biologie-der-baume.html.

[37] T. T. Kozlowski and S. G. Pallardy, Eds., *Physiology of woody plants [Internet]*, Academic Press, San Diego, USA, 1997.

[38] D. O. Otieno, M. W. T. Schmidt, S. Adiku, and J. Tenhunen, "Physiological and morphological responses to water stress in two Acacia species from contrasting habitats," *Tree Physiology*, vol. 25, no. 3, pp. 361–371, 2005.

[39] H. Ellenberg and C. Leuschner, *Vegetation Mitteleuropas mit den Alpen*, Ulmer, Stuttgart, Germany, 6th edition, 2010.

[40] D. Hölscher, O. Koch, S. Korn, and C. Leuschner, "Sap flux of five co-occurring tree species in a temperate broad-leaved forest during seasonal soil drought," *Trees - Structure and Function*, vol. 19, no. 6, pp. 628–637, 2005.

[41] N. Legner, S. Fleck, and C. Leuschner, "Within-canopy variation in photosynthetic capacity, SLA and foliar N in temperate broad-leaved trees with contrasting shade tolerance," *Trees - Structure and Function*, vol. 28, no. 1, pp. 263–280, 2014.

[42] C. A. Jaleel, P. Manivannan, G. M. A. Lakshmanan, M. Gomathinayagam, and R. Panneerselvam, "Alterations in morphological parameters and photosynthetic pigment responses of *Catharanthus roseus* under soil water deficits," *Colloids and Surfaces B: Biointerfaces*, vol. 61, no. 2, pp. 298–303, 2008.

[43] A. Mafakheri, A. Siosemardeh, B. Bahramnejad, P. C. Struik, and E. Sohrabi, "Effect of drought stress on yield, proline and chlorophyll contents in three chickpea cultivars," *Australian Journal of Crop Science*, vol. 4, no. 8, pp. 580–585, 2010.

[44] R. Gholamin and M. Khayatnezhad, "The effect of end season drought stress on the chlorophyll content, chlorophyll fluorescence parameters and yield in maize cultivars," *Scientific Research and Essays*, vol. 6, no. 25, pp. 5351–5357, 2011.

[45] I. Aref, H. El Atta, M. El Obeid, A. Ahmed, P. Khan, and M. Iqbal, "Effect of water stress on relative water and chlorophyll contents of Juniperus procera Hochst. ex Endlicher in Saudi Arabia," *Life Science Journal*, vol. 10, pp. 681–685, 2013.

[46] J. K. Mensah, B. O. Obadoni, P. G. Eruotor, and F. Onome-Irieguna, "Simulated flooding and drought effects on germination, growth, and yield parameters of sesame (Sesamum indicum L.)," *African Journal of Biotechnology*, vol. 5, no. 13, pp. 1249–1253, 2006.

5

Laboratory Assessment of Forest Soil Respiration Affected by Wildfires under Various Environments of Russia

Evgeny Abakumov,[1] Ekaterina Maksimova,[1] Anna Tsibart,[2] and George Shamilishviliy[1]

[1] *Department of Applied Ecology, Saint Petersburg State University, 16th Line 29, Vasilyevsky Island, Saint Petersburg 199178, Russia*
[2] *Department of Landscape Geochemistry and Soil Georgaphy, Moscow State University, Vorobyovy Gory 1, Moscow 119991, Russia*

Correspondence should be addressed to Evgeny Abakumov; e_abakumov@mail.ru

Academic Editor: L. M. Chu

Pyrogenic carbon emission rates were estimated in the soils of three natural zones in Russia: forest-tundra, south-taiga, and forest-steppe. Postfire soils were found to be characterized by essential losses of soil C due to the combustion fire effect. Soils lost 3 or 5 parts of initial carbon content and showed an essential decrease in the C/N ratio during the fire effect. The pH values increased due to soil enrichment by ash during the fire events. CO_2 emission rates were highest in natural soil samples, because the amount of organic matter affected by mineralization in those soils was higher than in natural ones. Simultaneously, the total values of mineralized carbon were higher in postfire soils because the SOM quality and composition were altered due to the fire effect. The only exception was in forest-tundra soils, where a high portion of dissolved organic compounds was released during the surface fire. The quality of initial SOM and intensity of the wildfire play the most important roles in the fate of SOM in postfire environments. Further study of CO_2 emissions is needed to better characterize postfire SOM dynamics and develop an approach to model this process.

1. Introduction

Great attention has been paid to soil organic matter (SOM) mineralization and possible CO_2 emissions from soil, due to their important effects on the global carbon cycle and ecosystem stability. Fire has an important impact on soil properties [1–13]. Soil processes in postfire soils are quite different from those in the soils of natural landscapes or in technogenous ones. After fires, there is an intensive accumulation of ash on the topsoil [14, 15], ash redistribution in landscapes and the leaching of nutrients and minerals into deeper horizons [16, 17], surface overcompaction and crust accumulation, and soil structure transformation of [4]. Nonetheless, the most expressed consequence of wildfires is uncontrolled CO_2 emission due to changes in soil organic matter stability that occur in high temperatures [18, 19]. The mass loss (i.e., gravimetric concentrations) of soil organic carbon in postfire environments is typically underestimated by researchers in terms of rates and balance units [20, 21]. Two main constituents of SOM loss in postfire environments are mineralization and erosion losses. The mass emission loss of organic carbon can be evaluated by field and laboratory experiments [22]. Postfire CO_2 emission rates have previously been shown to be strongly related to the intensity of the fire impact [1, 22–24]. Furthermore, the mineralization rate is affected by the type of burned organic material [9]. SOM mineralization rates are higher in burned postfire soils than in unburned ones, and alterations in organic matter in postfire soils can lead to alteration of total SOM stability [22, 25].

Controlled laboratory experiments provide data on possible mineralization losses from soils that face different fire impacts. Such experiments are needed to evaluate the apparent stability of SOM mineralization losses and to predict the possible stability and loss of organic carbon in the soil of postfires environments. Data obtained in controlled laboratory experiment are not comparable with data from field experiments (e.g., closed chambers method). Nevertheless,

FIGURE 1: Location of study sites.

data obtained in unified conditions can be used for further simulation modeling of SOM transformation.

The aim of this study is to examine the mineralization of soil organic carbon (SOC) from different types of postfire soils from different environments in Russia, using laboratory incubation with special attention to carbon mineralization rates and CO_2 emissions.

2. Materials and Methods

2.1. Study Sites. Three types of zonal environments (ecoregions) were selected to investigate soil properties and CO_2 emission intensity: forest-tundra, south-taiga, and forest-steppe. Plots situated in the areas surrounding the cities of Nadym (N), Shatura (S), and Togliatti (T) were investigated, using the scenario of surface fires. Sampling was conducted in the one months after surface fires in 2012 (S and T) and in 2016 (N). The location of the sites are shown in Figure 1. The environments studied differ in terms of soil profile organization and the nature of organic matter that accumulated in the soil profiles. Nevertheless, these three plots were studied to understand SOM stabilization rates and biodegradability because there is a lack of data on the quality and alteration of SOM quality in various climatic environments and different soil horizons.

Podzols of forest-tundra were sampled from all soil horizons because of the possible redistribution of organic matter due to eluviation processes. Peatland soils from the Shatura plots were also sampled from all the horizons of solum; two types of soil horizons were studied: burned and unburned. In the case of forest-tundra and south-taiga, only the surface fire scenario was accounted for investigation. In case of forest-steppe, we only compared SOM quality in the superficial horizons (A), because solums were undeveloped and only presented by a few centimeters of humus-enriched topsoils.

2.1.1. Forest-Tundra. Recent data on the natural fire effect on polar (arctic) ecosystems were gathered between July 2016 and August 2016 in the Nadym subregion of the Yamal-Nenets Autonomous Region, Northwest Siberia, Russian Federation (N 65°32′00″; E 72°31′00″, 40–60 m asl). Forest fires were reported during this time period in the northeast area of the studied region. The affected territory was approximately 15,000 hectares and included the areas surrounding the Pangodi settlement (i.e., location of the compressor Pangodinkaya station, which is a key part of the Siberian gas pipeline system). The reason of the fire occurrence was the abnormal high temperatures established since the beginning of the July (27–33°C). The fire arose in studied tundra-forest ecosystem was characterized by medium to high severity, which is proved by thickness of the burned litter and obtained data on the soil respiration rates in fire-affected and control soil. In accordance with the investigations in soils, subjected to fires of medium intensity, the release of carbon dioxide is larger in comparison with the control, whereas in case of strong fire this parameter is smaller than in unaffected soil. Thus the high severity of the wildfire is supposed.

According to geographical zoning assessments of the Tumensky region, the studied territory belongs to the Nadym province of tundra-forest width-zonal Champaign area [26]. From a geomorphological point of view, the studied area is characterized by a distribution of lake-alluvial terraces and represents a subhorizontal surface with absolute altitudes ranging from 30 to 60 m. The following landscape-geomorphological complexes are well distinguished in the modern floodplain of the Nadym River: a strip of ancient above-floodplain terraces forming the base of the root bank, remnants of segmented meanders, and a modern river floodplain with active channel processes. Forest ecosystems of the territory are characterized by three periods and stages of vegetation succession on alluvium. The first stage is the primary colonization of alluvial sediments with sedges and willow that forms separate bushes and small thickets where the accumulation of alluvium continues. The second stage is a homogenous dense willow tree (up to 45 trunks per 10 m^2) with a height of up to 4 m, under the canopy of which there are single specimens of sedges, reed grass, and floodplain grass. The thickness of litter is small, as it is periodically overlapped by alluvial deposits applied by spring-summer spills. Serried stand is formed by willow and birch on plot related to the third stage of the first period. Here, under the canopy, there are single specimens of spruce growth with 5–10 cm in height, ground cover is rare, represented by reed, sedge, and horsetail, and mosses are found only in depressions of the relief.

Described area is related to the second period of succession represented by two variants of the second stage, differing in composition of the stand. On these plots, spruce (sometimes larch) forms the second tree floor of the community, and under the canopy there are many cedar and spruce undergrowth and a small number of larches. In the surveyed areas, the age of spruce and cedar was 80 to 120 years and only for single trees was about 160 years. This indicates a relatively short duration of the period of reorganization of riverine landscape complexes in the lower floodplain.

A study of the renewal potential of the main forest-forming species showed that shoots and young coniferous trees appeared under the canopy of the birch and willow communities when they reached the age of 45–60 years. By this time, a diluted grass cover with an abundance of forest species, a litter, and a thin humus horizon of the soil had already been formed. Further, as the development of woody vegetation changes, the species composition of coniferous growth was changing correspondingly.

The landscapes of the most humid areas of the territory are characterized by the alternation of hilly permafrost-affected peat bogs (with shrub, moss, and lichen vegetation on the hillocks and sedge-cotton grass-moss vegetation in the depressions) and hummocky permafrost-affected shrub-lichen-moss bogs combined with hummock-hollow bogs. The landscapes of the drained areas are covered with coniferous forests containing birch and larch [26].

The burned ecosystem located by lowlands with tundra-forest landscape and was characterized by low soil diversity due to the predomination of homogeneous lacustrine sand parent material and general flatness of the relief. The vegetation cover was composed of highly stratified larch-pine forest (*Larix sibirica*, *Picea obovata*, *Pinus sibirica*, and *P. sylvestris*) with shrubs (*Ledum palustre, Empetrum nigrum, Vaccinium vitis-idaea,* and *V. uliginosum*) and moss-lichen underbrush (*Pleurozium schreberi, Hylocomium splendens, Polytrichum alpestre, Cladonia arbuscula, C. rangiferina,* and *C. stellaris*). The pine forest ecosystems were represented by stands consisting of 10–15 trees and were surrounded by peat land cover. In this study, only soils affected by fires under the pine stands were investigated. Wildfires in Nadym area have become more frequent during last decades and ecosystems of this particular area are faced to wildfires every 5 years. The fire arose in described tundra-forest ecosystem of Nadym region was characterized by medium to high severity, which was evidenced by the thickness of the burned litter. A study of the renewal potential of the main forest-forming species showed that shoots and seedlings of coniferous trees appear under the canopy of the birch and willow communities when they reach the age of 45–60 years. By this time, a diluted grass cover with an abundance of forest species, a litter, and a thin humus horizon of the soil have already been formed. Further, as the development of woody vegetation changes, the species composition of coniferous growth was changing correspondingly.

The soil cover of the investigated area is characterized by a complex mosaic of Podzols that belong to drained sandy positions and Histosols formed in swampy hollows [27]. Gleysols and Cryosols are also typical for the described territory, especially to sites with presence of permafrost, causing stagnic, moisture conditions. The soil is characterized by a sandy texture and represents a Podzol with well-expressed spodic horizons of yellow-ocherous tones underlying albic horizons that were affected by the fire.

2.1.2. South-Taiga. The research was conducted in the West Meshchera fluvioglacial plain of the Moscow Region (Shatura District, N 55°39, E 39°36′, 120 m asl). The investigated area is covered by peatlands and lies on ancient alluvial deposits [16, 17]. This plot belongs to the Petrovsko-Kobelevskoe high-moor peat, which is a part of the Shatura wetland area. Its total area is 6,443 ha. Prior to peat extraction, the peat layer was found to be 7.5 m and its average thickness was 2.5–4.0 m [18]. The large areas of the study site were transformed by drainage and peat mining. Mining began in the peat deposits of the Shatura district in the 1920s [28]. Nowadays the excavation has stopped and the most of peat-hags are in a renewal stage [29]. Because of peat excavation, the natural bog complex changed to large open pits, which were connected with the numerous channels and distributaries and the fields of peat mining at different stages of exploitation. The birch-aspen forests (*Betula pendula* Roth. and *Populus tremula* L.) grow on banks. These types of tree stands were selected for sampling of postfire soils and mature soils. Because of changes in the hydrological regime, the number of peat fires and areas impacted by peat fires increased dramatically [29]. Zonov and Konstantinovitch [30] identified a repeating pattern of fires in drained peatlands, whereby large fires occur nearly every 10 years (i.e., 1972, 1982, 1992, and 2002). In 2010 the wildfires in Russia damaged large areas, and in the Moscow region they caused significant air pollution and ecosystem degradation [31–33]. The Gleyic Histosols, Histic Podzols, and Sod Podzols [34] represent the soil cover of the studied area. This work investigates the peat soils affected by the fires of 2002, 2010, and 2012. The field work and sampling procedures were both conducted in 2012. Histosols and Histic Podzols represent mature soils. The Histosols have the following soil horizons: O-H-Hp-He-Ha-C. The typical soil profile of the Histic Podzols is O-A-E-EBhs-Bhs-Cs. After the fires smoldered and burned out the peat horizon, new ash horizons (Cpir) formed up to 5 cm deep; they formed over peat charry horizons (Hpir, 2-3 cm) and pyrogenic muck peat horizons (Ha,pir, 2-3 cm). Soils under the birch-aspen tree stands were sampled for further laboratory analyses.

2.1.3. Forest-Steppe. The study was conducted in the Samara region near the Volga River, in the central part of European Russia (N 58°39′44″; E 39°17′48, 179 m asl). The extreme summer temperatures in Russia that occurred in 2010 played a large role in the emergence of these wildfires. The extreme weather started in the middle of July and expanded to all of European Russian Europe, Ukraine, and Eastern Europe. Ultimately, the extreme heat led to drought and wildfires in the vast territories of European and Siberian Russia. The forest fire in question occurred in 2010 and affected more than 8,000 hectares. The fire severity was very high according to one of the most applied methods of identifying fire severity (that is the ash colour and ash $CaCO_3$ content [12, 13]).

The study plots were the so-called "steppe islands of pine forests" near Togljatty city, Samara region, and were exposed to widespread forest fires in 2010. It is necessary to emphasize that these "steppe islands of pine forests" are very important parts of forest-steppe ecosystems which conserve the boreal species of plants and animals as well as some soils which are untypical for a steppe zone.

The parent material is composed of old (Pleistocene) alluvial-dune landscape. A forest-steppe environment with higher pedodiversity characterizes the affected ecosystem. The vegetation is composed of pine forests *Pinus sylvestris* L. and other xerophytes species (*Veronica spicata* L., *Sedum acre* L., *Antennaria dioica* L., *Calamagrostis epigeios* (L.) Roth., and *Centaurea marschalliana* Spreng). Hardwoods (*Quercus robur* L., *Betula pendula* Roth., and *Populus tremula* L.) form one or two synfoliums and are added to a pine forest in more humid conditions. Woody and shrubby plantations were burned out, and the trunks of trees were charred to different extents in fire-affected areas. These charred trees were cut down or remained uncleaned for 1-2 years after the wildfires. The grassy cover of the postfire plots was formed by rhizomatous and loose-bunch gramen: *Bromus inermis* Leyss., *Elytrigia repens* L., *Poa* spp., and *Agrostis canina* L. The plants forming the lower forest synfolium (a shrubby, grassy, and mossy cover) burn down in case of surface fire, which indicates a surface fire. The crown fire covers the entire forest from the soil surface to the tops of tree crowns or passes through the trees and the underbrush, herbage, and moss layer. Fruits and seeds which are on the surface and in the solum (the upper soil horizons) can avoid the influence of the fire. Therefore, a crown fire is a type of surface and crown fire.

There is gramineous and herb succession stage at crown forest fire plot. The foliage cover was 60–70%. The following species of plants were presented at that plot: *Calamagrostis epigejos* (L.) Roth., *Bromopsis inermis* (Leyss.) Holub, *Artemisia absinthium* L., *Artemisia austriaca* Jacq., *Artemisia campestris* L., *Nonea pulla* (L.) DC., *Sisymbrium loeselii* L., and *Chamaecytisus ruthenicus* (Fisch. ex Wolosz.) Klask. A shrub steppe will be forming at surface forest fire plot. The foliage cover is 90–95%. The following species of plants were presented at that plot: *Chamaecytisus ruthenicus* (Fisch. ex Wolosz.) Klask., *Veronica chamaedrys* L., *Taraxacum officinale* Wigg. s.l., *Saponaria officinalis* L., *Berteroa incana* (L.) DC., and *Calamagrostis epigejos* (L.) Roth.

A variety of Luvisols and Chernozems prevail in the watershed [35–39]. The soil of the investigated area is sandy and sandy loam textured. In the studied area, the soil was classified as sandy loam soil on Late Pleistocene alluvial Volga sands [35], and it has weak features of alluvial phenomena without the formation of separate horizons. This soil was classified as Eutric Fluvic Arenosols with weak features of illuviation process development and sand and clay contents of 70.5–86.4% and 0.3–2.6%, respectively. The plots were selected for sampling: mature plot with developed old pine forest, crown fire plot where dead pine trees were cut and removed from the plot, and surface fire plot with accumulation of wood remnants on the soil surface. This pine stand was not faced to fires during previous 150 years, and wildfire has appeared for the first time in 2010. Sampling was conducted for plots, situated on the dune hills with flat relief without expressed mesoforms [8, 10, 11].

2.2. Laboratory Analyses. All chemical and biological parameters were determined using a fine earth of soil that had been ground and then sifted through a 2 mm sieve. Routine soil analyses (i.e., pH, C, and N) were conducted according to standard procedures: C and N were determined using a C-H-N analyzer, and pH was measured using a soil to water ratio of 1 : 25 [40]. The mineralization experiment was conducted with controlled laboratory incubation, using plastic cylinders with a diameter of 10 cm and a height of 20 cm. During controlled laboratory incubation, 10 g of mineral soil and 5 g of organic fine earth were put into a cylinder. The experiment was conducted in five replications for every soil horizon (they were collected from polypedons with the square 2 × 2 m). The cylinders were then put in a beaker that had a water table that was close to the bottom of the soil, so as to simulate the field conditions of paddy (i.e., approximately 60% of water-holding capacity). The water content (%) was controlled during soil incubation. CO_2 emission values were determined using the protocol outlined by Anderson [41]: aliquots (5 g) of fresh soil were adjusted to a moisture content that was 60% of the water-holding capacity and were then incubated at 20°C in sealed plastic bottles with 1 M NaOH. The amount of CO_2 trapped in the alkaline solution was measured using titration after 7 days of incubation. In total, the incubation was conducted for 112 days. The experiment was performed in analytical triplicates. For the Nadym and Shatura samples, only the scenario of surface fire was used for comparison of burned and unaffected soil samples. For the Shatura samples, the pyrogenically affected materials (i.e., Cpir and T pir) were compared with deeper horizons of unaffected solum (T). For the Togliatti plots samples, the soil presented weak Arenosols with undeveloped profile. Thus, the sampling design for Shatura and Togliatti differed from the Nadym and Shatura samples, and only the superficial soil horizons in case of Togliatti were sampled for further analyses.

2.3. Statistic. The normal distribution of the data was verified previously, and analysis of variance (ANOVA) and a post hoc test were conducted. Differences were considered significant at $p < 0.05$. Only superficial horizons of burned and unburned soils were examined with the use of post hoc test (Fisher's least significant difference).

3. Results

3.1. General Soil Properties. The soils from the Nadym plot differ from each other in terms of its C and N content (Table 1). The C content decreased due to the impact of the fire on the topsoil horizon. This alteration can be related directly to the combustion effect. An accumulation of organic matter of illuviation origin in the Epir soil horizon was also observed. This accumulation is confirmed by both the increased concentration of total organic carbon (TOC) compared with mature soil samples and the extended C/N ratio in this horizon, which might be result of dissolved low molecular compounds of illuviation. Fire-affected soil samples showed small increases in pH values, which is typical for these types of impacted soils.

The TOC content in the Shatura plots showed greater concentrations in the mature soil samples (S12) than in the topsoil horizons, which were affected by fire (i.e., S10, S11, and S13). The decreasing TOC content is the topsoils recognized alteration of soil under the surface fire. The C/N ratio changes were the most pronounced in postfire soils. These ratios were 4 or 5 times lower in fire-affected soil samples than in the control. The greatest change in C/N ratios was seen in the C pir and Tpir horizons of the fire-affected soil samples. In contrast, the lowest T horizon, namely, those unaffected by fires, had C/N values that were comparable to those of mature

TABLE 1: General soil properties.

Plot, depth, cm	C, %	N, %	C/N	pH_{H2O}	Color, Munsell color chart
Nadym surface fire					
O ash, 0–10	8,03 ± 0,93	0,51 ± 0,02	15.74	4,79	10 YR 4/2
Epir, 10–17	1,30 ± 0,23	0,06 ± 0,03	21.66	5,02	10 YR 4/1
BF, 17–26	1,20 ± 0,25	0,06 ± 0,01	20.00	5,21	7,5 YR 4/6
BC, 26–45	1,25 ± 0,32	0.05 ± 0,01	25.00	4,98	10,5 YR 5/8
Nadym mature soil					
OT, 0–10	13,83 ± 0,98	0,62 ± 0,05	22.30	4,22	10 YR 5/6
E, 10–18	0,26 ± 0,09	0,02 ± 0,01	8.67	4,37	10 YR 6/3
BF, 18–27	1,47 ± 0,05	0,09 ± 0,01	16.33	4,71	7,5 YR 4/6
BC, 27–36	0,40 ± 0,05	0,03 ± 0,01	13.33	5,03	10,5 YR 5/8
Shatura 10					
W, 0–5	6,90 ± 0,45	0.18 ± 0,02	39.33	6,7	10 YR 4/4
Cpir, 5–12	6,10 ± 0,67	0.78 ± 0,03	7.82	7,4	10 YR 4/2
Tpir, 12–40	24,50 ± 0,78	1.78 ± 0,05	13.76	6,1	2,5 YR 2,5/1
T, 40–50	36,79 ± 0,24	0.99 ± 0,05	37.61	5,2	10 YR 3/1
Shatura 11					
Cpir, 0–10	2,73 ± 0,15	0.56 ± 00,3	4.87	7,4	10 YR 4/2
Tpir, 10–15	10,5 ± 0,12	1.12 ± 0,07	9.37	5,7	2,5 YR 2,5/1
T, 15–30	25,3 ± 0,67	0.98 ± 0,08	25.81	5,3	10 YR 3/1
Shatura 12, mature soil					
Mature T, 0–10	35,40 ± 1,23	0.63 ± 0,02	56.19	7,8	10 YR 3/1
Mature T, 10–20	28,11 ± 0,54	0.78 ± 0,03	36.02	8,4	10 YR 3/1
Shatura 13					
Cpir, 0–10	5,03 ± 0,21	0.65 ± 00,4	7,73	7,5	10 YR 4/2
Tpir, 10–20	11,5 ± 0,45	1.05 ± 0,09	10,95	6,6	2,5 YR 2,5/1
T, 20–30	24,53 ± 0,65	1.23 ± 0,11	19,94	6,5	10 YR 3/1
Togliatti					
Crone fire, 0–15	2,94 ± 0,23	0,57 ± 0,06	5,16	7,5	10 YR 3/2
Surface fire, 0–15	2.58 ± 0,25	0,40 ± 0,03	6,45	6,4	10 YR 3/2
Mature soil, 0–15	16.64 ± 0,88	1,36 ± 0,12	12,23	6,5	10 YR 4/2

soil samples. The Shatura soil samples showed increasing pH values under fire effect.

The SOM content in the crown fire ash of the Togliatti plot was less than that observed in the surface fire ash. A similar trend was observed in the humus pyrogenic horizons: the crown fire carbon content was 1.21 ± 0.50% and the surface fire sample was 1.42 ± 0.31%. Thus, more intensive organic matter losses occurred as a result of crown fire where there is a complete burning of forest vegetation. As for the mineral horizons, the carbon content was generally comparable to similar indicators in the control plots. High temperatures had little impact on the mineral horizons due to the very low thermal conductivity of the litter. Thus, a carbon loss in the surface soil horizons was observed due to the destruction of organic horizons, the mineralization of root residues, and the nearly complete absence of new litter that would serve as material for humification. The C/N ratio suggests that humification processes were more prevalent than mineralization processes in the control plot. The C/N ratio decreased twofold in the postpyrogenic plots, due to the fact that high temperatures induced the mineralization of organic residues.

3.2. CO_2 Mineralization Rates. Data from the Nadym plots (Table 2) showed higher weekly mineralization rates in the topsoil of mature soil than in postfire soil. In contrast, mineralization rates were higher in all of the mineral horizons of the postfire soil than in the unaltered ones. This can be caused by the movement of dissolved organic material down the soil profile and further by the mineralization of SOM in these soil horizons. The differences between organic topsoils of burned and unburned soils were statistically significant ($p < 0.05$).

In the Shatura plots, the highest weekly mineralization rates were observed in the topsoil of mature plots, and this rate decreased rapidly within the time period observed. In the case of the postfire plots, the highest mineralization rates were typically observed in lower T (Hisitc) horizons that were unaffected by the fire. This was common for the lowest horizons of T-unaffected soils and the topsoils

TABLE 2: Average CO_2 emission rates, mg CO_2/100 g soil/day.

Plot	1–3	4–6	7–9	10–13	14–16
Nadym surface fire					
O ash	109,23	76,67	67,21	58,24	58,73
Epir	19,20	20,34	22,82	25,98	18,78
BF1	20,40	32,17	24,41	20,16	21,97
BF2	45,64	38,64	38,70	37,36	12,53
BF3	33,82	33,77	34,95	30,86	21,99
BC	27,62	27,73	37,75	32,96	25,11
C	29,83	24,25	33,75	31,76	18,77
Nadym mature soil					
OT	421,28	306,64	403,25	451,81	341,58
E	17,80	24,471	34,363	35,92	21,86
BF	47,01	38,82	39,08	43,15	34,39
BC	26,58	34,14	31,88	34,25	15,64
Shatura 10					
W	151,15	146,32	144,16	255,21	242,19
Cpir	222,40	204,92	180,09	164,23	119,44
Tpir	193,71	171,48	181,93	175,48	170,31
T	246,87	134,78	152,08	127,23	111,33
Shatura 11				0	
Cpir	108,05	66,53	75,41	83,70	58,70
Tpir	70,35	51,66	44,79	52,37	47,61
T	139,10	68,93	59,92	36,24	49,41
Shatura 12, mature soil				0	
Mature T	400,35	274,57	289,02	289,02	269,75
Mature T	253,25	150,25	158,87	178,08	165,52
Shatura 13				0	
Cpir	106,81	64,32	82,50	94,46	67,12
Tpir	181,91	131,55	140,78	193,57	123,18
T	326,95	231,79	261,17	261,17	239,41
Togliatti	0	0	0	0	
Crone fire	44,41	30,11	33,91	28,04	15,71
Surface fire	40,80	29,82	33,35	28,87	18,83
Mature soil	63,37	40,85	40,96	35,90	28,28

Post hoc test Nadym surface, Nadym mature: $p < 0{,}05$; post hoc test Shatura Mature, Shatura 10: $p < 0{,}03$; post hoc test Shatura Mature, Shatura 11: $p < 0{,}03$; post hoc test Shatura mature, Shatura 13: $p < 0{,}03$; post hoc test Togliatti, Crone-Mature: $p < 0,04$; post hoc test Togliatti, Surface-Mature: $p < 0{,}04$.

of mature soils. Fire-affected horizons (i.e., Cpir and Tpir) showed gradually decreasing mineralization rates within the experiment period. The post hoc test showed essential differences between topsoil horizons of mature and forest fire-affected plots ($p < 0.03$).

The topsoil samples of the Togliatti plots showed essential differences ($p < 0.05$) in weekly mineralization rates if one compare natural soil with both fires affected (surface and crown).

3.3. CO_2 Emission and SOM Stability to Mineralization. Data on TOC mineralization rates (Table 3) show that the total amount of CO_2 evolved from the soil samples decreased with time. This trend is more pronounced for topsoil horizons than for lower organomineral ones. If we compare the amount of CO_2 evolved from topsoil samples of postfire Nadym plots with those from mature Nadym soil samples, we can see that mature soil provides more intensive mineralization than postfire soil, but that this emission remains steady over the time. The potential for TOC mineralization is higher in mature topsoil than in the topsoil of fire-affected soil. This is caused by the direct fire effect. The SOM of postfire soil is initially more destroyed before the incubation experiment, while the organic matter of mature soil is initially stable and presents the unit system of the natural origin. Thus, the soil organic matter of the surface postfire soil of the Nadym plots is more stable to mineralization, because the most decomposable fractions of soil organic matter were removed by combustion during the fire event. In the mineral horizons of postfire soils the mineralization rate is lower than that of the same horizons for natural soil profiles that are situated nearby. The differences between C-CO_2/TOC

TABLE 3: Portions of C mineralized on weeks and portion of total minerlized carbon to TOC content.

Plot, horizons	Depth, cm	C mineralized, $gC\text{-}CO_2$/kg soil fine earth, weeks					% $C\text{-}CO_2$/TOC
		1–3	4–6	7–9	10–13	14–16	
				Nadym surface fire			
O ash	0–3	6,26	4,39	3,85	3,95	3,19	26,99
Epir	3–10 (20)	1,10	1,16	1,30	1,34	1,37	57,26
BF1	10–20 (25)	1,17	1,84	1,40	1,24	1,13	55,29
BF2	25–40	2,61	2,21	2,22	2,00	1,69	86.08
BF3	40–50	1,94	1,93	2,00	2,15	1,37	84,05
				Nadym mature soil			
OT	0–20	24,17	17,59	23,13	29,05	19,83	82,28
E	20–30	1,02	1,40	1,97	2,15	1,66	70,81
BF	30–50	2,69	2,22	2,24	2,51	2,04	79,81
BC	50–70	1,52	1,99	1,82	2,06	1,31	86,99
				Shatura 10			
W	0-1 (3)	8,67	7,72	8,71	11,02	14,03	72,71
Cpir	1 (3)–3 (4)	12,76	10,27	11,17	10,27	9,42	88,39
Tpir	3 (4)–5 (7)	11,11	9,18	11,93	9,47	9,77	21,01
T	7–18	14,16	6,69	9,47	8,82	6,38	12,37
				Shatura 11			
Cpir	0,5–2 (4)	6,20	2,99	4,62	5,23	3,92	89,8
Tpir	2 (4)–7	4,03	2,55	3,46	2,82	2,45	14,60
T	7-8	7,98	4,34	3,33	3,49	1,70	8,24
				Shatura 12, mature soil			
T	0–5	22,97	13,26	18,97	16,39	16,21	24,81
T	5–10	14,53	7,53	10,83	8,66	9,99	18,35
				Shatura 13			
Cpir	0–3	6,12	3,42	4,52	5,13	4,84	47,84
Tpir	3–5 (6)	10,43	6,05	9,05	10,09	9,08	38,88
T	7 (9)–15	18,75	12,48	14,81	16,23	13,73	30,99
				Togliatti			
Crone fire	0–4	2,54	1,72	1,94	1,96	1,13	31,70
Surface fire	0–3	2,34	1,71	1,91	1,77	1,35	35,25
Mature soil	0–4	3,63	2,34	2,35	1,90	2,03	7,37

Post hoc test Nadym surface, Nadym mature: $p < 0,05$; post hoc test Shatura Mature, Shatura 10: $p < 0,04$; post hoc test Shatura Mature, Shatura 11: $p < 0,04$; post hoc test Shatura mature, Shatura 13: $p < 0,04$; post hoc test Togliatti, Crone-Mature: $p < 0,05$; post hoc test Togliatti, Surface-Mature: $p < 0,05$.

values were statistically significant for the Oash and Ot topsoil (burned and unburned) horizons for Nadym plot ($p < 0.05$).

Data from the Shatura soil samples showed a general trend of declining CO_2 weekly evolving rates, with the exception of the topsoil of Shatura 10 soil. In general, weekly amount of evolving was higher in mature soil samples if the data were expressed in g/kg of fine earth ($p < 0.05$, $p < 0.04$ and $p < 0.04$ for pare comparison of topsoils of mature soil with Shatura 10, Shatura 11, and Shatura 13 soils correspondingly). But if one compares $C\text{-}CO_2$/TOC in percentages to initial TOC, then the organic horizons of postfire soil samples provide higher mineralization losses than unaffected soil. Within the soil profiles of the investigated soils, this trend of decreasing rates was observed in all 3 postfire locations. This overall trend indicates that the SOM of upper organic horizons has been altered, as expressed by the decrease in sustainability of organic matter to mineralization. This is the direct effect of fire on SOM quality and stability.

The Togliatti plot soil samples also showed this trend of decreasing of weekly evolved amount of CO_2 during the experiment. The weekly evolved amounts, expressed in g/kg of fine earth, were higher in mature soil samples than in fire-affected ones. The $C\text{-}CO_2$/TOC values were significantly high when comparing mature plot topsoil with soil from both crown and surface fires ($p < 0.05$). This difference is caused by higher initial TOC levels in mature soil that were present before the incubation experiment.

4. Discussion

Soil samples from three regions that were affected by fires showed essential differences in their general chemical properties. Firstly, there were essential changes in TOC content due to the combustion effect. The postfire soil samples are characterized by decreased TOC content and increased Nitrogen content, which in turn decreased the C/N ratio compared to unaffected soil. Furthermore, the pH values increased relative to unaffected soil due to the accumulation of ash compounds [42–46].

The total amount of initial TOC that was present before the incubation experiment played a significant role in the mineralization rate. Thus, the sum of evolved CO_2 was higher in those horizons that were unaffected by fire. This finding corresponds to data of Michelloti and Miesel [9], who showed that control samples provide higher CO_2 emissions than those amended with postfire remnants. At the same time the portion of TOC mineralized during the experiment, expressed in percentages, was lower in topsoil samples and the lowest horizons of unaffected soils than in fire-altered materials. This finding corresponds with the data of Maksimova and Abakumov [8], who showed that there is an essential alteration of organic matter quality in postfire soil. Previously it was shown that the aromaticity of soil affected by fire increases, due to declining concentrations of carboxylic and aliphatic compounds [6, 8]. This can affect the difference in mineralization rates of the SOM in postfire and mature soils. Thus, structural changes in SOM systems play an essential role in stabilizing SOM for postfire soil [47], and the degree of SOM oxidation regulates the mineralization rates and other indexes of postfire SOM quality [1, 8]. Comparing the zonal trend of organic matter stability to mineralization, we can conclude that the most stabilized organic matter is found in forest-steppe mature and postfire environments, where the degree of humification and TOM transformation is highest within the ecoregion sequence. The soil of forest-tundra environments is characterized by the highest alteration in organic matter and decreased sustainability of mineralization because of initial quality of SOM, as was previously substantiated [47] for polar ecosystems. Namely, the aliphatic compounds and dissolved organic carbons that experience fire events mineralize immediately or during the first 3 weeks [1, 8] and prevail in polar soils. In any case, the drought frequency increases in Polar and Siberian landscapes, and further estimates of SOM quality and its stability to mineralization processes should be provided in the near future. The Russian peatlands play a critical role in the formation of global and regional storages of carbon dioxide; thus the organic matter stabilization rates and possible risk of C pools huge losses from soil due to wildfires should be precisely estimated [48–51]. In this context, the data obtained for the Shatura south-taiga peatlands provide essential information about SOM alteration during the fire effect. It was shown that fire leads to the destabilization of peat material, which results in increased mineralization rates. In contrast, the unaffected topsoil and lowest horizons showed little mineralization activity, which indicates the essential stabilization rate of SOM in the peatlands. Our data corresponds well with those of Zhao et al. [22], who showed that mineralization rates in peatland soils increase immediately after fire events. Finally, our findings suggest that additional laboratory incubation experiments are needed for various Russian postfire environments, with the aim of specifying the role of SOM origins, quality, and soil type on SOM stabilization rates and to better understand possible postfire mineralization risks.

5. Conclusions

The soil processes in postfire soils differed from those in the soils of natural landscapes. After the fire event, there were (1) an intensive accumulation of ash on the topsoil, (2) SOM alteration, and (3) mineralization. Soil C emission rates were estimated in the soil of three ecoregions of Russia: forest-tundra (Nadym city, Yamal Region), south-taiga (Shatura city, Moscow region), and forest-steppe (Togliatti city, Samara region). Our findings demonstrated that postfire soil was characterized by the essential loss of soil carbon, due to the burning fire effect. In our samples, soil lost 3 or 5 of initial C content, decreased C/N ratios, and increased pH values during the fire effect. The CO_2 emission rates were highest in the natural soil samples, because the amount of organic matter faced to mineralization in those soils is higher than in natural ones. The total values of mineralized C were higher in postfire soil samples because the SOM quality and composition were altered due to fire effect; SOM destruction under the fire effect results in decreasing of organic matter stability. The quality of initial SOM and type of the wildfire both play important roles in the fate of SOM in postfire environments. Thus, further study of CO_2 emissions is needed, with the aim of better characterizing postfire SOM dynamics creating an approach to model this process.

Conflicts of Interest

The authors declare that there are no conflicts of interest regarding the publication of this paper.

Acknowledgments

This work was partially supported by the Russian Foundation for Basic Research, Projects nos. 15-34-20844 and 16-34-60010.

References

[1] G. Certini, "Effects of fire on properties of forest soils: a review," *Oecologia*, vol. 143, no. 1, pp. 1–10, 2005.

[2] G. Certini, "Fire as a soil-forming factor," *AMBIO*, vol. 43, no. 2, pp. 191–195, 2014.

[3] R. Guénon and R. Gros, "Frequent-wildfires with shortened time-since-fire affect soil microbial functional stability to drying and rewetting events," *Soil Biology and Biochemistry*, vol. 57, pp. 663–674, 2013.

[4] J. Mataix-Solera, A. Cerdà, V. Arcenegui, A. Jordán, and L. M. Zavala, "Fire effects on soil aggregation: a review," *Earth-Science Reviews*, vol. 109, no. 1-2, pp. 44–60, 2011.

[5] F. Oris, H. Asselin, A. A. Ali, W. Finsinger, and Y. Bergeron, "Effect of increased fire activity on global warming in the boreal forest," *Environmental Reviews*, vol. 22, no. 3, pp. 206–219, 2014.

[6] A. A. Dymov and D. N. Gabov, "Pyrogenic alterations of podzols at the North-east European part of Russia: morphology,

carbon pools, PAH content," *Geoderma*, vol. 241-242, pp. 230–237, 2015.

[7] E. A. Zharikova, "Post pyrogenic transformation of soil of Lower Amur region," *Agroecology*, vol. 8, pp. 57–60, 2015.

[8] E. Maksimova and E. Abakumov, "Wildfire effects on ash composition and biological properties of soils in forest–steppe ecosystems of Russia," *Environmental Earth Sciences*, vol. 74, no. 5, pp. 4395–4405, 2015.

[9] L. A. Michelotti and J. R. Miesel, "Source material and concentration of wildfire-produced pyrogenic carbon influence post-fire soil nutrient dynamics," *Forests*, vol. 6, no. 4, pp. 1325–1342, 2015.

[10] E. Maksimova and E. Abakumov, "Micromorphological characteristics of sandy forest soils recently impacted by wildfires in Russia," *Solid Earth*, vol. 8, no. 2, pp. 553–560, 2017.

[11] E. Y. Maksimova, A. G. Kudinova, and E. V. Abakumov, "Functional activity of soil microbial communities in post-fire pine stands of Tolyatti, Samara oblast," *Eurasian Soil Science*, vol. 50, no. 2, pp. 239–245, 2017.

[12] B. R. Goforth, R. C. Graham, K. R. Hubbert, C. W. Zanner, and R. A. Minnich, "Spatial distribution and properties of ash and thermally altered soils after high-severity forest fire, southern California," *International Journal of Wildland Fire*, vol. 14, no. 4, pp. 343–354, 2005.

[13] X. Úbeda, P. Pereira, L. Outeiro, and D. A. Martin, "Effects of fire temperature on the physical and chemical characteristics of the ash from two plots of cork oak (*Quercus Suber*)," *Land Degradation and Development*, vol. 20, no. 6, pp. 589–608, 2009.

[14] P. Pereira, X. Úbeda, and D. A. Martin, "Fire severity effects on ash chemical composition and water-extractable elements," *Geoderma*, vol. 191, pp. 105–114, 2012.

[15] P. Pereira, A. Cerdà, X. Úbeda et al., "Spatial models for monitoring the spatio-temporal evolution of ashes after fire—a case study of a burnt grassland in Lithuania," *Solid Earth*, vol. 4, no. 1, pp. 153–165, 2013.

[16] M. B. Bodí, D. A. Martin, V. N. Balfour et al., "Wildland fire ash: production, composition and eco-hydro-geomorphic effects," *Earth-Science Reviews*, vol. 130, pp. 103–127, 2014.

[17] H. Knicker, "How does fire affect the nature and stability of soil organic nitrogen and carbon? A review," *Biogeochemistry*, vol. 85, no. 1, pp. 91–118, 2007.

[18] C. Wüthrich, D. Schaub, M. Weber, P. Marxer, and M. Conedera, "Soil respiration and soil microbial biomass after fire in a sweet chestnut forest in southern Switzerland," *Catena*, vol. 48, no. 3, pp. 201–215, 2002.

[19] R. Guénon, M. Vennetier, N. Dupuy, S. Roussos, A. Pailler, and R. Gros, "Trends in recovery of mediterranean soil chemical properties and microbial activities after infrequent and frequent wildfires," *Land Degradation and Development*, vol. 24, no. 2, pp. 115–128, 2013.

[20] J. A. González-Pérez, F. J. González-Vila, G. Almendros, and H. Knicker, "The effect of fire on soil organic matter—a review," *Environment International*, vol. 30, no. 6, pp. 855–870, 2004.

[21] G. Certini, C. Nocentini, H. Knicker, P. Arfaioli, and C. Rumpel, "Wildfire effects on soil organic matter quantity and quality in two fire-prone Mediterranean pine forests," *Geoderma*, vol. 167-168, pp. 148–155, 2011.

[22] H. Zhao, D. Q. Tong, Q. Lin, X. Lu, and G. Wang, "Effect of fires on soil organic carbon pool and mineralization in a Northeastern China wetland," *Geoderma*, vol. 189-190, pp. 532–539, 2012.

[23] G. Almendros, F. J. González-Vila, and F. Martin, "Fire-induced transformation of soil organic matter from an oak forest: an experimental approach to the effects of fire on humic substances," *Soil Science*, vol. 149, no. 3, pp. 158–168, 1990.

[24] S. P. Bergeron, R. L. Bradley, A. Munson, and W. Parsons, "Physico-chemical and functional characteristics of soil charcoal produced at five different temperatures," *Soil Biology and Biochemistry*, vol. 58, pp. 140–146, 2013.

[25] M. A. Alexis, C. Rumpel, H. Knicker et al., "Thermal alteration of organic matter during a shrubland fire: a field study," *Organic Geochemistry*, vol. 41, no. 7, pp. 690–697, 2010.

[26] E. Gvozdetskiy, "Physical-geographical regionalization of the Tyumen region," Under the edition of Gvozdetskii N.A., M.: MGU, 1973. p. 246.

[27] "World reference base for soil resources. International soil classification system for naming soils and creating legends for soil maps," 2014. FAO. Rome, 2015.

[28] Wetlands of Moscow Region, Natural, economic and historical-cultural aspects, Wetland International-Russia Programme, KMK, Scientific Press, Tech. Rep., Moscow, Russia, p. 89, 2008.

[29] N. L. Kudravtseva, *Geostructural location and conditions of peat bogs and deep lakes nourishment in Meshchera downfold and on the Oka-Tsna bar, The Aspects of ground water and engineer-geological processes studying*, Moscow, Russia, 1973.

[30] N. T. Zonov and A. E. Konstantinovitch, *Geologic structure, mineral resources and ground waters of Egorievskii, Shaturskii, Korobovskii and Kurovskii Districts of Moscow Region*, vol. 11, Transactions of Moscow Geological Survey Bureau, Moscow, Russia, 1932.

[31] A. Timashev, *Shatura Power Plant*, Moscow, Russia, 1932.

[32] F. R. Zaidel'man, D. I. Morozova, and A. P. Shvarov, "Changes in the properties of pyrogenic formations and vegetation on burnt previously drained peat soils of poles'ie landscapes," *Eurasian Soil Science*, vol. 36, no. 11, pp. 1159–1167, 2003.

[33] A. Van Donkelaar, R. V. Martin, R. C. Levy et al., "Satellite-based estimates of ground-level fine particulate matter during extreme events: a case study of the Moscow fires in 2010," *Atmospheric Environment*, vol. 45, no. 34, pp. 6225–6232, 2011.

[34] A. Z. Shvidenko, D. G. Shchepashchenko, E. A. Vaganov et al., "Impact of wildfire in Russia between 1998–2010 on ecosystems and the global carbon budget," *Doklady Earth Sciences*, vol. 441, part 2, pp. 1678–1682, 2011.

[35] E. V. Abakumov and E. I. Gagarina, *The Soils of Samara Luka: Diversity, Genesis, Protection*, St. Petersburg State University, St. Petersburg, Russia, 2008.

[36] E. V. Abakumov, N. Fujitake, and T. Kosaki, "Humus and Humic Acids of Luvisol and Cambisol of Jiguli Ridges, Samara Region, Russia," *Applied and Environmental Soil Science*, vol. 2009, Article ID 671359, 5 pages, 2009.

[37] V. A. Nosin, *Soils of Kujbyshev Region*, OGIZ, Kujbyshev, Russia, 1949.

[38] I. S. Urusevskaja, J. L. Meshalkina, and O. S. Hohlova, "Geographic and genetic features of Luvisols' humus status," *Eurasian Soil Science*, vol. 11, pp. 1377–1390, 2000.

[39] D. I. Vasil'eva and M. N. Baranova, *Natural Resources of Samara Region*, Samara municipal management institute, Samara, Russia, 2007.

[40] L. A. Vorobyova, *Theory and Practice of the Chemical Soil Analysis*, GEOS, Moscow, Russia, 2006.

[41] J. P. E. Anderson, "Soil respiration. Methods of Soil Analysis," in *Chemical and Microbiological Properties*, R. H. Miller, Ed., part

2, pp. 831–871, American Society of Agronomy, Madison, WI, USA, 1982.

[42] M. B. Bodí, S. H. Doerr, A. Cerdà, and J. Mataix-Solera, "Hydrological effects of a layer of vegetation ash on underlying wettable and water repellent soil," *Geoderma*, vol. 191, pp. 14–23, 2012.

[43] S. W. Woods and V. N. Balfour, "The effects of soil texture and ash thickness on the post-fire hydrological response from ash-covered soils," *Journal of Hydrology*, vol. 393, no. 3-4, pp. 274–286, 2010.

[44] J. León, M. B. Bodí, A. Cerdà, and D. Badía, "The contrasted response of ash to wetting. The effects of ash type, thickness and rainfall events.," *Geoderma*, vol. 209-210, pp. 143–152, 2013.

[45] P. Pereira, X. Úbeda, L. Outeiro, and D. Martin, "Factor analysis applied to fire temperature effects on water quality," in *Forest Fires: Detection, Suppression and Prevention*, E. Gomez and K. Alvarez, Eds., Series Natural Disaster Research, Prediction and Mitigation, pp. 273–285, Nova Science Publishers, New York, NY, USA, 2009.

[46] P. Pereira, A. Cerdà, X. Úbeda, J. Mataix-Solera, V. Arcenegui, and L. M. Zavala, "Modelling the impacts of wildfire on ash thickness in a short-term period," *Land Degradation and Development*, vol. 26, no. 2, pp. 180–192, 2013.

[47] E. Ejarque and E. Abakumov, "Stability and biodegradability of organic matter from Arctic soils of Western Siberia: insights from ^{13}C-NMR spectroscopy and elemental analysis," *Solid Earth*, vol. 7, no. 1, pp. 153–165, 2016.

[48] N. T. Roulet, "Peatlands, carbon storage, greenhouse gases, and the kyoto protocol: prospects and significance for Canada," *Wetlands*, vol. 20, no. 4, pp. 605–615, 2000.

[49] D. G. Zamolodchikov, A. I. Utkin, G. N. Korovin, and O. V. Chestnykh, "Dynamics of carbon pools and fluxes in Russia's forest lands," *Russian Journal of Ecology*, vol. 36, no. 5, pp. 291–301, 2005.

[50] V. L. Gavrikov, R. A. Sharafutdinov, A. A. Knorre et al., "How much carbon can the Siberian boreal taiga store: a case study of partitioning among the above-ground and soil pools," *Journal of Forestry Research*, vol. 27, no. 4, pp. 907–912, 2016.

[51] E. Y. Maksimova, A. S. Tsibart, and E. V. Abakumov, "Soil properties in the Tol'yatti pine forest after the 2010 catastrophic wildfires," *Eurasian Soil Science*, vol. 47, no. 9, pp. 940–951, 2014.

6

Environmental Determinants Influencing Fish Community Structure and Diversity in Two Distinct Seasons among Wetlands of Northern Region (Ghana)

Collins Ayine Nsor[1] and Edward Adzesiwor Obodai[2]

[1] *Department of Ecotourism and Forest Recreation, Kwame Nkrumah University of Science and Technology, Kumasi, Ghana*
[2] *Department of Fisheries and Aquatic Sciences, University of Cape Coast, University Post Office, Cape Coast, Ghana*

Correspondence should be addressed to Collins Ayine Nsor; ayineus@yahoo.com

Academic Editor: Jean-Guy Godin

Fish community structure was assessed in six wetlands using cast nets, to correlate with environmental variables with diversity and distribution patterns, from 2010 to 2012. A total of 2,239 individuals belonging to 44 species and 1,938 individuals belonging to 40 species were sampled in the dry and wet seasons. Mochokid and Mormyrid families dominated fish community and constituted 14.8%, respectively, followed by Alestids (12.9%) and Chlariids (11.1%). Rarer taxons were centropomids, channids, malapteruds, and oesteoglossids and represented 1.9%, respectively. Overall, CPUE per net did not vary significantly (*Tukey HSD test*, $p = 0.27$) in the dry and wet seasons. Wuntori marsh consistently showed dominance in mean monthly CPUE per net (dry = 115 ± 4.5; wet = 107 ± 7.7 seasons), while Bunglung constructed wetland was the least recorded (dry = 56.5 ± 6.2; wet = 58.3 ± 4.1 seasons). Fish diversity and richness differed significantly ($F = 0.11$, $p = 0.03$) among seasons. Environmental disturbances were season-specific and did not differ significantly ($F = 0.16$, df = 14, $p = 0.97$) among sites. A DCA ordination explained 69% variability in fish distribution patterns, while PCA showed that 81.8% of nitrate-nitrogen, phosphate, and grazing intensity on axis 1 and conductivity, temperature, and turbidity on axis 2 influenced fish community structure. Wetland conservation must be promoted to sustain fish abundance and overall ecosystem stability.

1. Introduction

The formation of freshwater systems thousands of years ago initially guided the fish assemblage and distribution patterns that we observe today [1]. However, human activities have largely altered aquatic landscapes through resource exploitation and other impacts. The Millennium Ecosystem Assessment Report [2] estimated that about 20% of the world's 10,000 freshwater fish species are listed as threatened, endangered, or extinct in the last few decades, partly due to human-led activities. The decline in fish populations is attributed to degradation in water quality [3–5] and periodic phenomena of low-flows and floods, habitat complexity, and agricultural and urban land use [6, 7]. Menzel et al. [8] explained how agriculture activities have had multiple effects on Great Plains streams in the US, including increased turbidity, siltation, and nutrients through soil erosion and irrigation return flows. Erosion and sediment alter substrate characteristics, leading to a decrease in fish species that require clean gravel for spawning [9]. These disturbances on wetland systems may lead to changes in fish community composition and spatial distribution. Variability in fish assemblages is generally influenced by landscape features [10], stream size, and elevation [11] and stocking of economically important exotic/invasive fishes (e.g., brown trout, grayling, carp, and pike) [12] (Pivnička and Humpl, 2004).

Ansari et al. [13] showed how fish species segregate into seasonal groups, as a result of changes in physico-chemical parameters, such as salinity, temperature, and dissolved oxygen concentration. The vulnerability of fish from the slightest disturbances in their environment is due to the intimate contact of the skin and gills with the surrounding water [14].

Ghana's water resources have been under increasing threat of pollution in recent years due to rapid demographic changes, which have coincided with the establishment of human settlements lacking appropriate sanitary infrastructure [15]. The rising demand for fish and other resources, due to increase in impoverished population in communities close to wetland areas of Northern Region of Ghana, has partly contributed to widespread environmental disturbances. For instance, the use of dynamite as one illegal method of fishing depletes water quality and renders it unwholesome for domestic use and impairs aquatic life. These disturbances are seasonal-specific (i.e., wet and dry seasons) and fish responses vary by season and spatiotemporal scales. Given the high conservation concern of wetlands in Northern Region of Ghana [16], the wise use of these ecosystems, which involves the maintenance of ecological integrity and sustainable exploitation of fish, for the continuous benefit to current and future generations, must be given critical attention.

Researchers in the fisheries field are acutely conscious of the overfishing problem associated with the unregulated or poorly regulated common-use conditions that have prevailed in most fisheries [17]. Many industrialized countries have management regimes for introduced fisheries designed to curb overexploitation and produce better net returns from the fishery, but with modest success. Attempts to manage fisheries in developing countries have been much weaker and most often have had little effect [17]. Thus, fisheries management must be of priority concern, especially to rural communities whose livelihood largely depends on the availability of fish resource.

Understanding the factors influencing seasonal variations on fish community structure, diversity and spatial distribution in Northern Region of Ghana, remains poor. Although native fishermen have observed reduction in fish composition and distribution in recent times, the observed seasonal changes have not been sufficiently investigated to establish the cause of these trends. This study attempts to apply multivariate techniques to establish the environmental determinants influencing the seasonal changes on fish community structure and distribution in six selected wetlands in the Northern Region of Ghana. Our main objective is to (1) quantify fish diversity and (2) examine the responses of fish community composition and spatial distribution to major environmental factors in the six different wetland types.

2. Materials and Methods

The study was in six wetlands located in the Northern Region of Ghana, with their coordinates as follows: (i) Wuntori (N09°08.335′ W00°1 09°.685′); (ii) Kukobila (N10°08.723′ W000°48.179′); (iii) Tugu (N09°22.550′ W000°35.004′); (iv) Bunglung (N09°35.576′ W000°47.443′); (v) Adayili (N09°41.391′ W000°41.480′); and (vi) Nabogo (N09°49.941′ W000°.51.942′) (Figure 1). All six sites lie on the extensive floodplain along the course of the White Volta River, which has become incised and modified through meandering and aligning along various topographic features. This has led to the development of streams that have diverted from the main White Volta [18]. All six wetlands were classified as close shallow marshes (Wuntori and Tugu wetlands), open deep marsh (Kukobila wetland), riparian wetlands (Adayili and Nabogo wetlands), and an artificial wetland (Bunglung wetland). The hydrological regimes of the six wetlands were typical of permanent wetlands, whose depth at low tide did not exceed 2 m on average. Area of the wetlands was quantified through on-screen digitizing of Landsat aerial images, obtained from Google Earth (October, 2012): (a) Wuntori = 7.7 ha; (b) Kukobila = 5 ha; (c) Tugu = 2.7 ha; (d) Nabogo = 7.9 ha; (e) Adayili = 6.7 ha; and (f) Bunglung = 11.5 ha. Annual rainfall ranges from 1000 to 1,300 mm per annum and the wet season from June to early October, while the dry season lasts from November to May. Average air temperature varies between 14°C and 40°C (Slaymaker and Blench, 2002). Altitude ranges between 108 and 138 meters above mean sea level. The vegetation cover is a mixture of grassland dominated by *Leersia hexandra* and woodland dominated by Mahogany (*Khaya senegalensis*) and shea tree (*Vitellaria paradoxa*) interspersed with shrubby communities of *Mitragyna inermis* and *Ziziphus abyssinica*. The trees are relatively short with thick bark and occlusions, signifying their adaptation to the cyclical dry season bush fires. Crop farming, livestock rearing, and fishing are the main activities among the inhabitants.

2.1. Fish and Water Quality Sampling Procedure. Since one of the objectives of this study was to quantify fish diversity, richness, and abundance, fish were sampled using cast nets of 3 cm mesh size and 5 m radius. Sampling commenced from November to February in the dry season and from July to October in the wet season (2010–2012). Each of the six wetlands (Wuntori, Kukobila, Tugu, Adayili, Nabogo, and Bunglung) was sampled four times each month (once per week), between 08 and 1300 hours. Five net throws per hour of sampling effort were carried out in each site. The total of sampling hours of net throws per site was four hours. Fish were identified *in situ* to species, with keys developed by Dankwa et al. [19] and Paugy et al. [20]. Species that could not be identified were placed in marked transparent wash bottles, containing formalin solution and sent to the Water Research Institute (WRI) in Tamale, Northern Region of Ghana, for identification.

A total of 11 physicochemical parameters and four anthropogenic factors were determined at all sites on the same day of fish collection. They included pH, turbidity, water temperature, conductivity, TDS, nitrate-N (NO_3-N), phosphate (PO_4), ammonia (NH_4-N), calcium, magnesium, dissolved oxygen, fire, grazing intensity, erosion, and local farming activities. Water samples were collected monthly (in dry and wet seasons), using a clean polyethylene bottle, at three sample stations: upper, mid, and downstream over a 2-year period. With the exception of pH and water temperatures that were measured *in situ* using a portable pH meter and mercury-in-glass thermometer, respectively, the remaining samples were stored in ice chest and transported

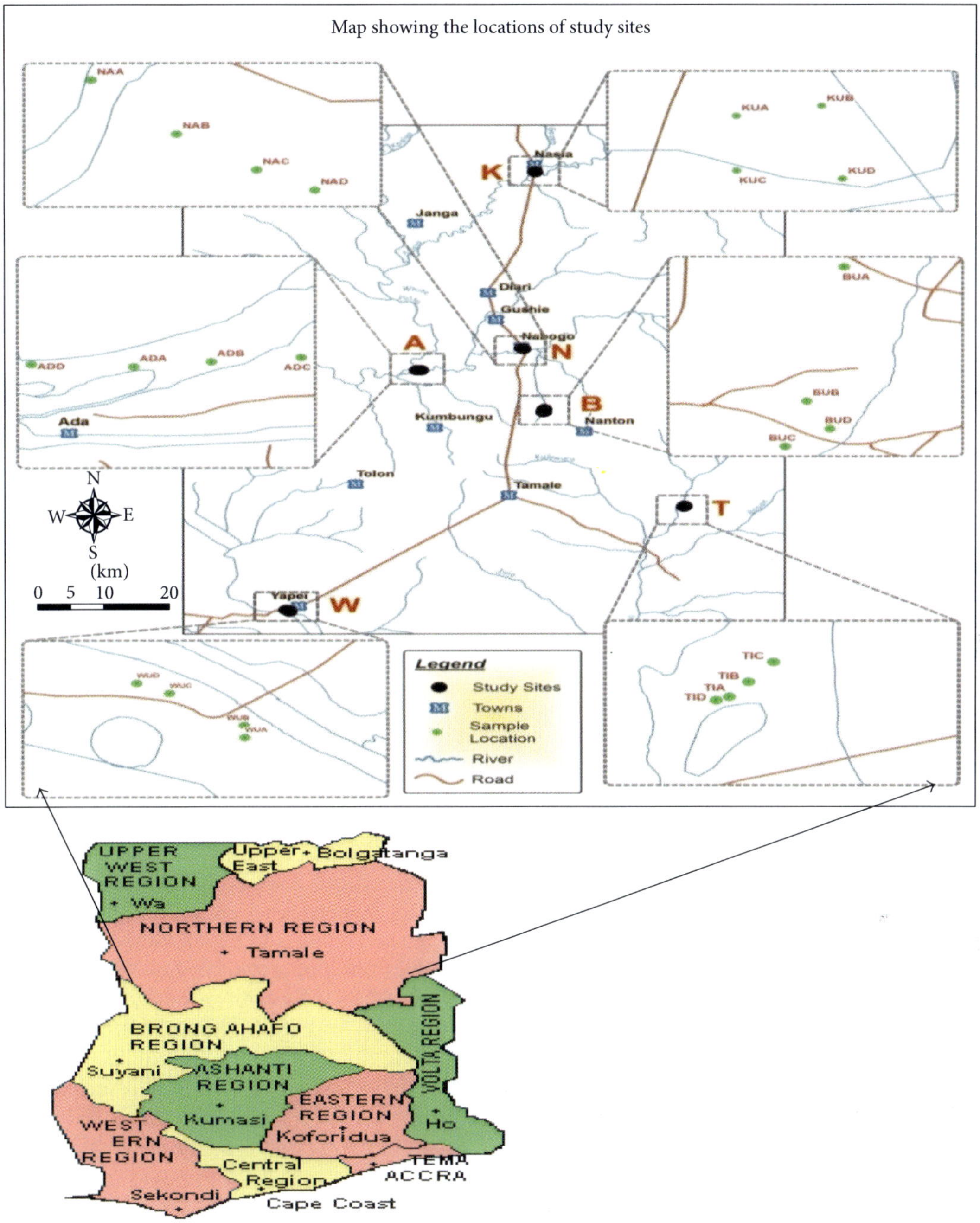

FIGURE 1: Map of the study areas, showing the location of the wetlands in the floodplains of the White Volta River catchment, Northern Region. The letters represent names of the wetlands; K = Kukobila, N = Nabogo, A = Adayili, B = Bunglung, T = Tugu, and W = Wuntori.

the same day to the Water Research Institute laboratory, for analysis of physicochemical constituents, following the APHA [21] Standard Methods for the Examination of Water and Wastewater.

Four environmental drivers of change or disturbance were bushfire, farming activities grazing, and erosion. Drivers were assessed to determine the severity of these threats on birds' habitat, since bird abundance and diversity are largely influenced by habitat quality. The hierarchical classification of these threats was comprehensive (containing all possible items, at least at higher levels of the hierarchy), consistent (ensuring that entries at a given level of the classification are of the same type), expandable (enabling new items to be added to the classification if they are discovered), and exclusive (allowing any given item to only be placed in one cell within the hierarchy) [22]. A score ranging from 1 to 4 (1 is the least impact and 4 the highest impact) was used to assess scope and severity of every threat. For "scope" we referred to the percentage ratio of the study area affected by a specific threat within the last 5 years (where 100% correspond to total

Table 1: Mean monthly catch per unit effort per net (CPUE) during the dry (November–February) and wet (July–October) seasons, in the six sites.

Sites	1st month	2nd month	3rd month	4th month	Total	Mean CPUE ± SE
			Dry season			
Wuntori	120	111	105	125	461	115.3 ± 4.5
Kukobila	85	101	81	99	366	91.5 ± 5.0
Tugu	93	77	60	75	305	76.3 ± 6.8
Nabogo	100	128	103	102	433	108.3 ± 6.6
Adayili	140	108	90	110	448	112 ± 10.4
Bunglung	71	59	41	55	226	56.5 ± 6.2
Total					*2,239*	
			Wet season			
Wuntori	102	103	95	130	430	107.5 ± 7.7
Kukobila	81	70	64	86	301	75.3 ± 5.0
Tugu	47	34	57	43	181	47.5 ± 3.3
Nabogo	115	97	105	88	405	101.3 ± 5.8
Adayili	71	117	93	107	388	97 ± 10.0
Bunglung	69	51	60	53	233	58.3 ± 4.1
Total					*1,938*	

site area: χ ha) [23]. The scores were assigned as follows: 4: the threat is found throughout (50%) the site area; 3: the threat is spread in 15–50% of the study area; 2: the threat is scattered (5–15%); and 1: the threat is localized (<5%).

Assessment of the disturbed area was carried out within a 1.2 km radius starting from the hydric delineated zone of the wetland. This was because all land use activities assessed were observed within the stated radius following a preliminary survey of the wetlands.

2.2. Statistical Analysis. The Shannon-Weiner index was calculated to determine the current status of fish community composition. Shannon-Weiner index equation is expressed as

$$H' = -\sum_{i=1}^{s} pi\,(\ln pi)\,, \qquad (1)$$

(see [24]) where s is the number of species and pi is the proportion of individuals or the abundance of the ith species expressed as a proportion of the total cover and ln is a natural logarithm (Shannon and Wiener, 1963). Margalef's index (D) of species richness that controls for sample size was calculated as

$$D = \frac{(S - 1)}{\ln N} \qquad (2)$$

(see [25]), where S is the number of species, ln is a natural logarithm, and N is the number of sites. A one-way ANOVA was applied to test for the differences in species diversity and species richness among the six wetlands, using SPSS version 20. A Kruskal-Wallis test was applied to test the differences in the mean of the diversity indices between riparian wetlands and the marshes.

We used detrended correspondence analysis (DCA) (an indirect ordination method of gradient analysis), to identify the strongest gradients of fish assemblage variation across the six wetlands. This was followed by principal component analysis (PCA) ver. 1.41 [26], to evaluate the variability in assemblage structure, diversity, and preferred wetland type in relation to the environmental variables. DCA was used primarily to determine the compositional variation of fish species in the respective wetlands, while PCA was performed to examine the relationship between fish species distribution and associated environmental gradients or variables. DCA techniques have the ability to handle large, complex data sets and uncover long ecological gradients, as well as help in data reduction and exploration [27, 28]. Environmental variables were $\log_{10}$ transformed when their variances were much larger than their means. This was followed by a one-way ANOVA test to determine whether environmental variables differed significantly among all sites. A Monte Carlo permutation test (9999 permutations) was used to test for the significance of the eigenvalues generated by the first three axes in the analyses of fish assemblage structure to environmental variables.

3. Results

Overall, 2,239 individuals from 44 species and belonging to 16 families were sampled in the dry season, while 1,938 individuals belonging to 40 species and from 12 families were identified during the wet season (Tables 1 and 2). Wuntori marsh contained the highest fish abundance in the dry season, closely followed by Adayili and Nabogo riparian systems and the lowest abundance was from Bunglung constructed wetland (Table 1). Similar patterns in species abundance were found in the six sites during the wet season. Fishes in the family Mochokidae were the most abundant and constituted 15.4% of the taxon, followed by Alestids (13.4%) and Clariids (11.5%) (Table 2). Rarer taxons were centropomids, channids, malapteruds, and osteoglossids and represented 1.9%, respectively. Mean monthly catch per unit

TABLE 2: Relative abundance of fish species recorded in the six sites, during the dry and wet seasons. (—) = absent.

Family	Species	Dry season	Wet season
Alestidae	*Alestes dentex*	129	108
	Alestes baremoze	64	71
	Brycinus nurse	160	155
	Brycinus macrolepidotus	92	24
	Hydrocynus brevis	17	—
	Hydrocynus forskali	4	—
	Micralestes occidentalis	—	83
Bagridae	*Bagrus bajad*	47	—
	Bagrus docmac	3	—
Cichlidae	*Oreochromis niloticus*	297	202
	Sarotherodon galilaeus	342	154
	Chromidotilapia guntheri	10	—
	Tilapia dageti	19	81
	Tilapia zilli	—	61
Citharinidae	*Citharinus citharus*	28	—
	Citharinus latus	5	—
Centropomidae	*Lates niloticus*	28	8
Channidae	*Parachanna obscura*	20	6
Clariidae	*Clarias gariepinus*	99	32
	Clarias anguillaris	38	80
	Heterobranchus bidorsalis	77	15
	Clarias laeviceps	—	19
	Clarias camerunensis	—	6
	Heterobranchus longifilis	—	7
Claroteidae	*Auchenoglanis occidentalis*	98	109
	Chrysichthys auratus	6	—
Cyprinidae	*Labeo coubie*	139	41
	Labeo senegalensis	103	99
	Labeo parvus	9	15
	Raimas nigeriensis	—	12
Distichodontidae	*Distichodus engycephalus*	25	—
	Distichodus rostratus	7	—
Mochokidae	*Synodontis clarias*	42	38
	Synodontis membranaceus	9	13
	Synodontis eupterus	35	—
	Synodontis schall	90	86
	Synodontis nigrita	12	36
	Synodontis velifer	85	8
	Hemisynodontis membranaceus	7	11
	Synodontis bastiani	—	12
Malapteruridae	*Malapterurus electricus*	6	2
Mormyridae	*Hyperopisus bebe*	13	18
	Hyppopotamyrus paugyi	—	2
	Marcusenius senegalensis	102	100
	Mormyrus hasselquistii	1	—
	Mormyrus rume	10	51
	Petrocephalus bovei	20	14
	Mormyrus delicious	—	10
	Mormyrops breviceps	—	4
Osteoglossidae	*Hetrotis niloticus*	14	12
Schilbeidae	*Schilbe intermedius*	131	90
	Schilbe mystus	17	33
Polypteridae	*Polypterus senegalus senegalus*	5	—
	Polypterus endlicheri	24	7
Total		*2,239*	*1,938*

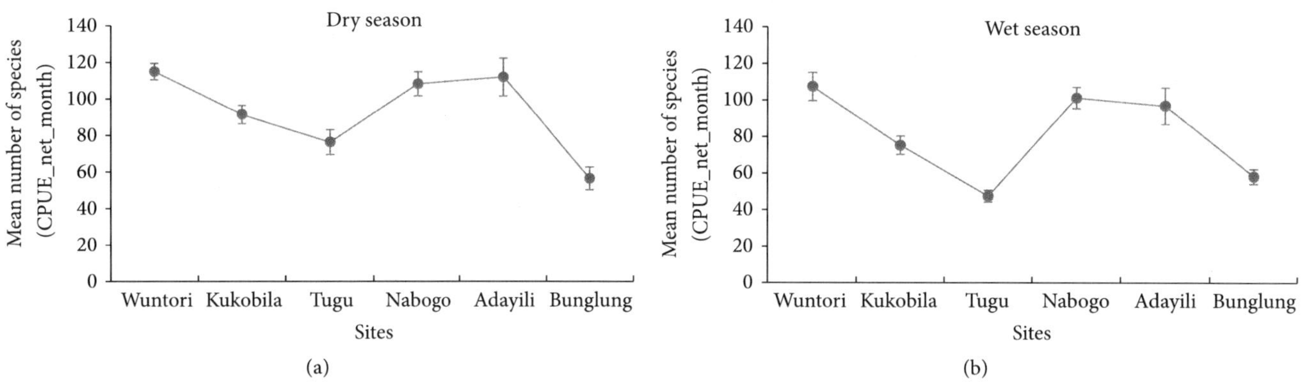

Figure 2: Variations in monthly catch per unit effort per net (CPUE) in the dry and wet season, across the six sites.

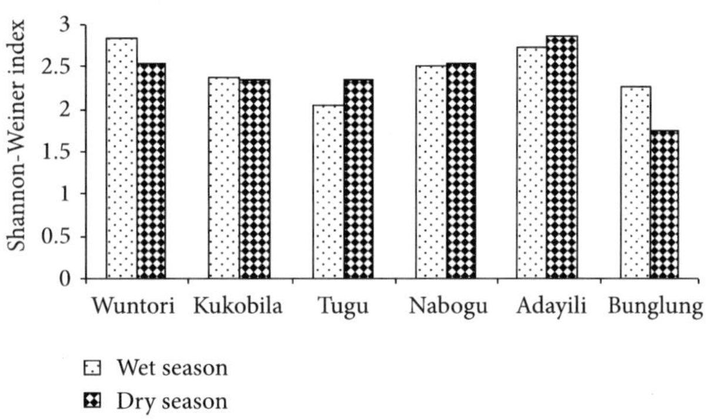

Figure 3: Variation in fish species diversity in the six wetlands for both dry season and wet season.

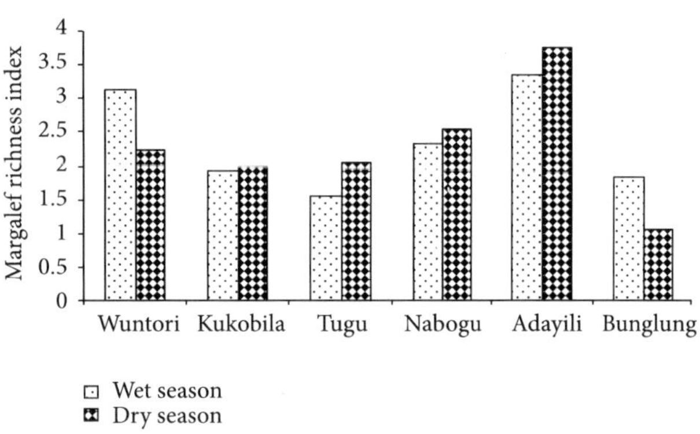

Figure 4: Seasonal changes of fish species richness in the six wetlands.

effort per net (CPUE) was consistently highest in Wuntori marsh, while the lowest CPUE was in Tugu and Bunglung constructed wetland (Figure 2).

Overall, CPUE did not vary significantly (*Tukey HSD test*, $p = 0.27$) in the dry and wet seasons (Figure 2). Fish diversity and richness marginally differed in the six sites ($F = 0.11$, $p = 0.03$) in the dry ($H' = 2.395 \pm 0.366$; $D = 2.395 \pm 0.366$) and wet seasons ($H' = 2.354 \pm 0.738$; $D = 2.354 \pm 0.738$) (Figures 3 and 4). Trends in diversity and richness reflect in the resilience of each site to environmental disturbances. However, from individual sites, Wuntori ($H' = 2.7$) and Adayilli ($H' = 2.6$) wetlands were significantly diverse and species rich ($p < 0.05$), compared to Bunglung constructed wetland ($H' = 2.0$).

3.1. General Trends in Fish Structural Distribution in the Six Wetlands. The first two axes of the DCA (axis 1 = 42% and axis 2 = 27%) jointly explained 69% of fish spatially varied distribution in the six sites (Figure 5). Fish community structure separated into three groups, as a result of habitat-specific conditions (Figure 5). Group 1 species from Bunlung man-made wetland (e.g., *Synodontis velifer, Malapterurus electricus,* and *Auchenoglanis occidentalis*) were generally less abundant in the wet and dry seasons and limited in their distribution. Farming activities, grazing intensity, water abstraction, and bush fires, were widespread in this site. Turbidity was equally high in both seasons, as were nutrient loads (phosphorus, ammonia, and nitrates). Fishes from this site survived low levels of nutrient loads, and showed a weak positive correlation with axis 1 (Figures 5 and 6). Fish communities in group II were from Nabogo and Adayili riparian systems and the shallow marshes of Tugu and Wuntori. Although species in this sites (e.g., *Mormyrus rume, Bagrus bajad, Petrocephalus bovei,* and *Lates niloticus*) were limited in their habitat distribution, their abundance were correlated with TDS ($r = -0.40$, $p < 0.05$), calcium ($r = -0.41$, $p < 0.05$), and conductivity ($r = -0.41$, $p < 0.05$) along axis 2 (Figure 6). High fish abundance at these sites reflects in their resilience to environmental disturbances like bushfires (Figures 5 and 6). Fish from group III were mainly from the deep marsh of Kubobila and dominated by Cichlids (e.g., *Sarotherodon galilaeus, Oreochromis niloticus*), Cyprinids (e.g., *Labeo coubie*), and Clariids (e.g., *Clarias gariepinus*). Surface water temperature did not significantly correlate with fish diversity and distribution, on axis 1 ($r = 0.19$, $p > 0.05$) or axis 2 ($r = 0.37$, $p > 0.05$) (Figures 5 and 6, Table 3).

The first two axes of the PCA jointly explained 81.84% of the seasonal variation in the structural distribution, abundance, and diversity of fishes in relation to the environmental variables (Table 2). Since axes 1 and 2 (81.84%) cumulatively explained >50% of variations in fish community structure and diversity [29], axis 3 correlation coefficient was not

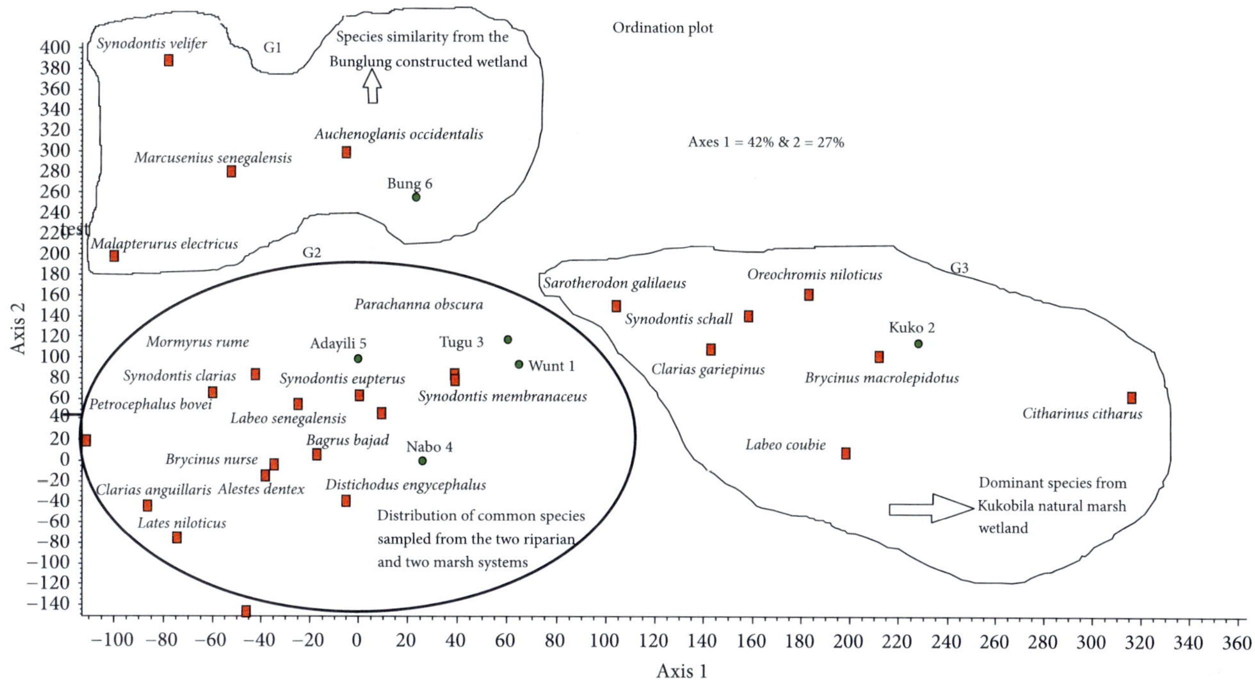

FIGURE 5: Detrended correspondence analysis (DCA) ordinations, showing fish community structure and distribution, into three groups. The red squares denote fish species, while the green circles denote sample sites. Environmental influences accounted for 69% of variations in fish community structure and distribution, while the remaining 31% was explained by a matrix of unidentified environmental and natural factors.

TABLE 3: Summary of PCA eigenvectors showing the levels of correlation with the first three axes of the environmental variables of the six sites. Asterisks (∗) indicate significance ($p < 0.05$), following Monte Carlo permutation procedures.

	Axis 1	Axis 2	Axis 3
Conductivity	−0.122	−0.414*	0.152
TDS	−1.41	−0.404*	0.156
pH	−0.202	−0.192	−0.530
Nitrate-nitrogen	−0.335	−0.165	0.080
Phosphate	−0.354	−0.072	0.148
Ammonia	−0.199	0.205	0.284
Calcium	−0.077	−0.408*	0.249
Magnesium	−0.324	−0.162	−0.194
Dissolved oxygen	−0.235	0.298	−0.258
Turbidity	0.205	0.374	0.0004
Temperature	0.190	0.374	−0.540
Fire	−0.308	−0.160	−0.181
Grazing intensity	−0.342	0.060	0.085
Erosion	−0.338	0.141	0.075
Farming activities	−0.275	0.225	−0.241
% total variance explained	7.4	4.9	1.8
Cum % total variance explained = 81.83% (first two axes)	49.1	32.8	11.9

interpreted. Eigenvectors generated by axes 1 and 2 were significant (Z-score = 1.931; $p < 0.05$; $n = 15$) in influencing fish assemblage and diversity, following Monte Carlo permutation test. Environmental variables that caused maximum change in the six sites did not differ significantly ($F = 0.158$, df = 14, and $p = 0.98$) (Figure 6).

4. Discussion

Our results revealed the direct and indirect influence of environmental disturbances on fish assemblages. High fish abundance and their structural distribution in Wuntori marsh and Adayili swamp forest systems were probably due to

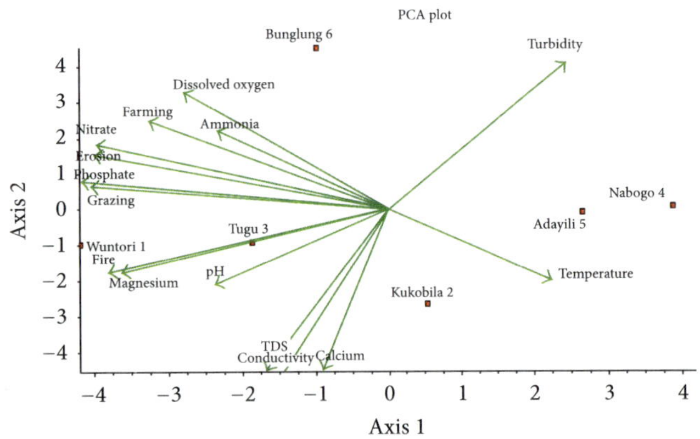

FIGURE 6: Principal component analysis (PCA) diagram, showing the significant correlations ($p < 0.05$) of environmental factors on fish community structure community and diversity in the six sites. The red squares represent the wetlands, while the arrows represent each of the environmental factors plotted pointing in the direction of maximum change of explanatory variables on a longitudinal gradient of the six sites. Axes 1 = 49.06% and Axes 2 = 32.77% jointly explained 81.83% influence of environmental variables on community structure, distribution, and diversity.

the migration of fish from the main course of the White Volta River into the wetlands via the tributaries, during cyclical inundation and the development of fish adaptive capacity to stress. We observed high diversity, richness, and abundance of the assemblage in Adayili forest swamp, although it contained lethal levels of ammonia (0.6–1.9 mg/L), which hitherto was lethal to fish survival and reproduction. Water temperature (29°C) and average pH of 6.4 were relatively low at this site. A toxic effect of high ammonia (NH_3^+) concentration on fish occurs only in prevailing higher pH values (>7.0) and warmer temperatures ~36°C in aquatic systems [30, 31]. Knepp and Arkin [32] observed that fish species can suffer gill damage even when ammonia levels are as low as 0.06 mg/L, while at 0.2 mg/L levels, sensitive fish like trout and salmon begin to die. Other factors like distance from the source, elevation, number of ponds, channel type, substratum type, and land use are key determinants that influenced littoral fish abundance [33].

However, sites with low fish abundance (e.g., Tugu marsh and Bunlung constructed wetland) were probably a result of overfishing and broad scale disturbances such as farming, grazing activities, bushfires, and water chemistry, although less severe human-led activities and optimal physicochemical factors occur in the dry season and largely influenced fish abundances. Fish harvest in the wet season was prohibited (close season) to allow for spawning activities. Thus, only regulated fishing activities were permissible at a lesser spatiotemporal scale. Ayoola and Kuton [5] reported similar high abundance and biomass composition of fish species in the dry season, relative to the wet season in a Lagos lagoon. But Abban et al. [34] suggested that the lower fish abundance encountered in the wet season could be linked to the physical expansions of the wetlands found within the catchment of the White Volta River, where the current study was conducted.

Distribution of *Bagrus bajad, Lates niloticus, Labeo coubie,* and *Brycinus macrolepidotus* in preferred habitats was influenced by varied water quality parameters, bushfire, and grazing intensity. While species in the riparian systems appear to respond positively to turbid condition, we observed that species assemblages from the marshes and the constructed wetland were distributed along a longitudinal profile, with TDS, conductivity, nitrates, dissolved oxygen, water temperature, grazing intensity, and fire occurrence (Figures 5 and 6). Using a similar ordination approach, Pires et al. [3] found that water depth, isolated pools, and cover segregated fish species in the streams of the Guadiana basin in Portugal. Flooding regime [6] and habitat size [35, 36] have also been reported as important natural drivers of change that structure fish assemblages in streams. These contrasting findings to the current study could be attributed to the varied response of species-specific to inherent ecological factors from different biogeographical zones (i.e., temperate to topical zones).

Though farming activities were widespread in this study, it did not directly influence fish distribution (as shown by a weak axes correlation in Table 1) but indirectly led to an increase in turbidity in the riparian and the constructed wetlands. Argent and Carline [37] observed the increased in agricultural activities as a major factor influencing fish structure subwatersheds in Pennsylvania. Levels of orthophosphate and nitrate were low across the sites and probably contributed to the high abundance of fishes. In spite of the low level of nutrient loads, they were spatially auto correlated with grazing intensity and farming activities, suggesting the direct influence of these human-led activities. Highly eutrophic conditions impair fish fecundity and abundance, compared to less eutrophic areas [38].

Fish diversity indices at the six sites ($H' = 2.395 \pm 0.366$ and 2.354 ± 0.738) reflected a highly diverse fish community compared to findings by Quarcoopome et al. [39] in Bontanga ($H' = 1.55$) and Libga ($H' = 2.36$) artificial wetlands, which are located within the same catchment of the current study. This appears to suggest that open natural habitats, characterized by a complex food web, support diverse fish species, compared to the artificial systems. But further comparison of the indices in this study to Fernandes and Achuthankutty [40] in the Salcete Taluka wetlands at Goa, India ($H' = 2.08$–4.27 and 2.23–4.28), shows that species in the Salcete Taluka wetlands are resilient to environmental perturbation and natural factors, or the systems are less impacted. The near 100% dominance of *Sarotherodon galilaeus* for the cichlids probably suggests their endemic nature in Northern wetlands, as a result of their broad scale tolerance to varying degrees of ecological stress conditions. This species may be considered as a suitable bioindicator candidate to impacts of wetland disturbances such as pollution and climate change. Similar observations on all year round dominance of Cichlids in the Lagos lagoon were documented by Ayoola and Kuton [5].

5. Conclusion

Although fish abundance and diversity were reasonably high in Wuntori, Adayili, and Nabogo, their spatial distributions were limited. Seasonal variability in diversity, richness, and abundance was as a result of varied habitat conditions mediated by environmental disturbances such as farming activities and bushfires. Most species appeared to be tolerant to low nutrient loads and this partly contributed in the high abundance and diversity. Finally, the low nutrient loads also suggest that the wetlands were not polluted. The implication of these results suggests that wetland managers must consider instituting conservation measures to check land use activities that has tendency to cause future eutrophication condition. The implementation of these conservation measures could contribute to the increase and sustenance of fish population, which serve as a major source of livelihoods for the rural dwellers.

Competing Interests

The authors declare that there are no competing interests regarding the publication of this paper.

Acknowledgments

The authors express their profound gratitude to the Ghana Education Trust Fund (GETFund), for their financial support of the research work.

References

[1] C. H. Hocutt and E. O. Wiley, Eds., *The Zoogeography of North American Freshwater Fishes*, John Wiley & Sons, New York, NY, USA, 1986.

[2] Millennium Ecosystem Assessment, *Ecosystems and Human Well-Being: Wetlands and Water Synthesis*, World Resources Institute, Washington, Wash, USA, 2005.

[3] A. M. Pires, I. G. Cowx, and M. M. Coelho, "Seasonal changes in fish community structure of intermittent streams in the middle reaches of the Guadiana basin, Portugal," *Journal of Fish Biology*, vol. 54, no. 2, pp. 235–249, 1999.

[4] K. J. Killgore and J. J. Hoover, "Effects of hypoxia on fish assemblages in a vegetated waterbody," *Journal of Aquatic Plant Management*, vol. 39, no. 1, pp. 40–44, 2001.

[5] S. O. Ayoola and M. P. Kuton, "Seasonal variation in fish abundance and physicochemical parameters of Lagos Lagoon, Nigeria," *African Journal of Environmental Science and Technology*, vol. 3, no. 5, pp. 149–156, 2009.

[6] K. D. Fausch and R. G. Bramblett, "Disturbance and fish communities in intermittent tributaries of a Western Great Plains river," *Copeia*, vol. 1991, no. 3, pp. 659–674, 1991.

[7] I. G. Cowx and R. L. Welcomme, Eds., *Rehabilitation of Rivers for Fish*, Blackwell Science, Oxford, UK, 1998.

[8] B. W. Menzel, J. B. Barnum, and L. M. Antosch, "Ecological alterations of Iowa prairie agricultural streams. Iowa State," *Journal of Science*, vol. 59, pp. 5–30, 1984.

[9] H. E. Berkman and C. F. Rabeni, "Effect of siltation on stream fish communities," *Environmental Biology of Fishes*, vol. 18, no. 4, pp. 285–294, 1987.

[10] P. L. Angermeier and M. R. Winston, "Characterizing fish community diversity across Virginia landscapes: prerequisite for conservation," *Ecological Applications*, vol. 9, no. 1, pp. 335–349, 1999.

[11] J. Lyons, "Patterns in the species composition of fish assemblages among Wisconsin streams," *Journal of Environmental Biology of Fishes*, vol. 45, no. 4, pp. 329–346, 1996.

[12] K. Pivnička and M. Humpl, "Fish assemblages in the Elbe River watershed—species richness, frequency, and clusters," *Acta Universitatis Carolinae Environmentalica*, vol. 18, no. 1-2, pp. 107–116, 2004.

[13] Z. A. Ansari, A. Chatterji, B. S. Ingole, R. A. Sreepada, C. U. Rivonkar, and A. H. Parulekar, "Community structure and seasonal variation of an inshore demersal fish community at Goa, West Coast of India," *Estuarine, Coastal and Shelf Science*, vol. 41, no. 5, pp. 593–610, 1995.

[14] A. Chezhian, N. Kabilan, T. Suresh-Kumar, and D. Senthamil-Selvan, "Influence of different calcium levels and low pH of water on the plasma electrolyte regulation of a fresh water teleost fish *Cyprinus carpio* Var. communies, (Linnaeus, 1958)," *Current Research Journal of Biological Science*, vol. 3, no. 2, pp. 147–154, 2011.

[15] A. Y. Karikari, K. A. Asante, and C. A. Biney, "Water quality characteristics at the Estuary of Korle Lagoon in Ghana," *West Africa Journal of Applied Ecology*, vol. 10, pp. 73–85, 2006.

[16] C. A. Nsor and E. A. Obodai, "Environmental determinants influencing seasonal variations of bird diversity and abundance in Wetlands, Northern Region (Ghana)," *International Journal of Zoology*, vol. 2014, Article ID 548401, 10 pages, 2014.

[17] A. T. Charles, T. R. Brainerd, M. A. Bermudez, H. M. Montalvo, and R. S. Pomeroy, *Fisheries Socioeconomics in the Developing World: Regional Assessments and an Annotated Bibliography*, International Development Research Centre (IDRC), Ottawa, Canada, 1994.

[18] T. Slaymaker and R. M. Blench, *Rethinking Natural Resource Degradation in Sub-Saharan Africa: Policies to Support Sustainable Soil Fertility Management, Soil and Water Conservation among Resource Poor Farmers in Semi-arid Areas: Country Studies*, University for Development Studies, Tamale, Ghana, 2002.

[19] H. R. Dankwa, E. K. Abban, and G. G. Teugels, *Freshwater Fishes of Ghana. Identification, Distribution, Ecological and Economic Importance*, vol. 283 of *Annales Science Zoologiques*, 1999.

[20] D. Paugy, C. Leveque, and G. G. Teugels, Eds., *The Fresh and Brackish water Fishes of West Africa*, Publications Scientifiques du Museum, MRAC. IRD Editions. Africa. Tervuren, 2004.

[21] APHA (American Public Health Administration), *Standard Methods for the Examination of Water and Wastewater*, APHA (American Public Health Administration), Washington, DC, USA, 20th edition, 1998.

[22] N. Salafsky, D. Salzer, A. J. Stattersfield et al., "A standard lexicon for biodiversity conservation: unified classifications of threats and actions," *Conservation Biology*, vol. 22, no. 4, pp. 897–911, 2008.

[23] C. Battisti, L. Luiselli, and C. Teofili, "Quantifying threats in a Mediterranean wetland: are there any changes in their evaluation during a training course?" *Biodiversity and Conservation*, vol. 18, no. 11, pp. 3053–3060, 2009.

[24] C. E. Shannon and W. Wiener, *The Mathematical Theory of Communication*, Urbana University Illinois Press, Champaign, Ill, USA, 1963.

[25] R. Margalef, *Perspective in Ecological Theory*, University of Chicago Press, Chicago, Ill, USA, 1968.

[26] P. A. Henderson and R. M. Seaby, *Community Analysis Package 1.14*, S041 8GN, Pisces Conservation Ltd. IRC House, Pennington, UK, 1999.

[27] M. Kent and P. Coker, *Vegetation Description and Analysis. A Practical Approach*, John Wiley & Sons, Chichester, UK, 1992.

[28] K. McGarigal, S. Cushman, and S. Stafford, *Multivariate Statistics for Wildlife and Ecology Research*, Springer, New York, NY, USA, 2000.

[29] C. J. F. terBraak, "Canonical correspondence analysis: a new eigenvector technique for multivariate direct gradient analysis," in *Multivariate Statistics for Wildlife and Ecology Research*, K. McGarigal, S. Cushman, and S. Stafford, Eds., p. 700, Springer, New York, NY, USA, 2000.

[30] US-EPA, "Ambient water quality criteria for ammonia," EPA Report 440/5-85-001, United States Environmental Protection Agency, Washington, DC, USA, 1984.

[31] Canadian Water Quality Guidelines for the Protection of Aquatic Life, Canadian Council of Ministers of the Environment, 2010.

[32] G. L. Knepp and G. F. Arkin, "Ammonia toxicity levels and nitrate tolerance of channel catfish," *The Progressive Fish-Culturist*, vol. 35, no. 4, pp. 221–224, 1973.

[33] M. Humpl and K. Pivnička, "Fish assemblages as influenced by environmental factors in streams in protected areas of the Czech Republic," *Ecology of Freshwater Fish*, vol. 15, no. 1, pp. 96–103, 2006.

[34] "Biodiversity, management and utilization of West African Fishes," in *Proceedings of the World Fish Center Conference Proceedings*, E. K. Abban, C. M. V. Casel, P. Dugan, and T. Falk, Eds., p. 63, Penang, Malaysia, 2004.

[35] L. R. Brown, "Fish communities and their associations with environmental variables, lower San Joaquin River drainage, California," *Environmental Biology of Fishes*, vol. 57, no. 3, pp. 251–269, 2000.

[36] G. Grenouillet, D. Pont, and C. Hérissé, "Within-basin fish assemblage structure: the relative influence of habitat versus stream spatial position on local species richness," *Canadian Journal of Fisheries and Aquatic Sciences*, vol. 61, no. 1, pp. 93–102, 2004.

[37] D. G. Argent and R. F. Carline, "Fish assemblage changes in relation to watershed," *Journal of Aquatic Ecosystem Health Management*, vol. 7, no. 1, pp. 101–114, 2004.

[38] A. Sandström and P. Karås, "Effects of eutrophication on young-of-the-year freshwater fish communities in coastal areas of the baltic," *Environmental Biology of Fishes*, vol. 63, no. 1, pp. 89–101, 2002.

[39] T. Quarcoopome, F. Amevenku, and O. Ansa-Asare, "Fisheries and limnology of two reservoirs in Northern Ghana," *West African Journal of Applied Ecology*, vol. 12, no. 1, 2008.

[40] B. Fernandes and C. T. Achuthankutty, "Seasonal variation in fishery diversity of some wetlands of the Salcete Taluka, Goa, India," *Indian Journal of Marine Sciences*, vol. 39, no. 2, pp. 238–247, 2010.

7

Impacts of Bush Encroachment on Wildlife Species Diversity, Composition, and Habitat Preference in Ol Pejeta Conservancy, Laikipia, Kenya

Cyrus M. Kavwele,[1] Johnstone K. Kimanzi,[2] and Mwangi J. Kinyanjui[1]

[1]*Department of Natural Resources, Karatina University, Karatina, Kenya*
[2]*Department of Wildlife Management, University of Eldoret, Eldoret, Kenya*

Correspondence should be addressed to Cyrus M. Kavwele; cyruskavwele@yahoo.com

Academic Editor: Daniel I. Rubenstein

Savannah ecosystems are currently facing a biome shift that changes grasslands to woody dominated landscapes, attributable to habitat degradation. In Ol Pejeta Conservancy (OPC), *Euclea divinorum*, an unpalatable and invasive woody species, is expanding to former savannah ecosystems with potential effects on herbivores key resources, wildlife species diversity, composition, and habitat use. We investigated wildlife species diversity, composition, and habitat preference or avoidance by wildlife in the conservancy. Infrared camera traps were deployed at the centroids of 2 km by 2 km, 50 cm above ground surface for 14 days and nights with 9 camera traps in each habitat type. Shannon wiener index revealed that wildlife species diversity was highest in *E. divinorum* dominated habitats and lowest in open grassland. Hierarchical Cluster Analysis revealed level of similarity in wildlife species composition between *E. divinorum* and mixed bushland. Jacobs index revealed that *E. divinorum* and mixed bushland were avoided by all guilds; however *E. divinorum* was significantly avoided while *A. drepanolobium* and open grassland were both preferred by all guilds. However, *A. drepanolobium* dominated habitats were significantly preferred compared to open grasslands. The findings are useful in management of sustainable ecosystems.

1. Introduction

Woody species encroachment has been described as gradual conversion of grasslands to woody and shrub dominated savannahs [1], a phenomenon that has been documented well in various parts of the world [2]. These increasingly changing grasslands [3] form persistent patches which alter composition and structure of flora in savannah ecosystems [4]. Overgrazing by exerting pressure on grazing lawns and increase in precipitation have shown positive correlation with woody species encroachment [5], hence touted as potential driver of the vegetation shifts. Further, soil conditions [6, 7] and fire suppression [8] among other perturbations are also attributable to this phenomenon.

According to Ward [9] an increase in woody species can significantly affect livestock and wildlife conservation efforts, as such, further indirectly or directly affecting livelihoods of the local people [10]. Particular facets of economy likely to be affected include but not limited to ecotourism, a key source of revenue in these ecosystems due to poor visibility, hence affecting game viewing experience [11]. From another ecological position, according to Wigley et al. [11], increase in woody species translates to increase in fuel wood, timber, and fencing material among other as well as increase in resource for browsing herbivores and structural diversity for avian biodiversity.

Even worse, encroachment in isolated ecosystems can result in decline and/or extinction of native species and can potentially affect species diversity, distribution, and abundance [12]. This also holds true according to Dalle et al. [13] where encroachment in savannah ecosystems poses a serious threat to ecosystems function especially tree-grass coexistence. Grass-tree balance influences grassland/rangeland

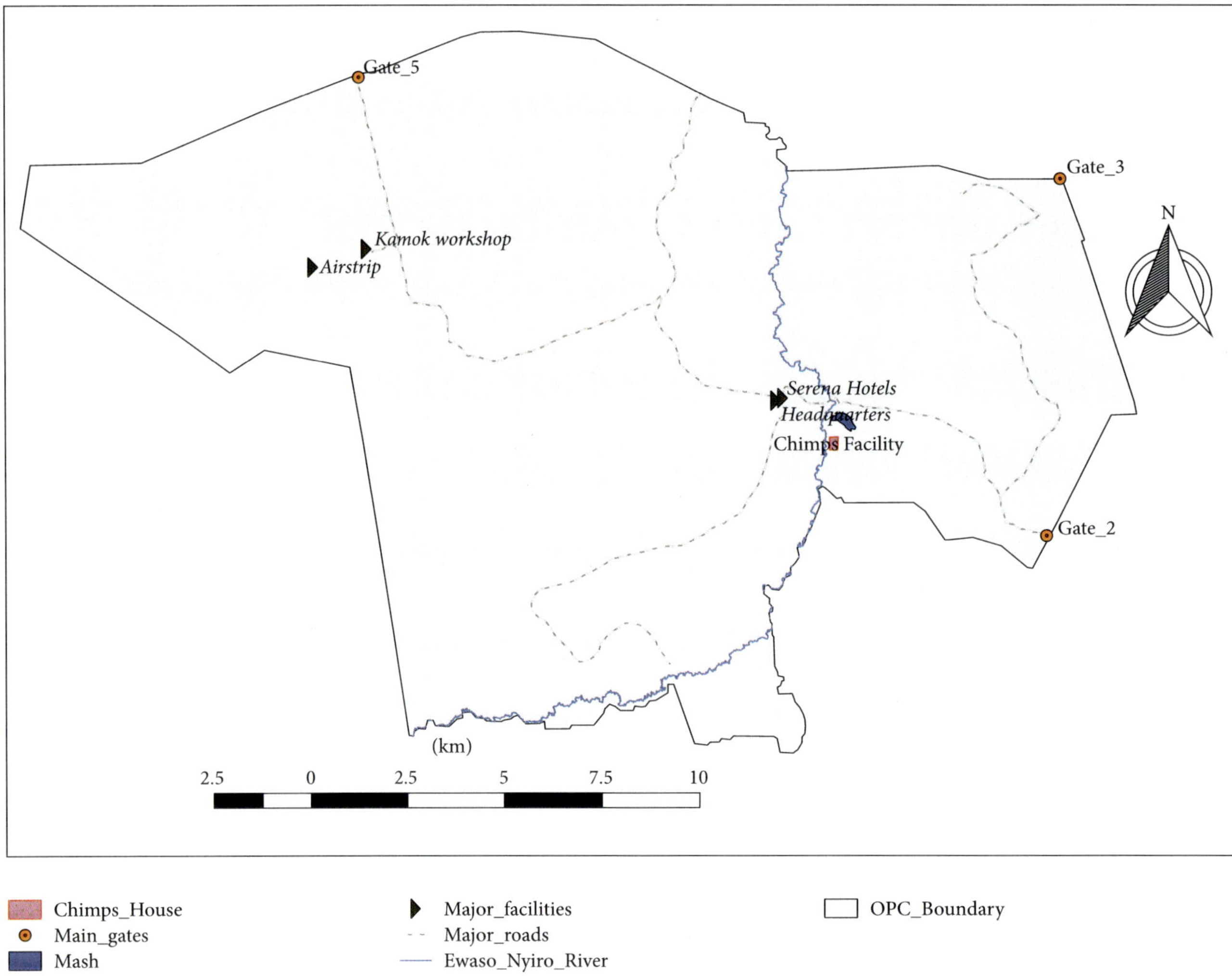

FIGURE 1: Map of the study site.

economic services, biodiversity conservation, and ecosystem function at local and landscape scales.

In Ol Pejeta Conservancy (OPC), encroachment by unpalatable *E. divinorum* encroachment towards *Acacia drepanolobium*, grasslands, and other open bushland vegetation cover types has become a concern for conservationists. Key resources for megafauna in these ecosystems especially the critically endangered Eastern black rhino, *Diceros bicornis*, International Union for Conservation of Nature (IUCN) red listing [14], and African elephants, *Loxodonta africana*, vulnerable (IUCN) red listing [15], among other herbivores can potentially be affected. This encroachment can further reduce the available ranging lands and to some extent exterminate some of the wild flora and fauna. Here, we investigated (a) wildlife species diversity and evenness in encroached and nonencroached habitats, (b) wildlife species composition in encroached and nonencroached habitats, and (c) habitat preference or avoidance by various feeding guilds within the conservancy. The findings demonstrate impacts of bush encroachment on biodiversity, hence, beneficial to conservationists.

2. Methods

2.1. Study Site. This study was undertaken in Ol Pejeta Conservancy in Laikipia County, Central Kenya, as shown in Figure 1. The property which covers 90,000 acres (360 km^2) lies in between Mt. Kenya and Aberdares (at 0° 7.288′N, 36°42.384′E and 0° 8.634′N, 37° 0.605′E) (0° 1.831′S, 36° 46.578′E and 0° 5.7025′S, 37° 2.492′E). It has an average altitude of 1810 m, mean annual rainfall of 739 mm, and mean maximum and minimum temperatures of 28°C and 12°C, respectively. It is a private owned conservancy, chiefly, a sanctuary for black rhinos *(Diceros bicornis)*. However, other wild animals are also conserved in the same property composed of various feeding guilds such as mixed feeders, carnivores, grazers, and browsers. It is also a sanctuary for chimpanzee *(Pan troglodytes)* providing refuge site for seized chimps from black markets. It has one permanent river flowing through the conservancy as well as various man-made water holes to supplement water scarcity in dry spells.

Habitat cover types include grasslands, *Acacia drepanolobium*, *A. xanthophloea*, *Euclea divinorum*, and mixed bushlands. The conservancy is surrounded by an electric fence

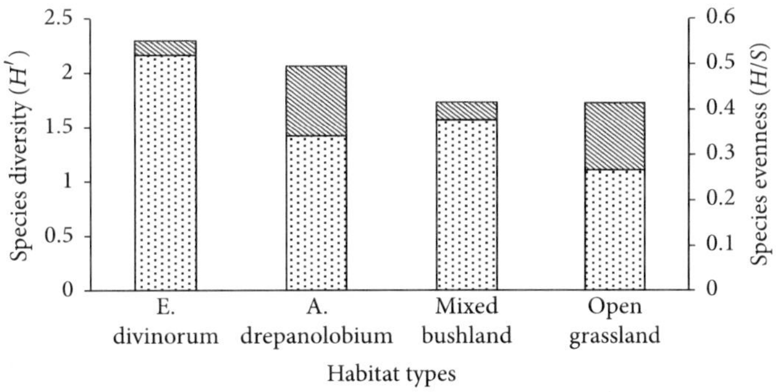

FIGURE 2: Wildlife species diversity (left y axis) and wildlife species evenness (right y axis) in different habitats in OPC.

with three "corridors" to allow movement of wild animals in and out of OPC (but movement of rhino species is restricted due to the risks involved). The conservancy is surrounded by agropastoral communities and towards the north by other adjoining conservancies.

2.2. Data Collection and Analysis. The entire conservancy was overlaid on 2 × 2 km grid as shown in Figure 1 (map) and centroids of each grid were identified. At every centroid, motion triggered infrared camera traps of Reconyx Rapid-fire RM[45] model were deployed systematically taking into cognisance of animal trails to capture wildlife species within given cover types. Camera traps were set to take photos with no delay and remain active for 24 hrs at 50 cm above the ground surface either attached to a tree using elastic strings (cords) or housed in a metal cage in case of grassland areas. They remained in the field for 14 consecutive days and nights though they were checked on the 7th day for battery charge level, remaining storage size in secure cards (SD), and any other malfunctions occasioned by knocking of the traps by rhinos or elephants. A total of 9 camera traps were deployed in each habitat type, namely, open grassland, *E. divinorum*, *A. drepanolobium*, and mixed bushland. Upon realising low battery charge or near filled up storage there were replacements of such to avoid loss of data due to nonperforming camera traps. Immediately after removal of camera traps from the field, all photographic data were downloaded and further sorted into species name, size, time, and habitat type the camera trap was deployed in. Further, duplicates were removed where thirty (30) minutes after the last photo was taken was considered as another photographic event.

After data cleaning, species diversity and composition using Shannon Weiner index (where if index is 5 it implies highest diversity and if it is 1 it implies lowest diversity) [16] and Hierarchical Cluster Analysis [17], respectively, were performed in Paleontological Statistics Software Package for Education and Data Analysis (PAST) software while habitat preference or avoidance was performed using Jacobs' index [18]. Further, Jacobs' index values were tested for significance preference or avoidance (where +1 implies maximum preference while −1 implies maximum avoidance) using t-tests against means of zero (0) in R version 3.1.2. Prior to these tests, normal distribution and homogeneity of variance were performed to ensure that conditions for parametric tests were not violated [19]. Overlay of grids on the spatial extent of the conservancy was performed in Quantum Geographic Information Systems (QGIS) platform version 2.8.2.

3. Results

3.1. Wildlife Species Diversity and Evenness. Shannon Weiner index revealed highest wildlife species diversity in *E. divinorum* dominated habitats ($H' = 2.291$), followed by *A. drepanolobium* habitats ($H' = 2.058$), mixed bushland ($H' = 1.728$), and least wildlife species diversity in open grassland ($H' = 1.715$). Wildlife species evenness (H/S) was highest in area under *E. divinorum* ($H/S = 0.5201$), followed by mixed bushland ($H/S = 0.3751$) and then closely followed by *A. drepanolobium* ($H/S = 0.3404$) and finally open grassland ($H/S = 0.2647$) as shown in Figure 2.

3.2. Wildlife Species Composition. Hierarchical Cluster Analysis (HCA) compared wildlife species composition (common wildlife species) across the four habitat types, namely, *E. divinorum*, *A. drepanolobium*, mixed bushland, and open grassland. Results revealed that *E. divinorum* and mixed bushland habitats shared 45% similarity in wildlife species composition. Additionally, *E. divinorum* habitat and mixed bushland shared 39% similarity in wildlife species composition with *A. drepanolobium* dominated habitat. On the other hand, open grassland habitat shared 27% similarity in wildlife species composition with three habitat types, namely, *E. divinorum*, *A. drepanolobium*, and mixed bushland as shown in Figure 3.

3.3. Habitat Preference or Avoidance by Various Feeding Guilds in OPC. Jacobs index revealed that carnivores preferred *A. drepanolobium* ($D = 0.469$) mostly followed by open grassland ($D = 0.327$) and least preferred mixed bushland ($D = 0.066$) but avoided *E. divinorum* dominated areas ($D = -0.698$) (Figure 4(a)). Grazers showed great preference for open grassland ($D = 0.773$) and *A. drepanolobium* ($D = 0.040$) but avoided *E. divinorum* ($D = -0.917$) and mixed bushland ($D = -0.192$) habitats (Figure 4(b)).

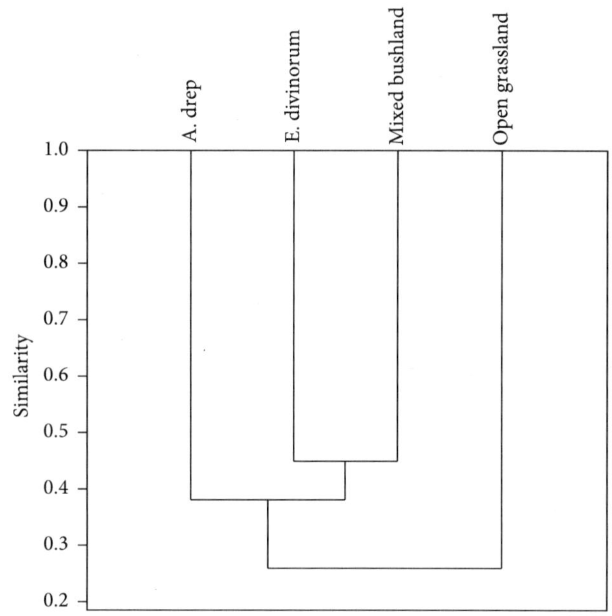

FIGURE 3: Percentage similarity in wildlife species composition of 4 habitats in OPC.

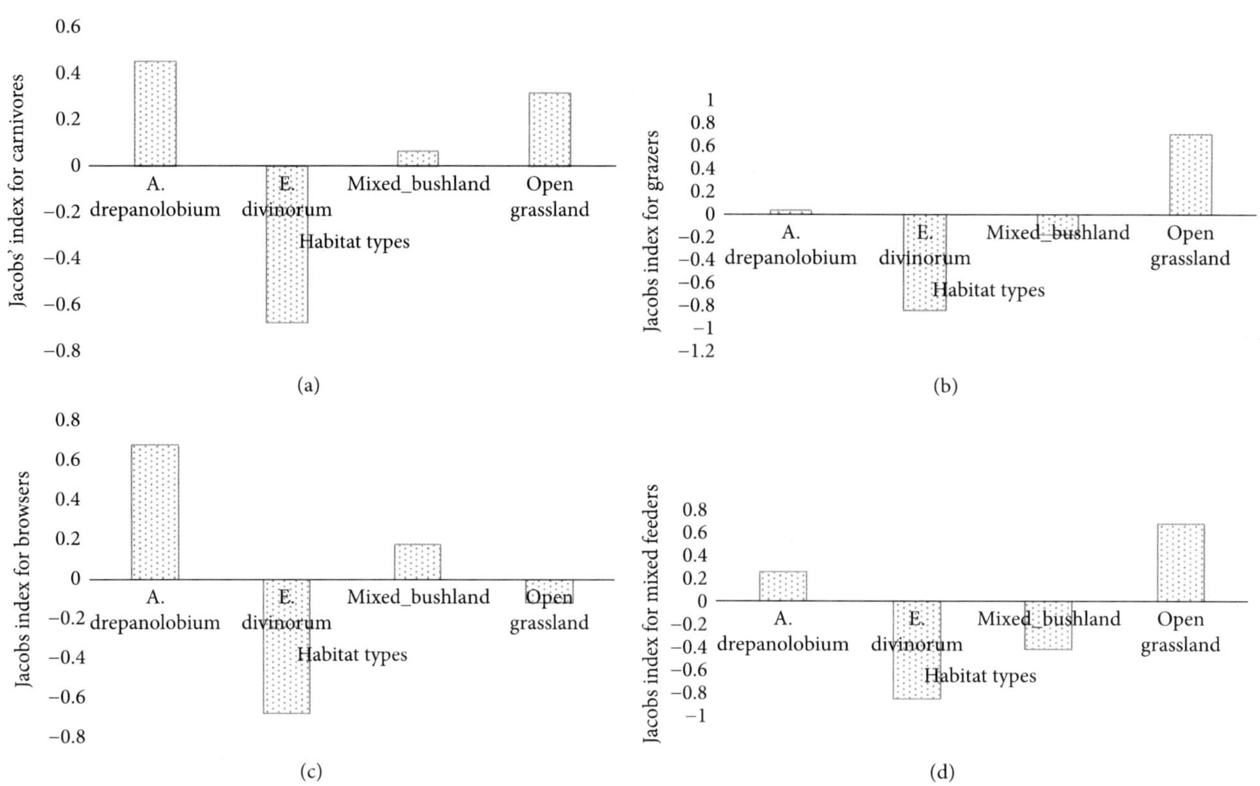

FIGURE 4: Habitat preference or avoidance (Jacobs' index) for (a) carnivores, (b) grazers, (c) browsers, and (d) mixed feeders in OPC.

Browsers showed preference for *A. drepanolobium* (D = 0.674) and mixed bushland habitat (D = 0.175) but avoided both *E. divinorum* (D = −0.673) and open grasslands (D = −0.116) dominated habitats (Figure 4(c)). However, mixed feeders preferred habitats dominated by open grassland (D = 0.688) and *A. drepanolobium* (D = 0.523) but avoided *E. divinorum* (D = −0.858) and mixed bushland (D = −0.420) dominated habitats (Figure 4(d)). A total of thirty (30) wildlife species were recorded as per the table in the supplementary material showing detection and nondetection of wildlife species in various habitat cover types.

Mean preference or avoidance revealed that *E. divinorum* and mixed bushland were avoided; however, *E. divinorum* was significantly avoided (t_1 = 2.253, d.f. = 3, and

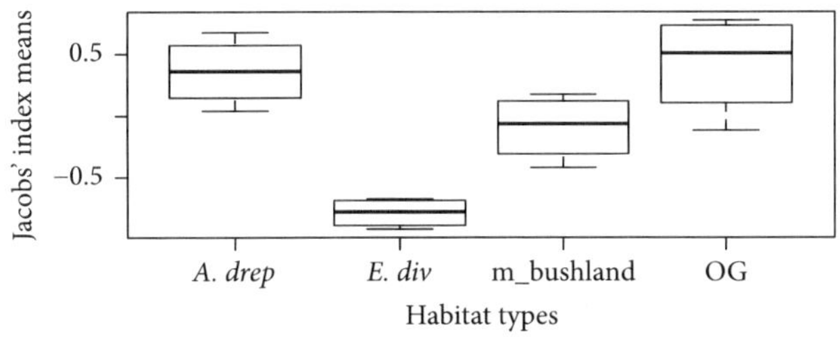

Figure 5: Jacobs index means for all guilds across the habitat types.

$p < 0.01$) compared with mixed bushland ($t_1 = 2.353$, d.f. = 3, and $p = 0.27$). On the other hand, *A. drepanolobium* and open grassland were both preferred by all feeding guilds; however, *A. drepanolobium* dominated habitats were significantly preferred ($t_1 = 2.353$, d.f. = 3, and $p = 0.03$) compared to open grasslands ($t_1 = 2.353$, d.f. = 3, and $p = 0.06$) as shown in Figure 5.

4. Discussions

Wildlife species diversity and evenness were highest in *E. divinorum* dominated areas compared to other habitat types. This is attributable to existence of resource rich patches that are hardly accessed by most wildlife species. As such, attraction of several wildlife species into this dense vegetation *(E. divinorum)* contributed to higher diversity chiefly small carnivores. Animals use cues such as landscape structure and vegetation structure like tree species phenology, resource availability (probability of encounter, quantity, and quality), and predator/parasite risks among others in order to explore a given habitat type. This implies that spatial and temporal habitat heterogeneity drives faunal species diversity and composition from local to regional to global scales. Hence, aggregation of resources may create dynamic impacts where these small sites/patches become a focus for higher proportion of faunal diversity and composition [20]. According to Sirami et al. [21] savannah ecosystems are diverse in structure and floral composition; hence different guilds will use different habitats based on their requirements.

Animals have different preference to certain habitats [22] as a function of direct and indirect effects of prey availability, detectability/cover, and resource availability [23]. Spatial heterogeneity of ecosystem is important for maintenance of diverse wild animal species and acts as buffer against changes in resources availability for wild animals in era of climate change [24]. On the other hand, woody encroachment can potentially change this heterogeneity and affect wide range of wild animals.

Similarity in wildlife species composition was highest between *E. divinorum* and mixed bushland habitats. These two habitat types have nearly homogeneous vegetation structure (dense canopy). This phenomenon is an illumination to higher percentage similarity in wildlife species composition in both *E. divinorum* and mixed bushland dominated areas due to nearly similar structure of the vegetation (canopy and dense bushland) providing ideal habitats for wildlife species which require such areas. According to Dupuch et al. [25], prey and predator can explore freely across habitats that differ in resource quantity, quality, and inherent habitat risks. As such, predators' movement and habitat selection are as a function of prey density, competition level, cover, and inherent risks while prey habitat selection and movement are balance of forage quality (benefits of foraging) and risk of predation [26].

On the other hand, habitats dominated by *A. drepanolobium* and open grassland were preferred by feeding guilds majorly herbivores due to their associated lesser predation risk. This is chiefly influenced by presence of relatively open landscapes allowing prey to detect predators from a distance and hence maintain safer flight distance (improved visibility). The phenomenon is consistent with other findings according to Dupuch et al. [25], where prey habitat selection is largely influenced by the need to minimise predation risks over forage quality. As such, these factors influence preference of habitats that offer significant benefits for most wildlife species.

5. Conclusion and Recommendations

The study revealed that bush encroachment by *E. divinorum* had impact on species diversity, composition, and habitat selection. In particular, species diversity was highest in *E. divinorum* dominated areas and least in open grasslands, a positive impact, albeit long term impacts are unknown. Likewise, percentage similarity in species composition was highest between *E. divinorum* and mixed bushland dominated habitats. Further, feeding guilds avoided areas dominated by *E. divinorum* but preferred *A. drepanolobium* and open grassland dominated habitats; hence, bush encroachment by *E. divinorum* affected habitat suitability; thus it has a negative impact. Based on this finding, active management of the encroaching species is recommended. Further studies on activity budget of wildlife species to ascertain how habitat types are being used are key. Additionally, long term study to provide insights into the effect of seasonality on species diversity dynamics across various land cover types in the conservancy is crucial. This information will be vital in designing habitat management protocols for sustainable ecosystem management.

Conflicts of Interest

The authors declare no conflicts of interest.

Acknowledgments

This study was funded by Rufford Foundation, United Kingdom (UK), Grant no. 19422-1. The authors are grateful to Carol Ngw'eno and Bernard Chira from Ol Pejeta Conservancy (OPC) for allowing them to carry out research in their property and Andrew Bradley and Dan Smith both from Bangor University for their help in camera trap image classification.

References

[1] A. T. Hudak, C. A. Wessman, and T. R. Seastedt, "Woody overstorey effects on soil carbon and nitrogen pools in South African savanna," *Austral Ecology*, vol. 28, no. 2, pp. 173–181, 2003.

[2] S. Archer, T. W. Boutton, and K. A. Hibbard, "Trees in grasslands: biogeochemical consequences of woody plant expansion," in *Global Biogeochemical Cycles in the Climate System*, pp. 115–137, Academic Press, San Diego, 2001.

[3] I. A. Dickie, S. A. Schnitzer, P. B. Reich, and S. E. Hobbie, "Is oak establishment in old-fields and savanna openings context dependent?" *Journal of Ecology*, vol. 95, no. 2, pp. 309–320, 2007.

[4] S. R. Wangen and C. R. Webster, "Potential for multiple lag phases during biotic invasions: Reconstructing an invasion of the exotic tree Acer platanoides," *Journal of Applied Ecology*, vol. 43, no. 2, pp. 258–268, 2006.

[5] D. F. Joubert, A. Rothauge, and G. N. Smit, "A conceptual model of vegetation dynamics in the semiarid Highland savanna of Namibia, with particular reference to bush thickening by Acacia mellifera," *Journal of Arid Environments*, vol. 72, no. 12, pp. 2201–2210, 2008.

[6] M. Sankaran, J. Ratnam, and N. Hanan, "Woody cover in African savannas: The role of resources, fire and herbivory," *Global Ecology and Biogeography*, vol. 17, no. 2, pp. 236–245, 2008.

[7] G. Oba, E. Post, P. O. Syvertsen, and N. C. Stenseth, "Bush cover and range condition assessments in relation to landscape and grazing in southern Ethiopia," *Landscape Ecology*, vol. 15, no. 6, pp. 535–546, 2000.

[8] O. W. Van Auken, "Shrub invasions of North American semiarid grasslands," *Annual Review of Ecology, Evolution, and Systematics*, vol. 31, pp. 197–215, 2000.

[9] D. Ward, "Do we understand the causes of bush encroachment in African savannas?" *African Journal of Range & Forage Science*, vol. 22, no. 2, pp. 101–105, 2005.

[10] A. T. Hudak and C. A. Wessman, "Textural analysis of high resolution imagery to quantify bush enroachment in Madikwe Game Reserve, South Africa, 1955-1996," *International Journal of Remote Sensing*, vol. 22, no. 14, pp. 2731–2740, 2001.

[11] B. J. Wigley, W. J. Bond, and M. T. Hoffman, "Bush encroachment under three contrasting land-use practices in a mesic South African savanna," *African Journal of Ecology*, vol. 47, no. 1, pp. 62–70, 2009.

[12] D. R. Towns, I. A. E. Atkinson, and C. H. Daugherty, "Have the harmful effects of introduced rats on islands been exaggerated?" *Biological Invasions*, vol. 8, no. 4, pp. 863–891, 2006.

[13] G. Dalle, B. L. Maass, and J. Isselstein, "Encroachment of woody plants and its impact on pastoral livestock production in the Borana lowlands, southern Oromia, Ethiopia," *African Journal of Ecology*, vol. 44, no. 2, pp. 237–246, 2006.

[14] R. Emslie, "Diceros bicornis, The IUCN Red List of Threatened Species 2012," 2012. e.T6557A16980917. Downloaded on 17April, http://dx.doi.org/10.2305/IUCN.UK.2012.RLTS.T6557A16980917.en. 2016.

[15] J. Blanc, "Loxodonta africana, The IUCN Red List of Threatened Species 2008,"2008. e.T12392A3339343. http://dx.doi.org/10.2305/IUCN.UK.2008.RLTS.T12392A3339343.en.Downloaded on 17 April 2016.

[16] C. E. Shannon and W. Weaver, *The Mathematical Theory of Communication*, University of Illinois Press, Chicago, 1949.

[17] H. G. Gauch, *Multivariate Analysis in Community Ecology*, Cambridge University Press, Cambridge, UK, 1982.

[18] J. Jacobs, "Quantitative measurement of food selection - A modification of the forage ratio and Ivlev's electivity index," *Oecologia*, vol. 14, no. 4, pp. 413–417, 1974.

[19] F. Palomares, M. Delibes, E. Revilla, J. Calzada, and J. M. Fedriani, "Spatial ecology of iberian lynx and abundance of european rabbits in southwestern spain," *Wildlife Monographs*, vol. 148, pp. 1–36, 2001.

[20] D. J. D. Natusch, J. A. Lyons, G. Brown, and R. Shine, "Communally nesting migratory birds create ecological hot-spots in Tropical Australia," *PLoS ONE*, vol. 11, no. 10, Article ID e0162651, 2016.

[21] C. Sirami, C. Seymour, G. Midgley, and P. Barnard, "The impact of shrub encroachment on savanna bird diversity from local to regional scale," *Diversity and Distributions*, vol. 15, no. 6, pp. 948–957, 2009.

[22] A. R. E. Sinclair, S. Mduma, and J. S. Brashares, "Patterns of predation in a diverse predator-prey system," *Nature*, vol. 425, no. 6955, pp. 288–290, 2003.

[23] W. J. Ripple and R. L. Beschta, "Wolves and the ecology of fear: Can predation risk structure ecosystems?" *Bioscience*, vol. 54, no. 8, pp. 755–766, 2004.

[24] G. Wang, N. T. Hobbs, R. B. Boone et al., "Spatial and temporal variability modify density dependence in populations of large herbivores," *Ecology*, vol. 87, no. 1, pp. 95–102, 2006.

[25] A. Dupuch, L. M. Dill, and P. Magnan, "Testing the effects of resource distribution and inherent habitat riskiness on simultaneous habitat selection by predators and prey," *Animal Behaviour*, vol. 78, no. 3, pp. 705–713, 2009.

[26] P. A. Abrams, "Habitat choice in predator-prey systems: Spatial instability due to interacting adaptive movements," *The American Naturalist*, vol. 169, no. 5, pp. 581–594, 2007.

8

Sacred Groves: Myths, Beliefs, and Biodiversity Conservation

Sushma Singh,[1] Mudasir Youssouf,[2] Zubair A. Malik,[3] and Rainer W. Bussmann[4]

[1]*Department of Environmental Sciences, HNB Garhwal University, Srinagar, Garhwal, Uttarakhand 246174, India*
[2]*Centre for Environmental Science and Technology, Central University of Punjab, Bathinda, Punjab 151001, India*
[3]*Department of Botany and Microbiology, HNB Garhwal University, Srinagar, Garhwal, Uttarakhand 246174, India*
[4]*Saving Knowledge, Casilla 13092, La Paz, Bolivia*

Correspondence should be addressed to Zubair A. Malik; malikmzubair081@gmail.com

Academic Editor: Béla Tóthmérész

Religious and traditional beliefs, cultural mores, and practices play a crucial role in the conservation of environment and biodiversity. The present paper describes a case study of two sacred groves in Western Himalaya. Sacred groves (SGs) are patches of land that are communally protected with religious zeal. A preliminary survey was conducted in these SGs to study their role in biodiversity conservation. The data collected included the general information regarding the SGs and the associated deity, nearest human habitation, access to them, and their floral and faunal diversity. Ethnomedicinal property of plants was collected from the indigenous communities. Many taboos are associated with both the SGs, which help in managing resources well through ritual representation. Different festivals are organized, where the local communities reaffirm their commitment to the forest and the deity. Sacred groves, in general, are a valuable tool of biodiversity conservation. But people's changing attitudes, erosion of traditional beliefs, and human impact have caused degradation of sacred groves over the years. Their conservation would not be possible without the active participation of the local people. By improving their living standards and by giving benefits of conservation to them, long-term conservation goals in these SGs can be achieved.

1. Introduction

Conservation of nature and natural resources has been an important part of cultural ethos, especially in remote rural and indigenous communities in many parts of the world, including India. These communities consider themselves connected with their biophysical environment in a web of spiritual relationship. These rural communities consider specific plants, animals, or even rivers and mountains as their ancestors and protect them. In India, nature worship dates back to the pre-Vedic period (5000 B.C.) and is based on the proposition that all creations of nature have to be protected. The forefathers of these communities were fully aware of the importance and significance of natural resources and the necessity of their conservation for the sustenance of future generations. They lived in harmony with nature and thereby played an important role in conservation of biodiversity [1]. One of the important traditions of nature reverence is to conserve those patches of forest that have been dedicated to a god or goddess or ancestral spirits as *"sacred groves."* According to Hughes and Chandran [2], sacred groves (SGs) are defined as *"segments of landscape containing vegetation, life forms and geographical features, delimited and protected by human societies under the belief that to keep them in a relatively undisturbed state is expression of an important relationship of humans with the divine or with nature."* In short, SGs are the relic forest segments preserved in the name of religion and culture. These groves are mostly associated with temples and are also culturally important. They manifest the spiritual and ecological ethos of rural indigenous communities. Various cultural and religious festivals are often arranged by local people within these patches, which they call *"Mela."* As a way of conservation of nature, SGs have proven to be a well-tried and tested method over thousands of years [3].

Mostly found in Africa and Asia, SGs also exist in Europe and the Americas. Around 100000 to 150000 sacred groves

TABLE 1: Locations of the studied sacred groves (SGs).

Site	Altitude (m asl)	Geographic coordinates
Hariyali Devi (HD)	1500–2800	N30°19′48.18″; E 79°00′24.77″
Tungnath (TN)	3000–4000	N30°29′13.07″; E 79°13′16.16″

have been reported from India [4]. In India, SGs are especially present in the Himalayan region, Western and Eastern Ghats, Coastal Region, Central Indian Plateau, and Western Desert.

The SGs play an important role in ensuring smooth ecosystem services such as clean environment, that is, air, soil, and water conservation, flora and fauna conservation, carbon sequestration, temperature control, and conservation of traditional knowledge. They are, therefore, of central importance as far as the ecological conservation and policy regarding conservation and management of forest at state and national levels are concerned [5]. Sacred groves serve as a home for birds and mammals, and hence they indirectly help in the conservation of biodiversity [3]. There are several studies carried out by various researchers on this subject, highlighting significant role and potential of the SGs [1, 5, 6]. The present paper presents a case study of two SGs of Western (Garhwal) Himalaya and the aim of the study was to document (i) the floral and faunal diversity of these SGs and (ii) the myths, beliefs, and taboos related to biodiversity conservation in these SGs.

2. Materials and Methods

2.1. Study Area. The present paper describes a case study of two sacred groves from the Western Himalaya. The study was carried out in Uttarakhand, also known as *Dev Bhumi* (abode or home of gods). Uttarakhand is very rich in biodiversity and there are many SGs for the conservation of this biodiversity. SGs like Chipla Kedar, Tarkeshwar, Hariyali Devi, Binsar, Kot, Kalimath, Goldev, Tapovan, Chandrabadani, Tungnath, and Triyuginarayan are some of the important SGs of *Dev Bhumi*. The present study was carried out in two SGs, Hariyali Devi and Tungnath, both of which are located in Rudraprayag district (Figure 1 and Table 1).

Hariyali Devi is located above Kodima village at an altitude of 1400 m in Rudraprayag. The temple contains a bejeweled idol of *Ma Hariyali Devi* astride a lion. The temple houses chiefly three idols, namely, *Ma Hariyali Devi*, *Kshatrapal*, and *Heet Devi*. The temple is open throughout the year but it is more festive at the time of Janmashtami, Navratri, and Deepawali.

Tungnath lies in the upper catchment of the Alaknanda River and the Mandakini River, two major tributaries of the Ganges at an altitude of about 2800–3300 m. Tungnath temple is dedicated to Lord Shiva and is the highest Shiva shrine among the *Panch Kedar*, 3 km uphill form Chopta.

2.2. Geology, Soil, and Climate. The rocks of study area are mainly mylonitized gneisses, augen gneisses, schists, and granites constituting Munsiyari Formation [8]. The weathering bedrocks, which provide the bulk of the loose material in these mountains, are crystalline and metamorphic, with sedimentary deposits of Paleozoic age [9]. The soil texture is sandy loam, light grey to brown in color and acidic in nature, with a pH range between 4 and 5 [10].

Four distinct seasons are observed in the study area: short summer (May-June), Monsoon (July–mid-September), autumn (mid-September-October), and long winter (November–April). The snow cover lasts for about 4-5 months and melts during April–May, which marks the arrival of favorable conditions for plant growth. The growth period lasts for about 5–7 months only [10].

2.3. Sampling Procedures (Methodology). To study the role of the SGs in biodiversity conservation mentioned in Table 1, a preliminary survey was conducted in these areas. Information about these SGs was collected by consulting the elderly people of the villages, governmental and nongovernmental agencies, after receiving prior informed consent, and literature sources (books and scientific journal articles). The data collected included the general information regarding the SGs and the associated deity, nearest human habitation, access to them, and their floral and faunal diversity. Information regarding the ethnomedicinal property of different plants was also collected from the surrounding indigenous communities.

3. Results and Discussion

3.1. Biodiversity of Hariyali Devi Sacred Grove. Hariyali Devi SG is rich in floral and faunal diversity (Tables 2, 3, 4, 5, and 6). A total of 80, 12, 9, and 7 species of plants, mammals, birds, and butterflies, respectively, were recorded from this SG (Tables 2, 3, 4, 5, and 6). The 80 plant species represented 75 genera and 44 families with different economic values. Rosaceae, with the highest number of species (8), was found to be the dominant family. The number of species varied in different life forms like herbs (37), shrubs (23), and trees (20). Most of the plant species in the SG had one or other ethnomedicinal importance (Tables 2, 3, and 4). The predominant vegetation is represented by *Quercus semecarpifolia* (kharsu), *Quercus leucotrichophora* (banj), *Rhododendron arboreum* (burans), and *Lyonia ovalifolia* (anyar).

Hariyali Devi forest harbors many sacred animal and butterfly species (Tables 5 and 6). *Capricornis sumatraensis*, *Cervus unicolor*, and *Felis bengalensis* are some common mammalian species. In addition to these, some reptile species were recorded from this SG.

3.2. Biodiversity of Tungnath Sacred Grove. Apart from being a sacred grove, Tungnath is also a part of Kedarnath Wildlife Sanctuary and hence is an important region as far as the conservation of biodiversity is concerned. Tungnath is a home to many rare, threatened, and endangered medicinal

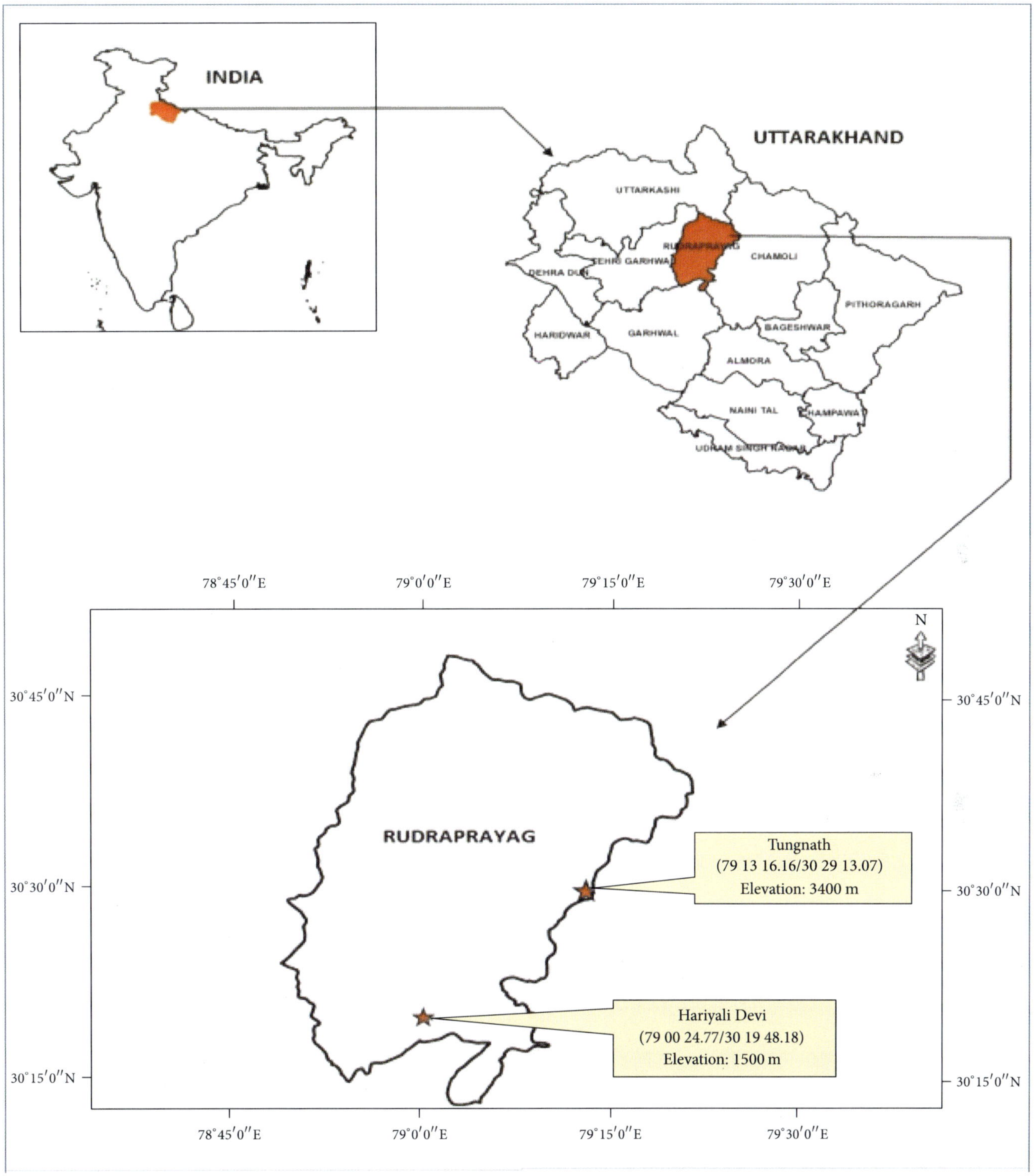

FIGURE 1: Location of the Hariyali Devi and Tungnath sacred groves in Uttarakhand.

plants like *Picrorhiza kurroa* and *Nardostachys jatamansi* (Tables 7 and 8). A total of 27 plant species were reported from this SG, which include 8 trees, 10 herbs, and 9 shrubs. Some endangered animals like musk deer, black bear, and so forth are also found there in good numbers (Table 9). Many rare and threatened birds and reptiles are also found there (Table 10).

3.3. Taboos as a Means of Plant and Animal Conservation. Taboos are the unwritten, orally transmitted traditional and

TABLE 2: List of *Tree* species (with ethnomedicinal properties) found in Hariyali Devi SG.

Botanical name	Vernacular name	Family	Ethnomedicinal property
Ilex dipyrena Wall.	Kandara	Aquifoliaceae	Agriculture implements, fuel, fodder
Alnus nepalensis D.Don	Utees	Betulaceae	Fuel, soil binder
Betula alnoides Buch.-Ham. ex D.Don	Bhuja patra	Betulaceae	Wood, fodder, medicine (rheumatism)
Benthamidia capitata (Wall.) H. Hara	Bhamora	Cornaceae	Edible (fruit), wood/tiber
Cupressus torulosa D.Don ex Lamb.	Surai	Cupressaceae	Wood, medicine (cough, cold, and bronchitis)
Lyonia ovalifolia (Wall.) Drude	Anyar	Ericaceae	Fuel, medicine (wounds and boils)
Rhododendron arboreum Sm.	Burans	Ericaceae	Fuel, edible flowers, medicinal (headache, diarrhea, and dysentery)
Quercus floribunda Lind. ex A. Camus	Tilonj	Fagaceae	Timber, fodder, fuel
Quercus leucotrichophora A. Camus	Banj	Fagaceae	Timber, fodder, fuel
Quercus semecarpifolia Sm.	Kharsu	Fagaceae	Fuel/timber and fodder
Aesculus indica (Wall. ex Cambess.) Hook.	Pangar	Hippocastanaceae	Fuel, fodder, medicine (rheumatism)
Juglans regia L.	Akhrot	Juglandaceae	Edible fruit, dye, medicine (antiseptic, astringent)
Lindera pulcherrima (Nees) Hook. f.	Cheri	Lauraceae	Wood/fuel, manure
Persea gamblei (King ex Hook. f.) Kosterm.	Kauwla	Lauraceae	Agricultural implements/fuel
Myrica esculenta Buch.-Ham. ex D.Don	Kaphal	Myricaceae	Edible fruit, medicine (cough, fever, and asthma)
Abies spectabilis (D.Don) Spach.	Morinda	Pinaceae	Timber/fuel, medicine (fever and antiseptic)
Picea smithiana (Wall.) Boiss.	Rai, spruce	Pinaceae	Wood/timber, medicine (cuts and sores)
Pinus wallichiana A. B Jacks.	Kail	Pinaceae	Fuel, paint, medicine (antiseptic)
Prunus cornuta (Wall. ex Royle) Steud.	Jamma	Rosaceae	Fuel, fodder, medicine (antipyretic)
Pyrus pashia Buch-Ham. ex D.Don	Mehal	Rosaceae	Fuel, fodder, edible, medicine (diabetes)

social rules that regulate human behaviour [11]. In rural areas of India like Garhwal (Uttarakhand), there are a number of plants, animals, and even lakes and rivers that are considered sacred and hence no felling or exploitation is being carried out. As a result of this consecration, different species of trees and animals that are economically important or threatened in other areas are preserved and can form a genetic reservoir and serve as a guide against extinction of these species [12]. *For example,* Dodital and Devariya Tal, two lakes in Uttarkashi and Rudraprayag district of Uttarakhand, are considered sacred, so fishing is completely restricted. There is a taboo that if fishing is done in these lakes, the fisherman will suffer from leprosy [13]. Therefore, the religious belief serves as an instrument of protection of rare and threatened species. Religious beliefs, tradition, and culture are the products of logical internalization of human experience and learning. These practices help in managing resources well through religious or ritual representation.

Trees have a very special role in the ethos of the people in Uttarakhand. Species of trees are worshipped as (1) manifestation of gods, (2) representatives of particular stars and planets, and (3) symbols of the natural elements (energy, water, land, and air), each of which has its own independent and rational meanings [12]. *For example,* in Garhwal Himalaya, *Ficus religiosa* is considered to be sacred. There are many instances where communities control the excess use of resource by confining the approach to resources and enforcing compliance through religious belief, rituals, and social convention which at last result in biodiversity conservation in such communities. The roles of religious and cultural beliefs in protecting trees have been observed by other researchers also [1, 12]. Species such as sacred fig *(Ficus religiosa)*, mountain lion *(Felis concolor)*, and southern pocket gopher *(Thomomys umbrinus emotus)* are protected by Hindus' taboo all across the Indian subcontinent [13]. In the present study also, we reported some taboos associated

TABLE 3: List of *Shrub* species (with ethnomedicinal properties) found in Hariyali Devi SG.

Botanical name	Vernacular name	Family	Ethnomedicinal property
Hedera nepalensis (K. Koch)	Laguli	Araliaceae	Medicinal (expectorant, whooping cough)
Berberis asiatica Roxb.	Kilmora	Berberidaceae	Fuel/fodder and medicinal (ophthalmic, conjunctivitis, and gastritis)
Sarcococca saligna (D.Don) Mull. Arg.	Piruli	Buxaceae	Sticks, soil binder, medicinal (joint pains and fever)
Abelia triflora R. Br.	Gogti	Caprifolaceae	Walking sticks, fodder
Lonicera quinquelocularis Hard.	Taknoi	Caprifolaceae	Edible, walking sticks
Viburnum cordifolium Wall. ex DC.	Bhatnoi, guya	Caprifolaceae	Edible fruits, medicinal (against menorrhagia-excessive menstruation)
Elaeagnus parvifolia Wall. ex Royle	Giwain, kanal	Elaeagnaceae	Medicine (treatment of bloody dysentery, cardiac tonic, cough, treatment of afflictions of lungs, cancer treatment), edible fruits, fodder
Indigofera heterantha Wall. ex Brandis	Sakina	Fabaceae	Medicinal (burns, skin diseases, and ulcers), fodder, edible
Desmodium elegans DC.	Chamali	Fagaceae	Medicine (antipyretic, vomiting)
Deutzia compacta Craib.	Mhujvar	Hydrangeaceae	Medicinal (diuretic)
Elsholtzia fruticosa (D.Don) Rehder	Pothi	Lamiaceae	Medicinal (abdominal pain and nausea)
Zanthoxylum armatum DC.	Timur	Rutaceae	Medicinal (toothache)
Myrsine africana L.	Chupra	Myricaceae	Medicinal (anthelmintic, antispasmodic, skin infections)
Boenninghausenia albiflora (Hook.) Rchb. ex Meisn.	Pishumar	Rutaceae	Medicinal (treatment of malaria, headache, treatment of scabies)
Rhamnus virgatus Roxb.	Chentuli	Rhamnaceae	Fuel, medicinal (eczema and ringworms)
Rosa brunonii Lindl.	Kunja	Rosaceae	Medicinal (cuts, wounds, and sprains), soil binder
Rosa sericea Lindl.	Dhurkunja	Rosaceae	Fodder, edible fruit rich in vitamin C
Rubus foliolosus D.Don	Kala hisar	Rosaceae	Edible fruits, medicinal (dysentery and whooping cough)
Spiraea bella Sims	Kuji	Rosaceae	Medicinal (wash sores and wounds), brooms
Leptodermis lanceolata Wall.	Padera	Rubiaceae	Medicinal (migraines), fodder
Randia tetrasperma (Wall. ex Roxb.) Benth. & Hook. f. ex Brandis	Kamoli	Rubiaceae	Fuel, walking sticks, medicinal (astringent, diuretic, and diarrhea)
Skimmia anquetilia Tayl. & Airy Shaw	Nairpatti	Rutaceae	Agricultural use, sticks, medicinal (treatment of headache and smallpox)
Debregeasia longifolia (Burm. F.)Wedd.	Tusara	Urticaceae	Fodder, used for making ropes, medicine (treatment of scabies)

with *Hariyali Devi* and *Tungnath* sacred groves. According to villagers, these taboos need to be followed by all. These include the following:

(a) Women are strictly prohibited from entering the sacred forest due to the belief that they are impure.

(b) Fetching/collection of fodder and fuelwood and the movement of women and *Shudras* (scheduled castes) have been strictly prohibited in this grove since the Mahabharata period. A temple of the goddess Hariyali Devi is located in this forest patch.

(c) Use of tools in any form (knife, sickle, etc.) on the plants and animals will be a step to hurt the sentiments of *Devi* (goddess). The forest fairies in turn are angered and their wrath can make person mad or deformed and also can lead to disaster in the family of offender.

(d) For a person who starts his journey, if a snake comes across his way, then he has to stop the journey and has to restart only after worshipping the god after an interval of a week.

(e) One week before pilgrimage, the villagers stop eating onion, garlic, egg, and meat.

(f) Anything that is made up of leather is prohibited in the temple and grove.

Table 4: List of *Herbaceous* species (with ethnomedicinal properties) found in Hariyali Devi SG.

Botanical name	Vernacular name	Family	Ethnomedicinal property
Barleria cristata L.	Kala bansa	Acanthaceae	Medicinal (anemia, toothache), soil binder
Peristrophe paniculata (Forssk.) Brummitt	Kaknado	Acanthaceae	Medicinal (used against TB)
Achyranthes aspera L.	Latjiri	Amaranthaceae	Medicinal (malarial fever and muscular sprains)
Heracleum lanatum Michx.	Kakriya	Apiaceae	Medicinal (nervine and tonic), edible
Pimpinella diversifolia DC.	Teroi	Apiaceae	Medicinal (respiratory diseases)
Arisaema intermedium Blume	Meen/magmungari	Araceae	Medicinal (burns)
Arisaema jacquemontii Blume	Khaprya	Araceae	Medicinal (antidote of poisonous mushrooms and snake bite, cough, kidney, skin diseases)
Impatiens sulcata Wall.	Chaul	Araliaceae	Medicinal (antirheumatic and burns), edible
Ageratum conyzoides L.	Gundrya	Asteraceae	Medicinal (anti-inflammatory, antibacterial)
Anaphalis triplinervis (Sims) C. B. Clarke	Bugla	Asteraceae	Medicinal (cuts and wounds, antiseptic)
Cynoglossum glochidiatum Wall. ex Benth.	Lichkura	Boraginaceae	Medicinal (dyspepsia and digestive disorder), vegetable
Silene edgeworthii Bocquet	Bakroyla	Caryophyllaceae	Medicinal (eye infections)
Stellaria media (L.) Vill.	Badyalu	Caryophyllaceae	Medicinal (antirheumatic, anti-inflammatory), vegetable, fodder
Bryophyllum pinnatum (Lam.) Oken.	Bish-khapura	Crassulaceae	Medicinal (burns, wounds, and swellings)
Dipsacus inermis Wall.	Phulee	Dipsacaceae	Medicinal (leucoderma and contusions), edible fruits
Lathyrus aphaca L.	Kurphail	Fabaceae	Fodder
Swertia angustifolia Buch.-Ham. ex. D.Don	Chirata	Gentianaceae	Medicinal (febrifuge)
Geranium nepalense Sweet	Phori	Geraniaceae	Medicinal (diarrhea, ulcers, and wounds), tennin
Hypericum elodeoides Choisy	Basanti	Hypericaceae	Medicinal (antidepressant, sedative, rheumatism)
Micromeria biflora (Buch.-Ham. ex D.Don) Benth.	Gorakhopan	Lamiaceae	Medicinal (carminative)
Origanum vulgare L.	Ban tulsi	Lamiaceae	Medicinal (antispasmodic, carminative), vegetable
Prunella vulgaris L.	Self-heal	Lamiaceae	Medicinal (wound healing, expectorant, antiseptic)
Salvia lanata Roxb.	Ghanyajhar	Lamiaceae	Vegetable and bee-forage source
Oxalis corniculata DC.	Bhilmoro	Oxalidaceae	Medicinal (headache, indigestion, and diarrhea), vegetable
Peperomia tetraphylla Hook. & Arn.	Tirpirya	Piperaceae	Medicinal (treatment of convulsions, skin diseases, cough, asthma, kidney disorders)
Rumex hastatus D.Don	Almoru	Polygonaceae	Medicinal (astringent)
Rumex nepalensis Spreng.	Khatura	Polygonaceae	Medicinal (purgative), vegetable
Anemone obtusiloba D.Don	Kanchphool	Ranunculaceae	Medicinal (nervine and sedative)
Thalictrum javanicum Blume	Mamiri	Ranunculaceae	Medicinal (febrifuge, antirheumatic, and antigout)
Fragaria nubicola (Hook. f.) Lindl. ex Lacaita	Gand-kaphal	Rosaceae	Medicinal (earache)
Potentilla fulgens Wall. ex Hook.	Bajardantu	Rosaceae	Medicinal (antidiarrheal, toothache), edible
Galium aparine L.	Khuskusa	Rubiaceae	Medicinal (diuretic and anti-inflammatory)

TABLE 4: Continued.

Botanical name	Vernacular name	Family	Ethnomedicinal property
Solanum erietinum D.Don	Ban-tambakhu	Solanaceae	Medicinal (vaginal discharges, inflammation), edible fruits
Solanum nigrum L.	Banbhatuja	Solanaceae	Medicinal (cough, cold, diuretic)
Selinum vaginatum C. B. Clark	Butkeshi	Spigeliaceae	Medicinal (nervine, sedative, and analgesic)
Girardinia diversifolia (Link) Friis	Bhainsya	Urticaceae	Medicinal (fever, headache, and swollen joints), fibers, ropes
Hedychium spicatum Buch.-Ham. ex Sm.	Banhaldi	Zingiberaceae	Medicinal (analgesic, anti-inflammatory)

TABLE 5: List of sacred animal species (with IUCN conservation status) found in Hariyali Devi.

Scientific name	Common name	Family	IUCN status*
Aonyx cinerea	Asian small-clawed otter	Mustelidae	Vulnerable
Capricornis sumatraensis	Serow	Bovidae	Vulnerable
Cervus unicolor	Sambar (jado)	Cervidae	Vulnerable
Felis bengalensis	Leopard cat	Felidae	Least concern
Felis chaus	Jungle cat	Felidae	Least concern
Hemitragus jemlahicus	Himalayan tahr	Bovidae	Near threatened
Martes flavigula	Himalayan marten (khursyala)	Mustelidae	Least concern
Panthera pardus	African leopard	Felidae	Vulnerable
Panthera uncia	Leopard (guldar)	Felidae	Endangered
Rattus nitidus	Himalayan field rat	Muridae	Least concern
Sus scrofa	Wild boar	Suidae	Least concern
Ursus arctos	Brown bear	Ursidae	Least concern

*http://www.iucnredlist.org/search.

TABLE 6: List of birds and butterflies (with IUCN conservation status) found in Hariyali Devi.

Scientific name	Common name	Family	IUCN status*
	Birds		
Columba eversmanni	Yellow-eyed pigeon	Columbidae	Vulnerable
Columba rupestris	Hill pigeon	Columbidae	Least concern
Dicrurus macrocercus	Black drongo	Dicruridae	Least concern
Gallus gallus	Red junglefowl	Phasianidae	Least concern
Streptopelia orientalis	Oriental turtle dove	Columbidae	Least concern
Pycnonotus leucogenys	Himalayan bulbul	Pycnonotidae	Least concern
Turdoides striata	Jungle babbler	Leiothrichidae	Least concern
Dendricitta vagabunda	Rufous treepie	Corvidae	Least concern
Urocissa flavirostris	Yellow-billed blue magpie	Corvidae	Least concern
	Butterflies		
Acraea issoria	Yellow coster	Nymphalidae	Least concern
Argynnis kamala	Common silverstripe	Nymphalidae	Least concern
Delias belladonna horsfieldi	Hill jezebel	Papilionidae	Least concern
Kallima inachus huegelii	Orange oakleaf	Lycaenidae	Least concern
Polyura dolon	Stately nawab	Lycaenidae	Least concern
Pseudergolis wedah	Tabby	Nymphalidae	Least concern
Ypthima sakra	Himalayan five-ring	Lycaenidae	Least concern

*http://www.iucnredlist.org/search.

Table 7: List of *Tree* species (with ethnomedicinal properties) found in Tungnath SG.

Botanical name	Vernacular name	Family	Ethnomedicinal property
Acer caesium Wall. ex Brandis	Indian maple	Aceraceae	Fuel, medicinal (for muscular swelling)
Ilex dipyrena Wall.	Himalayan holly	Aquifoliaceae	Fuel, fodder, agricultural implements
Betula utilis D.Don	Bhojpatra	Betulaceae	Medicinal (diuretic, skin infections)
Euonymus tingens Wall.	Spindle tree (kasuree)	Celastraceae	Fuel, also used as dyes
Rhododendron arboreum Sm.	Burans	Ericaceae	Fuel, flowers for squash
Quercus semecarpifolia Sm.	Kharsu	Fagaceae	Fuel, fodder, and timber
Prunus cornuta (Wall. ex Royle) Steud.	Himalayan bird cherry	Rosaceae	Fuel and fodder
Taxus wallichiana Zucc.	Himalayan yew	Taxaceae	Medicinal (anticancerous), fuel, timber

Table 8: List of *Herbaceous* and *Shrub* species (with ethnomedicinal properties) found in Tungnath SG.

Botanical name	Vernacular name	Family	Ethnomedicinal property
	Herbs		
Selinum candolle Edgew.	Muur	Apiaceae	Medicinal (analgesic, cough, fever)
Silene conoidea L.	Chota takla, thumriya	Caryophyllaceae	Medicinal (eye infections, treatment of ophthalmia)
Morina longifolia Wall.	Kathi, kathoj, sakina	Caprifoliaceae	Medicinal (wounds and incense, burns, and boils)
Polygonatum verticillatum (L.) All.	Mitha dudhiya	Asparagaceae	Medicinal (used in treatment of emaciation, senility, gastric diseases)
Corydalis govaniana Wall.	Inderajatta	Papaveraceae	Medicinal (fever, liver diseases, and eye infections)
Cynodon dactylon (L.) Pers.	Dhub	Poaceae	Medicinal (cuts, wounds, piles, inflammation, skin diseases)
Rubus nepalensis (Hook. f.) Kuntze	Hisar	Rosaceae	Medicinal (cuts and wounds)
Picrorhiza kurroa Royle ex Benth	Kutki	Scrophulariaceae	Medicinal (fever, hepatitis, chronic dysentery)
Nardostachys jatamansi (D.Don.) DC.	Jattamaansi	Valerianaceae	Medicinal (tranquilizer, sedative, high blood pressure, used in dysmenorrhoea for pain relief and smooth menstrual flow)
	Shrubs		
Berberis aristata DC.	Kingor	Berberidaceae	Medicinal (used in ophthalmia, conjunctivitis, ulcers)
Juniperus indica Bertol.	Guugal	Cupressaceae	Medicinal (cough, cold, and fever)
Rhododendron anthopogon D.Don	Taalisri, burans	Ericaceae	Medicinal (against respiratory diseases)
Rhododendron campanulatum D.Don.	Chimal/burans	Ericaceae	Medicinal (chronic rheumatism and sciatica)
Rhododendron lepidotum Wall. ex. G.Don	Burans	Ericaceae	Medicinal (respiratory and digestive ailments)
Clematis buchananiana DC.	Belkangu	Ranunculaceae	Medicinal (skin ailments, sinus inflammation, wounds)
Cotoneaster acuminatus Lindl.	Ruins	Rosaceae	Medicinal (diarrhea and dysentery)
Cotoneaster microphyllus Wall. ex Lindl.	Ruins	Rosaceae	Anti-inflammatory, cuts, and wounds
Leptodermis lanceolata Wall.	Koo-basya	Rubiaceae	Medicinal (migraines)

TABLE 9: List of animal species (with IUCN conservation status) found in Tungnath SG.

Scientific name	Common name	Family	IUCN status*
Canis aureus	Jackal	Canidae	Least concern
Capricornis sumatraensis	Serow	Bovidae	Vulnerable
Hemitragus jemlahicus	Himalayan tahr	Bovidae	Near threatened
Macaca mulatta	Rhesus macaque	Cercopithecidae	Least concern
Moschus chrysogaster	Alpine musk deer	Moschidae	Endangered
Ochotona roylei	Himalayan mouse-hare	Ochotonidae	Least concern
Panthera pardus	Common leopard	Felidae	Vulnerable
Presbytis entellus	Common langur	Cercopithecidae	Least concern
Pseudois nayaur	Bharal	Bovidae	Least concern
Pteropus giganteus	Indian flying fox	Pteropodidae	Least concern
Ursus thibetanus	Himalayan black bear	Ursidae	Vulnerable

*http://www.iucnredlist.org/search.

TABLE 10: List of birds and reptiles (with IUCN conservation status) found in Tungnath SG.

Scientific name	Common name	Family	IUCN status*
	Birds		
Aquila nipalensis	Steppe eagle	Accipitridae	Endangered
Gypaetus barbatus	Bearded vulture	Accipitridae	Near threatened
Lophophorus impejanus	Himalayan monal	Phasianidae	Least concern
Megalaima viridis	White-cheeked barbet	Megalaimidae	Least concern
Neophron percnopterus	Egyptian Vulture	Megalaimidae	Least concern
Pucrasia macrolopha	Koklass	Phasianidae	Least concern
Tragopan melanocephalus	Western tragopan	Phasianidae	Vulnerable
Zoothera monticola	Greater long-billed thrush	Turdidae	Least concern
	Reptiles		
Calotes versicolor	Indian garden lizard	Agamidae	Not evaluated
Hemidactylus brookii	Spotted Indian gecko	Gekkonidae	Least concern
Naja naja	Spectacled cobra	Elapidae	Near threatened
Orthriophis hodgsoni	Himalayan trinket snake	Colubridae	Not evaluated
Scincella himalayanum	Himalayan ground skink	Scincidae	Least concern

*http://www.iucnredlist.org/search.

(g) Killing/hunting of animals and plucking/uprooting of plants are strictly forbidden in the SGs.

3.4. Myths Associated with the Hariyali Devi and Tungnath SGs. According to Hindu Mythology, when Devi Mahamaya was conceived in the form of Devaki's seventh child, the cruel brother of Devaki, Kansa threw Devi Mahamaya aggressively on the ground. Immediately multiple body parts of Devi got scattered all over the earth. The hand fell at Hariyali Devi. Since then, it has become a revered place as *Siddha Peeth*. The temple is open for all seasons but it is more celebratory at the time of *Janmashtami, Navratri, and Deepawali*. The myth which prevails according to the *Bhagwat Puran* is the following: Yogmaya was the sister of Lord Krishna, and she replaced him in the cell of his parents during his birth. When Kansa threw her against the wall, she turned into lightning and came to Hariyali Parvat to make her abode. Since then, the adjoining forest is known as "Hariyali" and is worshipped by people.

The Tungnath temple is the highest Hindu shrine and is believed to be 1000 years old. It has a rich legend linked to the Pandavas, Heroes of Mahabharata epic. According to Hindu Mythology, Vyas Rishi advised the Pandavas that since they were culpable of slaying their own relatives (Kauravas, their cousins) during the Mahabharata war, their act could be pardoned only by Lord Shiva. Consequently, the Pandavas went in search of Shiva who was convinced of the guilt of Pandavas. In order to keep away from them, Shiva took the form of a bull and went into hiding in an underground safe haven of Guptakashi, where Pandavas chased him. But later Shiva's body in the form of bull's body parts rematerialized at five different locations that represent the "Panch Kedar," where Pandavas built temples of Lord Shiva at each location to worship and venerate, seeking his pardon and blessings. Each location is identified with a part of his body; Tungnath is identified as a place where his "Bahu (hands)" were seen. Legend also states that Lord Ram, the chief icon on the Ramayana epic, meditated at the Chandrashila peak, which is close to Tungnath.

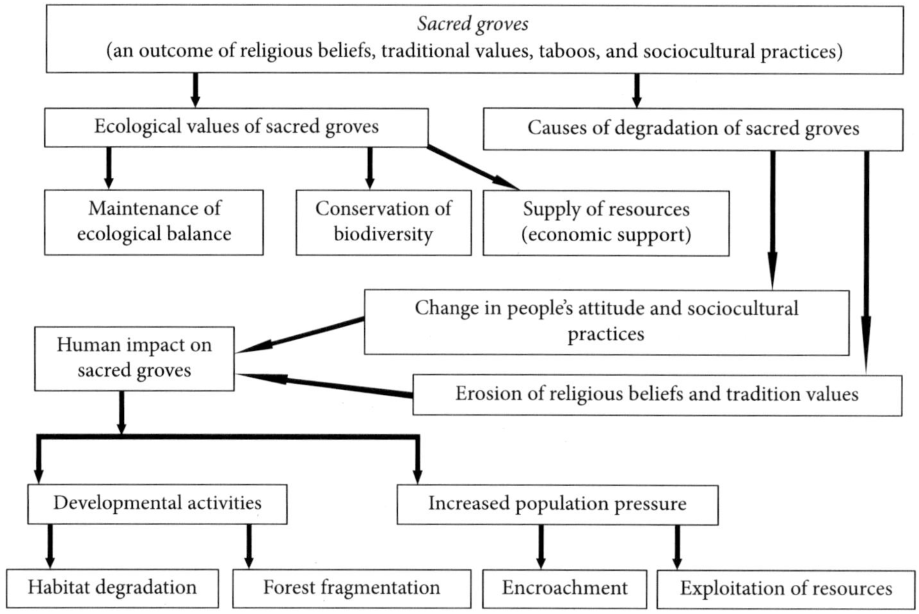

FIGURE 2: Relationships between ecological values, religious beliefs, and traditional values and causes of degradation of sacred groves (courtesy: [7]).

3.5. Festivals (Melas) Associated with the Sacred Groves. SGs are associated with religious rites, festivals, and recreation. The organization and celebration of fairs and festivals have preserved the traditional and sociocultural heritage of Garhwal to a great extent. In Hariyali Devi and Tungnath SGs, festivals (locally called as *melas*) are organized during April and October every year on the occasion of *Navratri, Shivratri, Holi*, and so forth. At these *melas* (festivals), the local communities reaffirm their commitment to the forest and the deity. The heads of the communities supervise the utilization and maintenance of the SGs to ensure that there is no deviation from the village-appointed rules. Anthwal et al. [14] also reported several festivals related to SGs in Uttarakhand. Many plant species have also been associated with religious festivals, namely, *Azadirachta indica* (Sheela Asthami, Nimb Saptami), *Ficus bengalensis* (Vat Savitri), *Aegle marmelos* (Bilvamengal sawan ke somvaar), *Musa paradisiaca* (Kadii Vrat), and *Ficus religiosa* (Somvati Amavasya), due to popular and common beliefs [14].

3.6. Conservation of Medicinal Plants. The traditional medical systems of northern India (such as Ayurveda and Tibetan) are a component of culture developed over long time [15]. Medicinally important plants have high importance for religious activities of north Indian native communities that worshiped the plants in the form of god, goddesses, and minor deities [16]. Thus, SGs are the valuable repositories of medicinal and aromatic plants.

Most of the denizens residing in the vicinity of groves are very simple, illiterate, and poor and are almost without any access to modern medicine systems. But they do have conscientiously nourished their traditional knowledge, customs, rituals, and ceremonies with great potency. Local traditional knowledge and the practice of plant-based medicine are still widespread in the rural areas of Garhwal and these play an important role in primary health care [17]. Even the local people prefer to stick to the traditional herbal remedies, and it is due to a situation of having no alternative choices, as well as poverty and belief in the effectiveness of folklore herbal remedies [17]. The denizens living around these SGs have conserved the medicinal plants of these regions for use in a sustainable way by themselves and by their future generations. Religious beliefs and traditional customs have played an important role in this conservation. They have deep faith that if someone from outside the village uproots the medicinal plants from their village, it is treated as an evil act that may bring misery of great order to the village folks.

3.7. Present Status of the HD and TN SGs. Sacred groves, in general, are good instruments of biodiversity conservation. As already mentioned, our ancestors were aware that the natural resources that sustained them should be conserved for the future generations. But, in the course of time, science and technology developed and industries were established and expanded to meet the increasing demands of the people. People's changing attitudes, the erosion of traditional beliefs, and human impact have caused degradation of sacred groves over the years (Figure 2). The same is true for the studied sacred groves. Various anthropogenic activities have altered the structure and function of different ecosystems all over the world [18]. One of the most noticeable effects of ecosystem perturbation has been the depletion of biodiversity [19]. Vanishing of species due to different anthropogenic disturbances like alteration of natural habitats, excessive utilization, pollution, universal climate change, and invasion of nonnative species is so fast that many precious taxa may disappear even before they are documented and identified and their scientific value is discovered [7]. Many scholars have worked on conservation of sacred groves through sociocultural practices in different parts of India [1, 12, 20–26].

Lack of awareness in terms of long-term future benefits has resulted in the destruction of SGs. No legislative protection has been implemented so far in India. This has caused considerable ecological damage. Sacred groves have become the victims of deteriorating faith. Such religiously protected areas provide a comprehensive and rich ecological niche as repositories of genetic diversity [12]. The increased threats to SGs can be related to the lack of an in-built conservation effort, higher demands for NTFPs, fuel wood collection, and decrease in the religious faiths along with the reduced commitment of the present generation toward such natural sacred places.

Encroachments of SGs areas by various government departments for different developmental projects, as well as migration and immigration of people, also have contributed to the extinction of SGs. These SGs need to be protected and managed wisely as was done a few decades ago. For providing necessary protection to the SGs and maintaining their natural identity and sanctity, it is imperative that the surrounding population is taken into confidence. The surrounding village communities need to be educated and guided for sustaining the sanctity of existing groves and strengthening them. Conservation of SGs is impossible without the active participation of the local people. Conservation without compensation is only conversation [17, 27]. By improving their living standards and by giving benefits of conservation to them, long-term conservation goals in these SGs can be achieved.

4. Conclusion

India has a very high number of scared groves that play an important role in biodiversity conservation because of various myths and religious beliefs associated with them. These SGs have been conserving the biodiversity for many decades. But, nowadays, the attitude of people has changed and this along with the mistrust of traditional beliefs has caused degradation of sacred groves all over India. For improving their degraded condition, it is suggested that the local people living inside and around the SGs need to be taken into confidence, so that long-term conservation goals can be achieved.

Conflicts of Interest

The authors declare that there are no conflicts of interest regarding the publication of this work.

Acknowledgments

The authors are thankful to the local people for their cooperation during the study period.

References

[1] L. S. Kandari, V. K. Bisht, M. Bhardwaj, and A. K. Thakur, "Conservation and management of sacred groves, myths and beliefs of tribal communities: a case study from north-India," *Environmental Systems Research*, vol. 3, no. 1, 2014.

[2] D. J. Hughes and S. M. D. Chandran, *Paper Presented in the Workshop on the Role of Sacred Groves in Conservation and Management of Biological Resources*, KFRI, Peechi, India, 1997, Paper presented in the workshop on, the role of sacred groves in conservation and management of biological resources.

[3] A. K. M. N. Islam, M. A. Islam, and A. E. Hoque, "Species composition of sacred groves, their diversity and conservation in Bangladesh," in *Conserving the Sacred for Biodiversity Management*, P. S. Ramakrishnan, Ed., pp. 163–165, KG Saxena & UM Chandrashekara, (UNESCO and Oxford-IBH Publishing), New Delhi, India, 1998.

[4] K. C. Malhotra, Y. Gokhale, S. Chatterjee, and S. Srivastava, "Sacred groves in India," in *Proceedings of the Aryan Books International*, p. 108, New Delhi, India, 2007.

[5] R. Ray and T. V. Ramachandra, "Small sacred groves in local landscape: are they really worthy for conservation?" *Current Science*, vol. 98, no. 9, pp. 1178–1180, 2010.

[6] H. Singh, T. Husain, and P. Agnihotri, "Haat Kali sacred grove, Central Himalaya, Uttarakhand," *Current Science*, vol. 98, no. 3, p. 290, 2010.

[7] M. L. Khan, A. D. Khumbongmayum, and R. S. Tripathi, "The sacred groves and their significance in conserving biodiversity an overview," *International Journal of Ecology and Environmental Sciences*, vol. 34, no. 3, pp. 277–291, 2008.

[8] N. K. Agrawala, *Working plan for the Kedarnath Forest Division 1972-73 to 1981-82*, Working plan circle, Nainital, Uttar Pradesh, India, 1973.

[9] R. K. Gupta, "Forest types of the Garhwal Himalaya in relation to edaphic and geological formations," *Indian Forestor*, vol. 4, no. 8, pp. 147–160, 1964.

[10] Z. A. Malik and M. C. Nautiyal, "Species richness and diversity along the altitudinal gradient in Tungnath, the Himalayan benchmark site of HIMADRI," *Tropical Plant Research*, vol. 3, no. 2, pp. 396–407, 2016.

[11] A. D. Banjo, G. A. Otufale, O. L. Abatan, and E. A. Banjo, "Taboo as a means of plant and animal conservation in South-Western Nigeria: a case study of Ogbe river and its basin," *World Applied Sciences Journal*, vol. 1, pp. 39–43, 2006.

[12] A. Anthwal, N. Gupta, A. Sharma, S. Anthwal, and K.-H. Kim, "Conserving biodiversity through traditional beliefs in sacred groves in Uttarakhand Himalaya, India," *Resources, Conservation & Recycling*, vol. 54, no. 11, pp. 962–971, 2010.

[13] J. Colding and C. Folke, "The relations among threatened species, their protection, and taboos," *Ecology and Society*, vol. 1, no. 1, 1997.

[14] A. Anthwal, R. C. Sharma, and A. Sharma, "Sacred groves: traditional way of conserving plant diversity in Garhwal Himalaya, Uttaranchal," *Journal of American Science*, vol. 2, no. 2, pp. 35–38, 2006.

[15] C. P. Kala, P. P. Dhyani, and B. S. Sajwan, "Developing the medicinal plants sector in northern India: challenges and opportunities," *Journal of Ethnobiology and Ethnomedicine*, vol. 2, article 32, 2006.

[16] C. S. Silori and R. Badola, "Medicinal plant cultivation and sustainable development: A case study in the buffer zone of the Nanda Devi biosphere reserve, Western Himalaya, India," *Mountain Research and Development*, vol. 20, no. 3, pp. 272–279, 2000.

[17] Z. A. Malik, J. A. Bhat, R. Ballabha, R. W. Bussmann, and A. B. Bhatt, "Ethnomedicinal plants traditionally used in health care practices by inhabitants of Western Himalaya," *Journal of Ethnopharmacology*, vol. 172, article no. 9563, pp. 133–144, 2015.

[18] Z. A. Malik, R. Pandey, and A. B. Bhatt, "Anthropogenic disturbances and their impact on vegetation in Western Himalaya, India," *Journal of Mountain Science*, vol. 13, no. 1, pp. 69–82, 2016.

[19] Z. A. Malik, J. A. Bhat, and A. B. Bhatt, "Forest resource use pattern in Kedarnath wildlife sanctuary and its fringe areas (a case study from Western Himalaya, India)," *Energy Policy*, vol. 67, pp. 138–145, 2014.

[20] B. Sinha and R. K. Maikhuri, "Conservation through socio-cultural-religious practice in Garhwal Himalaya: A case study of Hariyali sacred site," in *Conserving the Sacred for Biodiversity Management*, P. Ramakrishnan, K. G. Saxena, and U. M. Chandrashekhara, Eds., 299, p. 289, Oxford and IBH Publishing, 1998.

[21] S. Sunitha and R. P. Rao, "Sacred groves in Kurnool District, Andhra Pradesh," in *Biodiversity, Taxonomy and Conservation of Flowering Plants*, M. Sivadasan and P. Mathew, Eds., pp. 367–373, Mentor Books, 1999.

[22] R. Basu, "Studies on sacred groves and taboos in Purulia District of West Bengal," *Indian Forester*, vol. 126, no. 12, pp. 1309–1318, 2000.

[23] C. G. Kushalapa, S. A. Bhagwat, and K. A. Kushalapa, "Conservation and management of sacred groves of Hodagu, Karnataka, South India-a unique approach in," in *Tropical Ecosystems: Structure. Diversity and Human Welfare*, K. N. Ganeshaiah, U. R. Shaanker, and K. S. Bawa, Eds., pp. 565–569, Oxford IBH Publishing, 2001.

[24] S. A. Bhagwat, C. G. Kushalappa, P. H. Williams, and N. D. Brown, "A landscape approach to biodiversity conservation of sacred groves in the Western Ghats of India," *Conservation Biology*, vol. 19, no. 6, pp. 1853–1862, 2005.

[25] V. Jaryan, S. K. Uniyal, Gopichand et al., "Role of traditional conservation practice: Highlighting the importance of Shivbari sacred grove in biodiversity conservation," *Environmentalist*, vol. 30, no. 2, pp. 101–110, 2010.

[26] M. Kala and A. Sharma, "Traditional Indian beliefs: A key toward sustainable living," *Environmentalist*, vol. 30, no. 1, pp. 85–89, 2010.

[27] Z. A. Malik, *Phytosociological behavoiur, anthropogenic disturbances and regeneration status along an altitudinal gradient in Kedarnath Wildlife Sanctuary (KWLS) and its adjoining areas [Ph.D. thesis]*, HNB Garhwal University, Srinagar Uttarakhand, Uttarakhand, India, 2014.

9

Lake Baikal Ecosystem Faces the Threat of Eutrophication

Galina I. Kobanova, Vadim V. Takhteev, Olga O. Rusanovskaya, and Maxim A. Timofeyev

Institute of Biology at Irkutsk State University, Irkutsk State University, Irkutsk 664003, Russia

Correspondence should be addressed to Maxim A. Timofeyev; m.a.timofeyev@gmail.com

Academic Editor: Ram Chander Sihag

Recently there have been reports about large accumulations of algae on the beaches of Lake Baikal, the oldest and deepest freshwater body on earth, near major population centers and in areas with large concentrations of tourists and tourism infrastructure. To evaluate the observations indicating the ongoing process of eutrophication of Lake Baikal, a field study in July 2012 in the two largest bays of Lake Baikal, Barguzinsky and Chivyrkuisky, was organized. The study of phytoplankton using the sedimentary method and quantitative records of accumulations of macrophytes in the surf zone was made. In Chivyrkuisky Bay, we found the massive growth of colorless flagellates and cryptomonads as well as the aggregations of *Elodea canadensis* along the sandy shoreline (up to 26 kg/m^2). Barguzinsky Bay registered abundantly cyanobacterial *Anabaena* species, cryptomonads, and extremely high biomass of *Spirogyra* species (up to 70 kg/m^3). The results show the presence of local but significant eutrophication of investigated bays. To prevent further extensions of this process in unique ecosystem of Lake Baikal, the detailed study and monitoring of the coastal zone, the identification of the sources of eutrophication, and the development of measures to reduce nutrient inputs in the waters are urgently needed.

1. Introduction

Lake Baikal, in southeastern Siberia, is the oldest and deepest freshwater body on earth, containing ca. 20% of the world's liquid freshwater (equivalent to all North American Great Lakes combined). It is known for its specific abiotic characteristics such as high oxygen content throughout the water column and a stable low water temperature with a long seasonal ice cover of the lake surface. The high rate of endemism and the exceptional biotic diversity of this ancient lake contributed to UNESCO's 1996 decision to designate Lake Baikal a World Heritage Site.

Lake Baikal is in one of three areas in the world experiencing the most rapid climate change. Despite the enormous temperature buffering capacity of the large water body, it has already been estimated that global and regional warming has caused a 1.21°C increase in the average surface water temperature in the past 60 years [1], a rate twice that of the global average, and the ice-free season has lengthened by 16.1 days between 1868 and 1995 [2].

In addition to temperature change, industrial pollution and cultural eutrophication are of particular concern for Lake Baikal. The Baikal region, containing an industrial corridor with chemical plants and aging industries, lies within the lake's airshed, and industrial pollutants are carried into the basin of the lake by prevailing winds, affecting the entire ecosystem of the lake [3, 4].

Furthermore, in the past two decades, the tourism industry has developed rapidly on the coasts of Lake Baikal. The number of people visiting Baikal has increased several times. A growing number of pleasure crafts are using the lake. Numerous hotels and tourist centers have been constructed on the banks of the lake. Many beaches are extensively used by individual tourists. Domestic waste effluents and sewage from these territories fall into Lake Baikal.

Researches of rivers Maksimikha and Barguzin, conducted in 2012, show that these rivers enrich the water of Barguzin Bay in nutrients. In the wellhead area of Maksimikha River quite high levels of nitrate (0.26 $mg\,L^{-1}$) and phosphate (0.22 $mg\,L^{-1}$) were found that significantly increased the value of total mineralization—135 $mg\,L^{-1}$ (K. S. Mikhalev and G. M. Speiser, personal communication). In the same year in the southern part of Lake Baikal in the west bank (Bolshie

FIGURE 1: Massive growth of green algae of the genus *Spirogyra* in Barguzinsky Bay of Lake Baikal (shoreline accumulations). Photo by V. Petrov.

Koty village) mineralization amounted to 98.5 mg/L; that is, it almost corresponds to the value registered here 50 years ago. Few literature data also indicate the supply of nutrients from the river Barguzin to the bay, which affects the composition of phytoplankton in the shallow zone [5].

It is well known that addition of organic matter to the aquatic medium is the major cause of eutrophication. A water bloom in a lake occurs when it becomes rich of nutrients (primarily phosphorus and nitrogen) [6].

The main focus of monitoring observations has traditionally been the deepwater pelagic zone of Lake Baikal [1, 7]. The shallow coastal zone has received much less attention. However, shallow zones are the places that are affected by the negative impacts of economic and recreational activity especially.

In recent years, public environmental organizations and local activists have regularly reported about accumulations of algae on the beaches of Lake Baikal near major population centers (Slyudyanka, Severobaykalsk, etc.) [8]. Information about abnormally high growth of filamentous algae of the genus *Spirogyra* in the coastal shallow waters of southern Baikal began to appear in the scientific literature [9, 10]. Local inhabitants have also reported large accumulations of algae on the shores of the largest Baikal bay, Barguzinsky Bay, in recent years during autumn storms. In 2012, we estimated that a number of these algal accumulations consisted primarily of green filamentous algae (Figure 1) [11].

With this background, we carried out investigations to study the status of algal flora and their abundance as measures of eutrophication in the coastal areas of two largest bays of the Lake Baikal.

2. Materials and Methods

2.1. Sampling Sites. The depths and morphometry of Barguzinsky Bay and Chivyrkuisky Bay are different. Both bays are located close but separated from each other and from the open waters of Baikal by the Svyatoy Nos Peninsula and the Svyatonossky Isthmus. Barguzinsky Bay, the largest bay, is located on the southern side of these geographical features. It has an almost-circular shape, with an area of approximately 700 square kilometers. Its length is close to 30 km, and its length exceeds the width by about maximum 5 km. The entrance to Barguzinsky Bay is wide (approximately 20 km) and deep (more than 1000 m), allowing the waters of Baikal to exchange freely with the bay water. The depths then gradually decrease in the direction of the central area of the bay. The largest tributary of the bay, the Barguzin River, empties into the bay in the middle portion of the coast, contributing a significant amount of organic matter. Barguzin is one of the largest tributaries of the Lake Baikal; its length is 480 km, with basin area of 21,100 square miles, and the average annual discharge at the mouth is 130 cu·m/s. In the upper reaches, it is a mountain stream with clean water, but the lower part flows through agricultural areas and quite large settlements (Barguzin, Ust-Barguzin).

The warm Maksimikha Cove, with sandy beaches, is located in the southern part of Barguzinsky Bay. The eponymous village of Maksimikha and the Maksimikha River are also located at this site. In the Maksimikha village, there are many tourist bases and hotels. A review of advertisements of travel companies shows that in summer these facilities are able to accommodate 750 to 1000 people. The number of unorganized tourists is difficult to account for, but probably even more than organized.

Chivyrkuisky Bay is located north of the Svyatonossky Isthmus and is much smaller in size and shallower than Barguzinsky Bay. It has an elongated shape, an area of approximately 270 square kilometers, a width of 6–12 km, and a length of approximately 27 km. Depths in excess of 200 m can only be observed at the entrance of the bay; the average depth does not exceed 20–25 m. The maximum depth in the southern part of the bay rarely exceeds 10 m, and the water is generally warm (more than 20°C in the coves) starting from June. These two bays are among the places that are most affected by the local fishing industry and by recreational activities on the Lake Baikal. In Chivyrkuisky Bay, local fishing industry has well developed. Although there are few stationary tourist camps, a large number of tourists arrive in cars and boats.

In Zmeevaya Cove, there are many hot springs. Preliminary estimates reveal that tourists and fishermen add about 160 tons of liquid waste to the Lake Baikal in one summer season [12].

2.2. Sampling. Field studies were conducted in the summer (July 2012) in the littoral zone of the east coast of Lake Baikal in the two largest bays of the lake, Barguzinsky Bay and Chivyrkuisky Bay (Figure 2). The principal sampling sites were in the southern part of Chivyrkuisky Bay, near Monakhovo village (53°39′53″ N, 109°00′33″ E); in the middle part of this same bay, in the Zmeevaya Cove ("Bay of Serpents," 53°46′00″ N, 109°01′61″ E); and in the southern part of Barguzinsky Bay, near village Maksimikha (53°16′13″ N, 108°42′59″ E). For comparison, we chose another point in the Barguzin Bay, north of the mouth of the Barguzin River at the Svyatonossky Isthmus, near the shore (53°31′45″ N, 108°57′32″ E).

Samples of phytoplankton were obtained by a simple scooping-up of the water near the shore or by the research vessel from the surface layers at a moderate distance (100 m)

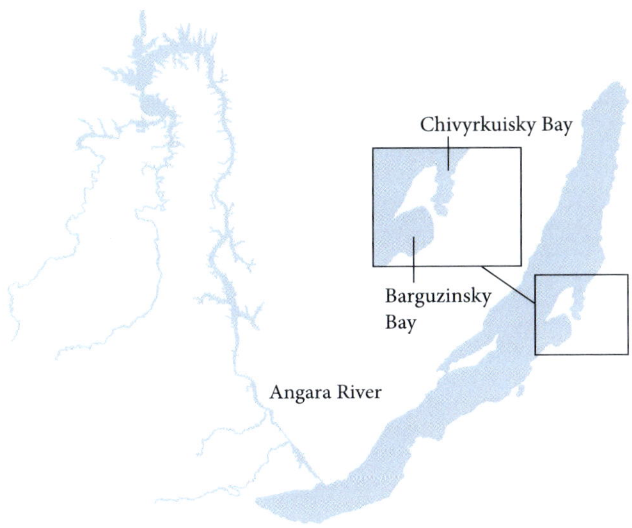

FIGURE 2: Map of the Lake Baikal sampling sites: Barguzinsky Bay and Chivyrkuisky Bay. The map was created using the "Paint" packages of Windows 7 software (Microsoft, USA).

from the shoreline. Simultaneously with the collection of the water samples, the shoreline zone was monitored to identify the quantity and composition of aquatic plants washed up on the shore by the waves. We filled glass bottles of 0.7 L with water and the samples were fixed with Utermöhl's solution. We processed samples using the sedimentary method [13]. We let them be settled previously for 20 days or more. After this, we drained off the top layer of water and estimated the number and the species diversity of algae in the concentrated samples (V = 10–15 mL) under a Leica DMLB microscope (Germany). Large-cell phytoplankton were counted in 0.1 mL under the microscope with 200x magnification. Small-cell phytoplankton were counted in 0.02 mL at a magnification of 1000. The biomass of dominant species was determined using the volume of their cells. The volume of algae was calculated based on the shape and size of their cells. Specific cells' weight was determined for 1 cm^3 = 1 g. Microphotographs of algae were obtained on the microscope Leica DMLB with a digital camera DC 300 (Germany).

3. Results and Discussion

3.1. Chivyrkuisky Bay. During the season of our study, the weather was generally cloudy, with frequent rain and wind. Thus, the temperature was not sufficiently high to warm the waters of Chivyrkuisky Bay, and the temperature did not exceed 19°C. We expected that the planktonic diatoms might go up from the bottom in the water thickness in such conditions. Planktonic diatoms are typical dominants of the spring and early summer phytoplankton in the Chivyrkuisky Bay [14]. However, the present study showed the dominance of Cryptophyta group in number as well as biomass of the phytoplankton close to Monakhovo village (Figure 3).

We observed a very high species abundance of the genus *Cryptomonas* in Chivyrkuisky Bay: *C. marssonii* Skuja was most abundant followed by *C. reflexa* (M. Marsson) Skuja and

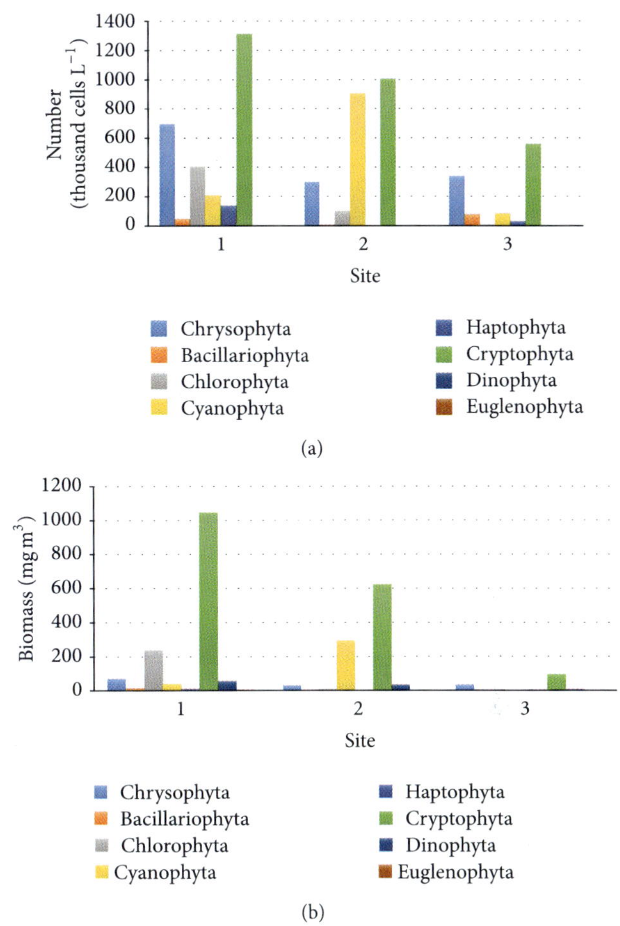

FIGURE 3: The ratio of algal divisions in the phytoplankton composition of the Chivyrkuisky Bay in July 2012: (a) by the number and (b) by the biomass. Sites: 1, Monakhovo near the shore; 2, Monakhovo, 100 m from shore; and 3, Zmeevaya, 100 m from shore.

C. erosa Ehrenberg (Figure 4). The total cells of this genus count exceeded 600000 cells L^{-1}, with a biomass greater than 1.2 g/m^3. This level was 20 times greater than that in the open waters of Baikal, where the average *Cryptomonas* cell counts range from 10000 to 20000 cells L^{-1}. This is the first report of such a massive development of the algal genus *Cryptomonas* in Lake Baikal.

At Chivyrkuisky Bay, we also observed small *Rhodomonas pusilla* (H. Bachmann) Javornicky, belonging to the Cryptophyta, with the highest estimated abundance of 690000 cells L^{-1} and a biomass of 0.07 g/m^3. A small flagellate alga, *Chrysochromulina parva* Lackey (Haptophyta), also reached a high abundance: more than 100000 cells L^{-1}. Besides these, there were a large number of small heterotrophic colorless flagellates and flagellates with chloroplasts from the division Chrysophyta, whose total cell count exceeded 400000 cells L^{-1}.

Thus, the cryptomonads and heterotrophic flagellates dominated in the algal plankton communities of Chivyrkuisky Bay. Intensive growth of these algae indicates that the water contains organic matter. Cyanobacteria (Cyanophyta

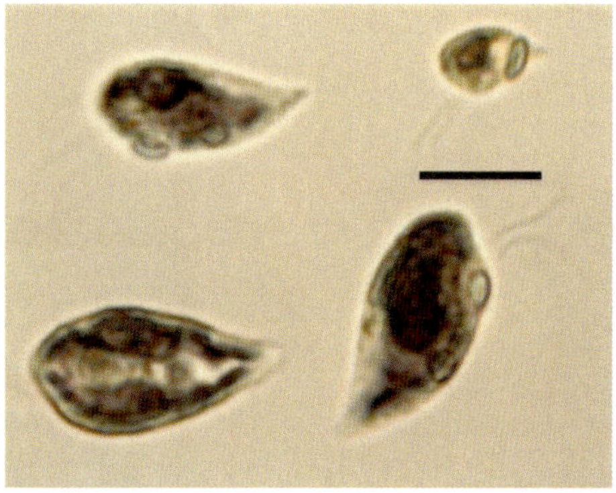

FIGURE 4: Dominant cryptophyte species in the phytoplankton of Chivyrkuisky Bay of Lake Baikal: *Cryptomonas marssonii* Skuja and *Rhodomonas pusilla* (Bachm.) Javor. (small cell). Scale bar: 10 μm.

or blue-green algae) were primarily represented by the genus *Anabaena*, among which *Anabaena lemmermannii* P.G. Richter was the most abundant (with total cell counts of 200000 cells L^{-1} and a biomass of 0.04 g/m^3).

In the coastal waters of Chivyrkuisky Bay, *Elodea canadensis* Michx. was highly abundant. *E. canadensis* first appeared in Lake Baikal in the late 1970s and developed extensively in a number of areas in the late 1980s. In our study, we found that *E. canadensis* now dominates the southern part of the bay, forming massive accumulations in the surf zone near the edge of the water, with a biomass of up to 26 kg/m^2 in wet weight (Figure 5).

Earlier researchers [15] reported that *E. canadensis* actively developed on soft ground in the shallow waters of Chivyrkuisky Bay and began to displace the native vegetation. We found that black sludge accumulates at the bottom of Chivyrkuisky Bay at a distance of as little as 5–10 m from the water's edge. It should be assumed that *E. canadensis* contributes to the siltation of the littoral zone of the bay, which has developed markedly in recent years.

In certain regions of Chivyrkuisky Bay, for example, in the upper strata of Zmeevaya Cove, we found entire colonies of the planktonic cyanobacterium *Gloeotrichia echinulata* (J.E. Smith) P. Richter. The colonies were shaped like spherical balls (with diameters up to 1.5 mm).

In Zmeevaya Cove, we also found a representative of the Dinophyta, *Peridiniopsis polonicum* (Woloszynska) Bourrelly, 1968, for the first time in Baikal (Figure 6). This is a potentially toxic dinoflagellate.

Euglenophytes were represented by *Trachelomonas volvocina* Ehrenberg, *T. hispida* (Perty) F. Stein emend. Deflandre, *T. ornata* (Svirenko) Skvortzov, and an unidentified *Euglena* species, each with an abundance of 500–1000 cells L^{-1}. All these three *Trachelomonas* are indicators of organic pollution and characterized as beta-mesosaprobionts [16].

Summarizing our observations at Chivyrkuisky Bay, we note clear signals of an ongoing trend of eutrophication. The most important of these signals are the massive development of heterotrophic colorless flagellates as well as the massive development of *E. canadensis* along the sandy shoreline

FIGURE 5: (a, b) Massive accumulations of *Elodea canadensis* Michx. on the shores of the Chivyrkuisky Bay of Lake Baikal. (c) Shoot of *E. canadensis* from the shore accumulation. Scale bar: 1 cm.

and its contributions to the siltation of the littoral zone. The number of registered dominant species in the bay has increased in comparison with the data for the years 1964–1973 [14]. Flora of summer plankton community was enriched by species of the genera *Cryptomonas*, *Rhodomonas*, and *Chrysochromulina*.

3.2. Barguzinsky Bay. In Barguzinsky Bay, near Maksimikha village (Maksimikha Cove), we observed the highest levels of phytoplankton growth, with a total biomass of up to 65 g/m^3. Near the shore, cyanobacteria clearly dominated both in abundance and in biomass (Figure 7); the most numerous cyanobacteria belonged to the genus *Anabaena*. They were dominated by *Anabaena lemmermannii* P. Richter. In open waters of bay, *A. lemmermannii* presented up to 21 million cells L^{-1} and a biomass of 4 g/m^3 (Figure 8). However, maximum growth of *A. lemmermannii* was recorded in a location sheltered from the wind in the waters in Maksimikha Cove with total cell counts up to 650 million cells L^{-1}, with a biomass of 65 g/m^3.

It is interesting to note that *A. lemmermannii* is a common species in lakes at all latitudes, where its massive growth usually produces a bloom of this species in eutrophic waters. The growth of the alga usually occurs during the period

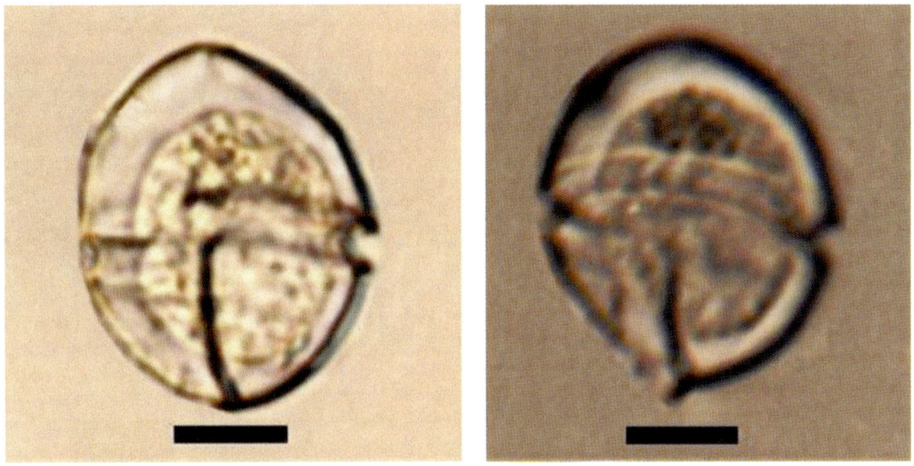

FIGURE 6: *Peridiniopsis polonicum* (Woloszynska) Bourrelly from the Chivyrkuisky Bay. The same cell in different focus of the microscope. Scale bar: 10 μm.

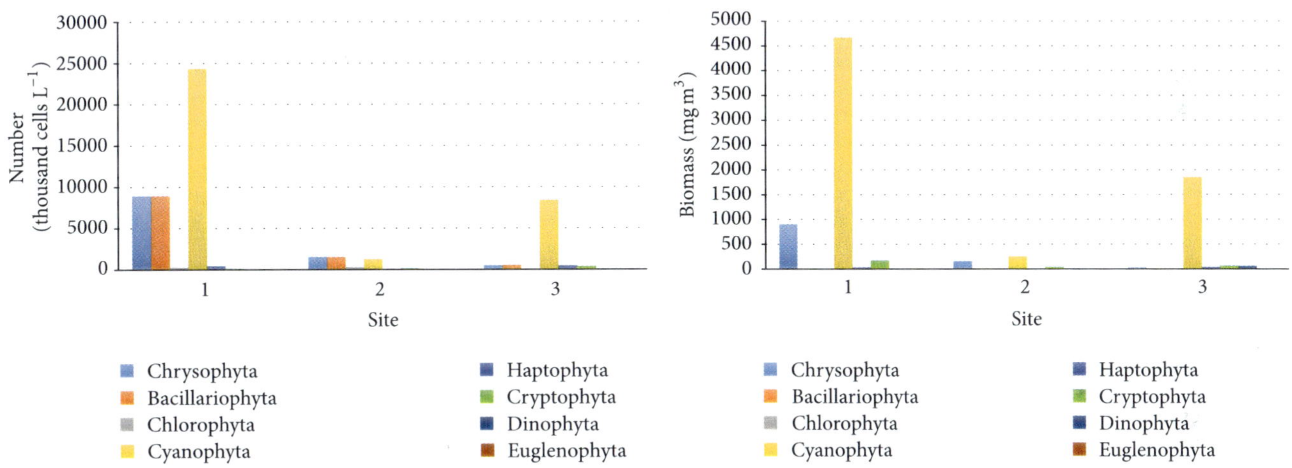

(a) (b)

FIGURE 7: The ratio of algal divisions in the phytoplankton composition of the Barguzinsky Bay in July 2012: (a) by the number and (b) by the biomass. Sites: 1, Maksimikha near the shore; 2, Maksimikha, 100 m from the shore; and 3, Svyatonossky Isthmus, near the shore.

of maximum warming of the water. The species is capable of nitrogen fixation and can grow luxuriantly regardless of the number and forms of mineral nitrogen in the water. This allows the alga to develop during the initial period of eutrophication, when the water is enriched with phosphorus but deficient in inorganic nitrogen [17, 18].

At the shore line, together with cyanobacteria, heterotrophic flagellates were observed in large numbers (up to 8 million cells L^{-1} and a biomass of 0.8 g/m^3). The cryptomonad species *Cryptomonas reflexa* (M. Marsson) Skuja, *C. marssonii* Skuja, and *C. erosa* Ehrenberg were also highly abundant (up to 100000 cells L^{-1} and a biomass of 0.22 g/m^3).

It is also a matter of interest that the observed composition of the dominant planktonic species communities in Maksimikha Cove is not typical for Lake Baikal. Such a community pattern is, most likely, characteristic of eutrophic aquatic environments with a high content of phosphorus. The extensive development of heterotrophic organisms such as small flagellates and various species of the genus *Cryptomonas* clearly shows that the bay environment is rich in dissolved organic compounds.

In the shallow waters of Maksimikha Cove, we found an extensive growth of periphyton. Numerous cloud-like clusters occurred on the sandy bottom or floated in the water, often at the surface. These clusters were formed by filamentous green algae and dominated by species of the genus *Spirogyra*. We also found species of the genera *Zygnema*, *Oedogonium*, *Cladophora*, and *Mougeotia*. To identify species of the genus *Spirogyra*, it is necessary to examine the zygospore, a product of sexual reproduction that is formed as a result of cell fusion [19]. However, the zygospores had not yet formed during the sampling period, and the *Spirogyra* cells in all analyzed samples were sterile. Only one pair of sexually mature conjugating cells were found, which were in the initial stage of cell fusion (Figure 9). The *Spirogyra* community of filamentous species was characterized by

(a)

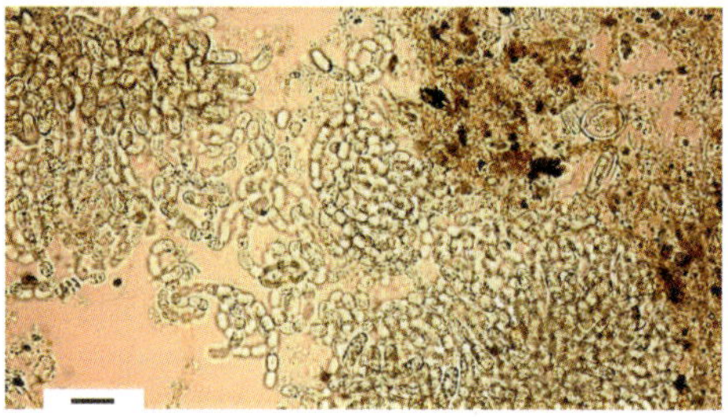

(b)

FIGURE 8: Maximal growth of cyanobacterial species *Anabaena lemmermannii* P. Richter near Maksimikha Cove in Barguzinsky Bay of Lake Baikal. (a) The water bloom near shoreline. (b) Colonies of *A. lemmermannii*. Scale bar: 20 μm.

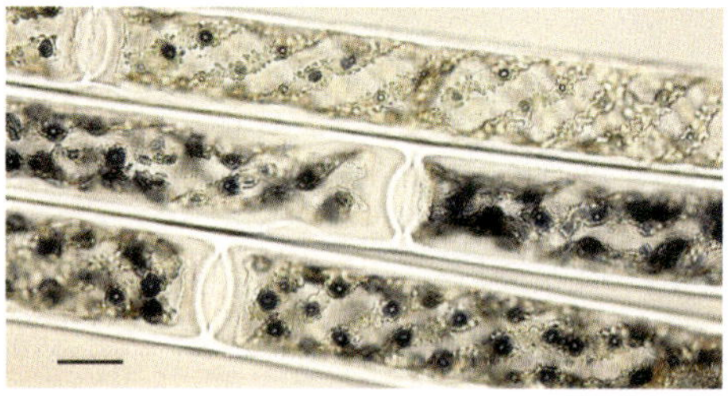

(a)

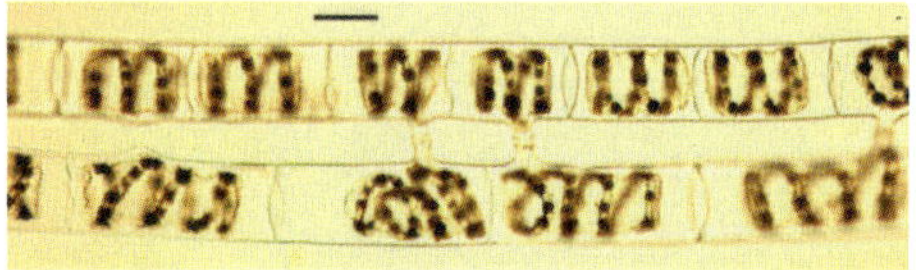

(b)

FIGURE 9: (a) Sterile filaments of *Spirogyra* species, dominant in the assemblages of algae in the Maksimikha Cove (Barguzinsky Bay of Lake Baikal, July 2012). Scale bar: 20 μm. (b) Sexually mature conjugating cells of *Spirogyra* species. Scale bar: 40 μm.

a high level of morphological diversity. In view of these morphological characteristics, we suggest that at least four species of the genus *Spirogyra* were present in Maksimikha Cove. All filamentous algae observed were active, in viable condition, and emerald-green in color. These algae have developed rapidly in the bay, where the current conditions are highly favorable for them. The species of *Spirogyra* have been noted previously in the Lake Baikal [20], but they have never been numerous. As already mentioned, only recently these algae have been reported in extreme abundance in the coastal zone, as, for example, near Listvyanka village (southern Baikal) [9, 10]. The current study presents the first report of a massive growth of *Spirogyra* in the largest bay of the Lake Baikal. The biomass of clusters of *Spirogyra* reached extremely high levels, up to 70 kg/m^3 (see Figure 1), with total filaments counts ranging from 17.5 to 25.5 thousand cells L^{-1}. The increased growth of these filamentous algae is a definitive sign of the ongoing changes in the environmental conditions of Barguzinsky Bay.

FIGURE 10: Colonial cyanobacteria of the *Stratonostoc* species on the coast of the Barguzinsky Bay of Lake Baikal. The diameter of the colonies is 8–20 mm.

It is also interesting to note that on the opposite side from the mouth of the Barguzin River, where there are no permanent tourist facilities, the deposits of algae on the beach were absent. In phytoplankton, cyanobacteria were dominated by the number as well as biomass (see Figure 7, site 3).

In contrast to Chivyrkuisky Bay, Barguzinsky Bay did not yield evidence of *Elodea canadensis* at the time of the study. We found profuse thickets of aquatic plants in a shallow estuary in an extended portion of the Maksimikha River. These thickets were formed primarily by *Equisetum fluviatile* L. and *Sagittaria natans* (Pallas). According to reports by the local inhabitants (fishermen and villagers) these aquatic plants have been intensively colonizing the shallow waters of the river during the past 10 years.

In Maksimikha Cove, the massive growth of filamentous algae was observed, accompanied by the release of these algae from the coastal strip during the autumn storms. Later, in September 2014, for the first time in the Lake Baikal, the massive development of colonial cyanobacteria of the genus *Stratonostoc* on the coast of the bay was reported by local inhabitants (Figure 10).

Occurrence of substantial colonies of *Stratonostoc* species freed from the substrate by the surf was also observed to the south of Barguzinsky Bay on a beach along the shore of the lake near Goryachinsk village (in September 2013). These *Stratonostoc* colonies covered up to 20% of the shoreline area of Maksimikha Cove, an additional clear signal of the ongoing trend of eutrophication. The most important of these signals are the massive growth of cyanobacteria species of the genus *Anabaena*, heterotrophic cryptomonads, and *Spirogyra* species.

4. Conclusion

The data collected in this study indicate the presence of local but significant eutrophication of the two largest shallow bays of the Lake Baikal. We conclude that the results of this field expedition and study clearly confirm that certain regions of the unique ultraclean aquatic ecosystem of this important lake face the threat of eutrophication. To prevent further extensions of this process of eutrophication, detailed study and monitoring of the coastal zone of the Lake Baikal is urgently needed. The identification of the sources of eutrophication and the development of measures to reduce nutrient inputs and organic matter in the waters of the lake should be given special consideration.

Competing Interests

The authors declare that they have no competing interests.

Acknowledgments

This research was funded by Russian Foundation for Basic Research (Grants 13-04-00614 and 14-04-00501); participation of Galina I. Kobanova and Maxim A. Timofeyev in this study was founded as a part of the projects within "Goszadanie" of the Ministry of Education and Science of RF (no. 6.382.2014/K). The support is gratefully acknowledged.

References

[1] S. E. Hampton, L. R. Izmest'eva, M. V. Moore, S. L. Katz, B. Dennis, and E. A. Silow, "Sixty years of environmental change in the world's largest freshwater lake—Lake Baikal, Siberia," *Global Change Biology*, vol. 14, no. 8, pp. 1947–1958, 2008.

[2] J. J. Magnuson, D. M. Robertson, B. J. Benson et al., "Historical trends in lake and river ice cover in the Northern Hemisphere," *Science*, vol. 289, no. 5485, pp. 1743–1746, 2000.

[3] H. Ishibashi, H. Iwata, E.-Y. Kim et al., "Contamination and effects of perfluorochemicals in baikal seal (*Pusa sibirica*). 1. Residue level, tissue distribution, and temporal trend," *Environmental Science and Technology*, vol. 42, no. 7, pp. 2295–2301, 2008.

[4] A. A. Mamontov, E. A. Mamontova, E. N. Tarasova, and M. S. McLachlan, "Tracing the sources of PCDD/Fs and PCBs to Lake Baikal," *Environmental Science & Technology*, vol. 34, no. 5, pp. 741–747, 2000.

[5] T. Katano, S.-I. Nakano, H. Ueno et al., "Abundance and composition of the summer phytoplankton community along a transect from the Barguzin River to the central basin of Lake Baikal," *Limnology*, vol. 9, no. 3, pp. 243–250, 2008.

[6] J. B. Shukla, A. K. Misra, and P. Chandra, "Modeling and analysis of the algal bloom in a lake caused by discharge of nutrients," *Applied Mathematics and Computation*, vol. 196, no. 2, pp. 782–790, 2008.

[7] M. V. Moore, S. E. Hampton, L. R. Izmest'eva, E. A. Silow, E. V. Peshkova, and B. K. Pavlov, "Climate change and the world's "sacred sea"-lake Baikal, Siberia," *BioScience*, vol. 59, no. 5, pp. 405–417, 2009.

[8] O. Timoshkin, M. Sakirko, V. Annenkov et al., "Baikal: ecology of the coastal zone. About the emergency sityation in the area of Severobaikalsk," *Eastern Siberia: The Arguments of the Week*, vol. 9, no. 401, pp. 16–17, 2014.

[9] L. S. Kravtsova, L. A. Izhboldina, I. V. Khanaev et al., "Nearshore benthic blooms of filamentous green algae in Lake Baikal," *Journal of Great Lakes Research*, vol. 40, no. 2, pp. 441–448, 2014.

[10] O. A. Timoshkin, N. A. Bondarenko, Y. A. Volkova, I. V. Tomberg, V. S. Vishnyakov, and V. V. Malnik, "Mass development of green filamentous algae of the genera *Spirogyra* and *Stigeoclonium* (Chlorophyta) in the littoral zone of the southern part of Lake Baikal," *Hydrobiological Journal*, vol. 51, no. 1, pp. 13–23, 2015.

[11] G. I. Kobanova, "The mass growth of filamentous macroalgae in the Barguzinsky Ba y of Baikal," in *Ecological Safety of the Coastal Regions (Seaports, Coastal Protection, Recreation, Mariculture): Proceeding of the International Scientific Conference Dedicated to 150-Years Anniversary of N.M. Knipovich. Rostov-on-Don*, pp. 115–118, SSC RAS, 2012.

[12] G. I. Kobanova, G. M. Speiser, M. P. Rikhvanova, K. S. Mikalyev, and E. A. Voronkova, "The results of the assessment of the ecological state of bays of Lake Baikal obtained during the Baikal social-science expedition in 2012," in *Materials of the VIII International Scientific-Practical Conference "Rivers of Siberia and the Far East", Irkutsk, June 2013*, pp. 108–110, Baikal Environmental Wave, Irkutsk, Russia, 2013.

[13] H. Utermöhl, *Zur Vervollkommung der Quantitativen Phytoplankton-Methodik*, vol. 9, Mitteilungen der Internationale Vereinigung für Theoretische and Angewandte Limnologie, 1958.

[14] G. I. Popovskaya, "The phytoplankton of coastal-lakes zone," in *Limnology of the Coastal-Lakes Zone of Baikal*, pp. 156–175, Nauka, Novosibirsk, Russia, 1977 (Russian).

[15] B. B. Bazarova and N. M. Pronin, "Spatial distribution of *Elodea canadensis* Michx. (Hydrocharitaceae) in Chivyrkuisky bay, Lake Baikal," *Inland Water Biology*, vol. 2, pp. 49–54, 2007.

[16] K. Wołowski and M. Grabowska, "Trachelomonas species as the main component of the euglenophyte community in the Siemianówka Reservoir (Narew River, Poland)," *Annales de Limnologie*, vol. 43, no. 3, pp. 207–218, 2007.

[17] I. S. Trifonova, *Ecology and Succession of Lacustrine Phytoplankton*, Nauka, Leningrad, Russia, 1990.

[18] C. Callieri, R. Bertoni, M. Contesini, and F. Bertoni, "Lake level fluctuations boost toxic cyanobacterial 'oligotrophic blooms'," *PLoS ONE*, vol. 9, no. 10, Article ID e109526, 2014.

[19] L. A. Rundina, *The Zygnematales of Russia (Chlorophyta: Zygnematophyceae)*, Nauka, St. Petersburg, Russia, 1998.

[20] L. A. Izhboldina, *Guide and Key to Benthonic and Periphyton Algae of Lake Baikal (Meio-and Macrophytes) with Short Notes on Their Ecology*, vol. 248, Nauka-Center, Novosibirsk, Russia, 2007.

Impact of Mass Bathing and Religious Activities on Water Quality Index of Prominent Water Bodies: A Multilocation Study in Haryana, India

Anita Bhatnagar,[1] Pooja Devi,[1] and M. P. George[1,2]

[1] *Department of Zoology, Kurukshetra University, Kurukshetra 136119, India*
[2] *Delhi Pollution Control Committee, ISBT Building, Delhi, India*

Correspondence should be addressed to Anita Bhatnagar; anitabhatnagar@gmail.com

Academic Editor: L. M. Chu

The present study was designed to assess the impact of mass bathing and religious activities on water quality index (WQI) of prominent water bodies (eight) in Haryana, India. Water quality characteristics revealed significant increase in the values of nitrate, biochemical oxygen demand (BOD), turbidity, total dissolved solids (TDS), conductivity, total hardness, total alkalinity, and MPN count after the religious activities. The computed WQI at all the eight selected sites varied from 47.55 to 211.42. The results revealed that there was a significant increase in the value of WQI after mass bathing or any other ritual performed. Out of eight water bodies studied three (sites 3, 4, and 5) were found under good water quality status; four sites (1, 2, 6, and 7) depicted medium water quality but site 8 was found under poor water quality after the religious activities. The good water quality status of water bodies was correlated with larger size of the water bodies and less number of pilgrims; however, the poor WQI values may be attributed to smaller size of the water body and heavy load of pilgrims on such sites. Therefore, water of these religious water bodies needed to be regularly changed after mass bathing to protect the aquatic component from different contaminations.

1. Introduction

Surface water bodies have been the basic resource to fulfill all kinds of human requirements but unfortunately today water quality of these water resources is under severe environmental stress and under threat due to various types of anthropogenic activities and the condition is more pathetic in lentic water bodies. Among various types of anthropogenic actions religious activities like immersion of flour, oil, soap, ash, detergents, floral offerings, and mass bathing are also one of the important causes that affect the water quality of a water body. The holy water bodies of state of Haryana, India, have their immense value in terms of religious aspect, as people of Haryana, Punjab, and other states also are associated emotionally and spiritually with these water bodies. The religious activities are deeply rooted in its cultural heritage; millions of people take holy bath and perform religious activities and also do *Aachman* (drinking water) regularly on occasions like solar eclipse and new moon day irrespective of thinking about the water quality of the aquatic bodies. According to WHO [1] about 80% of all diseases in human beings are caused by water; therefore a regular monitoring of such water bodies is very essential for physicochemical and microbiological analysis to know the suitability of water under use not only to check the outbreak of diseases and occurrence of hazards but also to prevent the water from further deterioration.

For taking a better picture of water quality the data of various quality indicating parameters data could be aggregated into an overall index called water quality index (WQI). WQI can be defined as "a rating reflecting the composite influence of different water quality parameters on overall quality of water" [2, 3]. WQI can be very important tool since it can be used for awareness of general public and policy makers to enhance the water quality. Therefore, the present study was conducted to depict the effect of religious activities

TABLE 1: The selected water bodies, their location, and the religious activities performed.

Site number	Name of water body	District	Activity
(1)	Kapalmochan tank	Yamunanagar	Full moon day (Kartik/Nov) on eve of Gurunanak birthday
(2)	Kirmach pond	Kurukshetra	Kulotarn Tirth, fair at Sept month
(3)	Ban Ganga Tirth tank	Kurukshetra	Vaishakhi (April)
(4)	Brahmsarovar tank	Kurukshetra	Amavasya (new moon day), solar eclipse
(5)	Jyotisar tank	Kurukshetra	Amavasya (new moon day), solar eclipse
(6)	Saraswati Tirth tank	Pehowa, Kurukshetra	Pind-dan and bathing during Sharadh (Sept-Oct), solar eclipse & Kartik amavasya
(7)	Phalgu Tirth pond	Kaithal	Somvati amavasya of Sharadh
(8)	Pandu-Pindara Tirth pond	Jind	Somvari amavasya (new moon day on Monday), solar eclipse, Sharadh

TABLE 2: Relative weight (Wi) values of each parameter.

S. number	Physicochemical parameters	Water quality standards for bathing or recreation	Assigned weight	Relative weight $Wi = wi/\sum wi$
1	pH	6.5–9.0 [4], 6.5–8.5 [5]	4	0.1
2	Conductivity (Ω/cm)	3000 [6]	4	0.1
3	TDS (mg L^{-1})	1500 [6]	4	0.12
4	Turbidity (NTU)	5 [7]	2	0.05
5	DO (mg L^{-1})	≥5 [5]	4	0.1
6	BOD (mg L^{-1})	≤3 [5]	3	0.07
7	Total alkalinity (mg L^{-1})	200 [6]	1	0.02
8	Total chloride (mg L^{-1})	350 [4]	3	0.07
9	Total hardness (mg L^{-1})	600 [8, 9]	2	0.05
10	Total sulphate (mg L^{-1})	350 [4]	4	0.1
11	Total nitrate (mg L^{-1})	5.6 [4]	5	0.12
12	Total coliform (MPN/100 mL of water sample)	≤500 [6, 10]	4	0.1

and mass bathing on water quality in terms of WQI of the eight prominent lentic water bodies of Haryana where such religious activities are carried out.

2. Materials and Methods

2.1. Study Area. A total of eight religious water bodies (lentic waters) in four districts of state of Haryana (Figure 1), India, were selected for the present study that were being used for the mass bathing and other religious activities (Table 1).

2.2. Sample Collection. Water samples for physicochemical analysis were collected two days before every religious activity and one day after religious activity from selected stations, in triplicate, in clean polythene bottles during February 2012 to March 2013 for physicochemical analysis. During the study period the religious activities were performed two times at stations Ban Ganga Tirth, Brahmsarovar, Jyotisar, Saraswati tank, and Pandu-Pindara Tirth; one time at stations Kapalmochan, Kulotaran Tirth, and Phalgu Tirth.

2.3. Physicochemical Analysis. Physicochemical parameters, namely, pH, conductivity, total dissolved solids (TDS), turbidity, dissolved oxygen (DO), biochemical oxygen demand (BOD), total alkalinity, chloride, hardness, total sulphate, and total nitrate, were determined using the standard procedures followed by APHA [11].

2.4. Microbiological Analysis. Concentration of total faecal coliform bacteria was reported as Most Probable Number per 100 mL (MPN/100 mL) following standard procedures APHA [11]. (i) Presumptive coliform test was done using multiple 5-tubes MPN dilution technique using MacConkey Broth. (ii) Conformation test for total coliform was done by inoculating 1 mL inoculum from positive tube from presumption test in Brilliant green broth. (iii) For completed test, EMB (eosin methylene blue) agar plates were prepared and inoculated with positive tubes of confirmed test by streak plate method and incubated at 37°C for 24 to 48 h and observed for faecal coliform colonies.

2.5. Calculation of Water Quality Index (WQI). For computing the WQI three basic steps were followed [12]. All the parameters were assigned a weight according to their effect on the public health (Table 2).

Maximum weight of 5 was assigned to parameter like nitrate due to their major importance in water quality assessment [13]. pH, conductivity, TDS, DO, and sulphate

Table 3: Status of water quality based on WQI.

S. number	WQI range	Status
1	<50	Excellent water quality
2	50–100	Good water quality
3	100–150	Medium or average water quality
4	150–200	Poor water quality
5	200–300	Very poor water quality
6	>300	Water unsuitable for drinking/bathing or any other human use

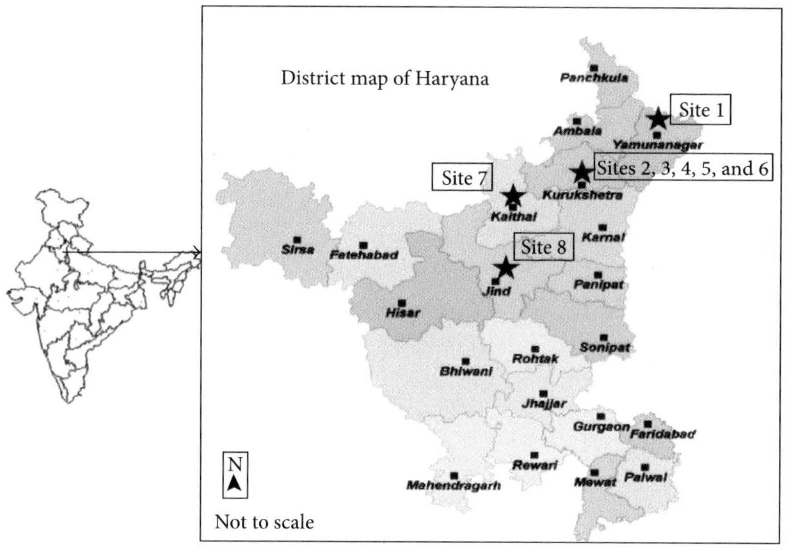

Site 1: Kapalmochan, Yamunanagar
Site 2: Kulotaran Tirth, Kurukshetra
Site 3: Ban-Ganga Tirth, Kurukshetra
Site 4: Brahmsarovar, Kurukshetra
Site 5: Jyotisar, Kurukshetra
Site 6: Saraswati Tirth, Kurukshetra
Site 7: Phalgu Tirth, Kaithal
Site 8: Pandu-Pindara Tirth, Jind

Figure 1: Map showing different sites of Haryana selected under the study area.

were assigned the weight 4 according to Al-Mashagbah [14] and Srinivasamoorthy et al. [15]. MPN coliform was assigned weight 4 since it is the most important factor in India, where most of the serious diseases are waterborne [16]. Turbidity as well as total hardness was given the weight of 2 as it plays a fewer role in the water quality assessment [17]. In the second step, the relative weight (Wi) of each parameter was computed using

$$Wi = \frac{wi}{\sum_{i=1}^{n} wi}, \quad (1)$$

where wi is the assigned weight of each parameter and n is the number of parameters. Weight (wi) and relative weight (Wi) values and the standards for bathing water quality for each parameter were given in Table 2. In the third step, quality rating scale (qi) was calculated for each parameter using

$$qi = \frac{ci}{si} \times 100, \quad (2)$$

where qi is the quality rating, ci is the observed concentration of each parameter in the water samples, and si represents the bathing standard value for each physicochemical parameter adopted from different recognized water quality organisations.

For the calculation of WQI, the SI (subindex) was computed for each given parameter as per the following equations:

$$\mathrm{SI} = Wi \times qi, \quad (3)$$

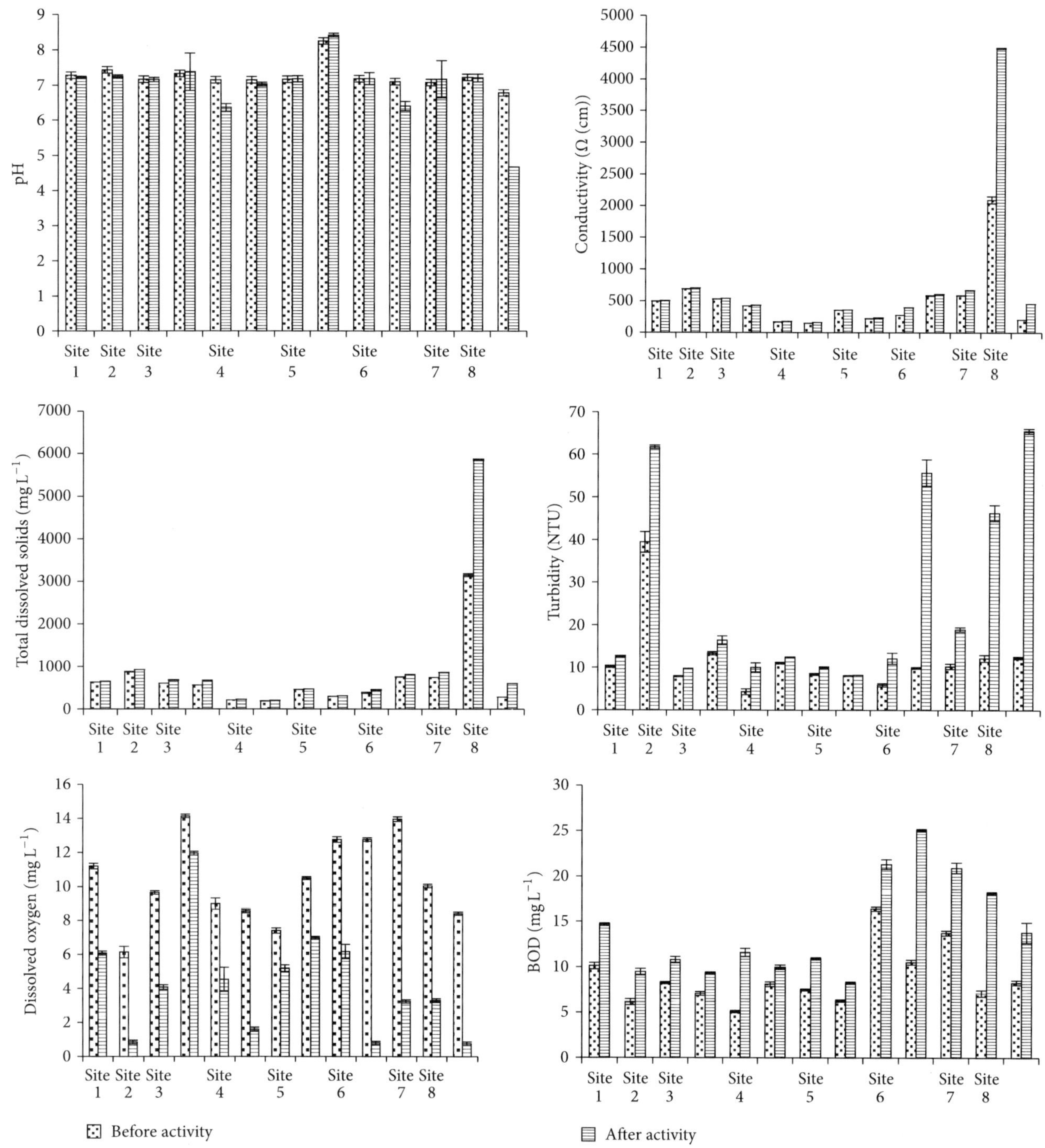

FIGURE 2: Variations in pH, conductivity, total dissolved solids, turbidity, dissolved oxygen, and biochemical oxygen demand (mean ± standard error of the mean) before and after the religious activities at all the selected sites.

$$\mathrm{WQI} = \sum_{i=1}^{n} \mathrm{SI}i, \qquad (4)$$

where SIi is the subindex of ith parameters. WQI values were classified into six categories (Table 3), Ramakrishnaiah et al. [18] and Mishra and Borah [4].

3. Results and Discussion

Mean values of all the eleven physicochemical parameters (mean ± SE) before and after the religious activities from February to March 2013 are represented in Figures 2 and 3. Significant ($P < 0.05$) increase in the values of conductivity, TDS, turbidity, BOD, total alkalinity, total chloride, total

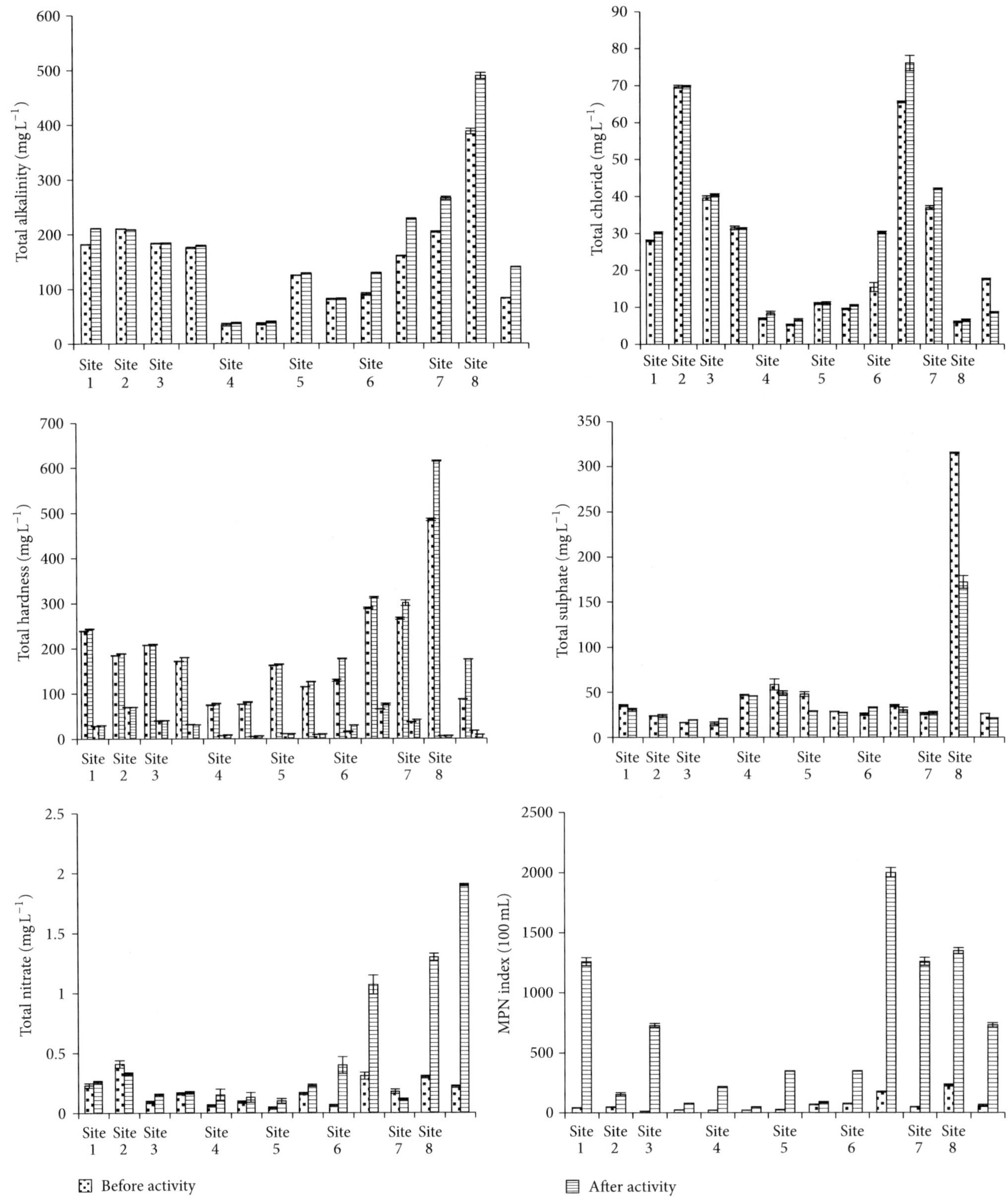

FIGURE 3: Variations in total alkalinity, total chloride, total hardness, total sulphate, and total nitrate (mean ± standard error of the mean) before and after the religious activities at all the selected sites.

hardness, and total nitrate and significant decrease in DO were observed at most of the sites after the mass bathing or other ritual activities. Similar observations were also reported by Vyas and Bajpai [19] and Sharma et al. [20]. The high organic matter and pollution load of pilgrims on the water bodies like bathing, adding ashes, flour, floral offerings, and urination may increase the chloride and nitrate contents of water in conformity with the findings of Kiran et al. [21], Gupta et al. [22]. Munawar [23] indicated that high concentration of chloride indicates high degree of organic pollution. Increase in the chloride may further be responsible for elevating the values of total dissolved solids and total hardness after the pilgrim's activities. Increase in the TDS further directly affects the conductivity as increase in the ion content will in turn increase the conductivity due to increase in the pollution load; similar observations were also recorded by [22]. Increase in total alkalinity may be attributed due to presence of basic salts of potassium and sodium along with those of calcium and magnesium as recorded by Jadhav et al. [24]. Overall minor differences were observed in pH before and after the religious activities. But the values of some parameters like conductivity, pH, TDS, turbidity, DO, BOD, and total alkalinity were recorded above or below the permissible limit at some sites suggested by water quality guidelines for bathing given by various organizations (Table 2). Among all the sites the values of maximum water quality indicating parameters were recorded beyond the permissible limits of water quality standards of bathing waters at sites 6 and 8. Minor fluctuations in the values of chloride and sulphate were observed after the religious activities; however, all the values were recorded within the permissible limits (Table 2). Bacteria of coliform group are considered the primary indicators of faecal contamination by Raina et al. [25] that have been correlated with the incidence of gastrointestinal disorders (Morace and McKenzie [26]). The link between poor microbiological water quality and infectious diarrhea is well established [27–29]. An increase in MPN index of total coliform population was also recorded at all the site after mass bathing or any other religious activity which may be due to high level of pollution caused in the form of immersion of flour, floral offerings, bathing, urination, and so forth during these peak bathing dates. A probable reason behind this may be the addition of organic materials and faecal waste due to mass bathing of millions of pilgrims and the religious activities performed by them, supporting the earlier findings of Sood et al. [30].

On the basis of physicochemical and bacteriological variables before and after the religious activities, WQI was calculated for all the selected sites, which is shown in Table 4. On the basis of WQI values the sites 3, 4, and 5 were found to have good water quality status before and after the religious activities. The status of sites 1, 2, 6, and 7 was recorded under the medium water quality after the religious activities. However, the water quality of site 8 was changed from good water quality to medium water quality after the Sharadh period and from medium water quality to very poor water quality after the solar eclipse period due to increase in the values of water quality index (Table 5). These results depicted that pollution load was maximum at sites 6 and 8 after the religious activities and these were not found suitable for the bathing purpose or for the holy dip. This may be attributed to smaller size of these water bodies and heavy pilgrims load on these sites.

TABLE 4: Water quality index values calculated at all the selected sites before and after the religious activities.

Sites		Before activity	After activity
Kapalmochan	Site 1	78.600	106.301
Kulotarn Tirth	Site 2	91.306	113.064
Ban Ganga	Site 3	67.187	79.101
		77.622	84.115
Brahmsarovar	Site 4	47.550	63.724
		60.890	53.696
Jyotisar	Site 5	58.934	70.686
		60.773	59.941
Saraswati Tirth	Site 6	87.625	101.256
		87.125	122.734
Phalgu Tirth	Site 7	94.278	125.427
Pandu-Pindara Tirth	Site 8	112.069	211.425
		63.727	133.061

Pollution status was found to be lower at the sites 1, 3, 4, and 5 indicating their good water quality status that was suitable for bathing purpose may be attributed due to larger sizes of the these water bodies and less number of bathers or pilgrims.

The higher values of water quality index at maximum sites after the religious activities were observed mainly due to increase in the values of physicochemical parameters, namely, nitrates, turbidity, BOD, TDS, total alkalinity, and total hardness, and also due to decrease in the values of DO.

4. Conclusions

The results of the present study clearly depict that there is an increase in the value of water quality index after the religious human activities due to increase in the values of nitrates, turbidity, BOD, TDS, total alkalinity, total hardness, and MPN and decrease in the values of DO. Out of eight water bodies studied three were found to be under good water quality status and four (1, 2, 6, and 7) were designated under the medium water quality status whereas site 8 fell under the category of poor water quality after the religious activities. The sites with good WQI status (sites 3, 4, and 5) may be considered suitable for bathing purpose.

Regular change or mixing of water after mass bathing to protect the water body from different contaminations is recommended. Educational campaigns for awareness of the public regarding suitable use of water bodies without impairing the water quality are also required. To ensure that water quality standards are not violated, there is need of regular monitoring and surveillance of water quality in terms of physicochemical, biological, and bacteriological parameters taking suitable remedial measures to control pollution and prevent the depletion of the quality of lentic waters.

TABLE 5: Water quality status of the selected sites on the basis of water quality index values before and after the religious activities.

Sites		Before activity	After activity
Kapalmochan	Site 1	Good water quality	Medium water quality
Kulotarn Tirth	Site 2	Good water quality	Medium water quality
Ban Ganga	Site 3	Good water quality	Good water quality
		Good water quality	Good water quality
Brahmsarovar	Site 4	Good water quality	Good water quality
		Good water quality	Good water quality
Jyotisar	Site 5	Good water quality	Good water quality
		Good water quality	Good water quality
Saraswati Tirth	Site 6	Good water quality	Medium water quality
		Good water quality	Medium water quality
Phalgu Tirth	Site 7	Good water quality	Medium water quality
Pandu-Pindara Tirth	Site 8	Medium water quality	Very poor water quality
		Good water quality	Medium water quality

Conflict of Interests

The authors declare that there is no conflict of interests regarding the publication of this paper.

References

[1] WHO (World Health Organisation), *Guidelines of Drinking Water Quality*, vol. 1, WHO (World Health Organisation, Geneva, Switzerland, 2nd edition, 1993.

[2] R. A. Deininger and J. J. Macinans, *A Water Quality of Environmental and Industrial Health*, School of Public,University of Michigan, Ann Arbor, Mich, USA, 1971.

[3] R. D. Harkins, "An objective water quality index," *Journal of Water Pollution Control Federation*, vol. 46, no. 3, pp. 588–591, 1974.

[4] U. Mishra and P. Borah, "Assessment of indicator parameters to investigate the variations in groundwater quality of west districts of tripura using entropy and correlation," *International Journal of Chemistry and Chemical Engineering*, vol. 3, no. 3, pp. 215–222, 2013.

[5] Organization for Economic Co-operation and Development (OECD), *Task Force for the Implementation of the Environmental Action Programme for Central and Eastern Europe, Caucasus and Central Asia: Regulatory Environmental Programme Implementation Network*, ENV/EPOC/EAP/REPIN(2011)1/FINAL, Organization for Economic Co-operation and Development (OECD), Paris, France, 2011.

[6] Central Pollution Control Board (CPCB), *Guidelines for Water Quality Management*, Central Pollution Control Board (CPCB), Delhi, India, 2008.

[7] Queensland Government, *Queensland Health Swimming and Spa Pool Water Quality and Operational Guidelines*, Communicable Diseases Unit Public Health Services, 2004.

[8] UNESCO/WHO/UNEP, *Water Quality Assessment—A Guide to the Use of Biota, Sediments and Water in Environmental Monitoring*, 2nd edition, 2001.

[9] Bureau of Indian Standards (BIS), *Indian Standard Drinking Water-Specifications*, First Revison, Bureau of Indian Standards (BIS), 1991.

[10] V. K. Tyagi, A. Bhatia, R. Z. Gaur et al., "Impairment in water quality of Ganges River and consequential health risks on account of mass ritualistic bathing," *Desalination and Water Treatment*, vol. 51, no. 10-12, pp. 2121–2129, 2013.

[11] APHA, *Standard Methods for the Examination of Water and Wastewater*, APHA, AWWA, WPCF, Washington, DC, USA, 2005.

[12] M. Kumar, K. Kumari, A. L. Ramanathan, and R. Saxena, "A comparative evaluation of groundwater suitability for irrigation and drinking purposes in two intensively cultivated districts of Punjab, India," *Environmental Geology*, vol. 53, no. 3, pp. 553–574, 2007.

[13] T. HemaLatha, G. N. Pradeep Kumar, P. Lakshminarayana, and A. Anil, "Assessment of groundwater quality index for upper Pincha Basin, Chittoor District, Andhra Pradesh, India using GIS," *International Journal of Scientific and Engineering Research*, vol. 3, no. 7, pp. 1–8, 2012.

[14] A. F. Al-Mashagbah, "Assessment of surface water quality of king abdullah canal, using physico-chemical characteristics and water quality index, Jordan," *Journal of Water Resource and Protection*, vol. 7, no. 4, pp. 339–352, 2015.

[15] K. Srinivasamoorthy, S. Chidambaram, M. V. Prasanna, M. Vasanthavihar, J. Peter, and P. Anandhan, "Identification of major sources controlling groundwater chemistry from a hard rock terrain—a case study from Mettur taluk, Salem district, Tamil Nadu, India," *Journal of Earth System Science*, vol. 117, no. 1, pp. 49–58, 2008.

[16] S. L. Dwivedi and V. Pathak, "A preliminary assignment of water quality index to Mandakini River, Chitrakoot," *Indian Journal of Environmental Protection*, vol. 27, no. 11, pp. 1036–1038, 2007.

[17] M. K. Rokbani, M. Guddari, and R. Bauhlila, "Use of Geographical Information System and Water Quality Index and assess ground water quality in EI Khairat Deep Aquifer (Enfidha, Tunisian Sahel)," *Iranica Journal of Energy and Environment*, vol. 2, no. 2, pp. 133–144, 2011.

[18] C. R. Ramakrishnaiah, C. Sadashivaiah, and G. Ranganna, "Assessment of water quality index for the groundwater in Tumkur Taluk, Karnataka State, India," *E-Journal of Chemistry*, vol. 6, no. 2, pp. 523–530, 2009.

[19] A. Vyas and A. Bajpai, "Water quality survey and monitoring of idol immersion in context of Lower lake," in *Proceedings of Taal 2007: The 12th World Lake Conference*, pp. 1818–1821, Bhopal, India, 2008.

[20] V. Sharma, S. Bhadula, and B. D. Joshi, "Impact of mass bathing on water quality of Ganga River during Maha Kumbh-2010," *Journal of Nature and Science*, vol. 10, no. 6, pp. 1–5, 2012.

[21] B. R. Kiran, K. Harish, M. Ravikumar, E. T. Puttaiah, and C. D. Kamat, "Water quality assessment of Bhadrariver with special reference to industrial pollution," *Indian Journal of Environmental protection*, vol. 26, no. 2, pp. 148–152, 2006.

[22] A. K. Gupta, K. Mishra, P. Kumar, C. Singh, and S. Srivastava, "Impact of religious activities on the water characteristics of prominent ponds at Varanasi (U.P.), India," *Plant Archives*, vol. 11, no. 1, pp. 297–300, 2011.

[23] M. Munawar, "Limnological studies on freshwater ponds of hyderabad-India," *Hydrobiologia*, vol. 36, no. 1, pp. 105–128, 1970.

[24] S. B. Jadhav, N. S. Chavan, and M. V. Gokhale, "Effect of ritual activity on the lentic water resources of Jotiba (Wadi-Ratnagiri), Kolhapur district, Maharashtra," *Ecology, Environment and Conservation*, vol. 15, no. 1, pp. 71–75, 2009.

[25] P. S. Raina, F. L. Pollari, G. F. Teare, M. J. Goss, D. A. J. Barry, and J. B. Wilson, "The relationship between *E. coli* indicator bacteria in well-water and gastrointestinal illnesses in rural families," *Canadian Journal of Public Health*, vol. 90, no. 3, pp. 172–175, 1999.

[26] J. L. Morace and S. W. McKenzie, "Faecal-indicator bacteria in the Yakima River Basin, Washington—an examination of 1999 and 2000 synoptic-sampling data and their relation to historical data," Water-Resources Investigations Report 02-4054, 2002.

[27] A. A. Cronin, D. Shrestha, N. Cornier, F. Abdalla, N. Ezard, and C. Aramburu, "A review of water and sanitation provision in refugee camps in association with selected health and nutrition indicators—the need for integrated service provision," *Journal of Water and Health*, vol. 6, no. 1, pp. 1–13, 2008.

[28] T. T. Do, T. T. Bui, K. Molbak, D. C. Phung, and A. Dalsgaard, "Epidemiology and aetiology of diarrhoeal diseases in adults engaged in wastewater-fed agriculture and aqua-culture in Hanoi, Vietnam," *Tropical Medicine and International Health*, vol. 12, no. 2, pp. 23–33, 2007.

[29] W. E. Oswald, A. G. Lescano, C. Bern, M. M. Calderon, L. Cabrera, and R. H. Gilman, "Fecal contamination of drinking water within peri-urban households, Lima, Peru," *American Journal of Tropical Medicine and Hygiene*, vol. 77, no. 4, pp. 699–704, 2007.

[30] A. Sood, K. D. Singh, P. Pandey, and S. Sharma, "Assessment of bacterial indicators and physicochemical parameters to investigate pollution status of Gangetic river system of Uttarakhand (India)," *Ecological Indicators*, vol. 8, no. 5, pp. 709–717, 2008.

Changes in the Ecological Parameters of Mixed Forests of Sal (*Shorea robusta* Gaertn.) are a Function of Distance from the Human Settlements

Ramesh P. Sapkota, [1,2] **Peter D. Stahl**, [1] **Krishna Hengaju**, [2] **and Kedar Rijal** [3]

[1] *Department of Ecosystem Science & Management and Program in Ecology, University of Wyoming, 1000 E University Avenue, Laramie, WY, USA*
[2] *Department of Environmental Science, Amrit Campus, Tribhuvan University, Kathmandu, Nepal*
[3] *Central Department of Environmental Science, Tribhuvan University, Kathmandu, Nepal*

Correspondence should be addressed to Ramesh P. Sapkota; rsapkota@uwyo.edu

Academic Editor: Ram Chander Sihag

Forests in Nepal are extremely important for supporting the livelihood of millions of people who collect forest products for their subsistence use and partly for income generation. Such inherent dependence is expected to cause disturbance in the forest ecosystem. We investigated changes in the structural assemblages caused by the interaction between anthropogenic disturbances and forest management activities in the mixed forests of Sal (*Shorea robusta* Gaertn.) of Terai, Central Nepal. We evaluated three buffer zone community forests (BZCFs), namely, Radha Krishna, Musharni Mai, and Janajagaran of Parsa Wildlife Reserve (PWR); the forest inside PWR was taken as a control. A transect of 2 km length was laid in each forest, and six plots, each of 1 ha size, were established at a successive interval of 300 m along the edge to the interior of the forests to count and record the diameter at breast height (DBH) of the studied plants. We observed that the species diversity increased linearly ($p < 0.05$) towards the forest interior in the BZCFs. Species other than *S. robusta* had significantly higher ($p < 0.05$) dominance and Importance Value Indices in the interior sites. We did not observe such trends in the control forest. Multivariate analysis showed that the sites of BZCFs had higher structural dissimilarity, but the control forest sites were closer to each other in composition. The forest sites near the settlements had undergone biotic homogenization (*S. robusta* mixed forest changed to *S. robusta* forest) due to the interaction between anthropogenic disturbances and forest management activities. On the basis of vegetation density, the edges of BZCFs appeared to be protected, but on the basis of diversity failed to do so. Future management strategies should be directed towards enhancing the diversity, heterogeneity, and forest quality, especially near the forest edges.

1. Introduction

Anthropogenic activities influence the ecological processes [1–3] and have extensive impacts on the forest ecosystems [4–7]. The fragmentation of natural forests creates a pattern in the landscapes and is regarded as one of the most important causes for the recent decline of forest species [8], and the human activities increase the risk of further degradation in the forests sharing borders with anthropogenic uses [9–12]. However, with an increasing concern towards forest degradation, management interventions are also underway with a focus on the restoration of degraded areas [13]. Human pressures from the nearby settlements and forest management activities interact to alter the forest structural and functional attributes [14]. Our concern was to determine the extent of such changes in the mixed forest of Sal (*Shorea robusta* Gaertn.; family: Dipterocarpaceae) of Terai, Central Nepal.

Terai forests of Nepal, dominated by *S. robusta*, are among the most disturbed ecosystems [15, 16]. Commonly occurring forests fires every year during dry months also kill young regenerations and burn the deciduous litter layer. In addition, these ecosystems are important for millions of people who depend on forest products for their subsistence

Table 1: List of predominant woody species (species with Importance Value Index > 10) found in the sampling sites and their local uses (all species are used as firewood).

Scientific name	Local name	Family	Local uses
Shorea robusta Gaertn.	Sal	Dipterocarpaceae	Fodder and animal bedding Religious plant Support for climbers (in gardens) Construction and furniture Handicrafts Cosmetics Seeds (oil) and resin Leaf plates and bowls Medicinal plant
Adina cordifolia (Willd. ex Roxb.) Benth. & Hook. f. ex Brandis	Karam	Rubiaceae	Fodder and animal bedding Support for climbers (in gardens) Construction and furniture Handicraft Medicinal plant
Lannea coromandelica (Houtt.) Merr.	Jinger	Anacardiaceae	Fodder and animal bedding Furniture Medicinal plant Living fences Gum
Lagerstroemia parviflora Roxb.	Botdhayaro	Lythraceae	Fodder and animal bedding Support for climbers (in gardens) Construction and furniture Handicrafts Tannins
Casearia graveolens Dalzell	Badkaule	Flacourtiaceae	Support for climbers (in gardens) Fruit extract used to poison fish in streams/rivers
Cassia fistula L.	Rajbrikshya	Fabaceae	Fruits/legume Veterinary medicine Construction and furniture Handicrafts Medicinal plant Religious plant Ornamental
Mallotus philippensis (Lam.) Müll.-Arg.	Sindure	Euphorbiaceae	Fodder and animal bedding Support for climbers (in gardens) Ornamental Construction and furniture Handicrafts Red dye Medicinal plant
Dillenia pentagyna Roxb.	Tatari	Dilleniaceae	Animal bedding Fruits Religious plant Vegetables Medicinal plant
Syzygium cumini (L.) Skeels	Jamuna	Myrtaceae	Fodder and animal bedding Fruits Religious plant Medicinal plant Support for climbers (in gardens) Construction and furniture

Table 1: Continued.

Scientific name	Local name	Family	Local uses
Semecarpus anacardium L.f.	Bhalayo	Anacardiaceae	Fodder and animal bedding Fruits and nuts Religious plant Support for climbers (in gardens) Medicinal plant
Miliusa velutina (Dunal) Hook. f. & Thoms.	Kalikath	Annonaceae	Fodder and animal bedding Fruits Support for climbers (in gardens) Construction materials Handicrafts
Terminalia alata Heyne ex Roth	Saj	Combretaceae	Fodder and animal bedding Construction and furniture Handicrafts Seeds Medicinal plant
Cleistocalyx operculatus (Roxb.) Meer. & Perry	Kyamuna	Myrtaceae	Fodder and animal bedding Fruits Construction Medicinal plant
Anogeissus latifolius (Roxb. ex DC.) Bedd.	Banjhi	Combretaceae	Fodder and animal bedding Support for climbers (in gardens) Construction Handicrafts
Terminalia bellirica (Gaertn.) Roxb.	Barro	Combretaceae	Fodder and animal bedding Fruits Construction and furniture Vegetable oil Medicinal plant
Phyllanthus emblica Linn.	Amala	Euphorbiaceae	Fruits and juice Religious plant Medicinal plant Support for climbers (in gardens) Construction and furniture Cosmetics Seed oil
Terminalia chebula Retz.	Harro	Combretaceae	Fodder and animal bedding Furniture Fruits Medicinal plant
Dalbergia latifolia Roxb.	Satisal	Fabaceae	Religious plant Ornamentals Handicrafts Construction and furniture Medicinal plant
Desmodium oojenense (Roxb.) Ohashi	Sandan	Fabaceae	Fodder Religious plant Construction and furniture Handicrafts Medicinal plant
Garuga pinnata Roxb.	Dabdabe	Burseraceae	Fodder and animal bedding
Buchanania latifolia Roxb.	Piyari	Anacardiaceae	Fodder and animal bedding Fruits Construction and furniture Handicrafts Medicinal plant

TABLE 1: Continued.

Scientific name	Local name	Family	Local uses
Dysoxylum gobara (Buch.-Ham.) Merr.	Lasune	Meliaceae	Fodder and animal bedding Construction and furniture Handicrafts Medicinal plant

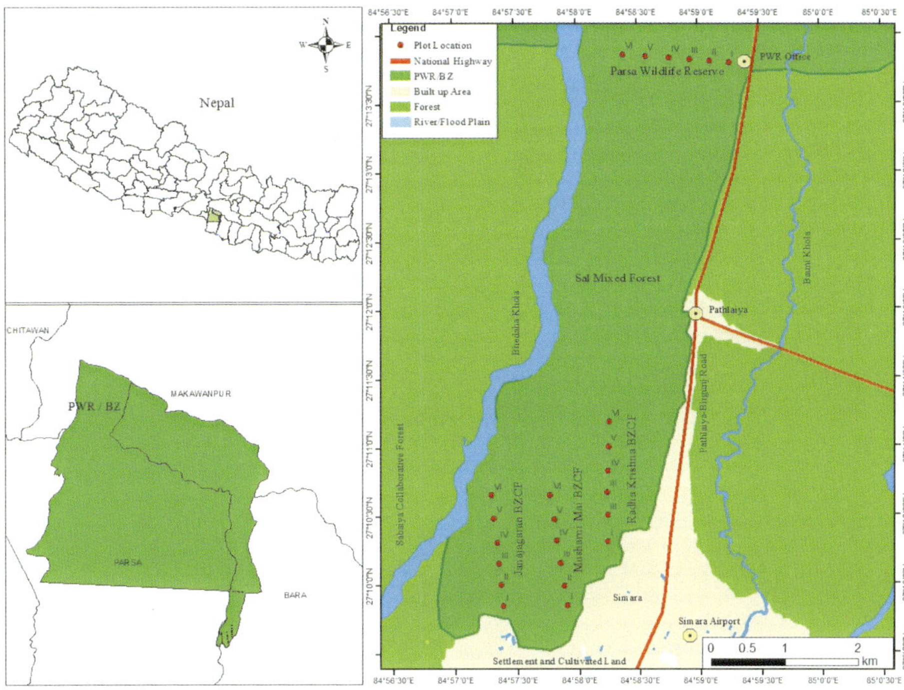

FIGURE 1: Location of studied BZCFs and PWR forest in Bara District, Central Terai, Nepal. In these BZCFs, the forest user groups are different and also vary in a number of households using forest resources. These BZCFs are the southernmost part of a large forest landscape also known as "Charkose Jhadi". People enter from the southern border of the BZCFs to collect/harvest forest resources.

(Table 1). Repeated extraction of forest resources to meet livelihood needs of adjoining human settlements can cause significant impacts on the forest regeneration, structure, and diversity, and these impacts often have been associated with the distance to the settlements [17–20]. In our studied buffer zone community forests (BZCFs), besides harvesting of the forest resources, the forests are also under regular management activities carried by Buffer Zone Management Committees, and the interaction between the disturbances and management interventions might have altered the ecological parameters differently. To our understanding, there are ample studies showing the impacts of forest disturbance on forest structure [16–19], but limited research exists on how management and disturbance interaction influences the composition, structure, richness, and taxonomic diversity of the forests. What ecological changes take place due to such interactions in the forests of Nepal are not known [21].

We attempted to study ecological changes in the Sal forest of Nepal as a result of interaction between inherent disturbances and ongoing management practices (for more than 20 years) in the BZCFs of Nepal. To understand responses of the woody plant species in the forests, our study aimed at elucidating (1) present condition of natural BZCFs in terms of forest community composition, distribution, density, richness, and species diversity as a function of distance from the human settlements, (2) changes in basal area (dominance) of woody species due to management and disturbance interventions from the edge to the forest interior, and (3) effect of anthropogenic disturbances and management interventions on the similarity and dissimilarities of the sites located along the distance from the human settlements. We hypothesized that forest structure differs when management interventions are applied in conjugation with the ongoing disturbances in the forests.

2. Materials and Methods

2.1. Study Area. Cross-sectional vegetation survey was conducted in three BZCFs of Parsa Wildlife Reserve (PWR) (PWR has been currently designated as a national park) at Gadimai Municipality, Central Terai Nepal, namely, Janajagaran BZCF (area: 371.10 ha), Musharni Mai BZCF (area: 231.31 ha), and Radha Krishna BZCF (area: 621.171 ha) (Figure 1). The forest in PWR was considered as a control.

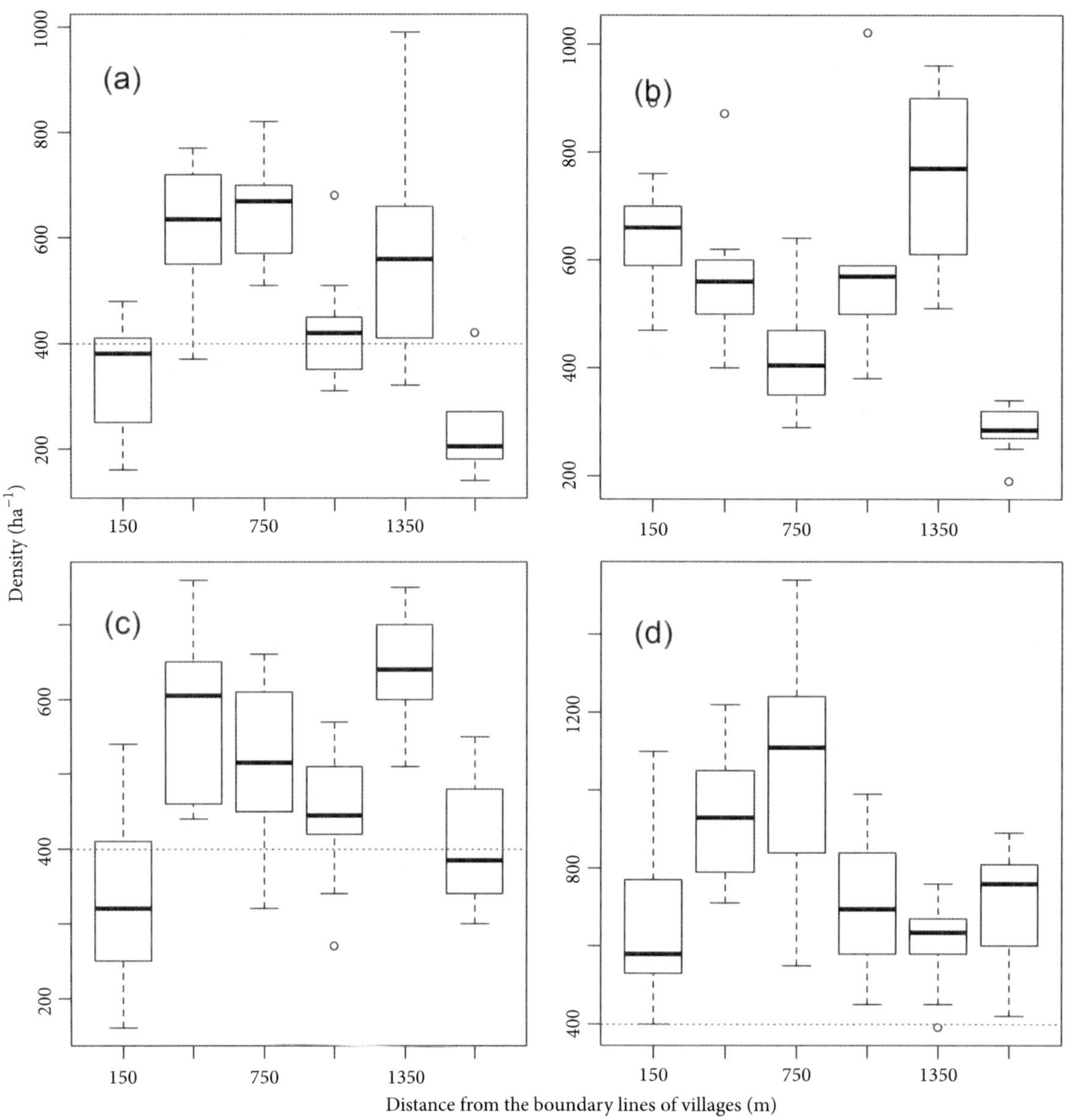

FIGURE 2: Box plot showing total density (stems ha^{-1}) of woody plants from the edge to the interior of the forests. (a) Janajagaran BZCF, (b) Musharni Mai BZCF, (c) Radha Krishna BZCF, and (d) Parsa Wildlife Reserve (control forest).

These forests are the part of the largest continuous forest landscape in Terai region, also known as "Charkose Jhadi" in Nepal. The BZCFs are situated in the southern part of PWR (Bhimeshwor BZCF is the other forest located in the southern part of PWR) on alluvial plains and composed of flood or river deposits of sand, silt, clay, coarser sediments, and conglomerates. The BZCFs are adjacent to each other with the connected boundaries between them. Since the forests are connected, the transects laid in the forests were kept at least at the distance of 1.5 km. The area harbors several protected faunal species such as Elephants (*Elephas maximus*), Royal Bengal Tiger (*Panthera tigris*), Gauri Gai (*Bos gaurus*), Wild Dogs (*Cannes aurous*), and many other common wild animals including spotted deer (*Axis axis*), barking deer (*Muntiacus muntjac*), hog deer (*Axis porcinus*), and wild boar (*Sus scrofa*). Common reptiles found in the area are golden lizard (*Calotes versicolor*), monitor lizard (*Varanus bengalensis*), krait (*Bungarus careleus*), branded krait (*Bungarus faciatus*), cobra (*Naja naja*), and common rat snake (*Ptyas mucosus*).

The climate in the area is tropical. A typical year can be divided into three main seasons: cold, hot, and rainy. April and May have the highest mean maximum temperature, while the coldest months are November, December, and January. Mean monthly minimum and maximum temperature are 18°C and 30°C, respectively. The average monthly rainfall in the area is 156 mm. Maximum rainfall occurs during the monsoon (June-September) season.

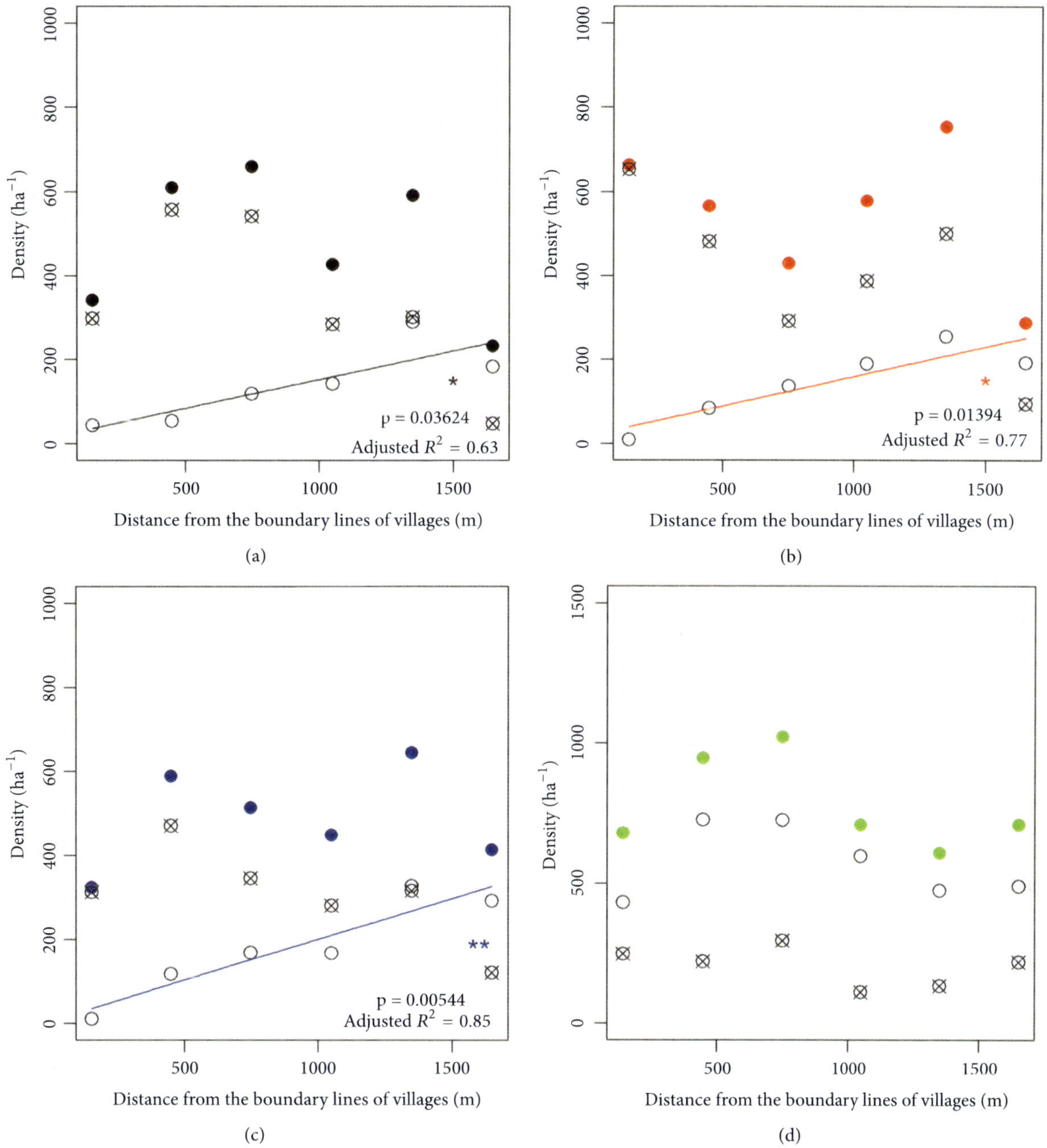

FIGURE 3: Density (stems ha^{-1}) of woody plant species as a function of distance from the village boundaries. In the figure, solid circles, hollow circles, and circles with crosses denote density of all woody species, density of species other than *Shorea robusta,* and density of *S. robusta,* respectively. For the density of stems other than *S. robusta,* the linear trend was significant for (a) Janajagaran BZCF ($p < 0.05$), (b) Musharni Mai BZCF ($p < 0.05$), and (c) Radha Krishna BZCF ($p < 0.01$), but the trend was nonsignificant for (d) Parsa Wildlife Reserve ($p > 0.05$).

2.2. Experimental Design and Sampling. A preliminary survey for the study was carried out during July 2015. Primary data collection and detail inventory of BZCFs and the protected forest inside PWR were carried out from July 2016 to October 2016. Using ArcGIS (10.2.2) and Google Earth Images (Google Earth 7.1.8.3036), an azimuth for each forest was oriented towards the forest interior, away from the corresponding settlement areas. A constant azimuth (355°, 350°, 1°, and 275° for Janajagaran BZCF, Musharni Mai BZCF, Radha Krishna BZCF, and PWR, respectively) was taken and a transect line was laid along the azimuth (Appendix A, Table 1). Similar sampling design was also adopted in other primary studies [7, 16, 17, 22].

Six plots each of 1 ha size were established on the left side of each transect at a successive interval (horizontal distance) of 300 m along the edge to the interior of each forest (150 m: Site I, 450 m: Site II, 750 m: Site III, 1050 m: Site IV, 1350 m: Site V, and 1650 m: Site VI). The farthest

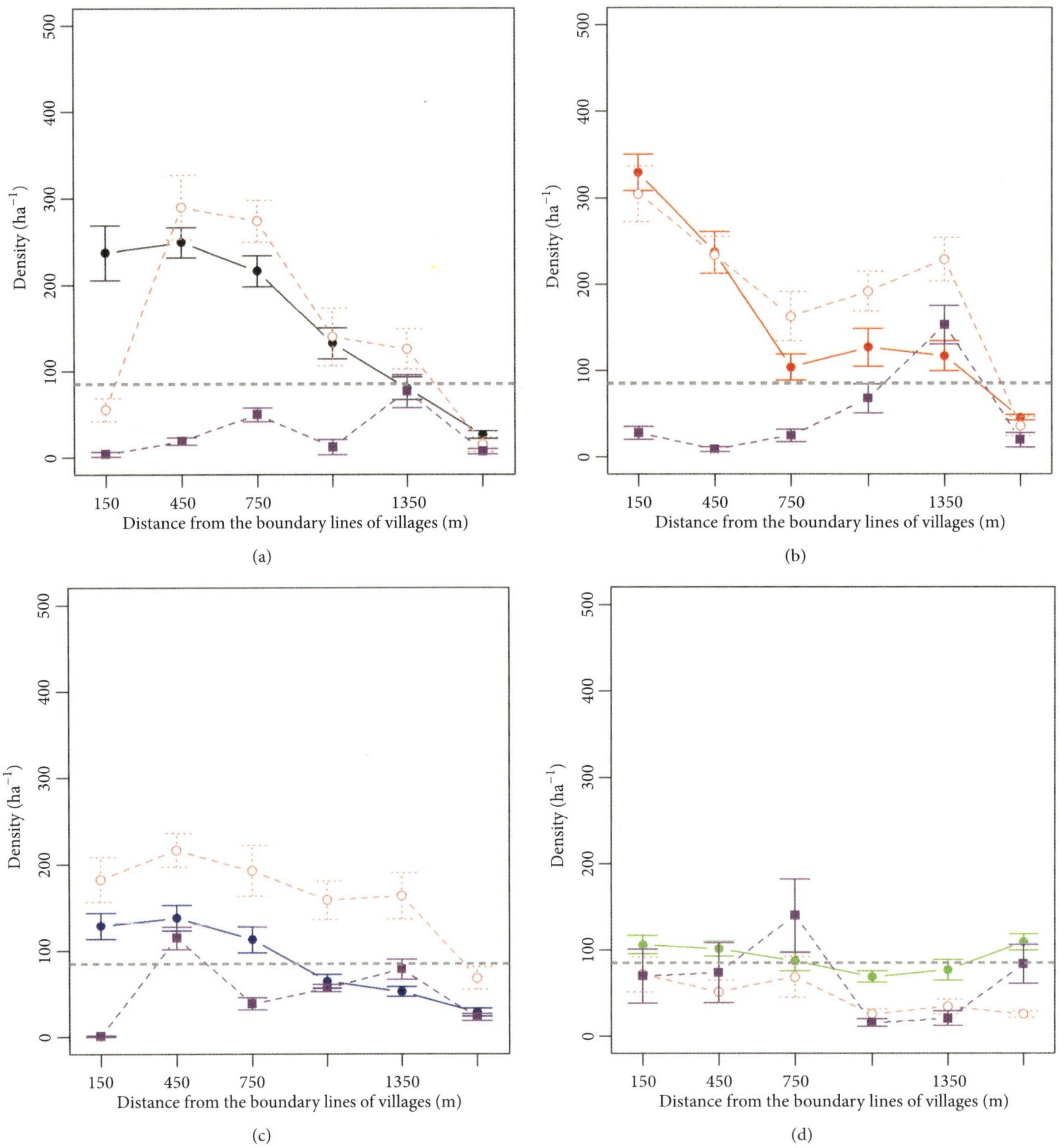

FIGURE 4: Density (stems ha^{-1}) of *Shorea robusta* (categorized as groups under saplings, poles, and trees) plotted against the distance from the village boundary lines. (a) Janajagaran BZCF, (b) Musharni Mai BZCF, (c) Radha Krishna BZCF, and (d) Parsa Wildlife Reserve. Solid lines in the figure indicate the density of the trees. Salmon colored dotted lines show the density of poles, and purple colored dotted lines show the density of saplings. The vertical lines are standard error bars. The horizontal lines are the reference density lines for the maximum number of mature trees (85 stems ha^{-1}) that 1 ha plot can withstand.

distance from the settlement (Site VI) consisted of relatively undisturbed stands (however, there were signs of illegal cutting of some individual stems). For the control forest, a transect was laid starting near the Central Office of PWR along the main entrance (fire line) towards the forest core. We used a handheld GPS (Garmin 60CSx) to determine the position of the plot-centers along the transect line (rope of 100 m was also stretched to ease transect layout). Each plot was then divided into 10 subplots each of 20 × 50 m for sampling [23]. The total number of sampling units (subplots) was sufficient to produce stabilizing species-area curves in the forests (Appendix B, Figure 13). In all

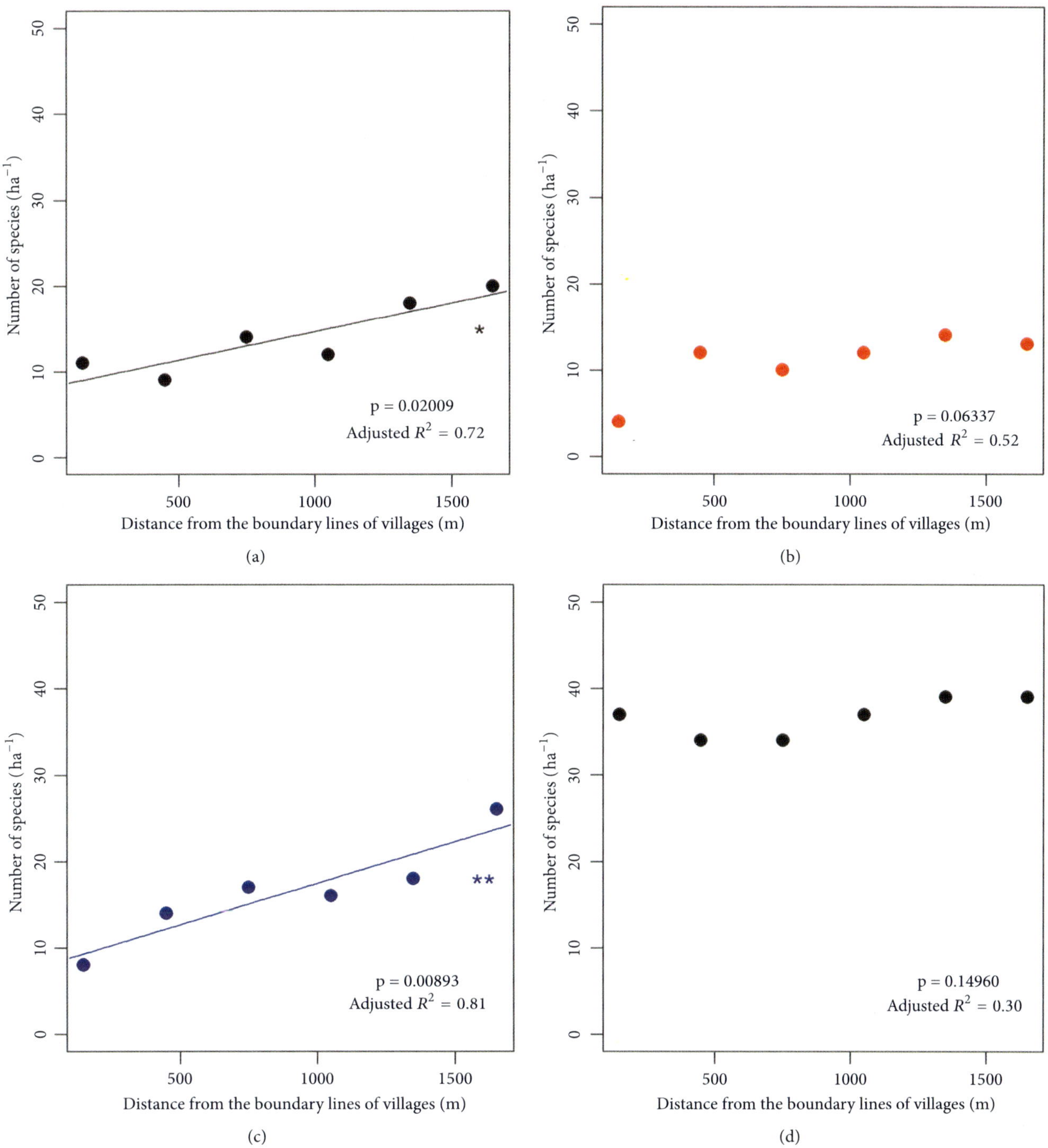

FIGURE 5: Observed species richness (species ha^{-1}) plotted against the distance from the village boundaries. (a) Janajagaran BZCF: significant ($p < 0.05$), (b) Musharni Mai BZCF: nonsignificant ($p > 0.05$), (c) Radha Krishna BZCF: significant ($p < 0.01$), and (d) Parsa Wildlife Reserve: nonsignificant ($p > 0.05$).

the studied forests, surface fire occurs at least once during the dry season of the year which usually destroys smaller seedlings, so we considered the woody species ≥ 1.5 cm in diameter at breast height (DBH) which have survived in moderate fire pressure. Structural attributes were collected from a total of 60 subplots in each studied BZCF. The woody species present were identified, counted, and tallied, and DBH of each individual ≥ 1.5 cm DBH was measured at 1.3 m above the ground with the help of a DBH tape (Kinglon Diameter Tape, Japan). Based on DBH, individuals were classified as saplings (1.5 to 10 cm DBH), poles (10 to 20 cm DBH), and trees (> 20 cm DBH). For a multistemmed tree at breast height, it was treated as a single individual whereby the diameters of all stems were measured separately and

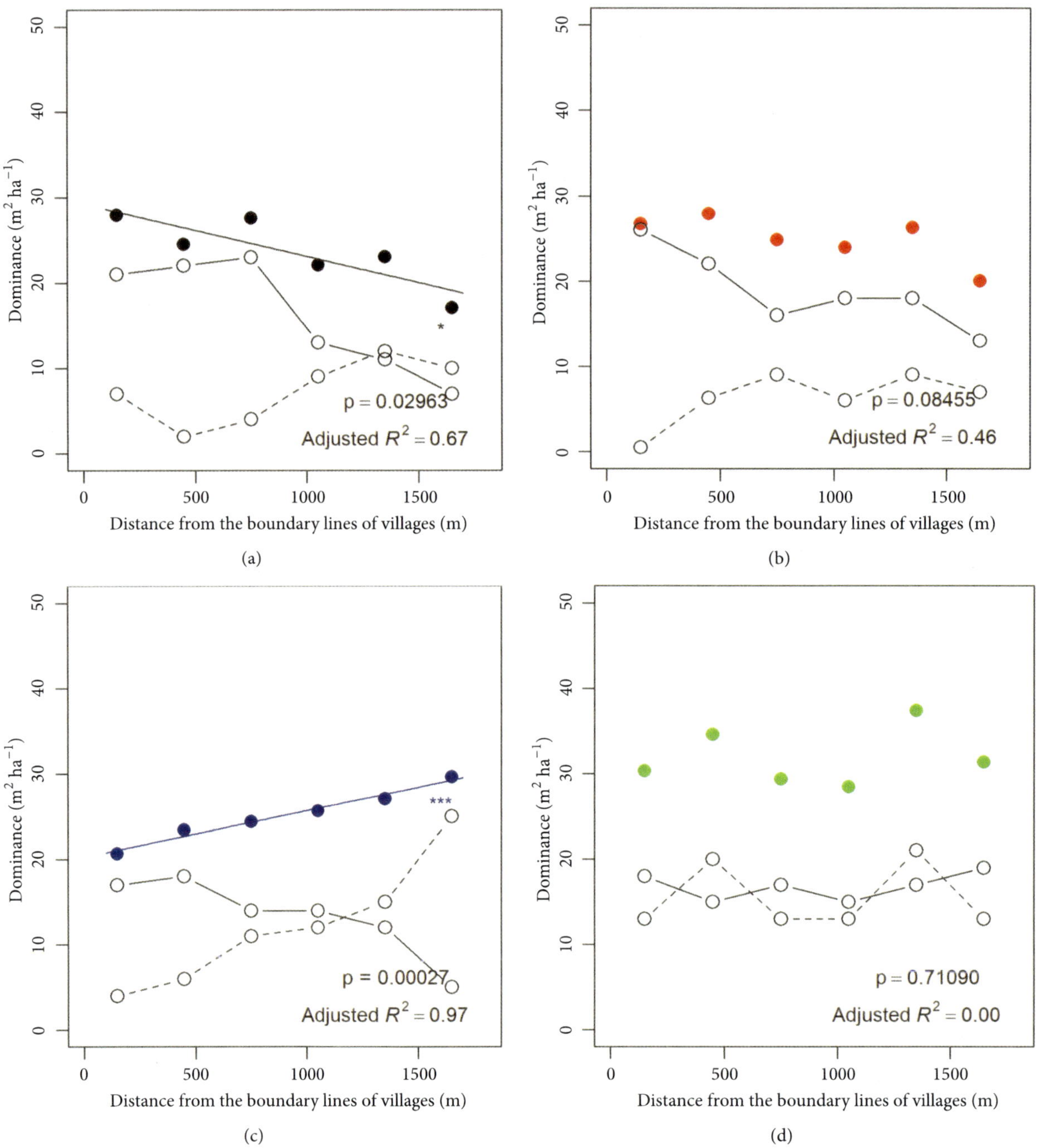

FIGURE 6: Dominance (basal area) of species as a function of the distance from the village boundaries. In the figure, solid circles with solid lines, hollow circles with solid lines, and hollow circle with dashed lines indicate total dominance, dominance of *Shorea robusta,* and dominance of other species except for *Shorea robusta,* respectively. (a) Janajagaran BZCF: significant ($p < 0.05$), (b) Musharni Mai BZCF: nonsignificant ($p > 0.05$), (c) Radha Krishna BZCF: significant ($p < 0.001$), and (d) Parsa Wildlife Reserve: nonsignificant ($p > 0.05$).

averaged. In case a tree had a buttress or an abnormality at 1.3 m height, the diameter was measured just above the abnormality.

Most plants were identified in situ; plants which were not identified to species in the field were identified by the National Herbarium and Plant Laboratories (KATH Herbarium Center) at Godavari, Lalitpur, Nepal (letter reference number 198 2073/2074). The species which could not be identified were considered as "unidentified".

2.3. Data Analysis. Basal area of live woody species (dominance) was calculated as $\Sigma \pi r^2$, where r is the radius (DBH/2) for all live woody species with DBH $\geq$ 1.5 cm. The density of a

TABLE 2: Geographical location of sampled sites∗ in the studied forests.

Forests		Site I	Site II	Site III	Site IV	Site V	Site VI
Janajagaran BZCF	Coordinates	27°09.822′N 084°57.253′E	27°10.018′N 084°57.215′E	27°10.172′N 084°57.216′E	27°10.326′N 084°57.205′E	27°10.500′N 084°57.357′E	27°10.675′N 084°57.396′E
	Elevation (m)	121	125	125	130	126	136
Musharni Mai BZCF	Coordinates	27°09.848′N 084°57.821′E	27°10.010′N 084°57.743′E	27°10.165′N 084°57.700′E	27°10.324′N 084°57.612′E	27°10.496′N 084°57.665′E	27°10.741′N 084°57.625′E
	Elevation (m)	109	113	108	118	120	135
Radha Krishna BZCF	Coordinates	27°10.334′N 084°58.101′E	27°10.525′N 084°58.103′E	27°10.678′N 084°58.121′E	27°10.821′N 084°58.132′E	27°11.022′N 084°58.126′E	27°11.203′N 084°58.105′E
	Elevation (m)	126	126	129	128	132	138
Parsa Wildlife Reserve	Coordinates	27°13.814′N 084°59.113′E	27°13.828′N 084°58.952′E	27°13.835′N 084°58.816′E	27°13.850′N 084°58.629′E	27°13.861′N 084°58.424′E	27°13.874′N 084°58.254′E
	Elevation (m)	200	199	203	200	210	202

∗Sites I, II, III, IV, V, and VI in the forests were 150 m, 450 m, 750 m, 1050 m, 1350 m, and 1650 m, respectively, far from boundary lines of villages. The gradient of increasing disturbance is from Site I to Site VI. Site I is nearer to settlements and prone to more resources (fuelwood, fodder, litter, etc.) collection as compared to Site VI. For the control forest (Parsa Wildlife Reserve), there was no such gradient of disturbances.

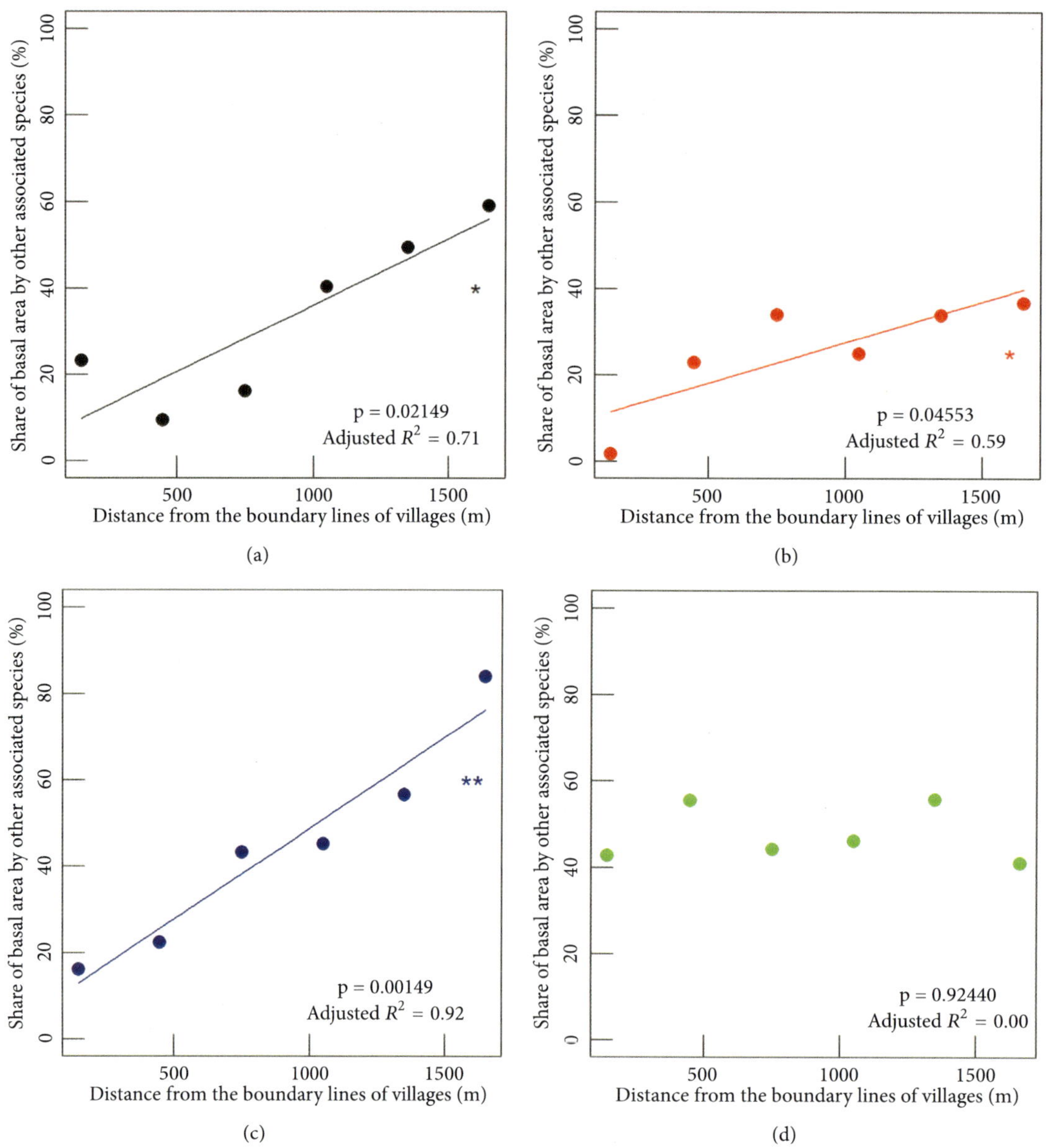

FIGURE 7: Dominance of species other than *Shorea robusta* as a function of the distance from the village boundaries. (a) Janajagaran BZCF: significant ($p < 0.05$), (b) Musharni Mai BZCF: significant ($p < 0.05$), (c) Radha Krishna BZCF: significant ($p < 0.01$), and (d) Parsa Wildlife Reserve: nonsignificant ($p > 0.05$).

TABLE 3: Observed richness, total density, and total basal area along the sites of Janajagaran BZCF.

Site	Observed richness (ha^{-1})	Density ± SD (individuals ha^{-1})	Basal area ± SD (m^2 ha^{-1})	Variance of frequency
I	11	340 ± 106.87	27.94 ± 10.99	761.82
II	9	609 ± 142.55	24.55 ± 6.66	1194.44
III	14	658 ± 97.27	27.61 ± 1.89	868.13
IV	12	426 ± 109.16	22.06 ± 2.69	1456.82
V	18	590 ± 227.50	23.02 ± 5.44	1287.91
VI	20	231 ± 78.94	17.10 ± 9.92	1231.58

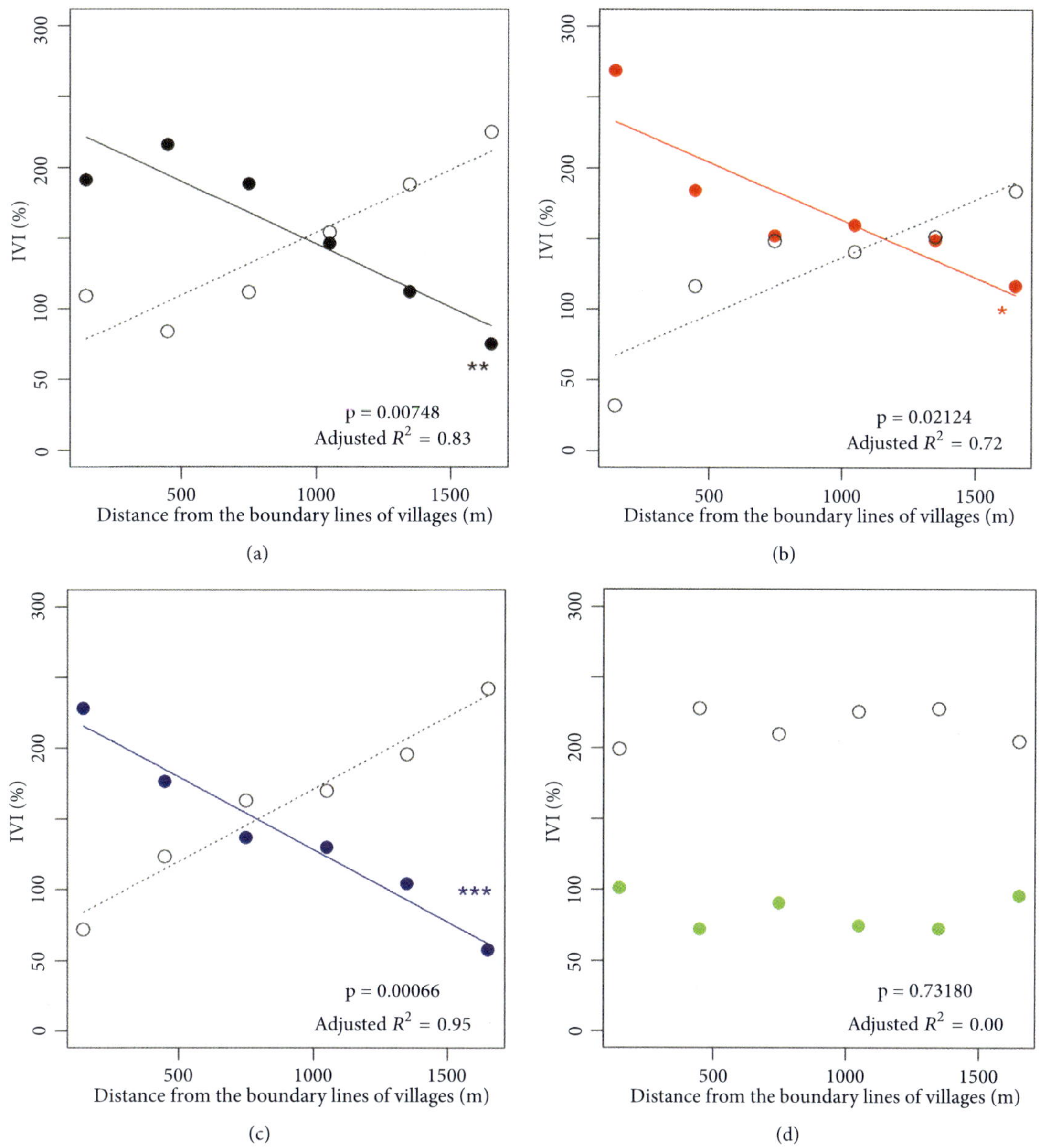

FIGURE 8: IVI (%) as a function of the distance from the village boundaries. (a) Janajagaran BZCF: significant ($p < 0.01$), (b) Musharni Mai BZCF: significant ($p < 0.05$), (c) Radha Krishna BZCF: significant ($p < 0.001$), and (d) Parsa Wildlife Reserve: nonsignificant ($p > 0.05$). The solid lines in the figure indicate IVI (%) of *S. robusta* and the dotted lines indicate total IVI (%) of other species except for *S. robusta*.

TABLE 4: Observed richness, total density, and total basal area along the sites of Musharni Mai BZCF.

Site	Observed richness (ha^{-1})	Density ± SD (individuals ha^{-1})	Basal area ± SD (m^2 ha^{-1})	Variance of frequency
I	4	662 ± 113.12	26.73 ± 5.53	1900
II	12	564 ± 131.17	27.92 ± 7.70	953.79
III	10	428 ± 109.52	24.87 ± 6.11	1262.22
IV	12	577 ± 169.84	23.94 ± 6.74	1384.09
V	14	753 ± 158.47	26.29 ± 4.42	1241.76
VI	13	286 ± 45.019	20.03 ± 6.52	1560.26

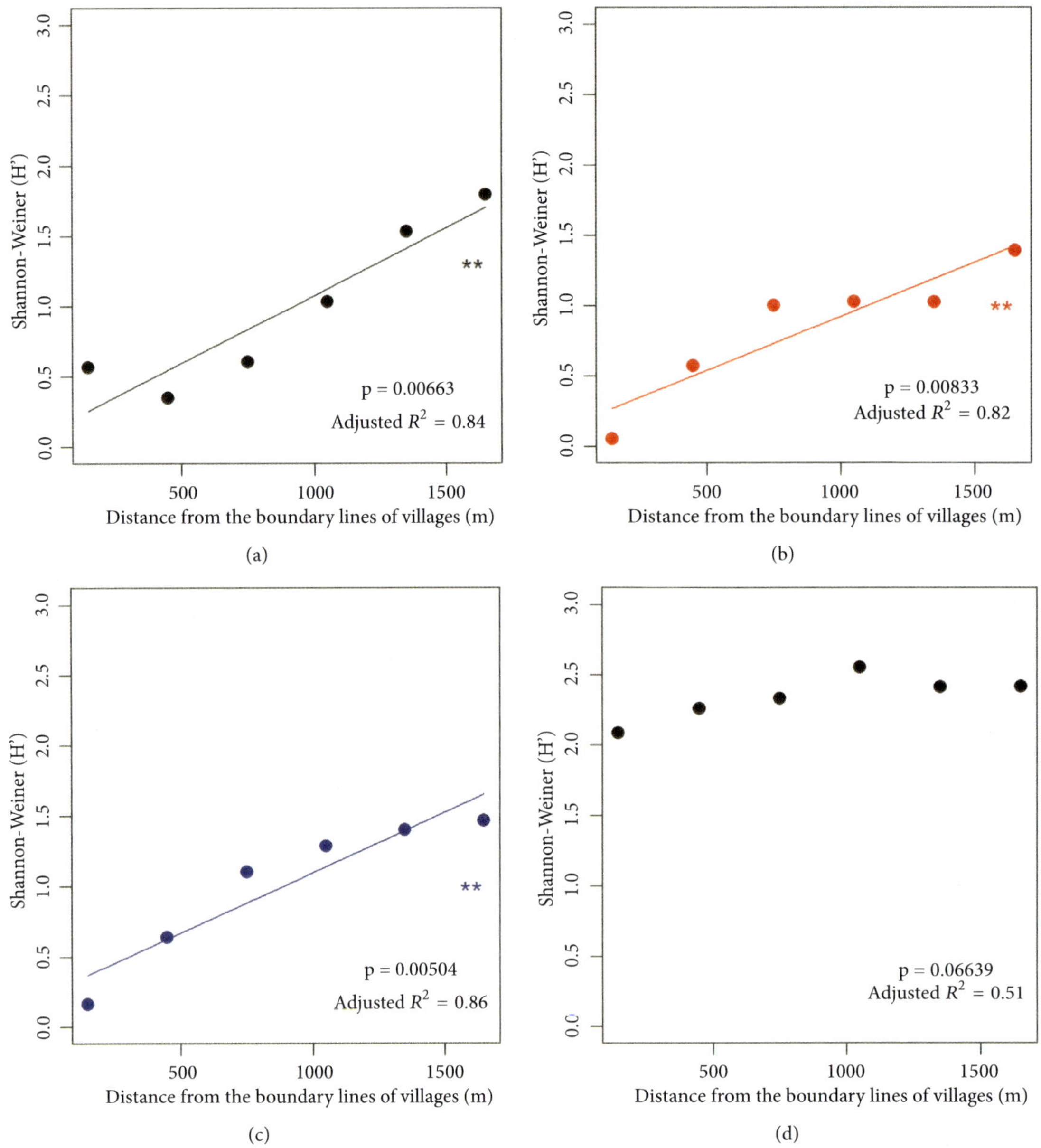

FIGURE 9: Shannon-Wiener H' diversity index as a function of the distance from the village boundaries. (a) Janajagaran Forest: significant ($p < 0.01$), (b) Musharni Mai Forest: significant ($p < 0.01$), (c) Radha Krishna Forest: significant ($p < 0.01$), and (d) Parsa Wildlife Reserve: nonsignificant ($p > 0.05$).

TABLE 5: Observed richness, total density, and total basal area along the sites of Radha Krishna BZCF.

Site	Observed richness (ha^{-1})	Density ± SD (individuals ha^{-1})	Basal area ± SD (m^2 ha^{-1})	Variance of frequency
I	8	323 ± 112.16	20.69 ± 6.74	942.86
II	14	587 ± 110.66	23.46 ± 4.53	965.39
III	17	513 ± 107.61	24.44 ± 4.25	1127.94
IV	16	448 ± 93.07	25.67 ± 6.00	1179.58
V	18	643 ± 71.81	27.09 ± 6.86	1379.41
VI	26	413 ± 89.20	29.68 ± 10.50	877.54

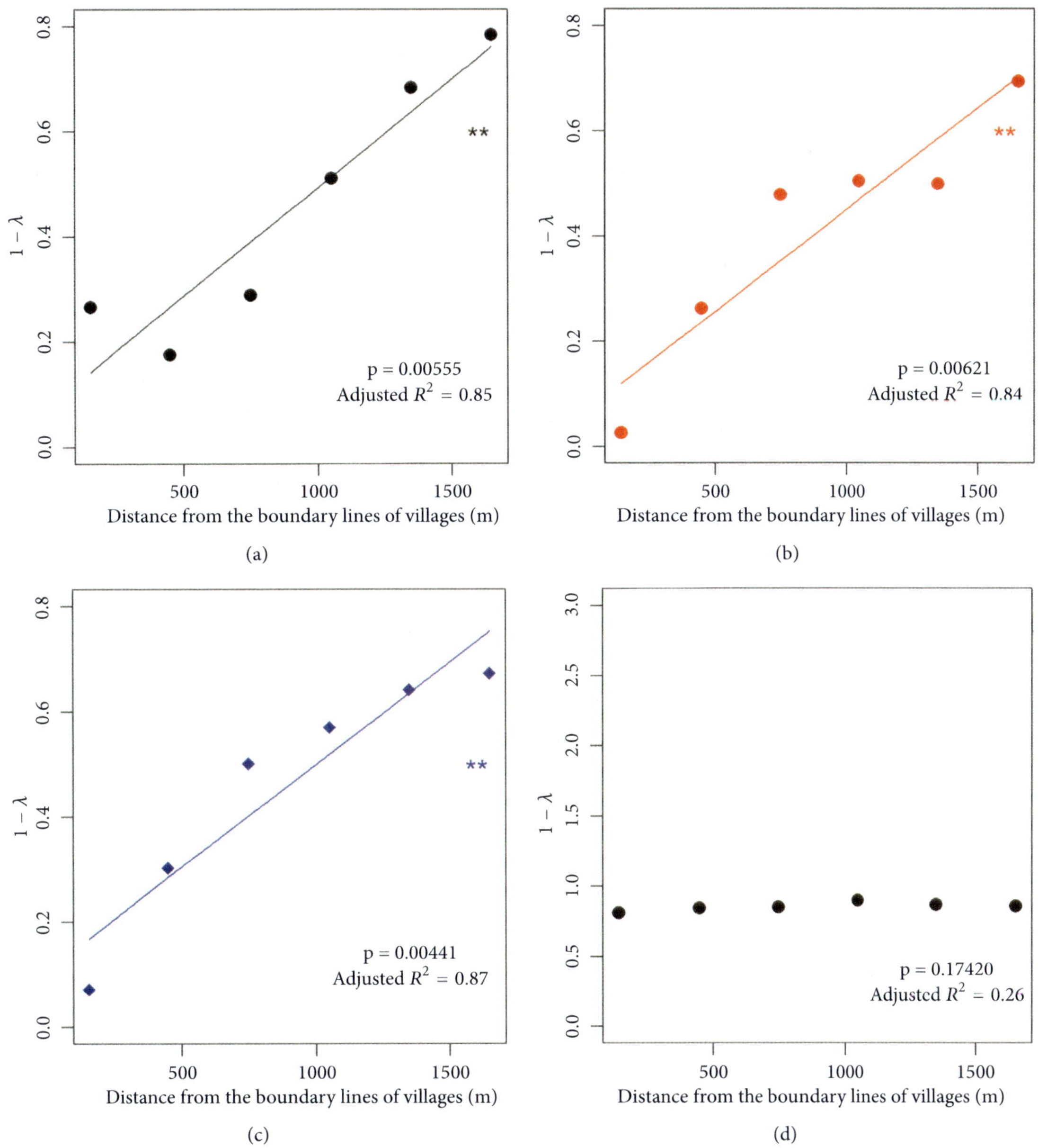

FIGURE 10: 1 – Simpson's λ ($1 - \sum pi^2$) plotted against the distance from the village boundaries. (a) Janajagaran Forest: significant ($p < 0.01$), (b) Musharni Mai Forest: significant ($p < 0.01$), (c) Radha Krishna Forest: significant ($p < 0.01$), and (d) Parsa Wildlife Reserve: nonsignificant ($p > 0.05$).

TABLE 6: Observed richness, total density, and total basal area along the sites of Parsa Wildlife Reserve.

Site	Observed richness (ha^{-1})	Density ± SD (individuals ha^{-1})	Basal area ± SD (m^2 ha^{-1})	Variance of frequency
I	37	679 ± 250.31	30.34 ± 6.42	1001.87
II	34	946 ± 182.04	34.59 ± 9.71	1208.91
III	34	1021 ± 311.43	29.40 ± 6.18	1109.80
IV	37	707 ± 160.56	28.46 ± 8.19	1091.59
V	39	607 ± 112.45	37.38 ± 8.60	968.15
VI	39	708 ± 139.43	31.34 ± 6.61	960.46

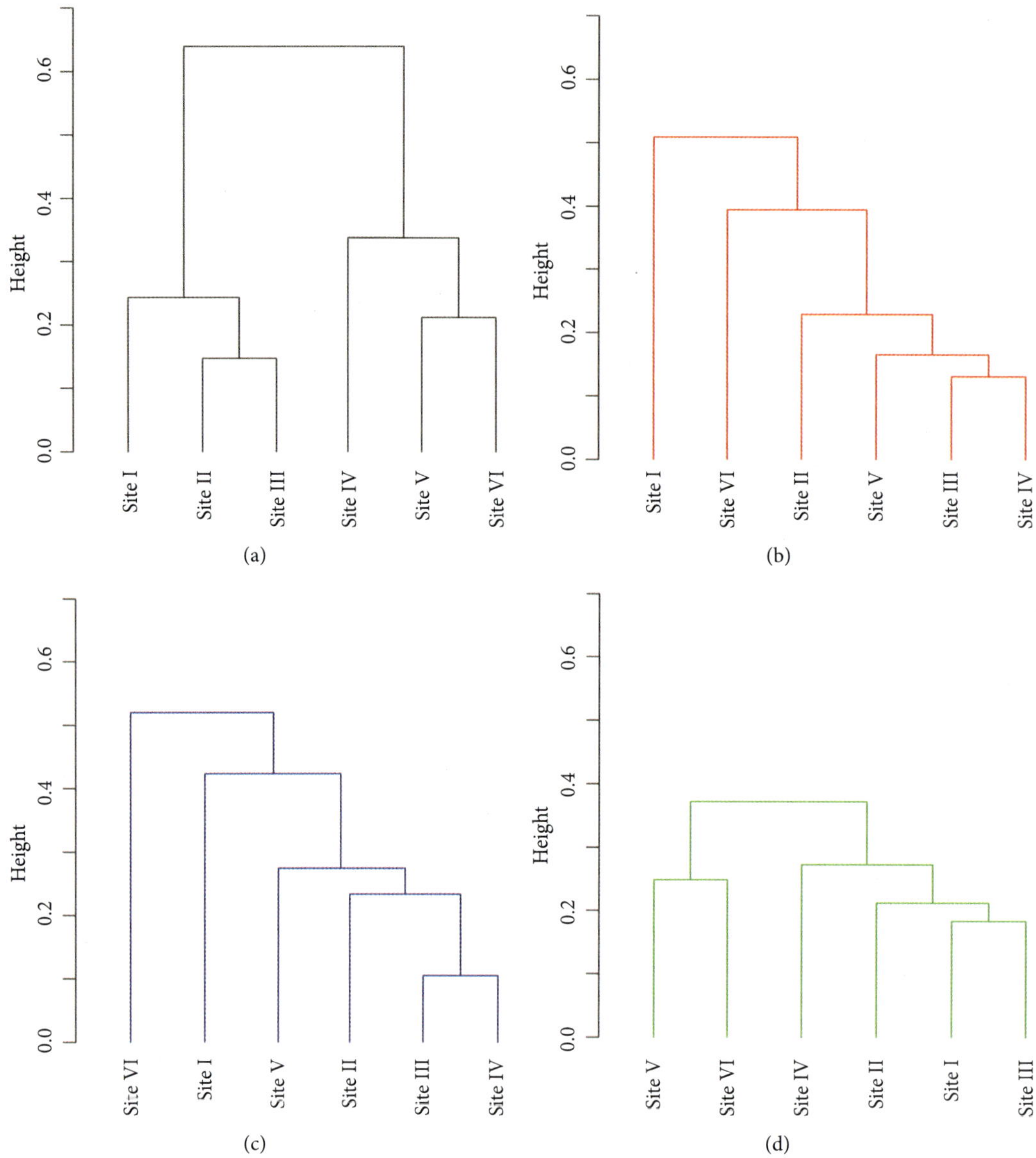

FIGURE 11: Multivariate agglomerative clustering (Ward D.2 group linkage) based on Bray-Curtis distance (species IVI dissimilarities) between the studied sites. (a) Janajagaran BZCF, (b) Musharni Mai BZCF, (c) Radha Krishna BZCF, and (d) Parsa Wildlife Reserve.

species was calculated as the number of individuals of species per unit area. Similarly, the estimated maximum density of mature *S. robusta* trees that 1 ha forest area can support was calculated with the mean value of the area covered by canopy projections of mature trees in natural growth conditions and considering that there will be no overlap of canopy between individuals of the species. The canopy projections (length) over cardinal and ordinal directions were measured based on the observation made vertically upwards and recorded as to whether or not the canopy of the trees obscures the sky [24]. The canopy cover area was calculated based on horizontal projections data and using readWKT and gArea functions in the rgeos package [25] and then was compared with 1 ha area.

The Importance Value Index (IVI) for each woody species was determined to evaluate the contribution of woody species in the forest composition, structure, and biomass. The IVI for a species was calculated by summing the relative frequency, relative density, and relative dominance [26–28]:

$$IVI_j = 100\left(\frac{n_j}{N} + \frac{d_j}{D} + \frac{x_j}{X}\right) \quad (1)$$

where IVI_j is the importance value of the jth species, n_j is the frequency of jth species, N is the sum of frequencies ($N = \sum n_j$), d_j is the number of individuals of the jth species present in the sample population, D is the total number of individuals in the sample population ($D = \sum d_j$), x_j is the sum of dominance for the jth species, and X is the total dominance across all species $X = \sum x_j$. Since relative frequency, relative density, and relative dominance are expressed in percentage, the maximum value of IVI is equal to 300.

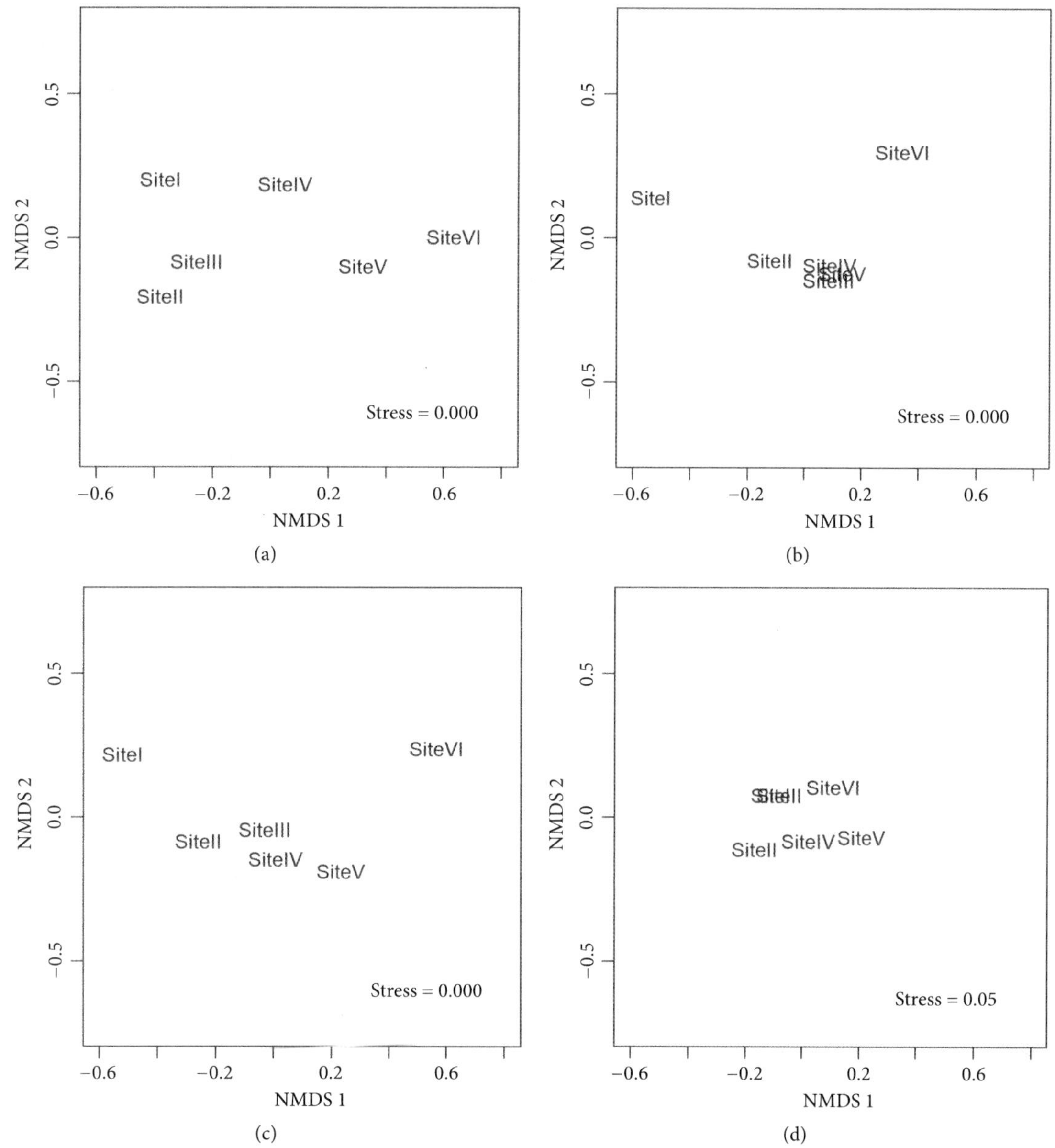

FIGURE 12: Nonmetric Multidimensional Scaling of the sites for (a) Janajagaran BZCF, (b) Musharni Mai BZCF, (c) Radha Krishna BZCF, and (d) Parsa Wildlife Reserve.

Species diversity (alpha diversity) was calculated by using Shannon-Wiener index [29] and it is defined as

$$H = -\Sigma pi \log (b)\, pi \qquad (2)$$

where pi is the proportional abundance of a species i and b is the base of the logarithm. The diversity function in the vegan package uses natural logarithms as the default while calculating Shannon-Wiener index, and 1– Simpson's alpha (which also describes overall alpha diversity) is calculated as 1 $-\Sigma pi^2$ [30]. All analyses for determining diversity (Shannon-Wiener, Simpson, and species accumulation curves) were performed using the diversity function in the vegan package [31].

To observe the functional association between forest community structure and distance from boundary lines of villages, a linear model was fit wherever applicable with the null hypothesis that there is no linear pattern between the structural attributes of forests and the distance from human settlements. Linear model assumptions were tested using the gvlma function in the gvlma package [32]. Residuals versus fitted and Q-Q plots were visualized.

Morisita's I_δ index of dispersion was used for determining the spatial pattern of individuals of the predominant woody species. I_δ, which is independent of sample size and diversity [33], was computed as follows [34]:

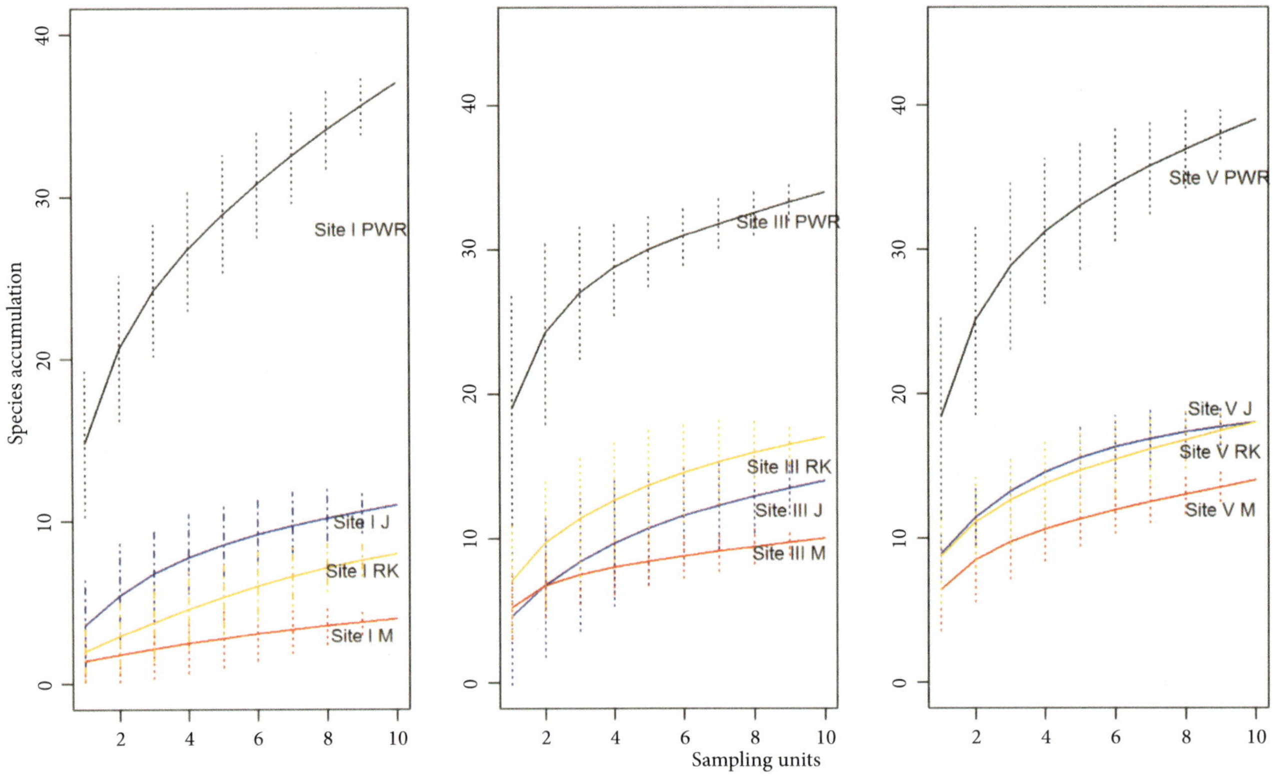

FIGURE 13: Species accumulation curves for different sites of studied forests plotted using Kindt's exact method. The curves are for Sites I, Sites III, and Sites V for Janajagaran (J), Musharni Mai (M), Radha Krishna (RK), and Parsa Wildlife Reserve (PWR) forests. The species accumulation curves reflected increase in number of species in the beginning with addition of sampling units. The vertical lines represent the standard deviations.

$$I_{\delta=}q\frac{\sum x_i (x_i - 1)}{N (N - 1)} \quad (3)$$

where q is the total number of subplots, N is the total number of individuals of the species in q subplots, and x is the number of individuals of a species in a subplot. Species were considered predominant if IVI > 10 observed in any one of the sites of the forests. We used dispindmorisita function in the vegan package to calculate intraspecific aggregation [31].

To examine the dissimilarity and similarity in species composition between the sites in the forests, Bray-Curtis Index (BCI) was used. BCI is a rank-order index and values were used to determine a standardized species score to combine the measurements of IVI in the woody plant species. The values of BCI range from 0 to 1. The value 1 reflects no similarity whereas 0 reflects complete similarity between plots. BCI of dissimilarity was calculated between the sites (from the edge to the interior of the forests) based on the IVI values determined for each species. Vegdist and hclust functions in the vegan package were used for determining dissimilarity and clustering (Ward.D2 group linkage), respectively [31]. Dissimilarity (DIS_{jk}) between sites j and k using BCI can be expressed as follows [35]:

$$DIS_{jk} = \frac{\sum_i^p abs\left(y_{ij} - y_{ik}\right)}{\sum_i^p \left(y_{ij} + y_{ik}\right)} \quad (4)$$

where y_{ij} and y_{ik} represent measures of species i in the sample sites j and k and abs indicates absolute value. To compare the dissimilarity in the forest community between sites, a matrix representation of the DIS_{jk} and the similarity ($S_{jk} = 1 - DIS_{jk}$) between the paired sites were tabulated and compared. A multivariate agglomerative clustering using Ward.D2 group linkage based on the Bray-Curtis distance of IVI values was performed.

We also determined the associated plant community composition and similarity between sites based on IVI of the species using Nonmetric Multidimensional Scaling (NMDS) ordination to collapse information from multiple species into just a few, so that they can be visualized and interpreted. In community ecology, NMDS is commonly regarded as the most robust unconstrained ordination method [36] and considered as an indirect gradient analysis approach for producing an ordination based on a distance or dissimilarity matrix. We chose Bray-Curtis distance measure and a maximum of 200 iterations for NMDS distance matrix. The analysis was formed with the metaMDS function in the vegan package [31]. To minimize the stress (usually the stress values more than 0.2 should be interpreted with caution and more than 0.3 are highly suspicious), NMDS was projected in K=3 dimensions (resulting in stress value for all our analysis less than 0.05; Shepard stress plots were visualized). All statistical analyses were performed in R [37].

TABLE 7: IVI of species in Janajagaran BZCF.

Species code	Name of the species	Site I	Site II	Site III	Site IV	Site V	Site VI
1	*Shorea robusta* Gaertn.	190.93	216.11	188.25	146.09	112.11	75.00
2	*Adina cordifolia* (Willd. ex Roxb.) Benth. & Hook. f. ex Brandis	32.81	4.62	8.21		9.47	
3	*Lannea coromandelica* (Houtt.) Merr.	14.96		3.84	21.80	4.44	9.96
4	*Lagerstroemia parviflora* Roxb.	10.61	35.60	32.71	41.19	36.87	36.11
5	*Schleichera oleosa* (Lour.) Oken	3.77		2.33	2.58	1.70	7.76
6	*Terminalia bellirica* (Gaertn.) Roxb.	5.18					3.07
7	*Careya arborea* Roxb.	4.36		3.20		7.56	7.18
8	*Casearia graveolens* Dalzell	13.19	7.39	10.40	13.16	25.42	28.89
9	*Cassia fistula* L.	7.62	3.78	12.87	2.44	2.76	1.69
10	*Mallotus philippensis* (Lam.) Müll.-Arg.	13.50	3.64	10.39	38.33	23.52	30.79
11	*Aegle marmelos* (L.) Correa	3.07		2.33			
12	*Dillenia pentagyna* Roxb.		5.24	6.31	22.48	24.16	28.92
13	*Syzygium cumini* (L.) Skeels		20.00	10.11		5.90	5.50
14	*Dysoxylum gobara* (Buch.-Ham.) Merr.		3.61				
15	*Semecarpus anacardium* L.f.			4.79		11.79	6.22
16	*Hymenodictyon excelsum* Wall.			4.26			
17	*Albizia* sps.				3.64		
18	*Miliusa velutina* (Dunal) Hook. f. & Thoms.				2.44	6.95	27.50
19	*Ehretia laevis* Roxb.				3.42	2.00	
20	*Cleistocalyx operculatus* (Roxb.) Meer. & Perry				2.45		
21	*Terminalia alata* Heyne ex Roth					17.03	15.13
22	*Terminalia chebula* Retz.					2.60	1.74
23	*Buchanania latifolia* Roxb.					1.30	
24	*Sterculia villosa* Roxb.					4.41	1.70
25	*Ficus religiosa* L.						2.41
26	*Anogeissus latifolius* (Roxb. ex DC.) Bedd.						6.03
27	*Dalbergia latifolia* Roxb.						2.15
28	*Wendlandia tinctoria* (Roxb.) DC.						2.26

3. Results

3.1. Community Structure and Distribution of Woody Plants. The studied BZCFs and PWR are *S. robusta* dominated mixed forests. The forests exhibit high total densities of woody plants (saplings, poles, and trees) but also high variance within forest sites (Figure 2). Total density (± SD) of woody species in Janajagaran BZCF ranged from 231±79 stems ha^{-1} to 658±97 stems ha^{-1} varying across different distances towards the forest interior from the boundary lines of the villages. Similarly, Musharni Mai BZCF and Radha Krishna BZCF also showed the wide range of variation in stem density (± SD) from 286±45 to 753±159 stems ha^{-1} and 323±112 stems ha^{-1} to 643±71.81 stems ha^{-1}, respectively. In PWR, the range of variation of densities (± SD) was from 607±113 to 1021±311 stems ha^{-1} (Appendix C, Tables 3–6). The median

Table 8: IVI of species in Musharni Mai BZCF.

Species code	Name of the species	Site I	Site II	Site III	Site IV	Site V	Site VI
1	*Shorea robusta* Gaertn.	268.26	183.83	151.92	159.47	148.73	116.36
2	*Adina cordifolia* (Willd. ex Roxb.) Benth. & Hook. f. ex Brandis	9.08	3.72	4.49		2.78	6.33
9	*Cassia fistula* L.	7.29	9.59	2.16	1.93	6.93	
8	*Casearia graveolens* Dalzell	15.36	22.60	29.91	30.03	34.10	20.51
12	*Dillenia pentagyna* Roxb.		22.19	25.87	15.06	15.47	17.52
13	*Syzygium cumini* (L.) Skeels		2.90	30.90	8.34		
20	*Cleistocalyx operculatus* (Roxb.) Meer. & Perry		9.80	3.10	2.04		
21	*Terminalia alata* Heyne ex Roth		7.16		9.47		2.75
4	*Lagerstroemia parviflora* Roxb.		21.11	24.64	34.56	25.39	54.35
17	*Albizia* sps.		4.98				
3	*Lannea coromandelica* (Houtt.) Merr.		7.34	12.16	11.88	18.46	
10	*Mallotus philippensis* (Lam.) Müll.-Arg.		4.79	14.85	19.66	23.20	48.78
16	*Hymenodictyon excelsum* Wall.				3.32	2.22	
14	*Dysoxylum gobara* (Buch.-Ham.) Merr.				4.24		2.70
7	*Careya arborea* Roxb.					2.27	
15	*Semecarpus anacardium* L.f.					2.06	2.10
5	*Schleichera oleosa* (Lour.) Oken					6.64	
29	*Ficus benghalensis* Linn.					2.88	
18	*Miliusa velutina* (Dunal) Hook. f. & Thoms.					8.88	2.46
26	*Anogeissus latifolius* (Roxb. ex DC.) Bedd.						18.75
56	*Pyrus pashia* Buch.- Ham. ex D.Don						4.47
31	UnidentifiedI						2.94

stem density was more than 400 individuals ha^{-1} (arbitrary value) in each site of PWR, but oscillations were observed across the sites in BZCFs. The variance of frequency indicates the abundance difference of woody species, within a site; high values indicate that the abundance of species in the plot differs from each other, whereas low values indicate that all the species have a similar pattern of abundance (Appendix C, Tables 3–6).

Significant pattern ($p > 0.05$) was not observed between the total density of species and the distance from the boundary lines of the villages in all the BZCFs (Figure 3). When stem densities were separated into the density of the dominant species *S. robusta* and the density of other woody species, a significant increasing linear trend was obtained between the density of other species (except *S. robusta*) and the distance from the boundary lines of the settlement areas (Figure 3) (Janajagaran BZCF: $p < 0.05$, Musharni Mai BZCF: $p < 0.05$, and Radha Krishna BZCF: < 0.01, but no significant difference was observed for PWR: $p > 0.05$). Similarly, the assemblage of *S. robusta,* when categorized as saplings, poles, and trees, reflected that the poles and trees had a high density near the human settlement areas in BZCFs (Figure 4). The BZCFs near the human settlements were receiving intense management and protection of the poles and tree-sized individuals of *S. robusta*. Densification has occurred in the area with very fewer plant species. The density of trees of *S. robusta* in the edges was very high above the estimated maximum density (85 stems ha^{-1}) of fully grown mature trees of *S. robusta* which 1 ha forest area can support. For PWR such difference in the density of *S. robusta* along the edge to the interior was not observed (Figure 4).

Table 9: IVI of species in Radha Krishna BZCF.

Species code	Name of the species	Site I	Site II	Site III	Site IV	Site V	Site VI
1	*Shorea robusta* Gaertn.	228.32	176.61	137.14	130.12	104.46	57.57
2	*Adina cordifolia* (Willd. ex Roxb.) Benth. & Hook. f. ex Brandis	7.39	10.17	14.76	8.87	4.61	10.04
3	*Lannea coromandelica* (Houtt.) Merr.	13.31	3.03	13.96	13.69	8.03	5.93
21	*Terminalia alata* Heyne ex Roth	19.56	7.36	16.06	9.72	10.51	19.41
12	*Dillenia pentagyna* Roxb.	6.81	15.13	22.69	21.04	17.26	23.97
6	*Terminalia bellirica* (Gaertn.) Roxb.	8.66	8.79	5.17	10.26	18.22	19.90
4	*Lagerstroemia parviflora* Roxb.	10.63	35.82	27.53	27.49	27.99	17.11
8	*Casearia graveolens* Dalzell	5.32	16.33	12.68	15.43	19.35	14.37
32	Baisake (Local name: Central Terai, Nepal)		2.38			1.31	
7	*Careya arborea* Roxb.		2.94	4.16	1.75	8.83	5.61
5	*Schleichera oleosa* (Lour.) Oken		2.55	4.13	5.92	6.10	3.18
10	*Mallotus philippensis* (Lam.) Müll.-Arg.		14.45	26.63	29.17	49.90	84.95
33	*Sapium insigne* (Royle) Benth. ex Hook. f.		2.22				
9	*Cassia fistula* L.		2.22	1.61		1.31	
18	*Miliusa velutina* (Dunal) Hook. f. & Thoms.			2.01	8.46	15.56	1.98
15	*Semecarpus anacardium* L. f.			5.27	11.14		1.47
26	*Anogeissus latifolius* (Roxb. ex DC.) Bedd.			2.30			
13	*Syzygium cumini* (L.) Skeels			2.29			1.64
34	*Lagerstroemia reginae* Roxb.			1.61			1.75
19	*Ehretia laevis* Roxb.				1.91		1.84
25	*Ficus religiosa* L.				3.51		1.89
35	*Madhuca longifolia* (Koeing) Macbride				1.52		
36	*Duabanga grandiflora* (Roxb. ex DC.) Walp.					2.62	
37	*Grewia eriocarpa* Juss.					1.31	
38	*Litsea monopetala* (Roxb.) Pers.					1.31	5.63
39	Unidentified II					1.31	
40	*Bombax ceiba* L.						6.34
17	*Albizia* sps.						3.57
14	*Dysoxylum gobara* (Buch.-Ham.) Merr.						1.64
41	*Toona ciliata* M. Roem.						3.67
42	*Trewia nudiflora* Linn.						1.56
43	*Premna barbata* Wall.						2.05
27	*Dalbergia latifolia* Roxb.						1.47
44	*Croton* sp.						1.46

Table 10: IVI of species in Parsa Wildlife Reserve.

Species code	Name of the species	Site I	Site II	Site III	Site IV	Site V	Site VI
8	*Casearia graveolens* Dalzell	16.20	18.69	12.30	15.48	4.96	4.95
4	*Lagerstroemia parviflora* Roxb.	28.03	35.81	23.25	18.65	21.96	15.84
45	*Bridelia retusa* (L.) Spreng.	0.85	3.15	6.74	4.04	1.81	3.51
18	*Miliusa velutina* (Dunal) Hook. f. & Thoms.	12.59	21.44	14.03	13.12	24.42	17.11
46	*Phyllanthus emblica* Linn.	6.00	1.19	4.30	13.76	5.99	6.46
15	*Semecarpus anacardium* L.f.	0.91				0.74	0.69
22	*Terminalia chebula* Retz.	8.02	6.92	6.03	0.99	0.82	
7	*Careya arborea* Roxb.	3.63	2.15	3.26	1.58	5.89	0.66
1	*Shorea robusta* Gaertn.	101.09	72.09	90.43	74.39	72.41	95.39
27	*Dalbergia latifolia* Roxb.	7.96	9.03	10.84	6.87	5.94	6.92
12	*Dillenia pentagyna* Roxb.	13.24	19.78	15.85	12.75	6.58	4.57
21	*Terminalia alata* Heyne ex Roth	13.69	15.55	11.17	10.55	12.71	13.78
2	*Adina cordifolia* (Willd. ex Roxb.) Benth. & Hook. f. ex Brandis	10.12	1.66	5.00	8.17	9.97	13.95
13	*Syzygium cumini* (L.) Skeels	5.67	12.24	7.78	0.99	0.81	2.51
47	*Desmodium oojenense* (Roxb.) Ohashi	15.74	12.51	5.83	24.32	13.45	13.93
26	*Anogeissus latifolius* (Roxb. ex DC.) Bedd.	8.07	7.92	4.87	13.34	4.44	10.80
48	*Garuga pinnata* Roxb.	6.97	8.47	15.50	7.70	11.72	13.53
3	*Lannea coromandelica* (Houtt.) Merr.	0.83	2.63	3.19	3.51	0.86	11.21
23	*Buchanania latifolia* Roxb.	10.68	9.14	12.54	11.12		0.67
19	*Ehretia laevis* Roxb.	0.83			5.32	3.01	3.82
54	*Lyonia ovalifolia* (Wall.) Drude	5.65	3.72	8.81	3.16	4.34	5.01
49	*Stereospermum personatum* (Hassk.) Chatterjee	2.83	3.92	3.30	7.07	2.84	0.67
10	*Mallotus philippensis* (Lam.) Müll.-Arg.	0.84				7.90	1.31
33	*Sapium insigne* (Royle) Benth. ex Hook. f.	1.69	3.44	4.60	3.40	6.36	2.05
50	*Bauhinia malabarica* Roxb.	0.83			0.69		3.21
51	*Bauhinia purpurea* L.	3.71	5.88	4.28	1.38	3.24	5.75
20	*Cleistocalyx operculatus* (Roxb.) Meer. & Perry	1.77	2.25				
17	*Albizia* sps.	0.96		3.04	4.67	3.77	2.52
52	Putalikath (Local name: Central Terai, Nepal)	1.82	5.84	0.66	0.76		1.36
14	*Dysoxylum gobara* (Buch.-Ham.) Merr.	1.95	1.62			22.96	4.36
5	*Schleichera oleosa* (Lour.) Oken	0.86			2.84	5.28	3.10
6	*Terminalia bellirica* (Gaertn.) Roxb.	0.98	2.20	2.99	3.66	6.12	1.32
55	*Tamarindus indica* L.		0.69				

TABLE 10: Continued.

Species code	Name of the species	Site I	Site II	Site III	Site IV	Site V	Site VI
56	*Pyrus pashia* Buch.- Ham. ex D.Don		0.71				
37	*Grewia eriocarpa* Juss.	1.70	4.94	5.94	6.93	6.99	6.44
16	*Hymenodictyon excelsum* Wall.	0.83	0.71	1.27	4.18	0.77	5.90
57	*Bauhinia variegata* L.		1.45	0.63			
58	*Ficus racemosa* L.		0.69			0.73	
9	*Cassia fistula* L.	0.82	0.68	0.75			
53	*Bauhinia vahlii* Wight & Arn	0.83	0.89		4.36		0.67
59	*Spatholobus parviflorus* (Roxb.) Kuntze			0.95			
60	*Nyctanthes arbor-tristis* Linn.			7.78			
61	*Viscum album* Linn.			0.66	1.71	3.15	6.26
62	*Gmelina arborea* Roxb.			0.79	4.60	2.40	1.77
63	*Antidesma acidum* Retz.			0.64		1.43	2.13
64	*Xeromphis spinosa* (Thunb) Keay				0.69		1.99
28	*Wendlandia tinctoria* (Roxb.) DC.				1.74	0.73	2.25
65	*Millettia extensa* (Benth.) Baker				0.81		
66	*Melia azedarach* Linn.				0.73		
67	*Holarrhena pubescens* (Buch.-Ham.) Wall. ex G. Don					1.15	1.66
40	*Bombax ceiba* L.					9.79	
68	*Alstonia scholaris* (L.) R. Br.					0.79	
69	*Cornus oblonga* (Wall.) Sojak					0.74	
34	*Lagerstroemia reginae* Roxb.	0.82					

We found a total of 69 woody species belonging to 30 families in all the sample plots of the studied forests. Plant species belonging to the Fabaceae family were the most abundant (12) followed by those belonging to the families Euphorbiaceae (7), Combretaceae (4), and Rubiaceae (4). The number of species varied within the sites from the forest edges to the interior. The numbers of woody species increased significantly as a function of the distance from the village boundaries for Janajagaran ($p < 0.05$) and Radha Krishna ($p < 0.05$) BZCFs. In the case of Musharni Mai BZCF, the number of species observed was the least for the site near the edge but a nonsignificant pattern was observed ($p>0.05$). For the control forest, such a trend was not observed ($p > 0.05$) (Figure 5). In BZCFs, species like *Dalbergia sissoo*, *Dalbergia latifolia*, *Wendlandia tinctoria*, *Bombax ceiba*, *Toona ciliata*, *Trewia nudiflora*, and *Premma barabata* were recorded only in the innermost sites.

The cross-sectional area (basal area) per unit sampling area is an indicator of woody species dominance. Basal area correlates with maturity and age of the woody plant species. Our study showed mixed patterns between total dominance and distance from the boundary lines of villages (Figure 6). A decreased linear trend was observed in Janajagaran BZCF ($p < 0.05$), a linear trend was absent in Musharni Mai BZCF, and an increased trend was observed in Radha Krishna BZCF ($p < 0.001$). Though no linear trend was present in the Musharni Mai BZCF, a decrease in the basal area moving from the edge to the interior of the forest was observed. As expected, PWR showed the oscillating pattern of increase and decrease of the total basal area along the sites. An interesting result was observed when the total basal area was segregated as contributed by *S. robusta* and the basal area of other species except for *S. robusta*. In BZCFs edges, the majority of the basal area was contributed by *S. robusta* but its value was observed decreasing when moving across the interior of the forests. In contrast, the basal area of other species showed an increasing trend ($p < 0.05$) in BZCFs (Figures 6 and 7) except for the control forest. Based on % basal area, the forest near

Table 11: IVI of predominant species and Morisita Index of Dispersion in Janajagaran BZCF.

Species code	Name of the species	Site I	Site II	Site III	Site IV	Site V	Site VI
1	*Shorea robusta* Gaertn.	190.93 (1.2)	216.11 (1.1)	188.25 (1.1)	146.09 (1.2)	112.11 (1.2)	75.00 (1.5)
2	*Adina cordifolia* (Willd. ex Roxb.) Benth. & Hook. f. ex Brandis	32.81 (10)	4.62 (10)	8.21 (4.7)		9.47 (2.3)	
3	*Lannea coromandelica* (Houtt.) Merr.	14.96 (3.1)		3.84 (10)	21.80 (3.8)	4.44 (4.7)	9.96 (3.1)
4	*Lagerstroemia parviflora* Roxb.	10.61 (3.8)	35.60 (1.6)	32.71 (1.4)	41.19 (1.3)	36.87 (1.2)	36.11 (1.4)
8	*Casearia graveolens* Dalzell	13.19 (2.5)	7.39 (5.4)	10.40 (2.9)	13.16 (2.7)	25.42 (1.2)	28.89 (1.9)
9	*Cassia fistula* L.	7.62 (5.0)	3.78 (10)	12.87 (2.5)	2.44 (10)	2.76 (5.4)	1.69 (10)
10	*Mallotus philippensis* (Lam.) Müll.-Arg.	13.50 (3.4)	3.64 (10)	10.39 (3.7)	38.33 (1.4)	23.52 (1.6)	30.79 (1.2)
12	*Dillenia pentagyna* Roxb.		5.24 (10)	6.31 (6.7)	22.48 (1.6)	24.16 (1.2)	28.92 (1.6)
13	*Syzygium cumini* (L.) Skeels		20.00 (2.3)	10.11 (3.1)		5.90 (4.3)	5.50 (4.7)
15	*Semecarpus anacardium* L.f.			4.79 (4.7)		11.79 (1.7)	6.22 (5.4)
18	*Miliusa velutina* (Dunal) Hook. f. & Thoms.				2.44 (10)	6.95 (3.6)	27.50 (1.7)
21	*Terminalia alata* Heyne ex Roth					17.03 (1.2)	15.13 (1.5)

Table 12: IVI of predominant species and Morisita Index of Dispersion in Musharni Mai BZCF.

Species code	Name of the species	Site I	Site II	Site III	Site IV	Site V	Site VI
1	*Shorea robusta* Gaertn.	268.26 (1.03)	183.83 (1.04)	151.92 (1.2)	159.47 (1.2)	148.73 (1.04)	116.36 (1.3)
8	*Casearia graveolens* Dalzell	15.36 (7.5)	22.60 (1.3)	29.91 (1.3)	30.03 (1.3)	34.10 (1.2)	20.51 (1.9)
12	*Dillenia pentagyna* Roxb.		22.19 (1.8)	25.87 (1.8)	15.06 (2.1)	15.47 (1.8)	17.52 (1.7)
13	*Syzygium cumini* (L.) Skeels		2.90 (10)	30.90 (1.6)	8.34 (3.1)		
20	*Cleistocalyx operculatus* (Roxb.) Meer. & Perry		9.80 (3.1)	3.10 (10.0)	2.04 (10.0)		
21	*Terminalia alata* Heyne ex Roth		7.16 (4.7)		9.47 (3.1)		2.75 (10.0)
4	*Lagerstroemia parviflora* Roxb.		21.11 (2.2)	24.64 (2.4)	34.56 (1.3)	25.39 (1.7)	54.35 (1.3)
3	*Lannea coromandelica* (Houtt.) Merr.		7.34 (4.7)	12.16 (5.4)	11.88 (3.1)	18.46 (1.3)	
10	*Mallotus philippensis* (Lam.) Müll.-Arg.		4.79 (5.4)	14.85 (2.2)	19.66 (1.8)	23.20 (2.3)	48.78 (1.1)
26	*Anogeissus latifolius* (Roxb. ex DC.) Bedd.						18.75 (1.7)

TABLE 13: IVI of predominant species and Morisita Index of Dispersion in Radha Krishna BZCF.

Species code	Name of the species	Site I	Site II	Site III	Site IV	Site V	Site VI
1	*Shorea robusta* Gaertn.	228.32 (1.1)	176.61 (1.02)	137.14 (1.1)	130.12 (1.1)	104.46 (1.1)	57.57 (1.1)
2	*Adina cordifolia* (Willd. ex Roxb.) Benth. & Hook. f. ex Brandis	7.39 (10.0)	10.17 (3.1)	14.76 (2.7)	8.87 (3.1)	4.61 (10.0)	10.04 (4.3)
3	*Lannea coromandelica* (Houtt.) Merr.	13.31 (4.7)	3.03 (10.0)	13.96 (2.1)	13.69 (2.0)	8.03 (2.3)	5.93 (3.6)
21	*Terminalia alata* Heyne ex Roth	19.56 (4.7)	7.36 (4.7)	16.06 (3.0)	9.72 (3.6)	10.51 (2.7)	19.41 (1.7)
12	*Dillenia pentagyna* Roxb.	6.81 (10.0)	15.13 (2.7)	22.69 (1.5)	21.04 (1.6)	17.26 (1.2)	23.97 (1.4)
6	*Terminalia bellirica* (Gaertn.) Roxb.	8.66 (10.0)	8.79 (3.1)	5.17 (4.7)	10.26 (2.3)	18.22 (1.9)	19.90 (2.8)
4	*Lagerstroemia parviflora* Roxb.	10.63 (4.7)	35.82 (1.8)	27.53 (1.4)	27.49 (1.3)	27.99 (1.2)	17.11 (1.5)
8	*Casearia graveolens* Dalzell	5.32 (10.0)	16.33 (2.2)	12.68 (1.8)	15.43 (1.8)	19.35 (1.4)	14.37 (1.7)
10	*Mallotus philippensis* (Lam.) Müll.-Arg.		14.45 (1.8)	26.63 (1.5)	29.17 (1.2)	49.90 (1.2)	84.95 (1.2)
18	*Miliusa velutina* (Dunal) Hook. f. & Thoms.			2.01 (10.0)	8.46 (2.1)	15.56 (2.8)	1.98 (10.0)
15	*Semecarpus anacardium* L.f.			5.27 (3.6)	11.14 (1.8)		1.47 (10.0)

the settlement areas was observed to have undergone biotic homogenization (*S. robusta* Terai Mixed Hardwood Forest to *S. robusta* Forest) due to disturbance and management interactions.

We calculated the Importance Value Index (IVI) for each species in every 1 ha plot. Though the forest was dominated by *S. robusta*, other species were identified as predominant species based on high IVI values (IVI > 10). Janajagaran BZCF, Musharni Mai BZCF, Radha Krishna BZCF, and Parsa Wildlife Reserve were observed to have 12, 9, 12, and 17 predominant species, respectively (Appendix D, Tables 7–10). In all the BZCFs, our results indicated that IVI of *S. robusta* was linearly decreasing (significant; $p < 0.05$ in all BZCFs) with moving farther from the villages, and consequently other species showed increasing IVI values with the distance. But the control forest did not show such an increasing or decreasing trend (Figure 8). The high IVI of *S. robusta* near the human settlement areas were due to the protection of *S. robusta* and thus increasing all the three parameters affecting IVI. Based on the variation of Morisita's dispersion index (I_δ), all these predominant species exhibited clumped distribution (Appendix E, Tables 11–14). No effect on the distribution pattern of species according to the distance from the settlements was observed (all species exhibited $I_\delta > 1$). The clumped distribution of the species reflects that the distance between neighboring individuals is minimized for each species.

3.2. Species Diversity along the Distance from the Human Settlements. To examine the variations in the species diversity of woody plants, Shannon-Wiener index (H') and Simpson's index (λ) were calculated for each plot and plotted against the distance from the village boundaries. The Shannon-Wiener index for the species diversity increased significantly ($p < 0.05$) for all the BZCFs (Figures 9(a), 9(b), and 9(c), respectively, for Janajagaran BZCF, Musharni Mai BZCF, and Radha Krishna BZCF) towards the forest edge to the interior. For PWR, this pattern was not observed (Figure 9(d)). At the nearer distance to the human settlements in BZCFs, all the sites showed very low diversity indicating a high dominance of *S. robusta* (the species receiving high management and restoration priority). For PWR, the index values were comparatively higher. Simpson's index of diversity, which ranges from 0 (highest diversity) to 1 (no diversity), also showed the similar pattern of increase when moving farther from the village boundary lines (Figure 10, 1 – Simpson's λ; $p < 0.01$ for BZCFs but insignificant for the control forest). This trend also indicates the conversion of natural mixed forests to the woodlands near the edges.

3.3. Community Similarity and Dissimilarity between the Sites. To demonstrate how the forest disturbances change the composition of the plant community as the distance gets closer to the human settlement areas, we determined similarity and dissimilarity between the sites. For Janajagaran BZCF, the maximum value of dissimilarity was 0.55 (similarity: 1 - 0.55 = 0.45) and the dissimilarity was between Site I versus Site VI and Site II versus Site VI. For Musharni Mai BZCF, the maximum value of dissimilarity was observed between Site I and Site VI (0.54) followed by Site I and Site V (0.42).

TABLE 14: IVI of predominant species and Morisita Index of Dispersion in Parsa Wildlife Reserve.

Species code	Name of the species	Site I	Site II	Site III	Site IV	Site V	Site VI
8	*Casearia graveolens* Dalzell	16.20 (1.2)	18.69 (1.6)	12.30 (1.3)	15.48 (1.1)	4.96 (1.8)	4.95 (1.9)
4	*Lagerstroemia parviflora* Roxb.	28.03 (1.1)	35.81 (1.2)	23.25 (1.1)	18.65 (1.3)	21.96 (1.1)	15.84 (1.2)
18	*Miliusa velutina* (Dunal) Hook. f. & Thoms.	12.59 (1.6)	21.44 (1.2)	14.03 (1.3)	13.12 (1.2)	24.42 (1.3)	17.11 (1.3)
46	*Phyllanthus emblica* Linn.	6.00 (3.0)	1.12 (10.0)	4.30 (2.8)	13.76 (1.3)	5.99 (1.6)	6.46 (2.1)
22	*Terminalia chebula* Retz.	8.02 (2.6)	6.92 (1.4)	6.03 (1.3)	0.99 (10.0)	0.82 (10)	
1	*Shorea robusta* Gaertn.	101.09 (1.4)	72.09 (1.4)	90.43 (1.3)	74.39 (1.1)	72.41 (1.3)	95.39 (1.1)
27	*Dalbergia latifolia* Roxb.	7.96 (1.8)	9.03 (1.2)	10.84 (1.2)	6.87 (1.8)	5.94 (2.5)	6.92 (1.4)
12	*Dillenia pentagyna* Roxb.	13.24 (1.8)	19.78 (1.3)	15.85 (1.8)	12.75 (1.5)	6.58 (2.1)	4.57 (2.3)
21	*Terminalia alata* Heyne ex Roth	13.69 (1.5)	15.55 (1.9)	11.17 (1.2)	10.55 (1.7)	12.71 (1.2)	13.78 (1.5)
2	*Adina cordifolia* (Willd. ex Roxb.) Benth. & Hook. f. ex Brandis	10.12 (1.8)	1.66 (4.7)	5.00 (1.7)	8.17 (2.1)	9.97 (1.9)	13.95 (2.1)
13	*Syzygium cumini* (L.) Skeels	5.67 (2.1)	12.24 (1.4)	7.78 (1.3)	0.99 (10.0)	0.81 (10.0)	2.51 (3.6)
47	*Desmodium oojenense* (Roxb.) Ohashi	15.74 (1.5)	12.51 (1.4)	5.83 (1.5)	24.32 (1.2)	13.45 (1.4)	13.93 (1.6)
26	*Anogeissus latifolius* (Roxb. ex DC.) Bedd.	8.07 (2.3)	7.92 (1.5)	4.87 (2.1)	13.34 (1.5)	4.44 (2.1)	10.80 (1.3)
48	*Garuga pinnata* Roxb.	6.97 (2.1)	8.47 (2.3)	15.50 (2.2)	7.70 (3.7)	11.72 (1.8)	13.53 (1.4)
3	*Lannea coromandelica* (Houtt.) Merr.	0.83 (10.0)	2.63 (5.5)	3.19 (4.0)	3.51 (3.4)	0.86 (10.0)	11.21 (1.4)
23	*Buchanania latifolia* Roxb.	10.68 (2.0)	9.14 (1.6)	12.54 (1.3)	11.12 (1.5)		0.67 (10)
14	*Dysoxylum gobara* (Buch.-Ham.) Merr.	1.95 (5.4)	1.62 (4.9)			22.96 (1.2)	4.36 (3.9)

TABLE 15: Index of similarity and dissimilarity among different sites for Janajagaran BZCF.

	Site I	Site II	Site III	Site IV	Site V	Site VI
Site I		**0.74**	**0.82**	**0.68**	**0.57**	**0.45**
Site II	0.26		**0.85**	**0.67**	**0.59**	**0.45**
Site III	0.18	0.15		**0.71**	**0.67**	**0.52**
Site IV	0.32	0.33	0.29		**0.74**	**0.65**
Site V	0.43	0.41	0.33	0.26		**0.79**
Site VI	0.55	0.55	0.48	0.35	0.21	

We observed similar trends for Radha Krishna BZCF with the highest dissimilarity between Site I and Site VI (0.59) followed by Site II and Site VI (0.49). Values showed the higher dissimilarity between the nearer plots and the farther plots in BZCFs. For PWR, the maximum dissimilarity was observed between Site II and Site VI, but the dissimilarity values across sites did not exceed 0.34. The results indicated that dissimilarity values and their clustering heights were higher between the sites of BZCFs than the control forest sites in PWR (Figure 11, Appendix F, Tables 15–18).

NMDS was conducted to ordinate the sites based on the Bray-Curtis distance of IVI values of the species. The NMDS results showed that the sites of BZCFs were ordinated in the farther distances as compared to PWR forest (Figure 12). In

TABLE 16: Index of similarity and dissimilarity among different sites for Musharni Mai BZCF.

	Site I	Site II	Site III	Site IV	Site V	Site VI
Site I		**0.70**	**0.58**	**0.59**	**0.58**	**0.46**
Site II	0.30		**0.80**	**0.81**	**0.77**	**0.62**
Site III	0.42	0.20		**0.87**	**0.84**	**0.66**
Site IV	0.41	0.19	0.13		**0.85**	**0.70**
Site V	0.42	0.23	0.16	0.15		**0.67**
Site VI	0.54	0.38	0.34	0.30	0.31	

TABLE 17: Index of similarity and dissimilarity among different sites for Radha Krishna BZCF.

	Site I	Site II	Site III	Site IV	Site V	Site VI
Site I		**0.75**	**0.67**	**0.64**	**0.53**	**0.41**
Site II	0.25		**0.80**	**0.78**	**0.70**	**0.51**
Site III	0.33	0.20		**0.90**	**0.76**	**0.63**
Site IV	0.36	0.22	0.10		**0.81**	**0.62**
Site V	0.47	0.30	0.24	0.19		**0.69**
Site VI	0.59	0.49	0.37	0.38	0.31	

BZCFs, Site I and Site II ordinated in the farthest distances from Site VI reflecting a high variation in the species composition due to varying gradients of anthropogenic pressures. However, in case of PWR, the sites were clumped (ordinated in the shorter distances) in the central region of the ordinated axes. The ordination distances, however, do not correspond to the original distances among the sites.

4. Discussion

In this paper, we assessed how the changes in the ecological parameters in the *S. robusta* mixed forest are a function of the distance from the human settlements, and the interaction between anthropogenic disturbances and the forest management. The changes in the structure and composition were determined by comparing their values from the edge to the interior sites of the BZCFs and also with the control forest conditions in PWR.

We did not observe a significant pattern between total stem density and the distance from the settlements, but when density was segregated into species, the sum density of species other than *S. robusta* witnessed a significant positive pattern (Figure 3). Human activities such as collection of fuelwood (leading to an absence of coarse woody debris in BZCFs that was present in the forest stands of PWR), fodder, litter, and other forest products, as well as grazing animals, had reduced the density of woody species in the forests. But the decrease had been compensated (regardless of species-wise contribution) by the forest management interventions in the BZCFs. Therefore, in overall, we observed neither an increase nor a decrease in the total stem density while moving towards the forest interior from the edges. Changes in the spatial and temporal heterogeneity as a result of anthropogenic activities have also been assessed in several other studies [8–11, 38].

Considering the total basal area, we observed mixed results in the three forests while moving from the edge to the interior. Basal area was observed with a significant decreasing trend for Janajagaran BZCF, an insignificant decreasing trend for Musharni Mai BZCF, and an increasing trend for Radha Krishna BZCF (Figure 6) along the edge to the interior of the forests. Previous research also demonstrated similar results of no effect on the basal area with a distance from the human settlements [17]. In contrast, another study has shown that biomass (a parameter indicating basal area) is reduced in the forest edge relative to the interior [39]. Consistent with this result, the basal area of the species other than *S. robusta* increased with the distance from the forest edges (Figure 6). Based on % basal area contributed, the forest edges were observed to have undergone biotic homogenization (*S. robusta* mixed stand to *S. robusta* stand) due to the disturbance and management interactions (Figure 6). The Ministry of Forests and Soil Conservation (MFSC) of Nepal has categorized the Sal (*S. robusta*) forest as a forest where *S. robusta* comprises more than 60% of the basal area and Sal Terai Mixed Hardwood Forest (STMH) as a forest where *S. robusta* comprises 33-60% of the total basal area [40]. In case of BZCFs, the edges had more than 60% *S. robusta* basal area, a percentage not observed in the interior stands of BZCFs and the control forest; in the latter forests the percent contribution of *S. robusta* was between 33 and 60% (Figure 7).

In our study, both Shannon-Wiener and 1 – Simpson's λ diversity values showed a positive trend with the distance from the boundary lines of villages in BZCFs (Figures 9 and 10). But in PWR such a trend was not observed. Some primary studies have determined a clear pattern of change in diversity, richness, density, and basal area of the tree species towards the forest interior from the edges [12, 20, 41], whereas other studies have obtained contrasting results with more species richness and abundancy of heliophilic species in the forest edge than the interior [42–44].

The increasing value of species diversity with the distance from the settlements in our study strongly supports that human interventions (interaction of both management and

TABLE 18: Index of similarity and dissimilarity among different sites for Parsa Wildlife Reserve.

	Site I	Site II	Site III	Site IV	Site V	Site VI
Site I		**0.81**	**0.82**	**0.77**	**0.69**	**0.76**
Site II	0.19		**0.78**	**0.73**	**0.68**	**0.66**
Site III	0.18	0.22		**0.75**	**0.69**	**0.73**
Site IV	0.23	0.27	0.25		**0.71**	**0.74**
Site V	0.31	0.32	0.31	0.29		**0.75**
Site VI	0.24	0.34	0.27	0.26	0.25	

disturbances) have fostered a decrease in the richness and evenness of plant community and have changed the *S. robusta* mixed stands to pure *S. robusta* stands near the forest edges due to overexploitation of other plant species. Consistent with observations of intense monoculture-driven forest management nearby human settlements, other researchers have also posited that intensification of silviculture threatens species diversity [45, 46] and leads to the development of a community dominated by a single species or even-aged stands [47]. Anthropogenic pressures which lead to changes in the biological diversity are also responsible for changing the composition and stand quality of the forest ecosystems, microclimate, and nutrient cycling [48]. Increase in the generalist woody species can be the site of food sources, nest sites, and roosting sites for a variety of other species [49]. Higher plant diversity enhances the stability of the physical, chemical, and biotic properties of soils [50]. The intermediate disturbance hypothesis explains that diversity is maximum at the intermediate level of disturbance [51]. In contrast, species diversity declined with increasing level of disturbance in our study (Figures 9 and 10). There are also views reflecting that stability increases with the complexity of the ecosystems [52, 53]. In addition, the similarity analysis between sites in BZCFs reflected that there is a very little similarity between the sites in the edges compared to the interior (Figure 11). The result was also supported by NMDS outcomes (Figure 12). In the case of the control forest sites, such variation in the similarity was not observed.

Our study, based on the structural attributes, showed loosely defined community assemblages in the edges of BZCFs primarily due to the anthropogenic activities. In the long run, the existing inherent disturbances and ongoing management in the Terai *S. robusta* mixed forests can potentially lead to a decrease in the stability of the ecosystems. Natural forest areas that have been disturbed or altered to a large spatial extent or degree may not reach the similar structural complex (richness and composition) as primary forest [54]. In this context, fostering heterogeneity of the woody plant species is essential for sustainability.

5. Conclusions

Our results clearly indicated the effects of distance from human settlements on the vegetative attributes in the BZCFs. Maximum disturbance on the structural characteristics was observed near the edges which gradually decreased along the interior of the forests. The stands near the edges were under active management; however, a decrease in the richness and species diversity of the woody species was observed when moving from the edge to the interior. The management interventions in the forest sites near the settlements have directed the stands of BZCFs towards biotic homogenization and densification. These conditions cannot support habitat requirements of multiple species, and if the homogenization continues towards the interior of the BZCFs, it ultimately will impact the species diversity of the landscape. These situations of ecological changes due to intervention between anthropogenic disturbances and management practices should be assessed in larger spatial and regional scales to frame the future forest management strategies.

Appendix

A. Predominant Species and Their Local Uses, and Geographical Location of the Sampled Sites

See Tables 1 and 2.

B. Species Accumulation Curves

The species accumulation curves for the BZCFs and Parsa Wildlife Reserve (control plot). The curves depend on the accumulation rates of new species over the increase in the sampling area. Comparatively more steep beginning rise of the curves showed that new species are increasing with addition of the sampling units. The accumulation rates of the new species in all forests were horizontal after the sampling area reached 1 ha (10 subplots each of 0.1 ha). This represents that 10, 0.1 ha plots are sufficient to represent the forest community at one site. See Figure 13.

C. Community Structure of the Studied Forests

See Tables 3, 4, 5, and 6.

D. Importance Value Index (IVI) for Woody Plant Species in Each Site in the Studied Forests. The Total Value of IVI Is 300 Which Is the Sum of Relative Frequency, Relative Density, and Relative Basal Area Values

See Tables 7, 8, 9, and 10.

E. IVI of Predominant Species and Morisita Index of Dispersion

Predominant woody species and their Importance Value Index (IVI) for each site. Total of 12 species were found predominant in Janajagaran BZCF, 10 species in Musharni Mai BZCF, 11 species in Radha Krishna BZCF, and 17 species in Parsa Wildlife Reserve. Predominant species are considered high IVIs (IVI > 10). The values in parenthesis show Morisita's index (I_δ) (a value of 1 indicates a random distribution, values >1 indicate clumped distribution, and values < 1 indicate a uniform or regular distribution). Chi-squared based probability shows that all the species showed clumped distribution (at critical level of 0.05) according to the distance from the boundary lines of villages (null hypothesis describes that the species have random distribution). See Tables 11, 12, 13, and 14.

F. Similarity and Dissimilarity among Different Sites of the Studied Forests

In the tables, the upper right section and the lower left section show similarity and dissimilarity values, respectively, between sites. See Tables 15, 16, 17, and 18.

Conflicts of Interest

The authors declare that they have no conflicts of interest.

Acknowledgments

The authors are especially thankful to the forest user groups and administrations of Janajagaran BZCF, Musharni Mai BZCF, and Radha Krishna BZCF. The authors would also like to express their gratitude to the Department of National Parks and Wildlife Conservation and Office of Parsa Wildlife Reserve (recently designated as Parsa National Park), Nepal, for granting permission (letter reference numbers ECO335 - 2071/2072 and ECO074 - 2073/2074) to study the forests. They are thankful to National Herbarium and Plant Laboratories, Nepal, and Central Department of Botany, Tribhuvan University for the identification of plant specimens. The technical and field support from Wyoming Reclamation and Restoration Center (WRRC), Amrit Campus, and Central Department of Environmental Science, Tribhuvan University, Nepal, is also thankfully acknowledged.

References

[1] D. Western, "Human-modified ecosystems and future evolution," *Proceedings of the National Acadamy of Sciences of the United States of America*, vol. 98, no. 10, pp. 5458–5465, 2001.

[2] C. Folke, S. Carpenter, B. Walker et al., "Regime shifts, resilience, and biodiversity in ecosystem management," *Annual Review of Ecology, Evolution and Systematics*, vol. 35, pp. 557–581, 2004.

[3] D. U. Hooper, F. S. Chapin III, J. J. Ewel et al., "Effects of biodiversity on ecosystem functioning: A consensus of current knowledge," *Ecological Monographs*, vol. 75, no. 1, pp. 3–35, 2005.

[4] G. J. Nowacki and M. D. Abrams, "Is climate an important driver of post-European vegetation change in the Eastern United States?" *GCB Bioenergy*, vol. 21, no. 1, pp. 314–334, 2015.

[5] R. J. Morris, "Anthropogenic impacts on tropical forest biodiversity: A network structure and ecosystem functioning perspective," *Philosophical Transactions of the Royal Society B: Biological Sciences*, vol. 365, no. 1558, pp. 3709–3718, 2010.

[6] P. F. Hessburg and J. K. Agee, "An environmental narrative of Inland Northwest United States forests, 1800-2000," *Forest Ecology and Management*, vol. 178, no. 1-2, pp. 23–59, 2003.

[7] M. Sassen and D. Sheil, "Human impacts on forest structure and species richness on the edges of a protected mountain forest in Uganda," *Forest Ecology and Management*, vol. 307, pp. 206–218, 2013.

[8] J. Kouki, S. Löfman, P. Martikainen, S. Rouvinen, and A. Uotila, "Forest fragmentation in Fennoscandia: Linking habitat requirements of wood-associated threatened species to landscape and habitat changes," *Scandinavian Journal of Forest Research*, vol. 16, no. 1, pp. 27–37, 2001.

[9] M. J. Lawes, B. C. C. Lamb, and S. Boudreau, "Area- but no edge-effect on woody seedling abundance and species richness in old Afromontane forest fragments," *Journal of Vegetation Science*, vol. 16, no. 4, pp. 363–372, 2005.

[10] C. M. Tucker, J. C. Randolph, T. Evans, K. P. Andersson, L. Persha, and G. M. Green, "An approach to assess relative degradation in dissimilar forests: Toward a comparative assessment of institutional outcomes," *Ecology and Society*, vol. 13, no. 1, 2008.

[11] T. G. Wade, K. Riitters, J. D. Wickham, and K. B. Jones, "Distribution and causes of global forest fragmentation," *Conservation Ecology*, vol. 7, no. 2, pp. 1–16, 2003.

[12] K. Majumdar and B. K. Datta, "Effects of anthropogenic disturbances on vegetation diversity and structure: A case study in the remnant forests surrounding the village ecosystems of Tripura, Northeast India," *Chinese Journal of Population Resources and Environment*, vol. 13, no. 4, pp. 332–340, 2015.

[13] S. R. Biswas and A. U. Mallik, "Species diversity and functional diversity relationship varies with disturbance intensity," *Ecosphere*, vol. 2, no. 4, 2011.

[14] J. Kimmins, *Forest Ecology: A Foundation for Sustainable Forest Management and Environmental Ethics in Forestry*, Prentice Hall, New Jersey, NJ, USA, 2004.

[15] E. L. Webb and R. N. Sah, "Structure and diversity of natural and managed sal (Shorea robusta Gaertn.f.) forest in the Terai of Nepal," *Forest Ecology and Management*, vol. 176, no. 1-3, pp. 337–353, 2003.

[16] I. P. Sapkota, M. Tigabu, and P. C. Odén, "Spatial distribution, advanced regeneration and stand structure of Nepalese Sal

(Shorea robusta) forests subject to disturbances of different intensities," *Forest Ecology and Management*, vol. 257, no. 9, pp. 1966–1975, 2009.

[17] A. Popradit, T. Srisatit, S. Kiratiprayoon et al., "Anthropogenic effects on a tropical forest according to the distance from human settlements," *Scientific Reports*, vol. 5, Article ID 14689, 2015.

[18] K. B. Shrestha, I. E. Måren, E. Arneberg, J. P. Sah, and O. R. Vetaas, "Effect of anthropogenic disturbance on plant species diversity in oak forests in Nepal, Central Himalaya," *International Journal of Biodiversity Science, Ecosystem Services & Management*, vol. 9, no. 1, pp. 21–29, 2013.

[19] S. Thapa and D. S. Chapman, "Impacts of resource extraction on forest structure and diversity in Bardia National Park, Nepal," *Forest Ecology and Management*, vol. 259, no. 3, pp. 641–649, 2010.

[20] D. S. Kacholi, "Edge-interior disparities in tree species and structural composition of the kilengwe forest in morogoro region, Tanzania," *ISRN Biodiversity*, vol. 2014, Article ID 873174, 8 pages, 2014.

[21] CBD, *Nepal Fourth National Report to the Convention on Biological Diversity*, Singha Durbar, Kathmandu, Nepal, 2009.

[22] K. A. Brown, D. F. B. Flynn, N. K. Abram, J. C. Ingram, S. E. Johnson, and P. Wright, "Assessing natural resource use by forest-reliant communities in madagascar using functional diversity and functional redundancy metrics," *PLoS ONE*, vol. 6, no. 9, 2011.

[23] G. Eilu, D. L. N. Hafashimana, and J. M. Kasenene, "Density and species diversity of trees in four tropical forests of the Albertine rift, western Uganda," *Diversity and Distributions*, vol. 10, no. 4, pp. 303–312, 2004.

[24] S. B. Jennings, N. D. Brown, and D. Sheil, "Assessing forest canopies and understorey illumination: Canopy closure, canopy cover and other measures," *Forestry*, vol. 72, no. 1, pp. 59–73, 1999.

[25] R. Bivand, C. Rundel, E. Pebesma, R. Stuetz, and K. O. Hufthammer, "Package "rgeos ." *R Packag. version 0.3-26*," 2017.

[26] H. Nguyen, D. Lamb, J. Herbohn, and J. Firn, "Designing mixed species tree plantations for the tropics: Balancing ecological attributes of species with landholder preferences in the Philippines," *PLoS ONE*, vol. 9, no. 4, 2014.

[27] B. Husch, T. Beers, and J. Kershaw, *Forest Mensuration*, Wiley, New York, NY, USA, 2003.

[28] A. C. Newton, *Forest Ecology and Conservation*, Oxford University Press Inc., New York, NY, USA, 2007.

[29] C. E. Shannon, "A mathematical theory of communication," *Bell Labs Technical Journal*, vol. 27, pp. 379–423, 1948.

[30] E. H. Simpson, "Measurement of diversity," *Nature*, vol. 163, article no. 688, 1949.

[31] J. Oksanen, F. G. Blanchet, R. Kindt et al., "Vegan: Community Ecology Package," *R Packag. ver. 2.4-3*, 2017.

[32] E. A. Peña and E. H. Slate, "Global validation of linear model assumptions," *Journal of the American Statistical Association*, vol. 101, no. 473, pp. 341–354, 2006.

[33] H. Wolda, "Similarity indices, sample size and diversity," *Oecologia*, vol. 50, no. 3, pp. 296–302, 1981.

[34] M. Morisita, "I-Index, a measure of dispersion of individuals," *Researches on Population Ecology (Kyoto)*, vol. 4, pp. 1–7, 1962.

[35] J. R. Bray and J. T. Curtis, "An ordination of the upland forest communities of southern wisconsin," *Ecological Monographs*, vol. 27, no. 4, pp. 325–349, 1957.

[36] P. R. Minchin, "An evaluation of the relative robustness of techniques for ecological ordination," *Plant Ecology*, vol. 69, no. 1-3, pp. 89–107, 1987.

[37] R Core Team R, "A language and environment for statistical computing," 2016.

[38] E. Ritter and D. Dauksta, "Human-forest relationships: Ancient values in modern perspectives," *Environment, Development and Sustainability*, vol. 15, no. 3, pp. 645–662, 2013.

[39] R. Chaplin-Kramer, I. Ramler, R. Sharp et al., "Degradation in carbon stocks near tropical forest edges," *Nature Communications*, vol. 6, 2015.

[40] DFRS, *Terai Forests of Nepal (2010-2012)*, Kathmandu, Babarmahal, 2010.

[41] J. A. Sayer and T. C. Whitmore, "Tropical moist forests: Destruction and species extinction," *Biological Conservation*, vol. 55, no. 2, pp. 199–213, 1991.

[42] J. W. Ranney, M. C. Bruner, and J. B. Levenson, "The Importance of Edge in the Structure and Dynamics of Forest Islands," in *Forest Island Dynamics in Man-Dominated Landscapes*, R. L. Burgess and D. M. Sharp, Eds., pp. 67–95, Springer-Verlag, New York, NY, USA, 1981.

[43] B. J. Palik and P. G. Murphy, "Disturbance versus edge effects in sugar-maple/beech forest fragments," *Forest Ecology and Management*, vol. 32, no. 2-4, pp. 187–202, 1990.

[44] S. Fraver, "Vegetation responses along edge-to-interior gradients in the mixed hardwood forests of the roanoke river basin, North Carolina," *Conservation Biology*, vol. 8, no. 3, pp. 822–832, 1994.

[45] K. J. Puettmann, K. D. Coates, and C. Messier, *A Critique of Silviculture*, Island Press, Washington, USA, 2009.

[46] J.-M. Carnus, J. Parrotta, E. Brockerhoff et al., "Planted forests and biodiversity," *Journal of Forestry*, vol. 104, no. 2, pp. 65–77, 2006.

[47] E. May, *At the Cutting Edge: The Crisis in Canada's Forests*, Key Porter Books, Toronto, Ontario, 2005.

[48] M. Kappelle, P. A. F. Kennis, and R. A. J. de Vries, "Changes in diversity along a successional gradient in a Costa Rican upper montane Quercus forest," *Biodiversity and Conservation*, vol. 4, no. 1, pp. 10–34, 1995.

[49] C. Martínez-Garza and H. F. Howe, "Restoring tropical diversity: Beating the time tax on species loss," *Journal of Applied Ecology*, vol. 40, no. 3, pp. 423–429, 2003.

[50] M. Pohl, D. Alig, C. Körner, and C. Rixen, "Higher plant diversity enhances soil stability in disturbed alpine ecosystems," *Plant and Soil*, vol. 324, no. 1, pp. 91–102, 2009.

[51] R. Abugov, "Species diversity and phasing of disturbance," *Ecology*, vol. 63, no. 2, pp. 289–293, 1982.

[52] M. I. Shafi and G. A. Yarranton, "Diversity, floristic richness, and species evenness during a secondary (post-fire) succession," *Ecology*, vol. 54, no. 4, pp. 897–902, 1973.

[53] E. G. Leigh, "On the relation between the productivity, biomass, diversity, and stability of a community," *Proceedings of the National Acadamy of Sciences of the United States of America*, vol. 53, no. 4, pp. 777–783, 1965.

[54] R. L. Chazdon, C. A. Peres, D. Dent et al., "The potential for species conservation in tropical secondary forests," *Conservation Biology*, vol. 23, no. 6, pp. 1406–1417, 2009.

12

Differences in Functional Trait Distribution between Inselberg and Adjacent Matrix Floras

John T. Hunter

School of Environmental and Rural Science, University of New England, Armidale, Australia

Correspondence should be addressed to John T. Hunter; jhunter8@bigpond.com

Academic Editor: Béla Tóthmérész

Inselbergs and the adjacent matrix represent extremes of different environmental conditions and should shed light on the changing allocation of plant resources across strong and abrupt resource gradients. Here I use collated life history trait data from 840 taxa found within typical insular inselberg and adjacent matrix floras from the New England Batholith region of eastern Australia. These species were sorted into guilds of specificity to the inselberg environment. Scored traits include life form, plant height, leaf area, fruit size, seed size, mono- or polycarpy, underground storage organs, regenerative/clonality, and flowering phenology. With reduced water and nutrient resources, typical of inselbergs, allocation of plant resources to vegetative reproduction and storage organs is a disadvantage. Plants restricted to inselbergs were shorter, usually polycarpic shrubs, with smaller leaves, fruits, and seeds. Flowering time was found to be earlier and reduced in length; diaspores often have dormancy and are dispersed locally in comparison to the matrix. The results show that with limited resources the creation of underground storage organs or vegetative reproduction becomes unviable on habitats characterised by shallow soil. Inselberg taxa of the study region are likely to be under greater threat than the matrix due to anthropogenic climate change.

1. Introduction

Central to functional ecology is the ability to link changes in species traits to ecosystem processes [1]. The distribution of functional traits is linked to the unique key evolutionary, environmental, predation, and competition pressures within a community. Functional traits determine how resources are secured and how those resources are allocated and/or stored, determining how species persist, compete, disperse, coexist, and reduce loss from herbivory [2–6]. Comparisons of trait differences between community types can allow for a more nuanced understanding of how traits and taxa may redistribute under global climate change [6]. However, selection acts on individuals and not on individual traits themselves indicating that it is a suite of coordinated traits that influence persistence, regeneration, and dispersal [7–9].

While functional trait investigations are increasingly common, comparisons between the floras of adjacent but contrasting environments are lacking. Granitic inselberg environments offer "habitat islands" that contrast dramatically from the surrounding vegetated matrix and have a worldwide distribution with replicate geomorphological structure [10–13]. Inselbergs are harsher and drier environments compared to the matrix in which they are embedded. Inselbergs shed up to 95% of precipitation, the soils are skeletal, humic, and highly acidic [10, 12, 14–22]. Temperatures up to 18°C higher than the ambient air and in excess of 60°C within the soil have been recorded; greater exposure to wind and solar radiation means that the thin soils rapidly dry out [16, 20, 22–25]. The vegetation around adjacent to the base of inselbergs receives increased moisture from runoff and nitrogen loads from inselberg biofilms [26]. Büdel et al. [26] showed that nitrogen content in the matrix soils adjacent to inselbergs was three times higher than the surrounding area. These characteristics are inherent features of inselbergs and their adjacent matrices worldwide [13].

As granitic inselbergs have a less favorable environment than the adjacent matrix in which they are imbedded, plants from such environments should develop persistence life history traits [13, 27]. Species may persist in situ by increasing the longevity of above ground parts, by increasing the longevity of below ground parts or via dormancy mechanisms of diaspores [5, 7]. Benwell [28] and Clarke [29] suggested that vegetative reproduction (resprouting and clonality) was a stress tolerant regeneration strategy compensating for high-risk recruitment and such strategies should be more common in drought affected environments and thus favored on inselbergs. In contrast, Halassy et al. [30] provided evidence that species that put greater resources into vegetative reproduction and/or storage organs were favored under mesic, resource rich, and competitive environments. Taxa with shorter stature with larger seeds, smaller long-lived leaves, longer lifespan, and denser wood are more likely to occur within areas of low resource availability compared to taxa from resource rich environments [5–7, 9, 31–33].

Although inselbergs are ideal "habitat islands," how permeable the barrier is between environments varies considerably, and thus some inselbergs are effectively more floristically insular than others [12, 34]. For example, within the New England Batholith of eastern Australia 76% of the granitic inselberg flora is shared with the surrounding matrix to varying degrees with overall endemism decreasing across the landscape with increasing landscape aridity [12]. McGann [12] created a set of guilds based on restriction to the inselberg environment and Hunter [13] used these in an investigation into the distribution of obligate seeding and resprouting life history traits on granitic inselbergs and within the adjacent matrix.

Many species are likely to be uninformative in cross community analyses as they are commonly shared taxa with similar abundances across both environments. Here, selected trait expressions between taxa with high specificity to the granitic inselberg environment of eastern Australia are compared to taxa with specificity to the immediate adjacent forested matrix based on the guilds of McGann [12]. Tested traits include plant height, specific leaf area, fruit size, seed size, life form, regenerative capacity, phenology, dormancy, and dispersal. Due to the contrasts between these adjacent vegetation types I hypothesized that inselberg endemics are likely to have long generation times, slow growth, and locally dispersed seeds and once established dominate spatial resources as occurring with other persistence niche specialists [13, 27].

2. Methods

2.1. Study Area. The New England Batholith lies in the north eastern corner of the state of New South Wales and the south eastern corner of the state of Queensland within Australia. The batholith extends for approximately 400 km north to south and 110 km east to west [35]. Elevations are highest on the eastern escarpment and fall towards the west. Surveyed areas varied in altitude from 310–1520 m above sea level. The vegetation on inselbergs in this region is largely of closed to open heaths, shrublands, or herbfields with the adjacent matrix comprised of rainforest, forest, and woodlands. Analysis of the relationships between the matrix and inselberg floras and detailed descriptions of the vegetation types, flora, and mapped distribution covered by this investigation are contained within Clarke et al. [36], Hunter and Clarke [37], Hunter et al. [38, 39], J. T. Hunter and V. H. Hunter [40], Hunter [41], Hunter [42], and Hunter and Sheringham [43].

Table 1: Inselberg insular guilds as defined by McGann [12].

Guild	Category of insularity
0	Never found on inselbergs, taxa restricted to the surrounding vegetation
1	Taxa always found in surrounding matrix but in low abundance on inselbergs
2	Ubiquitous taxa found equally on inselbergs or the matrix
3	Taxa common on inselbergs but occasionally found in the surrounding matrix
4	Taxa restricted to inselbergs on the New England Batholith but found in other habitats elsewhere
5	Taxa restricted to inselbergs and other rock outcrops throughout their distribution
6	Taxa restricted to granitic inselbergs throughout their distribution

Rainfall is seasonal and occurs mostly in summer (60–70%) with snow occurring occasionally at higher altitudes (Bureau of Meteorology http://www.bom.gov.au/). Rainfall ranges from 600 to 1000 mm annually in the west and from 1000 to 2500 mm annually along the eastern escarpment (Bureau of Meteorology http://www.bom.gov.au/).

2.2. Survey Design. The survey was conducted using 32 m × 32 m (*c.* 0.1 ha) nested plots (9 nests; hence a relative abundance score out of 10 was derived for each species within each plot) [44] and collected solely by the author. A total of 399 plots were allocated to 216 inselbergs and 123 plots were placed within the surrounding forested matrix. Full details of the survey procedure and the stratification of sites are given in Hunter and Clarke [37]. In total 840 vascular plant taxa were found within plots during April 1994 to June 1996.

2.3. Soil pH Determination. The pH of both matrix and inselberg soils were tested at each site. The "1 : 5 Soil/Water suspension method" of Raymont and Higginson [45] was used for determining soil pH. 20.00 g of air-dry soil and 100 mL of deionized water were placed within jars and tumble within a mechanical tumbler for one hour at room temperature. Solutions were allowed to settle for 30 mins. Electrodes of a standardized pH meter were placed within solutions that were mechanically stirred until the reading settled. After each measurement electrodes were washed with deionized water. Replicate determinations were periodically made on the same solutions to ensure that readings did not vary within samples by more than 0.1 pH.

2.4. Statistical Methods. Taxon related tables were created that included information on species relative abundance (out of 10) within each quadrat, a table of life form and life history traits for each species and a table scoring each species within one of the seven inselberg insular guilds (as scored by McGann [12], Table 1). The community based technique of Community Weighted Means of traits [46] was applied to assess the response of functional traits within floristic inselberg insular guilds. Trait expressions of all species were weighted by their relative abundances from each plot. A redundancy analysis was then performed on the plot by trait matrix constrained by the table created from the summed relative abundances of each insular guild via the CANOCO5 package [47]. Thus, functional traits are assessed across the gradients of insularity/endemicity to the inselberg environment as expressed by the insular guilds. Traits recorded for each of the 840 flora taxa included the following: trait scores were based on maximum values along with life form type scored from localized collections obtained by the author and those from the region housed within the NCW Beadle Herbarium at the University of New England for each taxon. Some additional information was also obtained from PlantNET (http://plantnet.rbgsyd.nsw.gov.au/; accessed October 2014). Values allocated to each of the 840 vascular species included plant height, leaf area (length by width), fruit size (length by width), fruit type (dry dehiscent, dry indehiscent, legume, and succulent), seed size (length by width), monocarpic or polycarpic, presence or absence of storage organs (e.g., bulbs, tubers, swollen taproots, corms, and lignotubers), and vegetative spread capacity (e.g., rhizomes, runners, stolons, suckering, and rooting at nodes).

Life form types were also attributed to each taxon and these included hemiparasites, lianas and vines, herbs, low shrubs, medium shrubs, tall shrubs, low trees, medium trees, and tall trees. Dispersal syndromes and propagule attachments were also recorded for each species and from these data each species was scored for the capacity to be transported more than 100. This was done based on the criteria of Elizabeth [7] whereby large, gravity dispersed propagules were considered to be locally dispersed only and scored as "none." Ant-dispersed or medium sized propagules were scored as "limited" due to their likely medial dispersal distance. Propagules were scored as "strong" for long distance dispersal if they possessed structures that enable wind or animal dispersal [7]. Similarly the capacity for diaspore dormancy was also scored in the same fashion for each taxon (i.e., no dormancy, limited dormancy, and strong dormancy). Comparisons between these life history attributes were made using percentage of the flora within each category. Flowering duration was coded by the maximum number of months for which a species has been recorded to flower. Comparisons between inselberg (guilds 5 or 6) and matrix floras (guilds 0 or 1) were made using univariate statistics after testing for normality of distribution and the differences tested using the Mann–Whitney test of significance using PAST 2.17b [48].

3. Results

For final presentation the RDA analysis was restricted to only the guild types (0-1 and 5-6). The explanatory variables accounted for 22.5% (test on all axes $p = 0.002$) of the variance of functional traits (Figure 1). This analysis indicates that overall the functional traits of smaller leaf size, lower height, a lack of underground storage organs, and lack of monocarpic life history are favored by inselberg endemic flora species on the New England Batholith of eastern Australia. All guild "6" taxa (inselberg endemics) were polycarpic species as opposed to guild type "0" (matrix restricted species) which had the second highest percentage of monocarpic species (13%). However, species from within the surrounding matrix are more likely to favor the traits of increased plant height, larger leaf area, larger seed, and fruit size.

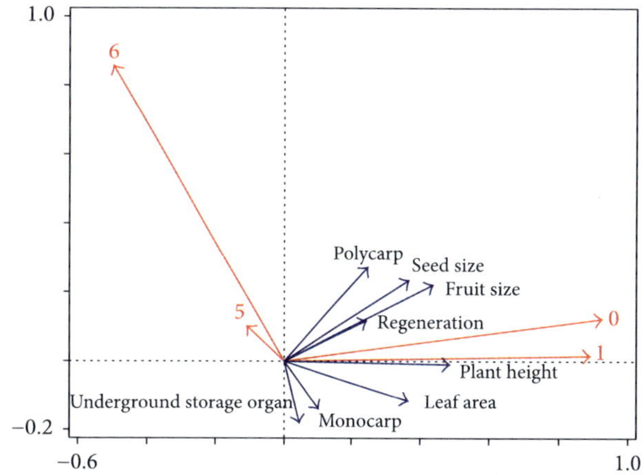

FIGURE 1: Redundancy analysis of the central weighted means of nine select functional traits of species found within the New England Batholith. Increasing number indicates increasing restriction to the inselberg environment (Table 1).

The inselberg soils were significantly higher in acidity than the adjacent matrix. Soil pH mean was 4.4 on inselbergs (Std. error 0.02; 3.5 minimum recorded); in contrast matrix soils had a mean of 5.6 (Std. error 0.06) ($p = 0.0001$).

Table 2 presents the percent distribution of life form types across the insular guild categories. Categories with increasing restriction to the inselberg environment are more likely to be low shrubs with no representation within the medium to tall tree categories, as hemiparasites or lianas and vines. Herbs dominate floristic guilds one to three which represent species that are found within inselbergs and matrix floras.

In terms of diaspore distance dispersal there was a relatively even distribution between types within matrix taxa (Table 3; 66% with some dispersal traits). In contrast, the majority of inselberg flora species (55%) had no distance dispersal mechanisms (Table 3). No diaspore dormancy was the most common strategy for matrix flora species (63%); however, there was an almost even distribution across the three dormancy types within the inselberg floras with most species having some form of diaspore dormancy (68%) (Table 3).

Flowering times for both inselbergs and the matrix floras were concentrated within the spring and summer months between August and February. June had the lowest number of species known to flower in both environments; however, the peak flowering month was September for inselberg species and October for matrix species. On average inselberg floras flowered over 4.5 months (Std. error 0.35) while matrix floras flowered for 5.1 months (Std. error 0.19) ($p = 0.048$).

TABLE 2: Percent of vascular flora within each major life form type from inselberg insularity guilds.

Insularity guild	Hemiparasites	Lianas and vines	Herbs	Low shrubs	Medium shrubs	Tall shrubs	Low trees	Medium trees	Tall trees
0	0	2.4	38.7	20.4	24.6	2.4	4.1	6.9	0.3
1	0.2	9.6	53.4	21.6	5.0	3.5	0.0	6.5	0.2
2	0.2	2.4	63.6	16.6	5.8	3.3	5.0	2.5	0.5
3	0	0	80.9	24.3	3.2	0.0	0.2	0.4	0.0
4	0	0	31.8	47.5	8.5	12.3	0	0	0
5	0	0	31.8	47.5	8.5	12.3	0	0	0
6	0	0	19.3	52.7	26.8	0.5	0.6	0	0

TABLE 3: Percentage allocation of matrix (guild 0 or 1) and inselberg (guild 5 or 6) floras to dispersal and dormancy categories.

Life history variable	None	Limited	Strong
Matrix flora diaspore distance dispersal capacity	33.8	36.7	29.5
Inselberg flora diaspore distance dispersal capacity	54.8	38.7	6.5
Matrix flora diaspore dormancy	62.5	29.1	8.7
Inselberg flora diaspore dormancy	33.9	33.9	32.3

4. Discussion

The granitic inselberg environment is an arid nutrient poor habitat compared to the surrounding matrix and as expected these environmental factors have favored persistence niche characteristics of (smaller leaves and lower stature). However other suites of functional traits are not consistent (smaller seeds, fruits, lack of dispersal, no vegetative spread/clonality, and no underground storage organs) [4, 5, 28, 29, 32, 49].

4.1. Resource Limitation. The soils on inselbergs had significantly higher acidity which would limit the capacity of plants to acquire nutrients. In low pH soil nutrients are quickly leached reducing soil fertility which has been linked to reduced leaf and fruit size [9]. As expected matrix taxa were taller, had larger leaf area scores, and produced larger seeds and fruits. Such features are typical of habitats with greater moisture and nutrient availability, where competition is high, light is limiting, and predation may be higher [8, 33, 50]. Hunter [13] suggested a model where competition for light was high in the matrix but that light was abundant on inselbergs and that predation/grazing was higher within the matrix of the New England Batholith; the results here support this model.

Resprouting/clonality traits were more commonly expressed within matrix species where nutrient and water resources were higher and climate more ameliorated. This provides support for Halassy et al. [30] who suggested that such species were favored in resource rich mesic environments with high competition and is at odds with Benwell [28] and Clarke [29] who proposed that such strategies were more likely to occur in locations of high stress and high recruitment risk.

4.2. Underground Storage Organs. The possession of underground storage organs was negatively associated with species specific to the inselberg environment, again indicating that such life history traits are of no advantage. Underground storage organs like diaspore dormancy can be a method of avoiding disadvantageous times by allowing above ground parts to die, yet allowing survival below ground enabling a rapid response when optimal conditions recur. Some form of diaspore dormancy was the norm for inselberg restricted taxa but lacking in most matrix species. However both fruits and seeds were smaller on inselbergs rather than larger as suggested by theory for harsh environments [49]. It is possible that the harsher conditions, lowered water, and nutrient availability on inselbergs or simply the reduced space on shallow soil basins do not favor the storage of additional resources (e.g., rhizomes, tubers, corms, bulbs, runners, and stolons) but instead there is selective pressure for diaspore dormancy in smaller rather than larger seeds. The combination of these traits would collectively reduce the need to capture greater resources from the impoverished soils. Alternatively, as inselbergs are deemed to have lower biomass removing disturbance events, yet juvenile mortality is still high (possibly through drought) [13] there is likely to be little advantage to the development of underground storage organs enabling quick growth following disturbances. The higher nutrient and water availability and ameliorated climate within the matrix along with more frequent biomass removing disturbances (fire and grazing) may select for vegetative reproduction and spread and more immediate regeneration as would be expected in highly competitive communities.

4.3. Persistence Niche. The shifting persistence niche hypothesis [51] proposes that the selective advantage of resprouting is bell shaped and changes along a gradient of disturbance. Hence where disturbances are infrequent resprouting is a disadvantage, under moderate disturbance it is advantageous but under frequent disturbance the ability to replenish bud resources diminishes, and it becomes a disadvantage [52, 53]. It is possible that the same mechanisms are at play across a gradient of resource availability whereby the ability to store additional resources becomes disadvantageous under very high and very low resource availability (nutrients/water, etc.) and the strategy is of greatest benefit under medial resource availability when competition between species is greatest. However to test this hypothesis further additional nutrient environments would need to be added. It is unlikely that any trait is advantaged in a continual linear fashion to any resource gradient [6, 54].

4.4. Carpy. Monocarpy (semelparity) in herbaceous species is generally favored in situations where juvenile mortality is low and in situations of episodic and short-lived opportunities is created by disturbances [7, 32, 55, 56]. Such taxa have small seeds and wider dispersal with larger geographic ranges [56, 57]. Monocarpic herbaceous taxa tend to have many small seeds and are favored in high resource environments [5, 7]. The lack of such taxa within inselberg restricted guilds suggests that juvenile mortality between reproductive events may be high and provides further evidence that dormancy in diaspores and local dispersal is of more selective advantage than smaller, dispersible seeds lacking longer term dormancy in harsher and resource limiting environments.

4.5. Phenology. Following the flowering phenology of two sister species, one restricted to inselbergs and the other within the adjacent matrix, Hunter and Williams [58] found that the inselberg restricted taxon flowered one month earlier than its matrix counterpart supporting the findings provided here. Hunter et al. [20] clearly showed that the temperature diurnal and seasonal range on inselbergs was much greater than the adjacent matrix and during summer periods could be over 10 degrees Celsius higher. Species may flower earlier on the inselbergs due to the earlier warming of these environments [59]. However, flowering is a physiologically expensive and the length of this phenophase is linked to optimal temperature ranges [60]. Thus, the significant reduced flowering period of approximately one month is potentially due to the physiological constraints on flowering being reached earlier on these exposed inselbergs with greater temperature amplitudes.

4.6. Adaptations of Inselberg Taxa. Overall the evidence supports the assumptions of Hunter [13] that inselberg restricted taxa are long-lived, locally dispersed and persist by dominating local limited resources. An unexpected discovery was that the lack of trait expressions associated with below ground storage and above ground vegetation spread suggesting that these traits are disadvantageous in these extreme and nutrient poor environments. Investigations by Poot et al. [61] suggest that inselberg endemics have adapted in a novel way by increasing their root mass and distributing it more evenly within the substrate. In agreement with Hunter [13] competition between inselberg flora species is strongest for the limited size of soil patches and the accessible nutrients and moisture within it.

4.7. Anthropogenic Climate Change. McGann [12] found the most significant correlated climatic factors with the occurrence of inselberg restricted species within the study region included a negative relationship with increased diurnal range, positive to increased precipitation of the driest quarter and negative to the maximum temperature of the wettest period. Predictions of global climate change indicate that within the study region the mean diurnal range will increase, precipitation of the driest quarter will decrease, and maximum temperature of the wettest period will increase [62–64]. All of which indicates that conditions are likely to cause potential extinctions of these rare, highly restricted inselberg species with long generation times and limited dispersal mechanisms in favor of more ubiquitous taxa shared by both environments. Thus, a significant reduction in local and regional diversity is lost increasing homogenization of the landscape. Other similar continental habitat islands (e.g., Kurgans of the Eurasian Steppe) are likely to share a similar fate.

5. Conclusion

Differences resource availability (nutrients and water) and climatic conditions have likely caused significant and different selective pressures on plant functional traits between species restricted to the inselberg environment and those within the adjacent matrix. The lack of resources and reduced habitability of the inselberg environment necessarily means that there is an inability or a disadvantage to the creation of underground storage organs and vegetative structures (rhizomes, runners, stolons, etc.) and that extended flowering is also problematic. Underground storage organs and vegetative structures can only be created where there are sufficient accessible resources (including space) and are likely favored under competitive environments with greater chances of biomass removing disturbances. The restrictions of the inselberg environment also limit the height of plants, size of leaves, fruits, and seeds. The effect of global climate change within the New England Batholith is likely to see a reduction or extinction in the unique and rare inselberg restricted species. This study has highlighted the need to address the differences between adjacent habitat types by removing shared species in order to better understand key divergence in functional traits. It is highly likely that endemic species of habitat islands with similarly specialized functional traits and poor dispersal will be under greater extinction threat than those within the surrounding matrix.

Competing Interests

The author declares that they have no competing interests.

Acknowledgments

The original survey sites were placed during a Ph.D. program funded by an Australia Postgraduate Award at the University of New England. The collation of additional life history data and the extension of the original research have been conducted independently of funding sources.

References

[1] S. Lavorel and E. Garnier, "Predicting changes in community composition and ecosystem functioning from plant traits: revisiting the Holy Grail," *Functional Ecology*, vol. 16, no. 5, pp. 545–556, 2002.

[2] M. Aiba and T. Nakashizuka, "Architectural differences associated with adult stature and wood density in 30 temperate tree species," *Functional Ecology*, vol. 23, no. 2, pp. 265–273, 2009.

[3] D. Carmona, M. J. Lajeunesse, and M. T. J. Johnson, "Plant traits that predict resistance to herbivores," *Functional Ecology*, vol. 25, no. 2, pp. 358–367, 2011.

[4] M. Kleyer, S. Dray, F. de Bello et al., "Assessing species and community functional responses to environmental gradients: which multivariate methods?" *Journal of Vegetation Science*, vol. 23, no. 5, pp. 805–821, 2012.

[5] P. B. Adler, R. Salguero-Gómez, A. Compagnoni et al., "Functional traits explain variation in plant life history strategies," *Proceedings of the National Academy of Sciences of the United States of America*, vol. 111, no. 2, pp. 740–745, 2014.

[6] J. T. Hunter, "Changes in allometric attributes and biomass of forests and woodlands across an altitudinal and rainfall gradient: what are the implications of increasing seasonality due to anthropogenic climate change?" *International Journal of Ecology*, vol. 2015, Article ID 208975, 10 pages, 2015.

[7] J. F. Elizabeth, "Plant life history traits of rare versus frequent plant taxa of sandplains: implications for research and management trials," *Biological Conservation*, vol. 136, no. 1, pp. 44–52, 2007.

[8] M. Kleyer, R. M. Bekker, I. C. Knevel et al., "The LEDA traitbase: a database of life-history traits of the Northwest European Flora," *Journal of Ecology*, vol. 96, no. 6, pp. 1266–1274, 2008.

[9] S. P. Bonser, B. Ladd, K. Monro, M. D. Hall, and M. A. Forster, "The adaptive value of functional and life-history traits across fertility treatments in an annual plant," *Annals of Botany*, vol. 106, no. 6, pp. 979–988, 2010.

[10] S. Porembski, W. Barthlott, S. Dörrstock, and N. Biedinger, "Vegetation of rock outcrops in Guinea: granite inselbergs, sandstone table mountains and ferricretes—remarks on species numbers and endemism," *Flora*, vol. 189, no. 4, pp. 315–326, 1994.

[11] A. Gröger and W. Barthlott, "Biogeography and diversity of the inselberg (Laja) vegetation of southern Venezuela," *Biodiversity Letters*, vol. 3, no. 6, pp. 165–179, 1996.

[12] T. D. McGann, "How insular are ecological 'islands'? An example from the granitic outcrops of the New England Batholith of Australia," *Proceedings of the Royal Society of Queensland*, vol. 110, pp. 1–13, 2002.

[13] J. T. Hunter, "Persistence on inselbergs: the role of obligate seeders and resprouters," *Journal of Biogeography*, vol. 30, no. 4, pp. 497–510, 2003.

[14] S. D. Hopper, "Honeyeaters and their winter food plants on granite rocks in the central wheatbelt of Western Australia," *Australian Wildlife Research*, vol. 8, no. 1, pp. 187–197, 1981.

[15] D. L. Phillips, "Succession in granite outcrop shrub-tree communities," *The American Midland Naturalist*, vol. 106, no. 2, pp. 313–317, 1981.

[16] J. M. Baskin and C. C. Baskin, "Endemism in rock outcrop plant communities of unglaciated eastern United States: an evaluation of the roles of the edaphic, genetic and light factors," *Journal of Biogeography*, vol. 15, no. 5-6, pp. 829–840, 1988.

[17] G. Houle and A. Delwaide, "Population structure and growth-stress relationship of *Pinus taeda* in rock outcrop habitats," *Journal of Vegetation Science*, vol. 2, no. 1, pp. 47–58, 1991.

[18] S. Ware, "Influence of interspecific competition, light and moisture levels on growth of rock outcrop *Talinum* (Portulacaceae)," *Bulletin of the Torrey Botanical Club*, vol. 118, no. 1, pp. 1–5, 1991.

[19] A. Burke, N. Jürgens, and M. K. Seely, "Floristic affinities of an inselberg archipelago in the southern Namib desert-relic of the past, centre of endemism or nothing special?" *Journal of Biogeography*, vol. 25, no. 2, pp. 311–317, 1998.

[20] J. T. Hunter, E. Fallavollita, and V. H. Hunter, "Observations on the ecology of *Muehlenbeckia costata* m.s. (Polygonaceae), a rare fire-ephemeral species occurring on the New England Batholith of northern New South Wales and southern Queensland," *Victorian Naturalist*, vol. 115, pp. 9–17, 1998.

[21] S. Lawler, S. Brown, G. Edney, S. Howelett, and P. Love, "Buffalo sallee at the back wall: an alpine species adapted to fire and drought," *Victorian Naturalist*, vol. 115, pp. 201–205, 1998.

[22] J. Szarzynski, "Xeric islands: environmental conditions on inselbergs," in *Inselbergs: Biotic Diversity of Isolated Rock Outcrops in Tropical and Temperate Regions*, S. Porembski and W. Barthlott, Eds., pp. 37–48, Springer, Berlin, Germany, 2000.

[23] G. S. Winterringer and A. G. Vestal, "Rock-Ledge vegetation in southern Illinois," *Ecological Monographs*, vol. 26, no. 2, pp. 105–130, 1956.

[24] W. H. Murdy, T. M. Johnson, and V. K. Wright, "Competitive replacement of *Talinum mangesii* by *T. teretifolium* in granite outcrop communities," *Botanical Gazette*, vol. 131, pp. 186–192, 1970.

[25] D. J. Shure and H. L. Ragsdale, "Patterns of primary succession on granite outcrop surfaces," *Ecology*, vol. 58, no. 5, pp. 993–1006, 1977.

[26] B. Büdel, U. Becker, G. Gollmann, and K. Sterflinger, "Algae, fungi and lichens on inselbergs," in *Inselbergs: Biotic Diversity of Isolated Rock Outcrops in Tropical and Temperate Regions*, S. Porembski and W. Barthlott, Eds., pp. 69–90, Springer, Berlin, Germany, 2000.

[27] W. J. Bond and J. J. Midgley, "Ecology of sprouting in woody plants: the persistence niche," *Trends in Ecology and Evolution*, vol. 16, no. 1, pp. 45–51, 2001.

[28] A. S. Benwell, "Post-fire seedling recruitment in coastal heathland in relation to regeneration strategy and habitat," *Australian Journal of Botany*, vol. 46, no. 1, pp. 75–101, 1998.

[29] P. J. Clarke, "Habitat insularity and fire response traits: evidence from a sclerophyll archipelago," *Oecologia*, vol. 132, no. 4, pp. 582–591, 2002.

[30] M. Halassy, G. Campetella, R. Canullo, and L. Mucina, "Patterns of functional clonal traits and clonal growth modes in contrasting grasslands in the central Appennines, Italy," *Journal of Vegetation Science*, vol. 16, pp. 29–36, 2005.

[31] M. Westoby, D. S. Falster, A. T. Moles, P. A. Vesk, and I. J. Wright, "Plant ecological strategies: some leading dimensions of variation between species," *Annual Review of Ecology and Systematics*, vol. 33, pp. 125–159, 2002.

[32] T. A. Forbis and D. F. Doak, "Seedling establishment and life history trade-offs in alpine plants," *American Journal of Botany*, vol. 91, no. 7, pp. 1147–1153, 2004.

[33] J. W. F. Slik, S.-I. Aiba, F. Q. Brearley et al., "Environmental correlates of tree biomass, basal area, wood specific gravity and stem density gradients in Borneo's tropical forests," *Global Ecology and Biogeography*, vol. 19, no. 1, pp. 50–60, 2010.

[34] S. K. Wiser and R. P. Buxton, "Context matters: matrix vegetation influences native and exotic species composition on habitat islands," *Ecology*, vol. 89, no. 2, pp. 380–391, 2008.

[35] C. Leigh, "The form and evolution of Bald Rock, New South Wales," *The Australian Geographer*, vol. 10, no. 5, pp. 333–345, 1968.

[36] P. J. Clarke, L. M. Copeland, J. T. Hunter, C. E. Nano, J. B. Williams, and K. E. Willis, *The Vegetation and Plant Species of Torrington State Recreation Area*, Division of Botany, University of New England, Armidale, Australia, 1998.

[37] J. T. Hunter and P. J. Clarke, "The vegetation of granitic outcrop communities of the New England Batholith of eastern Australia," *Cunninghamia*, vol. 5, pp. 547–618, 1998.

[38] J. T. Hunter, J. Kingston, and P. J. Croft, "Vegetation and floristics of Kwiambal National Park and surrounds, Ashford, New South Wales," *Cunninghamia*, vol. 6, pp. 351–378, 1999.

[39] J. T. Hunter, A. Wyatt, D. Hofmeyer, L. Brown, N. Barkwell, and N. J. Beresford-Smith, "Vegetation and floristics of the demon nature reserve, Tenterfield, New South Wales," *Cunninghamia*, vol. 6, pp. 331–350, 1999.

[40] J. T. Hunter and V. H. Hunter, "Vegetation of ironbark nature reserve and *Bornhardtia* voluntary conservation agreement, Northern Tablelands, New South Wales," *Cunninghamia*, vol. 8, pp. 93–110, 2003.

[41] J. T. Hunter, "Vegetation of basket swamp National Park, Northern Tablelands, New South Wales," *Cunninghamia*, vol. 8, pp. 453–466, 2004.

[42] JT. Hunter, "Vegetation and floristics of Warra National Park and Wattleridge Northern Tablelands," *NSW. Cunninghamia*, vol. 9, pp. 255–274, 2005.

[43] J. T. Hunter and P. Sheringham, "Vegetation and floristic diversity in Gibraltar Range and part of Washpool National Parks, New South Wales," *Cunninghamia*, vol. 10, pp. 434–479, 2008.

[44] R. K. Outhred, "Semi-quantitative sampling in vegetation survey," in *Survey Methods for Nature Conservation*, K. Myers, C. Margules, and I. Musto, Eds., vol. 1, pp. 87–100, CSIRO Division of Water and Land Resources, Canberra, Australia, 1984.

[45] C. E. Raymont and F. R. Higginson, *Australian Laboratory Handbook of Soil and Water Chemical Method*, Inkata Press, Melbourne, Australia, 1992.

[46] E. Garnier, S. Lavorel, P. Ansquer et al., "Assessing the effects of land-use change on plant traits, communities and ecosystem functioning in grasslands: a standardized methodology and lessons from an application to 11 European sites," *Annals of Botany*, vol. 99, no. 5, pp. 967–985, 2007.

[47] C. J. F. ter Braak and P. Šmilauer, *CANOCO Reference Manual and Users Guide Software for Ordination (Version 5.0)*, Microcomputer Power, Ithaca, NY, USA, 2012.

[48] Ø. Hammer, D. A. T. Harper, and P. D. Ryan, "Past: paleontological statistics software package for education and data analysis," *Palaeontologia Electronica*, vol. 4, article 9, 2001.

[49] P. B. Adler, A. Fajardo, A. R. Kleinhesselink, and N. J. B. Kraft, "Trait-based tests of coexistence mechanisms," *Ecology Letters*, vol. 16, no. 10, pp. 1294–1306, 2013.

[50] A. L. Mack, "An advantage of large seed size: tolerating rather than succumbing to seed predators," *Biotropica*, vol. 30, no. 4, pp. 604–608, 1998.

[51] P. J. Clarke, D. M. Bell, and M. J. Lawes, "Testing the shifting persistence niche concept: plant resprouting along gradients of disturbance," *The American Naturalist*, vol. 185, no. 6, pp. 747–755, 2015.

[52] P. Croft, J. T. Hunter, and N. Reid, "Depletion of regenerative bud resources during cyclic drought. What are the implications for fire management?" *Ecological Management & Restoration*, vol. 8, no. 3, pp. 187–192, 2007.

[53] P. J. Croft, J. T. Hunter, and N. Reid, "Threat of frequent fire and drought for the rare wattle *Acacia williamsiana* J. T. Hunter: an experimental burn highlights implications for fire management," *Ecological Management and Restoration*, vol. 11, pp. 217–220, 2010.

[54] L. Baastrup-Spohr, K. Sand-Jensen, S. V. Nicolajsen, and H. H. Bruun, "From soaking wet to bone dry: predicting plant community composition along a steep hydrological gradient," *Journal of Vegetation Science*, vol. 26, no. 4, pp. 619–630, 2015.

[55] J. Silvertown, *Introduction to Plant Population Ecology*, Longman Scientific and Technical, New York, NY, USA, 2nd edition, 1987.

[56] R. Snell and L. W. Aarssen, "Life history traits in selfing versus outcrossing annuals: exploring the 'time-limitation' hypothesis for the fitness benefit of self-pollination," *BMC Ecology*, vol. 5, article 2, 2005.

[57] J. T. Hunter, "Factors affecting range size differences for plant species on rock outcrops in eastern Australia," *Diversity and Distributions*, vol. 9, no. 3, pp. 211–220, 2003.

[58] J. T. Hunter and J. B. Williams, "A new species of *Brachyloma* and three new subspecies of *B. daphnoides* (Epacridaceae) from south-eastern Australia," *Telopea*, vol. 6, pp. 1–7, 1994.

[59] D. Primack, C. Imbres, R. B. Primack, A. J. Miller-Rushing, and P. Del Tredici, "Herbarium specimens demonstrate earlier flowering times in response to warming in Boston," *American Journal of Botany*, vol. 91, no. 8, pp. 1260–1264, 2004.

[60] R. L. Specht and A. Specht, *Australian Plant Communities: Dynamics of Structure, Growth and Biodiversity*, Oxford University Press, Melbourne, Australia, 1999.

[61] P. Poot, S. D. Hopper, and J. M. H. van Diggelen, "Exploring rock fissures: does a specialized root morphology explain endemism on granite outcrops?" *Annals of Botany*, vol. 110, no. 2, pp. 291–300, 2012.

[62] L. Hughes, "Climate change and Australia: trends, projections and impacts," *Austral Ecology*, vol. 28, no. 4, pp. 423–443, 2003.

[63] K. Hennessy, C. Page, K. McInnes, R. Jones, and A. Bathols, *Climate Change in New South Wales. Part 1: Past Climate Variability and Projected Changes in Average Climate*, CSIRO, Australian Government Bureau of Meteorology, Canberra, Australia, 2004.

[64] PMSEIC Independent Working Group, "Climate change in Australia: regional impacts and adaptation—managing risk for Australia," Tech. Rep., Prime Minister's Science, Engineering and Innovation Council, Canberra, Australia, 2007.

13

Species Richness, Community Organization, and Spatiotemporal Distribution of Earthworms in the Pineapple Agroecosystems of Tripura, India

Animesh Dey and P. S. Chaudhuri

Earthworm Research Laboratory, Department of Zoology, Tripura University, Suryamaninagar, West Tripura 799022, India

Correspondence should be addressed to P. S. Chaudhuri; priya_1956@rediffmail.com

Academic Editor: L. M. Chu

The impact that plant communities may have on underground faunal diversity is unclear. Therefore, understanding the links between plants and organisms is of major interest. Earthworm population dynamics were studied in the pineapple agroecosystems of Tripura to evaluate the impact of monoculture plantation on earthworm communities. A total of thirteen earthworm species belonging to four families and five genera were collected from different sampling sites. Application of sample-based rarefaction curve and nonparametric richness estimators reveal 90–95% completeness of sampling. Earthworm community of pineapple agroecosystems was dominated by endogeic earthworms and *Drawida assamensis* was the dominant species with respect to its density, biomass, and relative abundance. Vertical distribution of earthworms was greatly influenced by seasonal variations. Population density and biomass of earthworms peaked during monsoon and postmonsoon period, respectively. Overall density and biomass of earthworms were in increasing trend with an increase in plantation age and were highest in the 30–35-year-old plantation. Significant decrease in the Shannon diversity and evenness index and increase in Simpson's dominance and spatial aggregation index with an increase in the age of pineapple plantation were recorded. Soil temperature and soil moisture were identified as the most potent regulators of earthworm distribution in the pineapple plantation.

1. Introduction

Earthworms are the most important soil biota of agroecosystem where they play a key role in regulating soil fertility and promoting plant growth through nutrient cycling and water infiltration [1, 2]. Since earthworms account for the highest biomass among tropical soil macrofauna [3], soil quality depends on population density, species diversity, and activity level of earthworm community. Under similar climate conditions, factors that determine distribution and community structure of earthworms may depend on the changes in land use pattern [4, 5] as well as plant species composition and the physical and chemical properties of soils within a plant community [6, 7]. Thus, earthworm population dynamics and the influence of farming practices on earthworms are of particular interest to organic growers [8].

Pineapple (*Ananas comosus* (L.) Merr.), native to Southern Brazil and Paraguay, was introduced to India by Portuguese in 1548 AD [9]. It is one of the most delicious tropical fruit crops of commercial importance and widely cultivated in the hill slopes of Tripura, checking soil erosion. Crude extracts from fruits, stems, and leaves of pineapples are good sources of various kinds of sugars, organic acids, vitamins, and several proteinase enzymes including bromelains and peroxidases with immunomodulatory, anti-inflammatory, antithrombotic, fibrinolytic, antihelminthic, and tumour growth inhibitory properties [10]. The agroclimatic conditions prevailing in Tripura are ideal for the commercial production of its three common varieties, namely, Queen, Kew, and Mauritius.

The role of earthworms in organic matter decomposition as well as building and maintenance of soil structure has been well documented for the soils of temperate regions [11]. Although, in tropics, studies on diversity, ecology, and role of earthworms have been carried out in different agroclimatic zones of Indian subcontinent [12–17], there are scanty records on the diversity and distribution of earthworms and their role in the fruit crop plantations [18] of India including its northeastern parts.

The present paper deals with the community composition and spatiotemporal distribution of earthworms in pineapple plantations with reference to the edaphic factors in Tripura. Our hypothesis is that higher density and dominance of earthworms would be associated with monoculture pineapple plantation. The underlying idea is that if a plant species affects earthworms positively (i.e., due to the matter input), presumably the presence of the resource in abundance will likely promote the establishment of earthworms that benefitted from the resource. Earthworms may use the exudates and metabolites of soil where pineapples are cultivated since they are useful in degrading pineapple wastes [19].

2. Materials and Methods

2.1. Study Area and Sites. The studies on the earthworm communities were conducted during April 2008–September 2011 in Tripura (22°51′–24°32′N and 90°10′–92°21′E) having a total area of 10,491 sq·km. The state is almost encircled by Bangladesh except in the north-east where it meets its neighboring states, Assam and Mizoram. The climatic features of the study area are summer (March–May), monsoon (June–September), autumn (October-November), and winter (December–February) with a mean annual rainfall of 2000 mm and temperature of 25°C. Sampling of earthworms was done in the pineapple (*Ananas comosus* var. *queen*) plantations (>15 years) at different localities, namely, Bamutia, Nandannagar, Shalbagan, Nutannagar, Bishalgarh, Bishramganj, Padmanagar, Jumerdhepa, and Boiragibazar of Tripura. The distance between the studied sites varied within 20–50 km.

Pineapple plantations of different age groups (1–5 years, 15–20 years, and 30–35 years old) were studied to find the variations in the earthworm community structure in different age groups of pineapple plantation because with increase in plantation age, the chemistry of the soils also changes with input of litter quality.

Pineapple cannot tolerate water-logged condition and is thus usually on undulating uplands, locally called tilla. The soils of well-drained pineapple plantations were acidic (pH 4.6–5.5) in nature with loamy sand, loam, or sandy loam texture. *Nephelium litchi* (Sapindaceae) was the most abundant tree of pineapple plantation. Besides this, presence of *Mangifera indica* (Anacardiaceae) and *Cassia tora* (Caesalpiniaceae) along with *Syzygium cumini* (Myrtaceae) and *Cassia sophera* (Caesalpiniaceae) in the pineapple plantations was remarkable.

2.2. Experimental Design. Earthworms were collected during summer (April–June) and monsoon (July–September) and postmonsoon period of autumn (October) of each year during the study period of 2008–2011 by TSBF soil monolith (25 cm × 25 cm × 40 cm) digging method [20]. This method works well in the soils of Tripura where earthworm communities are dominated mainly by endogeic species. A total of 40 widely separated 10 m × 10 m plots were randomly selected for sampling in study site. A composite sample comprising five TSBF soil monoliths (25 cm × 25 cm × 40 cm) were taken from the centre and 4 corners of each sampling plot [21]. Thus, a total of 120 samples (600 TSBF monoliths) were taken from each sampling site. All earthworms were collected directly by hand-sorting from vertically stratified four successive strata of 10 cm depth from the top to the bottom of each soil monolith [22]. Sampling was done only from plain plots above and below the stiff slopes due to difficulties of sampling in the latter. Later, individuals were identified based on taxonomic keys provided by Gates [23] and Julka [24], counted, weighed, and preserved in the laboratory. Results were expressed in terms of biomass (fresh weight in g m^{-2}) and density (ind m^{-2}). Using the data available, relative abundance, frequency, Simpson's index of dominance, Menhinick's species richness index, Shannon's index of general diversity, and species evenness of earthworm communities were calculated [25] and the ecological categories of earthworms [26] of the studied sites were also determined.

2.3. Sample-Based Rarefaction Curve and Nonparametric Richness Estimators. Sample-based data were used for the calculation of sample-based rarefaction curve, which is actually the statistical expectation of the corresponding species accumulation curve [27] to plot the cumulative number of species recorded as a function of sampling effort [28]. This curve is a classic but informal way to assess the completeness of an inventory [29]. It may provide an estimate of the total species richness of an assemblage, unless sampling has been exhaustive [30].

For the calculation of nonparametric richness estimators, sample-based data were used with ESTIMATES 8.2.0 [31] and graphs were generated by ORIGIN 6.0 Professional [32]. A total of eight richness estimators were compared based on two types of data, namely, incidence and abundance based data [33] in order to verify which fitted best the obtained data set. Estimators like Chao 1 [34], abundance based coverage estimator (ACE) [35], and first-order Jackknife richness estimators (Jackknife 1) [36, 37] are based on abundance data, whereas Chao 2 [38], incidence based coverage estimator (ICE) [35], second-order Jackknife richness estimators (Jackknife 2) [36, 37], boot strap [39], and Michaelis-Menten estimators [40] are based on incidence data. Detailed descriptions of these algorithms are available in Toti et al. [41] and Colwell [42].

2.4. Spatial Pattern Analysis. The horizontal distribution of earthworms in pineapple plantation was studied by analyzing the spatial pattern [43]. Discrete probability distributions like Poisson and Negative Binomial Distribution (NBD) models were applied to study the spatial patterns of earthworms [44]. Goodness of fit of the observed probabilities and cumulative distribution of earthworm densities were compared with the expected binomial (random distribution) and NBD (clumped distribution) probabilities using the Kolmogorov-Smirnov test [45–47] and Anderson Darling test [48]. Differences between relative cumulative frequencies and critical values decide whether the maximum differences between the observed and expected cumulative frequency distribution are significant [7].

Three indices of dispersion for quadrate count of earthworms were employed: the variance (s^2) to mean (m) ratio [49], Morisita's index of dispersion (I_d) [50], and Negative Binomial (NB) parameter (k) [44]. The variance to mean ratio is also called index of dispersion (I). The I and I_d will give expected values <1 and >1 for uniform and clumped distribution, respectively [7]. Lower value of k indicates pronounced clumping whereas its higher value indicates slight clumping. If k equals zero, the clumping is maximum [44].

2.5. Soil Analysis. Soil samples were collected at 0–15 cm depth from the localities of maximum earthworm occurrence with a metal shovel and composite soil samples (30 samples per sampling site) comprising 5 subsamples were prepared for physicochemical analysis. Soil samples were air-dried, ground with mortar and pestle, and sieved with 1 mm and 2 mm sieves. Sieved soil samples were analyzed for their moisture (gravimetric wet weight method), pH (1 : 2.5 dilution method), soil organic matter [51], and texture [52]. Soil temperatures were recorded in situ at each sample plot at a depth of 15 cm.

2.6. Multivariate Analysis. Pairwise comparison of earthworm species composition between three age groups of pineapple plantations was identified by calculating Bray-Curtis similarity coefficient (BSC). Based on BSCs, cluster analysis was performed. A dendrogram was constructed by means of unweighted paired-group method with arithmetic averages (UPGMA) [53]. Euclidean distance (ED) was also applied in cluster analysis according to soil physicochemical properties [54]. Following cluster analysis, principal coordinates analysis (PCorA) was computed to investigate the overall variations in the present data set from different age groups of pineapple plantations and to provide an additional representation of similarity [55].

A canonical correlation analysis (CCorA) was performed to analyze the nature of relationship between soil physicochemical properties (independent variables) and overall earthworm population density and biomass (dependent variables) [56]. CCorA is a multivariate statistical model that facilitates the study of linear interrelationship between two sets of variables. One set of variables is referred to as independent variables and others are considered dependent variables; a canonical variate is formed for each set [57]. This analysis develops a canonical function that maximizes the canonical correlation coefficient between two canonical variates and determines the strength and nature of intra- and intervariable relationships [56].

Canonical correspondence analysis (CCA) [58] was performed to explore species specific associations among earthworm abundances relative to soil physicochemical parameters (temperature, moisture, pH, and organic matter). CCA results are displayed graphically with biplot scaling focused on interspecies distances, where vectors depict environmental variables and taxa (species) are represented as points. The biplot approximates the weighted averages of each taxon (species) with respect to each of the environmental variables (soil properties).

3. Results

3.1. Species Composition and Community Organization. A total of 13 species of earthworms with a minimum of five and a maximum of eleven earthworm species were collected from different sampling sites of pineapple plantations. Among them, 4 species belonged to the family Megascolecidae (*Metaphire houlleti* (Perrier), *Metaphire posthuma* (Vailant), *Kanchuria* sp., and *Kanchuria sumerianus* Julka), 5 species to the family Octochaetidae (*Eutyphoeus gigas* Stephenson, *Eutyphoeus scutarius* Michaelsen, *Eutyphoeus comillahnus* Michaelsen, *Eutyphoeus gammiei* (Beddard), and *Eutyphoeus* sp.), 3 species to the family Moniligastridae (*Drawida assamensis* Gates, *Drawida papillifer papillifer* Stephenson, and *Drawida nepalensis* Michaelsen), and one species to the family Glossoscolecidae (*Pontoscolex corethrurus* (Muller)). In live condition, *E. gammiei* was the largest (length 210–320 mm, diameter 8–12 mm) earthworm and *P. Corethrurus* was the smallest (length 50–70 mm, diameter 3–5 mm) earthworm of pineapple plantation in Tripura (Table 1). *M. houlleti*, *M. posthuma*, and *P. corethrurus* are exotic, whereas the rest are endemic to the Indian subcontinent. Density, biomass, and relative abundance of different earthworm species are shown in Table 1. In respect to their density, biomass, and relative abundance, *D. assamensis* was the dominant earthworm species and *M. posthuma*, *Kanchuria* sp., *K. sumerianus*, *E. gigas*, *E. scutarius*, *E. comillahnus*, *E. gammiei*, *Eutyphoeus* sp., and *D. nepalensis* were the rare species (Table 1) of pineapple plantations. Among 13 species, two were epianecic (*M. houlleti* and *D. papillifer papillifer*), two were endoanecic (*E. gigas* and *E gammiei*), and the rest were of endogeic categories.

Density, biomass, relative abundance, and frequency of 13 earthworm species of pineapple plantations are given in Table 1. *D. assamensis* topped the list followed by *P. corethrurus*, *D. papillifer papillifer*, and *M. houlleti*. Density of earthworms ranged from 52 ind m^{-2} to 194 ind m^{-2} (mean 142.3 ± 4.43) and biomass ranged from 12 g m^{-2} to 51 g m^{-2} (mean 37.19 ± 1.97). *Drawida assamensis* was the dominant earthworm species of pineapple plantations, representing 81.2% density and 73.4% biomass of total earthworm species population. *P. corethrurus*, *D. papillifer papillifer*, *M. houlleti*, and *Kanchuria* sp. contributed 15.1%, 3.6%, 2.9%, and 0.9% to total earthworm biomass.

3.2. Sampling Optimization and Nonparametric Richness Estimation. The sample-based rarefaction curve (Figure 1(a)) generated from our complete sampling revealed the observed species richness (S_{obs}). Initially, the curve rose rapidly but the rate of rise slowed down later on. Thirteen earthworm species (S_{obs}) were observed in the pineapple plantations of West Tripura during the present study (Table 2; Figure 1(a)).

Table 1: Ecological categories, habitat, and morphological and population characteristics of the earthworm species of pineapple plantation in Tripura.

Family and earthworm species	Ecological categories	Habitat	Size ($D \times L$) mm	Biomass ($g \cdot m^{-2}$)	Density ($ind \cdot m^{-2}$)	Relative abundance (%)	Frequency (%)
Megascolecidae							
*M. houlleti**	Epianecic	Surface soil	5–7 × 190–220	1.09 ± 0.50	3.68 ± 0.87	2.52 ± 0.43	14.72 ± 4.64
*M. posthuma**	Endogeic	0–10 cm	3–5 × 85–90	0.06 ± 0.01	0.16 ± 0.05	0.12 ± 0.09	0.56 ± 1.07
Kanchuria sp.	Endogeic	0–20 cm	4-5 × 180–210	0.32 ± 0.09	0.64 ± 0.06	0.49 ± 0.30	2.78 ± 0.83
K. sumerianus	Endogeic	0–15 cm	2-3 × 70–140	0.05 ± 0.01	0.05 ± 0.03	0.03 ± 0.01	0.28 ± 0.03
Octochaetidae							
E. gigas	Endoanecic	0–30 cm, in vertical burrows	8–12 × 195–290	0.72 ± 0.01	0.48 ± 0.02	0.37 ± 0.02	2.78 ± 0.02
E. scutarius	Endogeic	0–15 cm	7–11 × 220–310	0.10 ± 0.03	0.09 ± 0.02	0.06 ± 0.01	0.56 ± 0.05
E. comillahnus	Endogeic	0–15 cm	3–5 × 120–255	0.27 ± 0.06	0.40 ± 0.04	0.28 ± 0.01	1.94 ± 0.07
Eutyphoeus sp.	Endogeic	0–15 cm, in horizontal burrow	4-5 × 160–210	0.09 ± 0.01	0.16 ± 0.07	0.12 ± 0.05	1.11 ± 0.05
E. gammiei	Endoanecic	0–30 cm, in vertical burrows	8–12 × 210–320	0.01 ± 0.03	0.05 ± 0.01	0.03 ± 0.02	0.28 ± 0.08
Moniligastridae							
D. assamensis	Endogeic	0–10 cm	3–5 × 50–80	27.29 ± 0.39	115.52 ± 4.96	81.23 ± 1.12	96.94 ± 3.01
D. papillifer papillifer	Epianecic	0–10 cm	2–4 × 50–110	1.33 ± 0.48	4.32 ± 0.93	2.99 ± 0.41	17.78 ± 4.17
D. nepalensis	Endogeic	0–15 cm	4–6 × 70–130	0.25 ± 0.02	0.64 ± 0.35	0.46 ± 0.21	1.67 ± 1.44
Glossoscolecidae							
*P. corethrurus**	Endogeic	0–15 cm soil	3–5 × 50–70	5.61 ± 2.27	16.16 ± 0.93	11.39 ± 0.48	72.22 ± 7.12
Total				**37.19 ± 1.97**	**142.35 ± 4.43**		

Values represent mean ± standard error (SE); *exotic species.

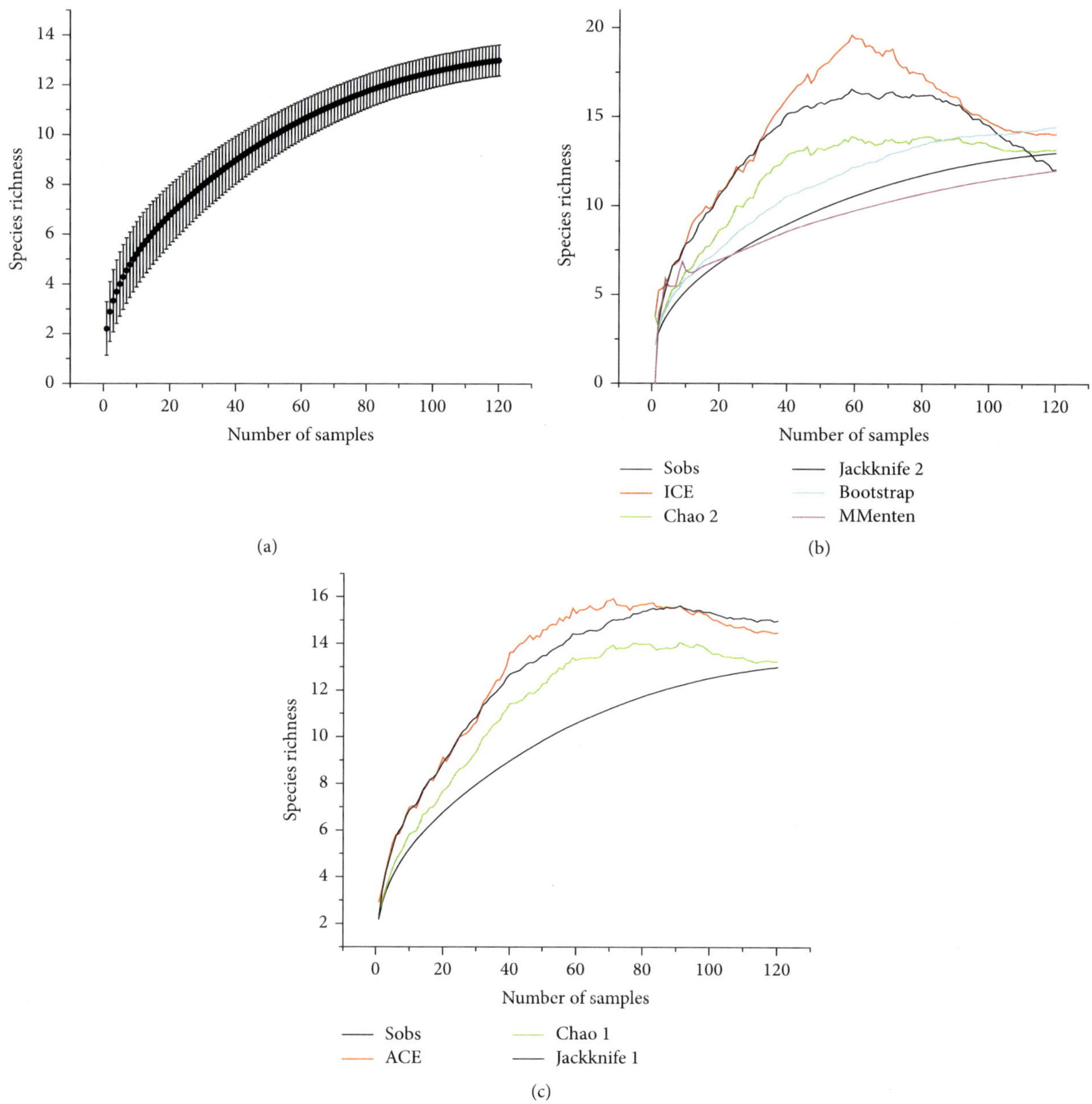

FIGURE 1: (a) Sample-based rarefaction curve revealed the observed species richness. (b) Showing the performance of incidence based nonparametric richness estimators for pineapple plantation. (c) Showing the performance of abundance based nonparametric richness estimators for pineapple plantation.

Sample-based rarefaction curve continued to rise as sample number increased and almost approached saturation while it reached up to 120 samples.

Among the richness estimators, the Jackknife 2 and Michaelis-Menten estimators underestimated the total species richness by producing the erratic and nonasymptotic curves (12.07 and 12.02) [59], whereas the Jackknife 1 and Bootstrap estimators overestimated the richness (14.98 and 14.46) [60] (Table 2; Figures 1(b) and 1(c)). The Chao 1 estimator curve closely resembled the species accumulation curve initially but leveled off in the middle and finished equally with the observed richness (Figure 1(c)). ICE and ACE curves, on the other hand, continued to rise with an increase in number of samples and appeared to fall as the number of samples approached maximum (Figures 1(b) and 1(c)). Unlike the other estimators, Chao 2 stabilized much before the total cumulative sampling effort and finished at 13 species (Figure 1(b)). This lack of consensus among the estimators is echoed in their total richness which did not cluster tightly but ranged from 12 (Michaelis-Menten) to 15 (Jackknife 1) (Table 2). Taking into consideration these curves, it appears that approximately 90–95% of species were discovered (Table 2) and another one or two earthworm species may yet be seen in the pineapple plantations of West Tripura.

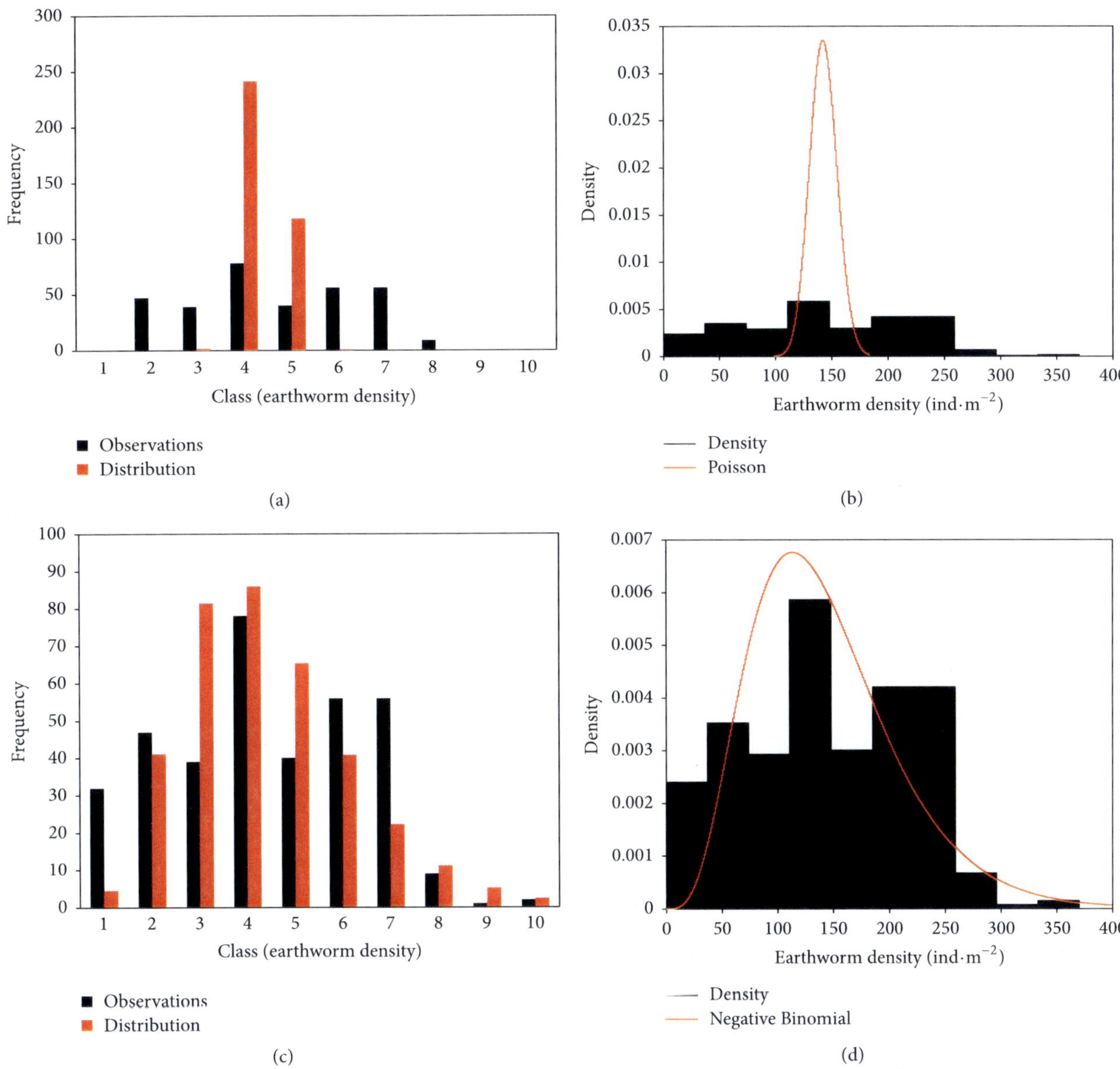

FIGURE 2: Showing the observed and expected frequency distribution and probability density function for (a, b) observed earthworm density, Poisson (random) distribution and (c, d) observed earthworm density, and Negative Binomial (clumped) distribution in pineapple plantations of Tripura.

TABLE 2: Performance of different nonparametric richness estimators.

Estimators	Observed richness (S_{obs})	Estimated richness (Est.)	Estimation percent (%)
Chao 1	13	13.25	98
ACE	13	14.49	90
Jackknife 1	13	14.98	87
Chao 2	13	13.17	99
ICE	13	14.06	92
Jackknife 2	13	12.07	—
Michaelis-Menten	13	12.02	—
Bootstrap	13	14.46	90

3.3. Spatiotemporal Patterns

3.3.1. Horizontal Distribution. The probability density function (PDF) for observed earthworm densities and expected Poisson distribution and Negative Binomial Distribution (Figure 2) were used for detecting the spatial pattern of earthworm distribution in pineapple plantations. NBD was well fitted to the observed pattern of community when compared to Poisson distribution model. Goodness of fit of NBD rather than Poisson distribution model is the indication of spatially clumped distribution of earthworm communities in the pineapple agroecosystems.

Means of counts was 142 individuals per sampling unit for pineapple plantation. The variance counts were high and

TABLE 3: Vertical distribution of earthworm species in different soil strata.

Earthworm species	Depth (cm)			
	0–10 cm	10–20 cm	20–30 cm	30–40 cm
M. houlleti	+	−	−	−
M. posthuma	+	−	−	−
Kanchuria sp.	+	+	−	−
K. sumerianus	−	+	−	−
E. gigas	+	+	+	+
E. scutarius	+	+	+	−
E. comillahnus	+	+	+	−
Eutyphoeus sp.	−	+	+	−
E. gammiei	−	+	+	+
D. assamensis	+	+	+	−
D. papillifer papillifer	+	+	−	−
D. nepalensis	+	+	+	−
P. corethrurus	+	+	+	−

(+) present; (−) absent.

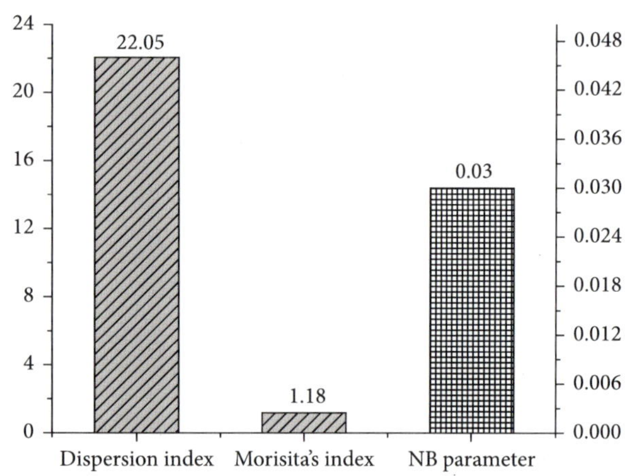

FIGURE 3: Showing the values of three different dispersion indices indicating the higher degree of clumped horizontal distribution of the earthworm community in pineapple plantation.

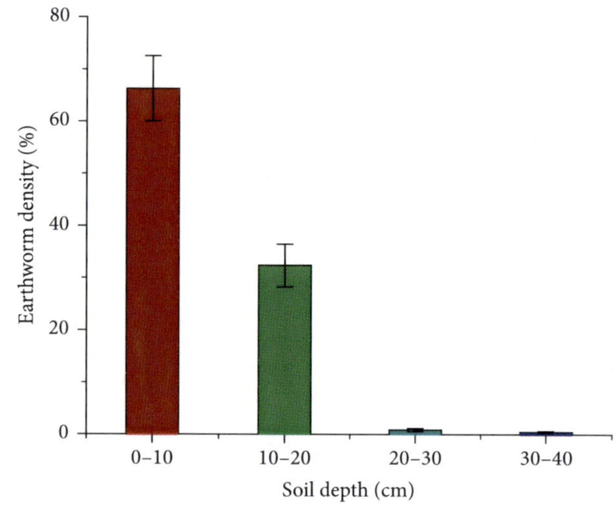

FIGURE 4: Showing concentration of earthworm density in different soil strata.

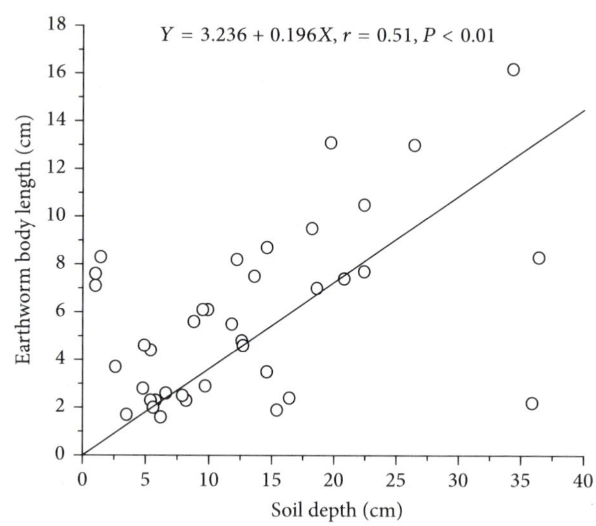

FIGURE 5: Linear regression analysis showing the overall relationship between earthworm body length and soil depth for the earthworms of pineapple plantation.

led to significant variance to mean aggregation (Figure 3). Morisita's index of dispersion (I_d) was higher (1.18) and the Negative Binomial parameter (k) was lower (0.03) in the pineapple plantations, indicating the higher clumping spatial pattern of earthworm community in the former (Figure 3).

3.3.2. Vertical and Temporal Distribution. The assessment of vertical pattern of distribution allowed us to determine those layers that were exploited by different earthworm species. The study during the summer, monsoon, and postmonsoon period of the year 2009-2010 revealed that about 90% of the total earthworm density was concentrated in the upper 20 cm of soils (Figure 4), among which a clear concentration was observed in the top soils of 10 cm depth.

In the earthworm communities, *E. gigas* was found to be the most versatile species being distributed in all the vertical strata of pineapple plantation (Table 3). On the other hand, *M. houlleti* and *M. posthuma* were found to be restricted to the top 10 cm of soil (Table 3). By analyzing the overall vertical distribution data, a significant positive correlation was found between earthworm length and soil depth (Figure 5); that is, the worms having larger body length generally inhabited a greater soil depth.

In the pineapple plantations of Tripura, vertical distribution of earthworm community was influenced by seasonality. During summer (dry season), earthworm density was greater in the deeper strata (10–20 cm, 20–30 cm), whereas the scenario was opposite in case of wet season (rain), that is, high worm density in the top soil layer (0–10 cm) (Figure 6). A decrease in worm density during postmonsoon period, compared to monsoon in the upper strata, was noteworthy.

The overall earthworm population density and biomass were greatly influenced by seasonality in the pineapple agroecosystems of Tripura. The highest population density and biomass of earthworms were recorded during the monsoon and postmonsoon period, respectively (Figures 7(a) and 7(b)).

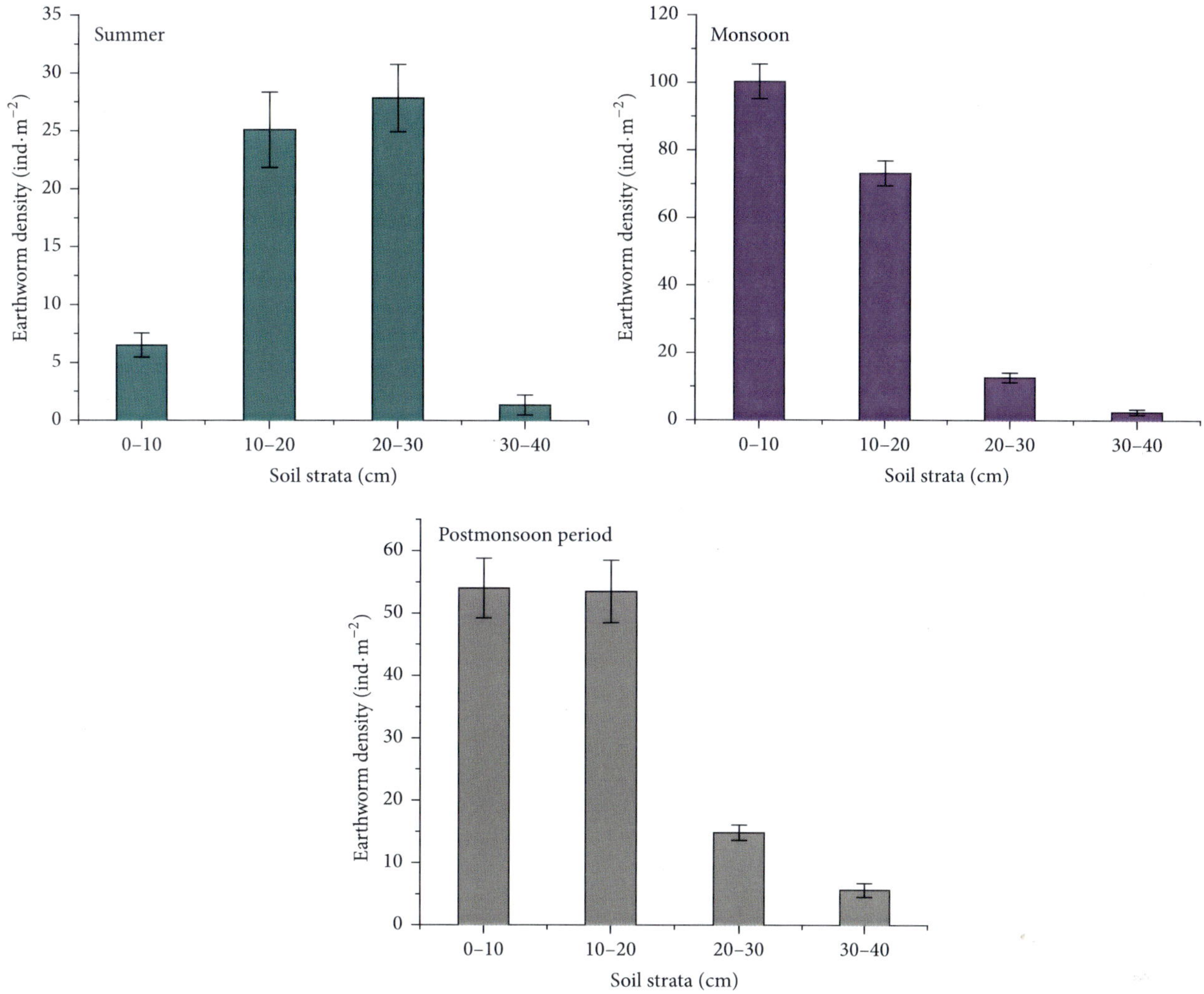

FIGURE 6: Showing the influence of seasonality on the vertical distribution of earthworm community in the pineapple plantations of Tripura.

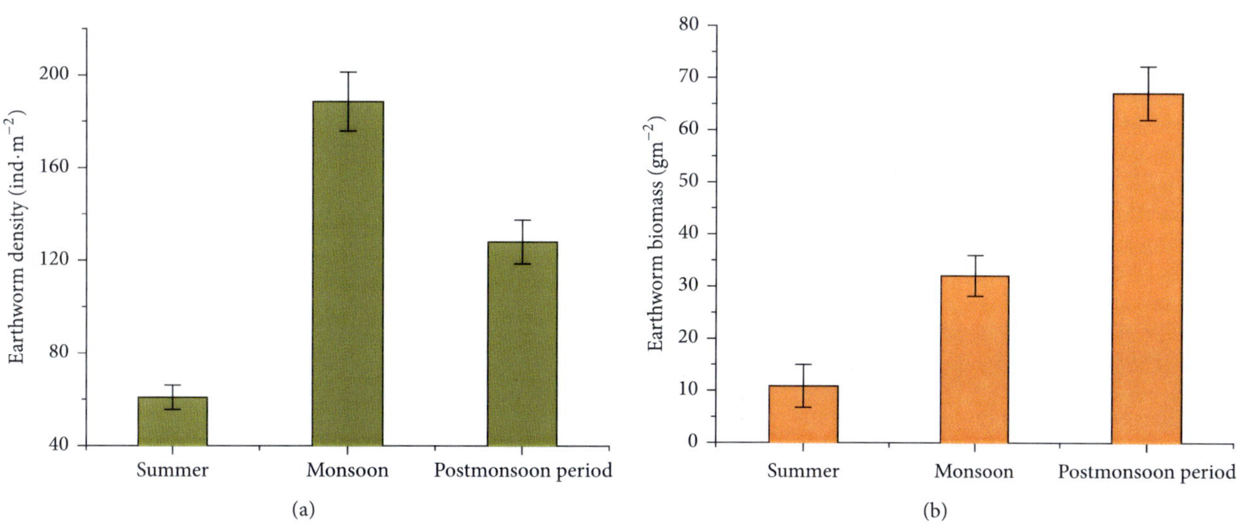

FIGURE 7: (a) Showing the influence of seasonality on the overall density of earthworm community in the pineapple plantation of Tripura. (b) Showing the influence of seasonality on the overall biomass of earthworm community in the pineapple plantation of Tripura.

TABLE 4: Comparison of soil physicochemical and earthworm community parameters among different age group of pineapple plantations of Tripura.

Parameters	1–5 years	15–20 years	30–35 years
Soil texture	Loamy sand	Loam	Sandy loam
Canopy cover	Absent	Minimum	Scattered
Temperature (°C)	26.62 ± 0.07^a	26.14 ± 0.07^b	25.83 ± 0.07^c
Moisture (%)	16.61 ± 0.29^a	18.84 ± 0.43^b	20.09 ± 0.35^c
pH	5.02 ± 0.04^a	4.81 ± 0.33^b	4.63 ± 0.04^c
Organic matter (%)	1.36 ± 0.02^a	1.93 ± 0.08^b	2.38 ± 0.06^c
Total species richness	07	07	11
Min. richness sample^{-1}	1	1	1
Max. richness sample^{-1}	3	4	5
Mean richness sample^{-1}	1.78 ± 0.08^a	1.90 ± 0.06^a	2.21 ± 0.06^b
Species name	*D. assamensis*	*D. assamensis*	*D. assamensis*
	D. papillifer papillifer	*D. papillifer papillifer*	*D. papillifer papillifer*
	E. gigas	*E. gigas*	*E. gigas*
	M. houlleti	*M. houlleti*	*M. houlleti*
	P. corethrurus	*P. corethrurus*	*P. corethrurus*
	M. posthuma	*Kanchuria* sp.	*Kanchuria* sp.
	E. gammiei	*Eutyphoeus* sp.	*Eutyphoeus* sp.
	—	—	*E. scutarius*
	—	—	*E. comillahnus*
	—	—	*K. sumerianus*
	—	—	*D. nepalensis*
Average biomass (gm^{-2})	15.04 ± 1.83^a	23.78 ± 2.06^b	41.92 ± 0.67^c
Average density ($\text{ind}\cdot\text{m}^{-2}$)	53.73 ± 2.57^a	77.20 ± 2.89^b	158.67 ± 8.17^c

Values represent mean ± standard error (SE); dissimilar letters indicate significant difference at 5% level of significance.

Interestingly, while earthworm biomass increased gradually with the seasonality from summer to postmonsoon period, earthworm density showed a sudden increase followed by a considerable decrease (Figures 7(a) and 7(b)).

3.4. Impact of Pineapple Plantation Age on Earthworm Community. Comparative studies on structure of earthworm communities in the three age groups of pineapple plantations (1–5 years, 15–20 years, and 30–35 years) revealed common occurrence of at least 5 species (*Drawida assamensis, Drawida papillifer papillifer, Eutyphoeus gigas, Metaphire houlleti*, and *Pontoscolex corethrurus*) of earthworms (Table 4). Species richness was significantly ($P < 0.05$) higher in the 30–35-year-old plantation with eleven species (mean species richness 2.21 ± 0.06), while in the other age groups of plantations it was only seven (mean species richness: 1.9 ± 0.06 (15–20 yrs); 1.8 ± 0.08 (1–5 yrs)).

Cluster analysis based on Euclidean distance and Bray-Curtis similarity coefficient revealed the distance and similarity between the three plantation types in terms of soil properties and earthworm species composition. Both dendrograms divide all the studied sites into two distinct categories (immature and old) based on plantation age (Figure 8(a)). The cluster of immature plantations again was subdivided into two definite age groups (juvenile and young). Immature and old plantation groups showed least similarity and highest distance (BSC 0.55 and ED 95). On the other hand, juvenile and young plantations showed BSC 0.72 and ED 25. This clustering almost perfectly classified pineapple plantation based on age structure.

Results of UPGMA clustering based on Bray-Curtis index were confirmed by the pattern of PCorA plotting (Figure 8(b)). A total of 94% of the variations among different age group of pineapple plantations was explained by the first two principal coordinates accounting for 84.2% and 9.9% of total variance, respectively. The first principal coordinate clearly divided studied sites into three groups, forming three clusters based on plantation age. This pattern corresponded entirely with the clusters of UPGMA dendrogram.

Interestingly, there was an increasing trend in the overall biomass and density of earthworms with increase in the plantation age. Maximum earthworm density (158 ind m^{-2}) and biomass (42 g m^{-2}) were recorded in 30–35-year-old pineapple plantation. Earthworm population density and biomass differed significantly ($P < 0.05$) among different age groups of plantations (Table 4).

A significant decrease ($P < 0.05$) in Shannon diversity index and species evenness and increase ($P < 0.05$) in Simpson's dominance index with increase in the age of pineapple plantations was noted (Figure 9). Morisita's index of dispersion (I_d) was greater than 1 in all the age groups of pineapple plantations that revealed aggregated nature of

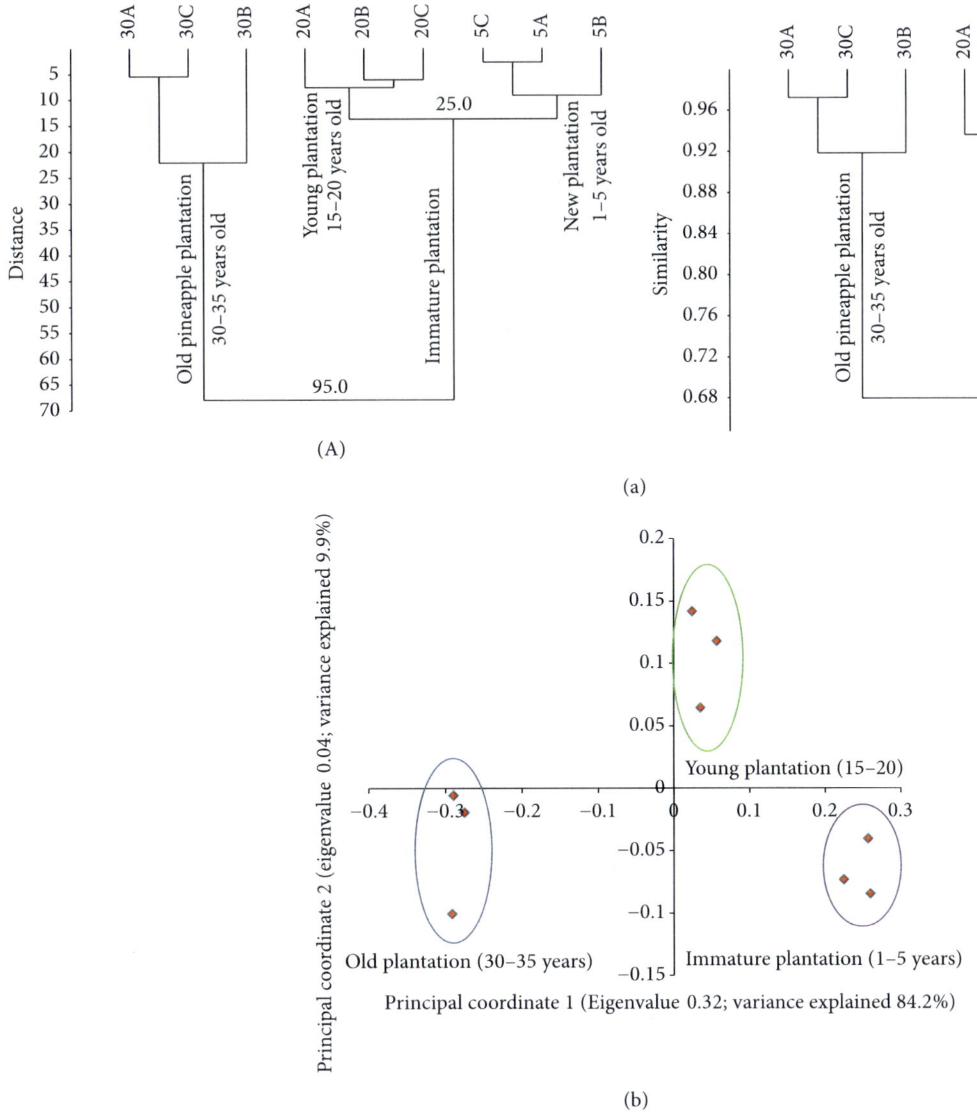

FIGURE 8: (a) Clustering of sampling plots: (A) according to soil properties (Euclidean distances); (B) according to characteristics of earthworm assemblages and species composition (Bray-Curtis index). (b) Principal coordinates analysis of pineapple plantation sampling plots. First and second principal coordinates explain 81.9% and 11.9% of total variance, respectively.

spatial distribution of earthworm communities (Figure 9). Highest value (1.28) of I_d in 15–20-year age group of plantation indicated the highest degree of clumping.

3.5. Analysis of Relationship between Soil Properties and Earthworms. Distribution of earthworm species in the pineapple agroecosystems with reference to their ecological parameters, namely, temperature, moisture, pH, and organic matter, is given in Table 5. Earthworms experienced a mean soil temperature of 25.8°C, moisture of 20.4%, pH of 4.6, and organic matter of 2.3% in the pineapple plantations of Tripura. In the studied sites, *D. assamensis*, *D. papillifer papillifer*, *P. corethrurus*, *M. houlleti*, and *E. gigas* had wide range and *E. comillahnus*, *Kanchuria* sp., *K. sumerianus*, and *D. nepalensis* showed narrow range of tolerance to edaphic factors (Table 5).

CCorA indicated that only a single relationship exists between the soil properties (independent variables) and the earthworm population characters (dependent variables) which is supported by a lack of statistical significance ($P > 0.05$) in the second canonical function (Table 6). Therefore, only the first canonical function ($P < 0.01$) explaining 75% of the total variance can be considered for drawing an inference on relationship between two variables. In this study, interpretation based on canonical weights is not considered due to lack of noncollinearity among the independent variables. Based on canonical and squared canonical loadings, soil temperature (0.96, 0.93) and pH (0.91, 0.83) were identified as most important independent variables followed by soil organic matter and moisture (Table 6). In studying cross- and squared cross-loadings for the first canonical function, soil temperature was found to be the most potent

TABLE 5: Occurrence of different earthworm species of pineapple plantation in different ecological conditions in Tripura.

Family and earthworm species	Temperature (°C)	Moisture (%)	Soil pH	Organic matter (%)
Megascolecidae				
M. houlleti	26.2 ± 0.28	19.78 ± 1.02	4.64 ± 0.10	2.03 ± 0.22
M. posthuma	24.8 ± 0.75	20.50 ± 1.43	5.72 ± 0.07	3.84 ± 0.06
Kanchuria sp.	26.2 ± 0.30	19.39 ± 0.85	4.63 ± 0.10	2.11 ± 0.17
K. sumerianus	26.8 ± 0.11	19.21 ± 0.75	4.79 ± 0.06	2.24 ± 0.05
Octochaetidae				
E. gigas	26.1 ± 0.58	20.35 ± 0.78	4.51 ± 0.15	2.00 ± 0.20
E. scutarius	25.9 ± 1.25	22.53 ± 2.14	5.23 ± 0.68	3.14 ± 0.88
E. comillahnus	27.0 ± 0.35	19.96 ± 0.42	4.81 ± 0.09	2.72 ± 0.19
Eutyphoeus sp.	25.3 ± 1.05	19.12 ± 0.97	5.02 ± 0.22	1.61 ± 0.16
E. gammiei	25.0 ± 0.40	23.29 ± 0.89	5.76 ± 0.13	2.33 ± 0.14
Moniligastridae				
D. assamensis	25.5 ± 0.31	19.03 ± 0.89	4.59 ± 0.14	2.16 ± 0.30
D. papillifer papillifer	25.4 ± 0.32	19.67 ± 1.27	4.67 ± 0.15	1.97 ± 0.21
D. nepalensis	26.3 ± 0.12	20.69 ± 0.18	4.41 ± 0.21	2.38 ± 0.10
Glossoscolecidae				
P. corethrurus	25.7 ± 0.70	18.38 ± 1.08	4.77 ± 0.18	2.16 ± 0.28

Values represent mean ± standard deviation (SD).

TABLE 6: Summary result of canonical correlation analysis relating soil properties and earthworm population characteristics.

(a) Measures of overall model fit for canonical correlation analysis

Canonical function	Canonical correlation	Eigen value	χ^2	*P* value
1	0.87	0.75	42.9	0.00
2	0.49	0.25	7.25	0.06

(b) Canonical weights, loadings, squared loadings, cross-loadings, squared cross-loadings, and redundancy index for the first canonical function

Variables	Canonical weight	Canonical loading	Squared canonical loading	Canonical cross-loading	Squared canonical cross loading	Redundancy index
(A) Independent						
(1) Temperature	1.98	0.96	0.93	−0.84	0.70	0.62
(2) Moisture	1.01	−0.87	0.76	0.76	0.57	
(3) Organic matter	−0.27	−0.87	0.76	0.76	0.57	
(4) Soil pH	−0.28	0.91	0.83	−0.79	0.63	
(B) Dependent						
(1) Worm density	0.85	0.83	0.69	−0.72	0.52	0.36
(2) Worm biomass	0.56	0.52	0.27	−0.45	0.21	

predictor of dependent variables with negative correlation explaining 70% variance of function 1, followed by soil pH (63%). Among dependent variables, earthworm density (cross-loading: −0.72) was mostly affected by independent variables compared to biomass explaining 52% and 21% of the variance, respectively (Table 6). Independent variables with high redundancy index (0.62) confirm their ability in predicting the values of dependent factors, namely, earthworm density and biomass.

According to CCA ordination, common earthworm species of pineapple plantation with greater relative abundance, namely, *D. assamensis*, *P. corethrurus*, *D. papillifer papillifer*, and *M. houlleti*, were mainly regulated by studied soil physicochemical properties (temperature, moisture, pH, and organic matter) (Figure 10). The dominant species, *D. assamensis*, was strictly associated with acidic pH, low temperature, and high moisture, whereas others preferred comparatively higher pH and temperature. Distribution of

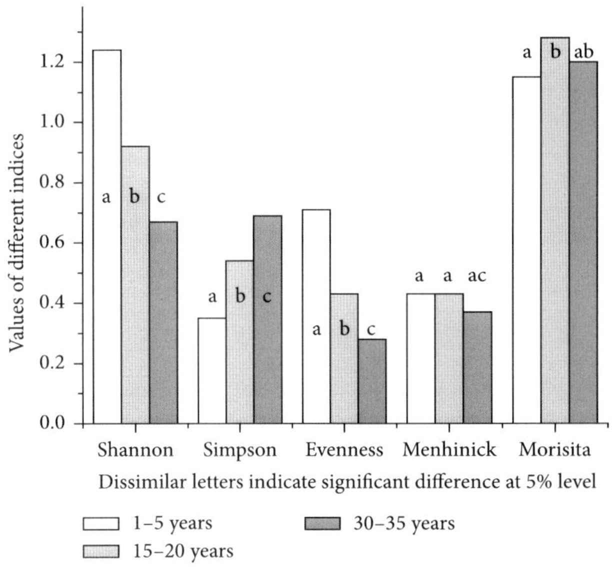

FIGURE 9: Showing the variation among diversity and dispersion indices in the three age groups of pineapple plantation.

most of the rare earthworm species was controlled by several unstudied factor combinations. However, *E. comillahnus* chiefly associated with greater organic matter and moisture content (Figure 10).

4. Discussion

The present study recorded 13 species of earthworms from the pineapple plantations of Tripura representing 5 genera and 4 families. Among the 13 species of earthworms, 3 species, namely, *M. houlleti*, *M. posthuma*, and *P. corethrurus*, are exotics and the rest are either native or endemic. Occurrence of a greater number of endemic species in the soils of pineapple plantations of Tripura indicates that the region belongs to biodiversity hot spot zone where the native species coexist with the exotics. This is in contrast to tropical countries like Peru, Brazil, and Mexico where native species of earthworms have largely been replaced by exotics [61, 62]. *Kanchuria* sp. and *Eutyphoeus* sp. are new species (yet to be described) of Tripura. Interestingly, *Eutyphoeus comillahnus* shows restricted distribution only in Tripura of India [63]. Eleven (*M. houlleti*, *Kanchuria* sp., *K. sumerianus*, *E. gigas*, *E. scutarius*, *E. comillahnus*, *E. gammiei*, *D. assamensis*, *D. papillifer papillifer*, *D. nepalensis*, and *P. corethrurus*) and five (*E. gigas*, *E. comillahnus*, *E. gammiei*, *D. nepalensis*, and *P. corethrurus*) species in the pineapple plantations of Tripura had also been reported from the soils of other habitats including rubber plantations in Tripura [64, 65] and its neighboring country Bangladesh [66]. At least eight earthworm species of pineapple plantations (*M. houlleti*, *M. posthuma*, *K. sumerianus*, *E. gammiei*, *D. assamensis*, *D. papillifer papillifer*, *D. nepalensis*, and *P. corethrurus*) had previously been reported from other northeastern states of India [18, 67, 68]. Recently, Lalthanzara and Ramanujam [69, 70] reported the presence of *M. houlleti*, *Drawida* sp., *Perionyx excavatus*, *Perionyx macintoshi*, and *Eutyphoeus mizoramensis* in the agroforestry system of Mizoram.

A minimum of five and maximum of eleven earthworm species distributed in the studied pineapple plantations corroborated the reported range of 4–14 species in the earthworm communities of tropical rain forest [11]. Occurrence of only 5 earthworm species in the pineapple plantations of East Khasi Hills in Meghalaya was reported by Tiwari et al. [18]. Such less species richness of earthworms was probably linked with altitude effect on faunal diversity [71, 72]. In fact, less number of earthworm species at higher altitude has been attributed to low temperature and unfavorable edaphic factors [73, 74]. Although pineapple plantation is a monoculture, its earthworm diversity is not remarkably less in comparison to the mixed fruit plantations of Tripura [75, 76]. This is due to the fact that mostly fallow lands were converted to pineapple plantations in order to check soil degradation, so that the original earthworm communities were retained [76, 77]. In addition, varieties of weeds, herbs, and shrubs found over the floors of the studied pineapple plantations under the tropical conditions with heavy rainfall in Tripura provide suitable physical habitat and trophic resource for a diverse earthworm community [75]. Dense arrangement of pineapple plants in the mature plantations that curtail solar radiation, protect surface soil from the thrust of rain, and reduce air temperature allows moisture conservation and retains soil organic carbon which creates favorable microclimatic conditions for the soil dwelling earthworms [78].

Drawida assamensis is the dominant species with respect to its biomass, density, relative abundance, and frequency in pineapple plantations. Interestingly Tiwari et al. [18] also reported *D. assamensis* as the dominant earthworm species in the pineapple plantations of East Khasi hills in Meghalaya. Dominance of *D. assamensis* in the studied plantations may be linked with "individual plant species effect" that favored *D. assamensis* over other species of earthworms [79]. Earthworm communities of monoculture agroecosystems dominated by single earthworm species were reported earlier by Chaudhuri and Nath [5] and Dey et al. [80]. Dominance of exotic earthworm *Pontoscolex corethrurus* in the rubber plantations of Tripura was reported by Nath [65]. Earthworm density and biomass (contributed mainly by *Drawida assamensis*) in pineapple plantation is comparable to other monoculture plantations like acacia, rubber [12, 81], and pine forests [4, 82], respectively.

Although sample-based rarefaction curve does not provide an estimation of asymptotic species richness [83], this can be used to evaluate sampling adequacy by assessing whether the cumulative number of species reaches a plateau and to compare observed species richness with nonparametric estimators [28]. In the pineapple plantation, inventory was not complete as revealed by the sample-based rarefaction curve [84, 85] after a reasonably large sampling effort, thus revealing that additional sampling effort would be needed to assess the true diversity of the earthworms in the study area. Less heterogeneous habitat conditions due to monoculture nature of plantation, dominance of *D. assamensis* in the assemblage, and patchy distribution of earthworms in

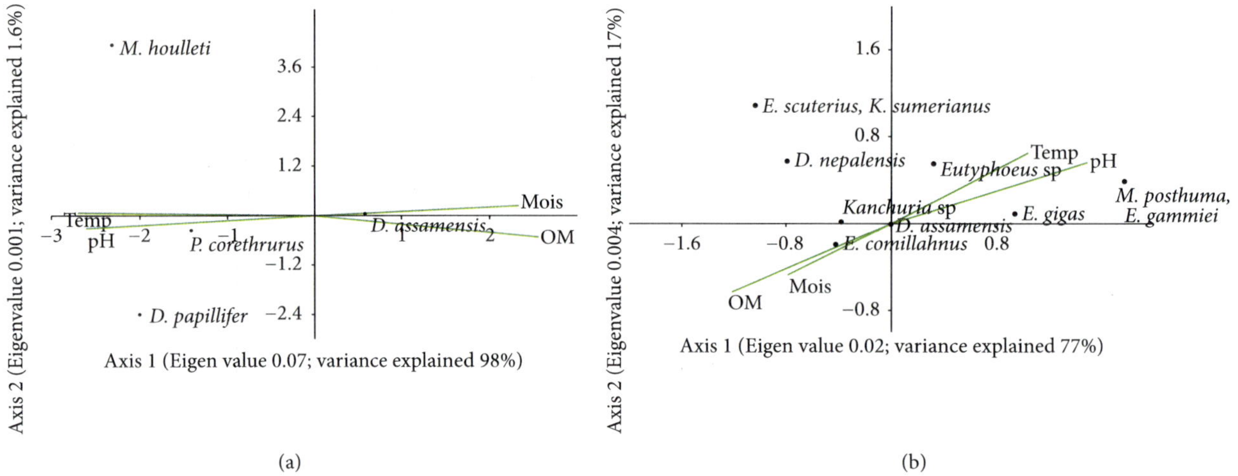

Figure 10: Canonical correspondence analysis (CCA) indicated the relationship between soil physicochemical parameters and (a) common and (b) rare earthworm species of pineapple plantation.

the pineapple plantations of Tripura are the possible factors for nonattainment of asymptotic species richness [86–88].

Among the selected eight nonparametric estimators, except for Jackknife 2 and Michaelis-Menten, others converged closely on the observed species richness. Since each estimator has been developed to work under the assumptions defined by a specific population model [89], performance tests must be carried out for the richness estimators for each case of inventory. In this case, Chao 2 estimator can be considered best suited for the present data set since it reached a stable asymptote earlier than the sample-based rarefaction curve [41] and the asymptote (13 species) was reached approximately after 83% of total sampling effort, the value of which was higher than other estimators [90]. Comparable results were obtained in other studies on both macro- and microorganisms with unequal catchability [21, 91, 92]. A great advantage of Chao 2 richness estimator is that it relies not on the precise abundance values of individuals and number of singletons and doubletons but on uniques and duplicates [60].

Earthworms are organisms with a highly contagious spatial distribution [93]. The present study agrees with Sileshi [44] and Chaudhuri and Dey [75] who found that earthworms were highly aggregated in the agroforestry assemblages of Eastern Zambia and West Tripura (India), respectively. According to Lloyd [94], aggregation indices are higher for rare species and *M. posthuma*, *Kanchuria* sp., *K sumerianus*, *D. nepalensis*, and members of the genus *Eutyphoeus* are the rare species of pineapple plantations [84]. Highly clustered horizontal distribution of earthworm communities in the pineapple plantations of West Tripura, dominated by the small sized endogeic earthworm *D. assamensis* [75, 76], corroborates the studies of Rossi and Lavelle [93] and Jiménez et al. [95], who also reported a negative correlation between earthworm size and horizontal distribution in case of earthworm communities from savannas of Africa and Columbia, respectively. Individual plant distribution as well as root architecture and activity [96] along with microtopography or intrinsic population processes like dispersal, reproduction, or competition [97] may also influence the spatial distribution of earthworms in the pineapple plantations.

Generally the replacement of original ecosystems by different land use systems of human origin has a deep impact on the functioning of the ecosystems [98]. The lack of clear relationship between the biometric variables employed, namely, length-diameter (L/D) and weight-diameter (W/D) ratios, and the vertical distribution of earthworm species studied was in line with other studies [99–101] in which larger species were found to be distributed deep in the soil profile. Lalthanzara and Ramanujam [70] and Piearce [102] also reported a positive correlation between earthworm size and depth of burrow. Higher density of earthworms in the upper soil strata (0–10 cm) of pineapple plantations corroborates Lalthanzara and Ramanujam [70], Kaushal et al. [103] and Bisht et al. [104], which may be attributed to higher percentage of organic matter, moisture, and availability of food. A gradual decrease in soil macroinvertebrate population with increase in soil depth was also recorded by Doblas-Miranda et al. [105] with small differences between microhabitat. Less density of earthworms in the deeper soil stratum during the activity period reveals the fact that the deeper stratum is not suitable for the majority of the species recorded.

Soil temperature and rainfall are the most potent regulatory factors for vertical distribution of earthworms [106]. Lower earthworm density in the upper stratum (0–10 cm) of soils during summer compared to 10–20 cm depth might be linked with lower soil moisture content and downward migration of earthworm to avoid extreme conditions [107]. Since the environmental conditions of tropical soil ecosystems are highly influenced by strong seasonality, earthworms migrate deeply to aestivate in different ways [99, 101]. The soil depth and physical characteristics of the soils are the most important factors influencing differential distribution of earthworm population exposed to seasonal variations [3, 106].

Unlike summer and postmonsoon period, a greater population density during monsoon in the pineapple plantations is probably related to higher breeding rate of earthworms at that period [11]. According to Chaudhuri and Bhattacharjee [108], *D. assamensis* is a discrete breeder and reproduces mainly during monsoon. However, earthworms in general favor monsoon for reproduction mainly due to favorable temperature that acts as cue for increased neurosecretion favoring cocoon production [109]. Lower density of earthworms with greater biomass during postmonsoon period is mainly attributed to higher monsoon mortality rate of juveniles compared to young and adults; these grow well in presence of suitable growth parameters like temperature, organic matter, and so forth [99, 110].

A minimum of seven (1–5-year-old and 15–20-year-old plantation) and a maximum of eleven (30–35-year-old plantation) earthworm species were distributed in the studied sites among the 13 earthworm species found during our survey. The difference in the species composition in the earthworm communities among the different studied age groups of pineapple plantations of West Tripura indicates the importance of habitat heterogeneity (β-diversity) in the diversity of earthworms as shown by Fragoso and Lavelle [111] in the forests of Mexico.

A gradual increase in average biomass and density of earthworms with plantation age is also reported in case of several other monocultures like rubber and *Eucalyptus* [112–114]. An approximate fourfold increase in density and threefold increase in biomass of earthworms in the 30–35-year age group plantation compared with the 1–5-year-old plantation were probably due to a significant increase ($P < 0.01$) in soil organic carbon and soil moisture and significant decrease ($P < 0.01$) in temperature with increasing plantation age [11]. Decaëns et al. [115] proposed that increased faunal activity with the aging of plots was due to availability of trophic resource (i.e., dead roots and decomposed leaves) that sustained a high carrying capacity.

The gradual decrease in the indices of Shannon diversity together with an increase in dominance is probably linked with the dramatic increase in the population density of dominant earthworm *D. assamensis* in pineapple plantations with increase in their age. According to Shakir and Dindal [116], population density is negatively correlated with species diversity. The lower population densities for the rare species were linked to high diversity, and the highest population densities for dominant species correlated with lower diversity. Thus, lower population densities of rare octochaetid species such as *E. gigas*, *E. comillahnus*, and *E. scutarius* in young plantations (1–5 year old) and higher population densities of dominant earthworm species, *D. assamensis*, in aged plantation (30–35 years old) were correlated with higher diversity in the former and lower diversity in the latter.

Highest value of Morisita's index (I_d) in 15–20-year plantation may be due to greater heterogeneity in soil conditions and food distribution [117] than in other plantation groups. A gradual decrease in the soil environmental heterogeneity [118] coupled with decline in the intensity of disturbance [95] with increasing plantation age may contribute to the decrease in Morisita's index (I_d), that is, less clumped earthworm community with increase in the pineapple plantation age.

A number of ecological factors are known to play a vital role in the distribution, diversity, and abundance of earthworms [11]. Earthworm species in pineapple plantation occurred in moist (20% moisture) and acidic soils (pH 4.6) with temperature 26°C and organic matter 2.3%. Because of their narrow ecological plasticity to soil edaphic factors, members of the largest native genus *Eutyphoeus* [119] can be considered as "stenoecious." On the other hand, exotic peregrine worms like *M. houlleti*, *M. posthuma*, and *P. corethrurus* and native peregrine species like *D. nepalensis*, *D. papillifer papillifer*, and so forth with wide ecological plasticity are called "euryoecious".

Soil temperature and moisture are usually inversely related and act synergistically to influence earthworm activity by affecting their metabolism, respiration, growth, reproduction, and osmoregulation [120]. In the present study, soil temperature and moisture had a significantly strong negative and positive correlation with the density and biomass of earthworms, respectively. Such correlations were also reported by several other workers from different natural and artificial agroecosystems across the tropics [18, 65, 77, 85]. According to Edwards and Bohlen [11], earthworm densities and biomasses are negatively affected by temperature due its inverse relationship with soil moisture content. However, optimum temperature increases biodiversity by improving access to soil nutrients and organic resources [121].

Earthworms are very sensitive to soil pH that limits their density and distribution [11]. Significant negative correlation between soil pH and overall earthworm density and biomass in the present study results from the domination of acidophilic (*D. assamensis*) and other acid-tolerant earthworm species in the pineapple plantations of Tripura [84]. Recently, Bhattacharjee [77] reported negative correlation between soil pH and earthworm population in rubber plantation. Spiers et al. [122] reported acid-tolerant earthworm species, *Arctiostrotus* sp., in the organic soil with pH ranging from 2.6 to 6.2 in the coniferous podzols of Vancouver Island, Canada. Earlier, Bachelier [123] reported that certain tropical species of *Megascolex* thrived in the acid soils with pH of 4.5 to 4.7. It has been advocated that acidophilic species of earthworm have a major role in the decomposer subsystem of soil ecosystem [11]. Staaf [124] suggested that pH and factors related to pH had very important influence on the distribution and abundance of earthworms in the acid beech forest soils in Sweden.

Organic matter has an important influence on earthworm population, activities, and their enzyme activities [125]. Significantly positive correlation ($P < 0.05$) between earthworm population density and soil organic matter content of the present study in pineapple plantation corroborates with the studies of Hendrix et al. [126] at 10 sites, which included conventional, nontillage agroecosystem, grass, meadows, and a mixed deciduous forest in the south eastern USA. Ghabbour and Shakir [127] reported that increased density and biomass of earthworms were associated with increase in organic carbon content of semiarid agricultural soils in Egypt.

Recently, Dey and Chaudhuri [85] reported strong positive correlation between soil organic matter content and density of earthworm in pineapple plantation of West Tripura. Thus, there is little doubt that the availability of organic matter is one of the most important factors in influencing earthworm abundance.

According to Blanchart and Julka [81], maximum population density and species diversity occur within a range of 2.4% to 4.5% organic matter content. High earthworm diversity in low organic matter is probably due to the domination of more geophagous species in the pineapple agroecosystem (9 endogeic species). Faster organic matter decomposition rate at higher temperature under tropical conditions results in decreased litter availabilities and thus litter feeding epigeic and anecic earthworm populations tend to be depleted in tropical soils compared to those in temperate ones. Fragoso et al. [61] also advocated that earthworm communities of tropical agroecosystems are mostly composed of endogeic species of earthworms. Texture, nutrient status, and moisture conditions of soil probably determine functional categories of earthworms in an ecosystem [12].

5. Conclusion

Native endogeic earthworm with dominance of acidophilic species, *Drawida assamensis*, is a characteristic of pineapple agroecosystems in Tripura. The increasing activity of *Drawida assamensis* in pineapple plantation, with increasing age of plantations, suggests that individual plant species with its soil microclimatic conditions favor *D. assamensis* over the other species of earthworms. Vertical distribution of earthworms is greatly influenced by seasonal variation.

Species richness is a fundamental measurement of community and regional diversity. Species accumulation and sample-based rarefaction curves are very useful techniques in analyzing inventory completeness and species richness. Based on our experience, we recommend Chao 2 richness estimator to be applied in a particular investigation site, with nonasymptotic species richness, to indicate the number of earthworm species yet to be discovered.

Soil temperature and pH are identified as the major regulatory factors for earthworm distribution under the pineapple agroecosystems.

Conflict of Interests

The authors declare that there is no conflict of interests regarding the publication of this paper.

Acknowledgment

The authors express their sincere thanks to the Head of the Department of Zoology, Maharaja Bir Bikram College, Agartala, for providing laboratory facilities.

References

[1] P. Lavelle, T. Decaëns, M. Aubert et al., "Soil invertebrates and ecosystem services," *European Journal of Soil Biology*, vol. 42, supplement 1, pp. S3–S15, 2006.

[2] A. Don, B. Steinberg, I. Schöning et al., "Organic carbon sequestration in earthworm burrows," *Soil Biology & Biochemistry*, vol. 40, no. 7, pp. 1803–1812, 2008.

[3] C. Fragoso and P. Lavelle, "Earthworm communities of tropical rain forests," *Soil Biology & Biochemistry*, vol. 24, no. 12, pp. 1397–1408, 1992.

[4] T. Bhadauria, P. S. Ramakrishnan, and K. N. Srivastava, "Diversity and distribution of endemic and exotic earthworms in natural and regenerating ecosystems in the central Himalayas, India," *Soil Biology & Biochemistry*, vol. 32, no. 14, pp. 2045–2054, 2000.

[5] P. S. Chaudhuri and S. Nath, "Community structure of earthworms under rubber plantations and mixed forests in Tripura, India," *Journal of Environmental Biology*, vol. 32, no. 5, pp. 537–541, 2011.

[6] G. Gonzalez, X. Zou, and S. Borges, "Earthworm abundance and species composition in abandoned tropical cropland: comparison of tree plantation and secondary forests," *Pedobiologia*, vol. 40, pp. 385–391, 1996.

[7] G. González, X. Zou, A. Sabat, and N. Fetcher, "Earthworm abundance and distribution pattern in contrasting plant communities within a tropical wet forest in Puerto Rico," *Caribbean Journal of Science*, vol. 35, no. 1-2, pp. 93–100, 1999.

[8] J. B. Kotcon, "Population dynamics of earthworms in organic farming systems," in *Biology of Earthworms*, A. Karaca, Ed., vol. 24 of *Soil Biology*, pp. 299–310, Springer, Berlin, Germany, 2011.

[9] D. P. Bartholomew, R. E. Paull, and K. G. Rohrback, *The Pineapple: Botany, Production and Used*, CABI Publishing, Wallingford, UK, 2003.

[10] C. Kole, *Wild Crop Relatives: Genomic and Breeding Resources. Tropical and Subtropical Fruits*, Springer, Berlin, Germany, 2011.

[11] C. A. Edwards and P. J. Bohlen, *Biology and Ecology of Earthworms*, Chapman and Hall, London, UK, 1996.

[12] P. S. Chaudhuri, S. Nath, and R. Paliwal, "Earthworm population of rubber plantations (*Hevea brasiliensis*) in Tripura, India," *Tropical Ecology*, vol. 49, no. 2, pp. 225–234, 2008.

[13] P. S. Chaudhuri, S. Nath, S. Bhattacharjee, and R. Paliwal, "Biomass, density of earthworm under rubber plantation (*Hevea brasilensis*) in Tripura, India," *The Bioscan*, vol. 4, no. 3, pp. 475–479, 2009.

[14] P. S. Chaudhuri, S. Nath, T. K. Pal, and T. K. Dey, "Earthworm casting activities under rubber plantations (*Hevea brasilensis*) in Tripura, India," *World Journal of Agricultural Sciences*, vol. 5, no. 4, pp. 515–521, 2009.

[15] R. D. Kale and N. Karmegam, "The role of earthworms in tropics with emphasis on Indian ecosystems," *Applied and Environmental Soil Science*, vol. 2010, Article ID 414356, 16 pages, 2010.

[16] I. A. Najar and A. B. Khan, "Earthworm communities of Kashmir Valley, India," *Tropical Ecology*, vol. 52, no. 2, pp. 151–162, 2011.

[17] M. C. Dash and K. G. Saxena, "Earthworms in the Himalaya and Western Ghats region of India: a review," *The Bioscan*, vol. 7, no. 1, pp. 1–12, 2012.

[18] S. C. Tiwari, B. K. Tiwari, and R. R. Mishra, "Relationship between seasonal populations of earthworms and abiotic factors in pineapple plantations," *Proceedings of the National Academy of Sciences, India Section B: Biological Sciences*, vol. 62, no. 2, pp. 223–226, 1992.

[19] N. O. K. Mainoo, S. Barrington, J. K. Whalen, and L. Sampedro, "Pilot-scale vermicomposting of pineapple wastes with earthworms native to Accra, Ghana," *Bioresource Technology*, vol. 100, no. 23, pp. 5872–5875, 2009.

[20] J. M. Anderson and J. S. I. Ingram, *Tropical Soil Biology and Fertility-A Handbook of Methods*, CAB International, Wallingford, UK, 2nd edition, 1993.

[21] A. Dey and P. S. Chaudhuri, "Quantifying earthworm species richness in the pineapple and mixed fruit plantations of West Tripura, India—a non-parametric approach," *European Journal of Soil Biology*, vol. 59, pp. 31–35, 2013.

[22] D. E. Bignell, R. Constantino, C. Csuzdi et al., "Macrofauna," in *A Handbook of Tropical Soil Biology: Sampling and Characterization of Below-Ground Biodiversity*, F. M. S. Moreira, E. J. Huising, and D. E. Bignell, Eds., pp. 43–58, Earthscan, Taylor & Francis, 2008.

[23] G. E. Gates, *Burmese Earthworms. An Introduction to the Systematics and Biology of Megadrile Oligochaetes with Special Reference to Southeast Asia*, vol. 62 of *Transactions of the American Philosophical Society*, American Philosophical Society, Philadelphia, Pa, USA, 1972.

[24] J. M. Julka, *The Fauna of India and Adjacent Countries: Megadrile Oligochaeta (Earthworms)*, Zoological Survey of India, Kolkata, India, 1988.

[25] M. C. Dash and S. P. Dash, *Fundamentals of Ecology*, Tata McGraw-Hill Education Pvt., New Delhi, India, 2009.

[26] P. F. Hendrix and P. J. Bohlen, "Exotic earthworm invasions in North America: ecological and policy implications," *BioScience*, vol. 52, no. 9, pp. 801–811, 2002.

[27] J. Valckx, G. Govers, M. Hermy, and B. Muys, "Optimizing earthworm sampling in ecosystems," in *Biology of Earthworms*, A. Karaca, Ed., Soil Biology, pp. 19–38, Springer, London, UK, 2011.

[28] R. K. Colwell and J. A. Coddington, "Estimating terrestrial biodiversity through extrapolation," *Philosophical transactions of the Royal Society of London Series: B*, vol. 345, no. 1311, pp. 101–118, 1994.

[29] J. A. Coddington, L. H. Young, and F. A. Coyle, "Estimating spider species richness in a southern Appalachian cove hardwood forest," *Journal of Arachnology*, vol. 24, no. 2, pp. 111–128, 1996.

[30] A. E. Magurran, *Measuring Biological Diversity*, Blackwell Science, Oxford, UK, 2004.

[31] R. K. Colwell, *EstimateS: Statistical Estimation of Species Richness and Shared Species from Samples*, Version 8.2.0, 2009, http://purl.oclc.org/estimates.

[32] Microcal Software, "Origin 6.0 Professional," 1999, http://www.microcal.com.

[33] N. J. Gotelli and R. K. Colwell, "Estimating species richness," in *Biological Diversity: Frontiers in Measurement and Assessment*, A. E. Magurran and B. J. McGill, Eds., Oxford University Press, London, UK, 2010.

[34] A. Chao, "Non-parametric estimation of the number of classes in a population," *Scandinavian Journal of Statistics*, vol. 11, no. 4, pp. 265–270, 1984.

[35] A. Chao, W.-H. Hwang, Y.-C. Chen, and C.-Y. Kuo, "Estimating the number of shared species in two communities," *Statistica Sinica*, vol. 10, no. 1, pp. 227–246, 2000.

[36] K. P. Burnham and W. S. Overton, "Estimation of the size of a closed population when capture probabilities vary among animals," *Biometrika*, vol. 65, no. 3, pp. 625–633, 1978.

[37] K. P. Burnham and P. S. Overton, "Robust estimation of population size when capture probabilities vary among animals," *Ecology*, vol. 60, no. 5, pp. 927–936, 1979.

[38] A. Chao, "Estimating the population size for capture-recapture data with unequal catchability," *Biometrics*, vol. 43, no. 4, pp. 783–791, 1987.

[39] E. P. Smith and G. van Belle, "Nonparametric estimation of species richness," *Biometrics*, vol. 40, no. 1, pp. 119–129, 1984.

[40] J. G. Raaijmakers, "Statistical analysis of the Michaelis-MENten equation," *Biometrics*, vol. 43, no. 4, pp. 793–803, 1987.

[41] D. S. Toti, F. A. Coyle, and J. A. Miller, "A structured inventory of Appalachian grass bald and heath bald spider assemblages and a test of species richness estimator performance," *Journal of Arachnology*, vol. 28, no. 3, pp. 329–345, 2000.

[42] R. K. Colwell, *EstimateS: Statistical Estimation of Species Richness and Shared Species from Samples*, Version 7.5, 2006, http://viceroy.eeb.uconn.edu/estimates.

[43] J. H. Connell, "Territorial behavior and dispersion in some marine invertebrates," *Researches on Population Ecology*, vol. 5, no. 2, pp. 87–101, 1963.

[44] G. Sileshi, "The excess-zero problem in soil animal count data and choice of appropriate models for statistical inference," *Pedobiologia*, vol. 52, no. 1, pp. 1–17, 2008.

[45] A. N. Kolmogorov, "Sulla determinazione empirica di una legge di distribuzione," *Giornale dell'Istituto Italiano degli Attuari*, vol. 4, pp. 83–91, 1933.

[46] N. V. Smirnov, "Sur les écarts de la courbe de distribution empirique," *Recueil Mathematics NS*, vol. 6, no. 1, pp. 3–26, 1939.

[47] N. V. Smirnov, "On the estimation of the discrepancy between empirical curves of distribution for two independent samples," *Bulletin of Moscow University*, vol. 2, pp. 3–16, 1939.

[48] M. A. Stephens, "EDF statistics for goodness of fit and some comparisons," *Journal of the American Statistical Association*, vol. 69, no. 347, pp. 730–737, 1974.

[49] E. C. Pielou, *Mathematical Ecology*, John Wiley & Sons, New York, NY, USA, 1977.

[50] M. Morisita, "Measuring of the dispersion of individual and analysis of distribution patterns," *Member Faculty of Science, Kyushu University, Series E (Biological Science)*, vol. 2, pp. 215–235, 1959.

[51] A. Walkley and I. A. Black, "An examination of the Degtjareff method for determining soil organic matter, and a proposed modification of the chromic acid titration method," *Soil Science*, vol. 37, no. 1, pp. 29–38, 1934.

[52] J. A. Daji, *A Text Book of Soil Science*, Media Promoters and Publishers, Mumbai, India, 1996.

[53] B. Heider, M. S. Andersson, and R. Schultze-Kraft, "RAPD variation among North Vietnamese *Flemingia macrophylla* (Willd.) Kuntze ex Merr. accessions," *Biodiversity and Conservation*, vol. 16, no. 6, pp. 1617–1631, 2007.

[54] J. F. Hair Jr., R. E. Anderson, R. L. Tatham, and W. C. Black, *Multivariate Data Analysis with Readings*, Macmillan, New York, NY, USA, 1992.

[55] B. F. J. Manly, *Multivariate Statistical Methods*, Chapman & Hall, London, UK, 1989.

[56] J. F. Hair Jr., R. E. Anderson, R. L. Tatham, and W. C. Black, *Multivariate Data Analysis*, Prentice Hall, London, UK, 5th edition, 1998.

[57] P. E. Green and J. D. Carroll, *Mathematical Tools for Applied Multivariate Analysis*, Academic Press, New York, NY, USA, 1978.

[58] C. J. F. ter Braak and P. Smilauer, *CANOCO Reference Manual and CanoDraw for Windows User's Guide: Software for Canonical Community Ordination (Version 4.5)*, Microcomputer Power, Ithaca, NY, USA, 2002.

[59] W. Fannes, D. De Bakker, K. Loosveldt, and R. Jocqué, "Estimating the diversity of arboreal oonopid spider assemblages (Araneae, Oonopidae) at Afrotropical sites," *Journal of Arachnology*, vol. 36, no. 2, pp. 322–330, 2008.

[60] M. Unterseher, M. Schnittler, C. Dormann, and A. Sickert, "Application of species richness estimators for the assessment of fungal diversity," *FEMS Microbiology Letters*, vol. 282, no. 2, pp. 205–213, 2008.

[61] C. Fragoso, P. Lavelle, E. Blanchart et al., "Earthworm communities of tropical agro ecosystem: origin, structure and influence of management practices," in *Earthworm Management in Tropical Agro Ecosystems*, P. Lavelle, L. Brusaard, and P. Hendrix, Eds., CAB International, Wallingford, UK, 1999.

[62] A. Feijoo, A. F. Carvajal, M. C. Zúñiga, H. Quintero, and C. Fragoso, "Diversity and abundance of earthworms in land use systems in Central-Western Colombia," *Pedobiologia*, vol. 54, pp. S69–S75, 2011.

[63] K. R. Halder, "Oligochaeta: earthworm," in *State Fauna Series 7: Fauna of Tripura, Part 4*, Zoological Survey of India, Calcutta, India, 2000.

[64] P. S. Chaudhuri and G. Bhattacharjee, "Earthworms of Tripura (India)," *Ecology, Environment & Conservation*, vol. 11, no. 2, pp. 295–301, 2005.

[65] S. Nath, *Earthworm resources in the rubber plantations of Tripura with special reference to their ecological characteristics [Ph.D. thesis]*, Tripura University, Agartala, India, 2012.

[66] J. W. Reynolds, J. M. Julka, and M. N. Khan, "Additional earthworm records from Bangladesh (Oligochaeta: Glossoscolecidae, Megascolecidae, Moniligastridae, Ocnerodrilidae and Octochaetidae)," *Megadrilogica*, vol. 6, no. 6, pp. 51–62, 1995.

[67] J. M. Julka, "Studies on the earthworm collected during the Daphabum expedition in Arunachal Pradesh, India," *Records of Zoological Survey of India*, vol. 69, pp. 229–239, 1976.

[68] K. R. Halder, "Oligochaeta: earthworm," in *State Fauna Series 4: Fauna of Meghalaya, Part 9*, Zoological Survey of India, Kolkata, India, 1999.

[69] H. Lalthanzara and S. N. Ramanujam, "Effect of fertilizer (NPK) on earthworm population in the agro-forestry system of Mizoram, India," *Science Vision*, vol. 10, no. 4, pp. 159–167, 2010.

[70] H. Lalthanzara and S. N. Ramanujam, "Vertical distribution of earthworms in agro-forestry system of Mizoram, India," in *Biology and Ecology of Tropical Earthworms*, P. S. Chaudhuri and S. M. Singh, Eds., pp. 125–140, Discovery Publishing House, New Delhi, India, 2014.

[71] J. Illig, R. A. Norton, S. Scheu, and M. Maraun, "Density and community structure of soil- and bark-dwelling microarthropods along an altitudinal gradient in a tropical montane rainforest," *Experimental and Applied Acarology*, vol. 52, no. 1, pp. 49–62, 2010.

[72] O. F. Palin, P. Eggleton, Y. Malhi, C. A. J. Girardin, A. Rozas-Dávila, and C. L. Parr, "Termite diversity along an Amazon-Andes elevation Gradient, Peru," *Biotropica*, vol. 43, no. 1, pp. 100–107, 2011.

[73] C. Salomé, C. Guenat, G. Bullinger-Weber, J.-M. Gobat, and R.-C. L. Bayon, "Earthworm communities in alluvial forests; influence of altitude, vegetation stages and soil parameters," *Pedobiologia*, vol. 54, supplement, pp. S89–S98, 2011.

[74] D. Verma and Sweta, "Earthworm resources of Western Himalayan region, India," *International Journal of Soil Science*, vol. 6, no. 2, pp. 124–133, 2011.

[75] P. S. Chaudhuri and A. Dey, "Earthworm communities in the pineapple (*Ananus comosus*) and mixed fruit plantations of West Tripura, India," *Proceedings of the Zoological Society*, vol. 66, no. 2, pp. 105–118, 2013.

[76] A. Dey and P. S. Chaudhuri, "Earthworm community structure of pineapple (*Ananas comosus*) plantations under monoculture and mixed culture in West Tripura, India," *Tropical Ecology*, vol. 55, no. 1, pp. 1–17, 2014.

[77] S. Bhattacharjee, *Impact of rubber plantations on the earthworm communities of West Tripura, India [Ph.D. thesis]*, Tripura University, Agartala, India, 2013.

[78] A. Dey, *Ecological studies on earthworms of pineapple (Ananas comosus) plantation with special reference to their community characteristics [Ph.D. thesis]*, Tripura University, Agartala, India, 2014.

[79] M. Sarlo, "Individual tree species effects on earthworm biomass in a tropical plantation in Panama," *Caribbean Journal of Science*, vol. 42, no. 3, pp. 419–427, 2006.

[80] A. Dey, S. Nath, and P. S. Chaudhuri, "Impact of monoculture (rubber and pineapple) practice on the community characteristics of earthworms in West Tripura (India)," *NeBIO*, vol. 3, no. 1, pp. 53–58, 2012.

[81] E. Blanchart and J. M. Julka, "Influence of forest disturbance on earthworm (Oligochaeta) communities in the Western Ghats (South India)," *Soil Biology & Biochemistry*, vol. 29, no. 3-4, pp. 303–306, 1997.

[82] T. Bhadauria and P. S. Ramakrishnan, "Population dynamics of earthworms and their activity in forest ecosystems of north-east India," *Journal of Tropical Ecology*, vol. 7, no. 3, pp. 305–318, 1991.

[83] C. X. Mao, R. K. Colwell, and J. Chang, "Estimating the species accumulation curve using mixtures," *Biometrics*, vol. 61, no. 2, pp. 433–441, 2005.

[84] A. Dey and P. S. Chaudhuri, "Community characteristics of earthworms in different age groups of pineapple plantations (*Ananus comosus*) in West Tripura, India," *Asian and Australasian Journal of Plant Science & Biotechnology*, vol. 6, no. 1, pp. 67–75, 2012.

[85] A. Dey and P. S. Chaudhuri, "Density, diversity and distribution of earthworms in pineapple plantation (*Ananus comosus*) plots of different ages in West Tripura, India," in *Climate Change and Biodiversity*, P. K. Bharti and A. Chauhan, Eds., Discovery Publishing House, New Delhi, India, 2013.

[86] C. X. Mao and R. K. Colwell, "Estimation of species richness: mixture models, the role of rare species, and inferential challenges," *Ecology*, vol. 86, pp. 1143–1153, 2005.

[87] J. Hortal, P. A. V. Borges, and C. Gaspar, "Evaluating the performance of species richness estimators: sensitivity to sample grain size," *Journal of Animal Ecology*, vol. 75, no. 1, pp. 274–287, 2006.

[88] A. Chao, R. K. Colwell, C.-W. Lin, and N. J. Gotelli, "Sufficient sampling for asymptotic minimum species richness estimators," *Ecology*, vol. 90, no. 4, pp. 1125–1133, 2009.

[89] B. A. Walther and J. L. Moore, "The concepts of bias, precision and accuracy, and their use in testing the performance of species richness estimators, with a literature review of estimator performance," *Ecography*, vol. 28, no. 6, pp. 815–829, 2005.

[90] A. E. Magurran, S. R. Baillie, S. T. Buckland et al., "Long-term datasets in biodiversity research and monitoring: assessing change in ecological communities through time," *Trends in Ecology & Evolution*, vol. 25, no. 10, pp. 574–582, 2010.

[91] J. B. Hughes, J. J. Hellmann, T. H. Ricketts, and B. J. M. Bohannan, "Counting the uncountable: statistical approaches to estimating microbial diversity," *Applied and Environmental Microbiology*, vol. 67, no. 10, pp. 4399–4406, 2001.

[92] B. J. M. Bohannan and J. Hughes, "New approaches to analyzing microbial biodiversity data," *Current Opinion in Microbiology*, vol. 6, no. 3, pp. 282–287, 2003.

[93] J. P. Rossi and P. Lavelle, "Earthworm aggregation in the Savannas of Lamto (Côte d' Noire)," *Applied Soil Ecology*, vol. 7, no. 2, pp. 195–199, 1998.

[94] M. Lloyd, "Mean crowding," *The Journal of Animal Ecology*, vol. 36, no. 1, pp. 1–30, 1967.

[95] J. J. Jiménez, J.-P. Rossi, and P. Lavelle, "Spatial distribution of earthworms in acid-soil savannas of the eastern plains of Colombia," *Applied Soil Ecology*, vol. 17, no. 3, pp. 267–278, 2001.

[96] J.-P. Rossi, "Clusters in earthworm spatial distribution," *Pedobiologia*, vol. 47, no. 5-6, pp. 490–496, 2003.

[97] C. H. Ettema and D. A. Wardle, "Spatial soil ecology," *Trends in Ecology & Evolution*, vol. 17, no. 4, pp. 177–183, 2002.

[98] J. J. Jiménez and T. Decaëns, "Vertical distribution of earthworms in grassland soils of the Colombian Llanos," *Biology & Fertility of Soils*, vol. 32, no. 6, pp. 463–473, 2000.

[99] P. Lavelle, *Les Vers de Terre de la Savane de Lamto, Côte d'Ivoire: Peuplements, Populations et Fonctions dans l'écosystème*, vol. 12, Publications du laborotoire de Zoologie de l'Ecole Normale Superieure, 1978.

[100] A. Nemeth and R. Herrera, "Earthworm populations in a Venezuelan tropical rain forest," *Pedobiologia*, vol. 23, no. 6, pp. 437–443, 1982.

[101] C. Fragoso, *Les peulements vers de terre dans l'est et sud'est du Mexique [Ph.D. thesis]*, Université Paris, Paris, France, 1993.

[102] T. G. Piearce, "Functional morphology of lumbricid earthworms, with special reference to locomotion," *Journal of Natural History*, vol. 17, no. 1, pp. 95–111, 2007.

[103] B. R. Kaushal, S. Kalia, and S. P. Bisht, "Growth and cocoon production by the earthworm *Drawida nepalensis* (Oligochaeta: Moniligastridae) in oak and pine litter," *Pedobiologia*, vol. 39, no. 5, pp. 417–422, 1995.

[104] R. Bisht, H. Pandey, D. Bharti, and B. R. Kaushal, "Population dynamics of earthworms (Oligochaeta) in cultivated soils of central Himalayan tarai region," *Tropical Ecology*, vol. 44, pp. 229–234, 2003.

[105] E. Doblas-Miranda, F. Sánchez-Piñero, and A. González-Megías, "Different microhabitats affect soil macro-invertebrate assemblages in a Mediterranean arid ecosystem," *Applied Soil Ecology*, vol. 41, no. 3, pp. 329–335, 2009.

[106] M. V. Reddy and M. Pasha, "Influence of rainfall, temperature and somes oil physico-chemical variables on seasonal population structure and vertical distribution of earthworms in two semi-arid tropical grassland soils," *International Journal of Biometeorology*, vol. 37, no. 1, pp. 19–26, 1993.

[107] G. Baker, J. Buckerfield, R. Grey-Gardner, R. Merry, and B. Doube, "The abundance and diversity of earthworms in pasture soils in the fleurieu peninsula, south australia," *Soil Biology & Biochemistry*, vol. 24, no. 12, pp. 1389–1395, 1992.

[108] P. S. Chaudhuri and S. Bhattacharjee, "Reproductive biology of eight tropical earthworm species of rubber plantations in tripura, india," *Tropical Ecology*, vol. 52, no. 1, pp. 49–60, 2011.

[109] G. Bhattacharjee and P. S. Chaudhuri, "Cocoon production, morphology, hatching pattern and fecundity in seven tropical earthworm species—a laboratory-based investigation," *Journal of Biosciences*, vol. 27, no. 3, pp. 283–294, 2002.

[110] T. A. McCredie, C. A. Parker, and I. Abbott, "Population dynamics of the earthworm *Aporrectodea trapezoides* (Annelida: Lumbricidae) in a Western Australian pasture soil," *Biology & Fertility of Soils*, vol. 12, no. 4, pp. 285–289, 1992.

[111] C. Fragoso and P. Lavelle, "The earthworm community of a tropical rain forest," in *On Earthworms*, A. M. Bonvicini-Pagliani and P. Omodeo, Eds., Mucchi Editore, Modena, Italy, 1987.

[112] C. Gilot, P. Lavelle, E. Blanchart, J. Keli, P. Kouassi, and G. Guillaume, "Biological activity of soil under rubber plantations in Cote d'Ivoire," *Acta Zoologica Fennica*, vol. 196, pp. 186–189, 1995.

[113] L. M. C. Mboukou-Kimbatsa and F. Bernhard-Reversat, "Effect of exotic tree plantations on invertebrate soil macro fauna," in *Effect of Exotic Tree Plantations on Plant Diversity and Biological Soil Fertility in Congo, Svana: With Special Reference to Eucalyptus*, F. Bernhard-Reversat, Ed., Centre for International Forestry Research, Bogor, Indonesia, 2001.

[114] P. S. Chaudhuri and S. Bhattacharjee, "Impact of rubber plantation on the earthworm communities in Tripura (India)," in *Earthworm Ecology and Environment*, S. M. Singh, Ed., International Book Distribution Co, Lucknow, India, 2009.

[115] T. Decaëns, F. Bureau, and P. Margerie, "Earthworm communities in a wet agricultural landscape of the Seine Valley (Upper Normandy, France)," *Pedobiologia*, vol. 47, no. 5-6, pp. 479–489, 2003.

[116] S. H. Shakir and D. L. Dindal, "Density and biomass of earthworms in forest and herbaceous microecosystems in central New York, North America," *Soil Biology & Biochemistry*, vol. 29, no. 3-4, pp. 275–285, 1997.

[117] J. K. Whalen, "Spatial and temporal distribution of earthworm patches in corn field, hayfield and forest systems of southwestern Quebec, Canada," *Applied Soil Ecology*, vol. 27, no. 2, pp. 143–151, 2004.

[118] J. P. Rossi, E. Huerta, C. Fragoso, and P. Lavelle, "Soil properties inside earthworm patches and gaps in a tropical grassland (la Mancha, Veracruz, Mexico)," *European Journal of Soil Biology*, vol. 42, no. 1, pp. S284–S288, 2006.

[119] P. S. Chaudhuri, A. Dey, G. Bhattacharjee, and S. Nath, "Earthworm diversity in Tripura—present status," *Science & Culture*, vol. 78, no. 7-8, pp. 343–346, 2012.

[120] C. A. Edwards, "The importance of earthworms as key representatives of the soil fauna," in *Earthworm Ecology*, C. A. Edwards, Ed., St. Lucie Press, Boca Raton, Fla, USA, 2004.

[121] P. Lavelle, L. Brussard, and P. Hendrix, *Earthworm Management in Tropical Agroecosystems*, CABI Publishing, New York, NY, USA, 1999.

[122] G. A. Spiers, D. Gagnon, G. E. Nason, E. C. Packee, and J. D. Lousier, "Effects and importance of indigenous earthworms on decomposition and nutrient cycling in coastal forest ecosystems," *Canadian Journal of Forest Research*, vol. 16, no. 5, pp. 983–989, 1986.

[123] G. Bachelier, *La Vie Animale dans les Sols*, Orston, Paris, France, 1963.

[124] H. Staaf, "Foliage litter turnover and earthworm populations in three beech forests of contrasting soil and vegetation types," *Oecologia*, vol. 72, no. 1, pp. 58–64, 1987.

[125] I. A. Najar and A. B. Khan, "Factors affecting distribution of earthworms in Kashmir Valley: a multivariate statistical approach," *Proceedings of the Zoological Society*, vol. 67, no. 2, pp. 126–135, 2014.

[126] P. F. Hendrix, B. R. Mueller, R. R. Bruce, G. W. Langdale, and R. W. Parmelee, "Abundance and distribution of earthworms in relation to landscape factors on the Georgia Piedmont, U.S.A," *Soil Biology & Biochemistry*, vol. 24, no. 12, pp. 1357–1361, 1992.

[127] S. I. Ghabbour and S. H. Shakir, "Population density and biomass of earthworms in agro-ecosystems of the Mariut coastal desert region," *Pedobiologia*, vol. 23, pp. 189–198, 1982.

14

Habitat Ecology and Ichthyofaunal Diversity of Two Creeks and their Associated Streams from Port Blair, South Andaman Islands

M. Arun Kumar, S. Venu, and G. Padmavati

Department of Ocean Studies and Marine Biology, Pondicherry University, Brookshabad Campus, P.O. Bag No. 26, Chakkaragaon (Post), Port Blair 744112, India

Correspondence should be addressed to M. Arun Kumar; arun4kavi@gmail.com

Academic Editor: L. M. Chu

Habitat ecology and ichthyofaunal diversity were analyzed from two creeks and their associated streams in Port Blair: one was perennial and the other one was seasonal. Various habitat types like riffles, pools, and major and minor bends in both streams were studied. The perennial stream has 12 riffles and 14 pools from head point to the entering point of the creek. The seasonal stream has 15 riffles and 13 pools from the starting point to the sink. The perennial stream was found to have higher ichthyofaunal diversity than the seasonal stream: 1701 individuals constituting 8 orders, 30 families, 42 genera, and 54 species were recorded. Among these, 395 specimens represented by 21 species were found in perennial stream, 291 specimens with 11 species were found in seasonal stream, 863 specimens by 48 species were noted in creek, and 152 specimens constituting 14 species in sink were recorded. The physicochemical parameters of the habitats showed more or less a similar trend. The average dissolved oxygen value of the perennial stream was higher than that of the seasonal stream. Higher temperature values were noted during the postmonsoon season. This study has indicated a wide lacuna in the knowledge of the fresh water habitats and their inhabitants on these islands.

1. Introduction

The Andaman and Nicobar group of islands are located in the Bay of Bengal between 6°45′–13°41′N and 92°12′–93°57′E approximately in a North–South direction in the shape of a crescent. These islands have unique ecological systems, mainly contributed by coral reefs, mangroves, seagrass, and seaweed beds [1]. The fish fauna of these islands consist of an assemblage of about 1450 species spread over all the diverse habitats representing 586 genera belonging to 175 families. The fish diversity in Andaman Islands and Nicobar Islands is of special interest in terms of marine zoogeography because of the confluence in the fishes of Andaman Sea with that of the Western Pacific and the Indian Ocean [2]. Apart from marine fishes, many fresh and brackish water species also come under the exploited fishery resources of these islands. The production of fresh water fishes from the inland waters in these islands is 90 tonnes per year. The exploitation and scope for further development are limited due to the lack of major riverine systems [3].

There is so much focus on the diversity of the marine environment around these islands and the fresh water ecology is significantly negligible. But it has to be noted that as much important as the marine environment is the rich rainforest ecosystem in these islands and the numerous water bodies nestled in the depths of the rainforest. Though not having a magnitude with regard to size, it is undisputable that they harbour a great percentage of wildlife yet to be discovered. Due to lack of accurate inventories, information on many of the taxa, in particular the fresh water fishes, is inadequate [4]. Though there were some early contributions on fresh water fish fauna, most of these were patchy or restricted to a particular group. These include studies on diversity [5–8], diversity and distribution [9, 10], and a study on Gobiidae [11]. Herre had recorded 112 species of fresh water and littoral fish and his study is the most comprehensive work so far [9, 10].

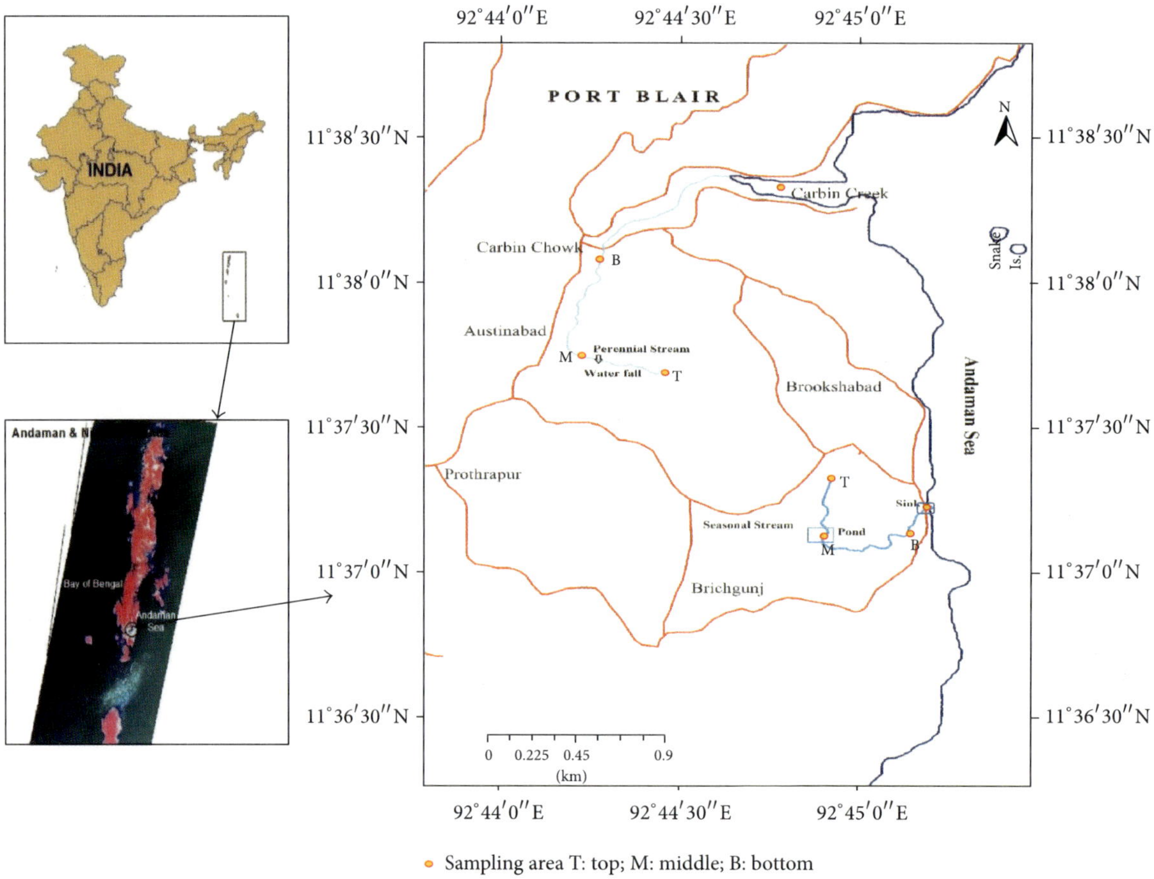

Figure 1: Map showing the study area.

Later studies on fresh water fishes of Andaman Islands were merely compilations of previous surveys [12]. Vijay and Priya have done a survey of the fishes in fresh water ecosystems and prepared a checklist [13], comparing it to that of Herre's.

Fish assemblages have widely been used as ecological indicators to assess and evaluate the level of degradation and health of rivers and streams at various spatial scales [14]. In streams, the structural complexity may be understood as a mosaic of mesohabitat patches (such as pools, runs or raceways, and riffles) which are clearly delimited by different combinations of current, depth, and substrate composition [15]. Thus, it is predictable that each mesohabitat holds a peculiar ichthyofauna in response to such habitat features. The current study focusses on the habitat ecology and fish diversity of two creeks in which two different streams empty their water.

2. Materials and Methods

2.1. Study Area. The study was conducted season-wise from the month of October 2012 to July 2013. The study area was from Brichgunj to Carbyn's Cove, southeastern side of Port Blair (Figure 1). Two different fresh water streams, one perennial and one seasonal, are flowing from the top of the hill and are emptying themselves into the sea. The study was conducted on four major habitats, namely, two streams and two creeks, in which these two streams empty their water. The creek in which the perennial stream emptied was bigger than the other one. The freshwater streams flow to the creeks which have seawater influx.

Table 1: Habitat types of fresh water streams [16].

Habitat type	Depth (m)	Current (m/sec)	Turbulence
Pool	Variable	<0.2	No
Side pool	<0.5 usually	Nil usually	No
Plunge pool	<2.0 usually	<0.20	Yes
Run	Variable	0.20–0.75+	No
Riffle	<0.5	>0.75	Yes
Cascade	2.0 usually	>2.0	Much

2.2. Sample Collection and Analysis. The environmental parameters such as dissolved oxygen, pH, salinity, and temperature of these habitats were studied season-wise using standard methods [17]. The habitat types were examined based on a combination of depth, current speed, and turbulence [16] (Table 1). The fishes were collected using cast nets of various sizes (2,540 mm × 7 mm and 2,032 mm × 10 mm), drag nets, baited hook, and lines and by hand-picking. The species-wise numbers were recorded during each sampling. The length and weight were measured for the collected fishes. The identification was done using standard identification keys [18–21]. Care was taken after obtaining the specimens so as not to kill them. Certain ambiguous fish species were brought to the laboratory for taxonomic identification and confirmation. The rest were released back into the waters immediately after identification. The collected specimens were preserved in formalin and deposited in the fishery museum, Department of Ocean Studies and Marine Biology, Pondicherry University, Port Blair.

2.3. Statistical Analysis. The data were processed and graphs were plotted using the Analytical Package of Microsoft Excel (2013 version). Four different indices were used to assess the diversity of fishes including the following.

2.3.1. Margalef's Diversity Index (D). Margalef's Diversity Index is a measure of the number of species present for a given number of individuals:

$$D = \frac{(S-1)}{\log N}, \quad (1)$$

where S is total number of species; N is total individuals.

2.3.2. Simpson's Index (D_s). Simpson's Index is considered as dominance index because it weighs the abundance of the most common species:

$$D_s = \sum_{i-1}^{s} \frac{(n_i (n_i - 1))}{(N (N - 1))}, \quad (2)$$

where n_i is the number of individuals in the ith species. N is total individuals. Since D_s and diversity are negatively related, Simpson's Index is usually expressed as the reciprocal $(1 - D)$ so that as the index goes up, so does diversity.

2.3.3. Shannon-Weiner Index (H'). The Shannon-Weiner Index is used for measuring diversity. It assumes that all species are represented in a sample and that the sample was obtained randomly:

$$H' = -\sum_{i=1}^{s} p_i \ln p_i, \quad (3)$$

where p_i is the proportion of individuals found in the ith species and ln is the natural logarithm.

2.3.4. Evenness (E). We can compare the actual diversity value to the maximum possible diversity by using a measure called evenness. The evenness of the sample is obtained from the formula

$$E = \frac{H'}{H_{\max}} = \frac{H'}{\ln S}. \quad (4)$$

The diversity indices were carried out using the software PAST [22].

3. Results

3.1. Physicochemical Parameters. The analysis of physicochemical parameters from these habitats showed more or less a similar trend. The pH, salinity, and temperature have shown little variation between the streams from the top to bottom. While dissolved oxygen varied between 3.39 mg·L^{-1} and 4.34 mg·L^{-1} in perennial stream (Table 2), it was 3.03 mg·L^{-1} to 4.12 mg·L^{-1} in seasonal stream (Table 3). The higher value for average dissolved oxygen recorded in the perennial stream when compared to the seasonal stream may be attributed to the continuous water flow throughout the year. The salinity and pH values showed less variation seasonally. High temperature (27–31°C) and moderate to low dissolved oxygen (3.03–3.89 mg·L^{-1}) values were recorded during postmonsoon in both streams (Tables 2 and 3).

The values for physicochemical parameters for the creek and the sink were comparable to each other but entirely different for the two streams. The dissolved oxygen in the creek was found to be between 3.89 mg·L^{-1} and 4.35 mg·L^{-1} (Table 4) while in the sink it ranged from 3.62 mg·L^{-1} to 4.23 mg·L^{-1} (Table 5). The temperature, salinity, and pH were more or less similar in both streams when compared to creek and sink (Tables 2–5). Among the four habitats (the two streams, creek, and sink), the values were more or less the same, and the perennial stream had the lowest average pH, temperature, and salinity (Table 2). Season-wise values of the physicochemical parameters of all the habitats showed significant variation ($p > 0.05$, one-way ANOVA) among themselves.

3.2. Habitat Types. One stream starts its flow from the head point situated near the top of the hill which is present between Brookshabad and Brichgunj and is flowing northeast. This stream has continuous water flow over the heavy rocky substratum on the hill, with a waterfall near Austinabad, and then it flows over a soft sandy bottom. It mixes with the canal from Austinabad and then with another drainage channel from the airport at Carbyn's Cove. Finally it mixes with the sea through the creek. The seasonal stream does not flow continuously, but the water is present as pools. This stream has a continuous flow only during the rainy season. The water is stored during times of overflow in a pond which is present near the Brichgunj quarry. Finally it mixes with the sea in the eastern side of the hill. There is a small sink near the shore to which the water will be emptied. Only during high tide does it mix with the sea because of the obstruction by a small road.

The results have shown that the major habitat types of both streams were side pools, riffles, pools, runs, cascades, and plunge pools, where their relative numbers play a major role in the faunal distribution. The perennial and seasonal streams both have almost the same number of riffles and pools (Figure 2). The perennial stream has 12 riffles and 14 pools from head point to the entering point of creek. The seasonal stream has 15 riffles and 13 pools from the starting point to the sink. A major difference was that the side pools, plunge pools, runs, and cascades are more in perennial stream than the seasonal stream. There were 7 plunge pools in

TABLE 2: Physicochemical properties of the water in the perennial stream.

Parameters	Premonsoon			Monsoon			Postmonsoon		
Area	Top	Middle	Bottom	Top	Middle	Bottom	Top	Middle	Bottom
Salinity (‰)	0	0	27	0	0	24	0	0	30
pH	7	7	7.6	7	7	7.3	7	7	7.9
Temperature (°C)	25	26	27	24	24	27	26	28	30
Dissolved oxygen ($mg{\cdot}L^{-1}$)	3.39	3.68	3.94	3.89	4.34	4.23	3.43	3.73	4.09

TABLE 3: Physicochemical properties of the water in the seasonal stream.

Parameters	Premonsoon			Monsoon			Postmonsoon		
Area	Top	Middle	Bottom	Top	Middle	Bottom	Top	Middle	Bottom
Salinity (‰)	0	0	28	0	0	24	0	0	30
pH	7	7	7.5	7	7	7.4	7	7	7.7
Temperature (°C)	25	25	26	23	25	26	27	28	31
Dissolved oxygen ($mg{\cdot}L^{-1}$)	3.2	3.38	3.76	3.67	4.12	4.01	3.03	3.58	3.9

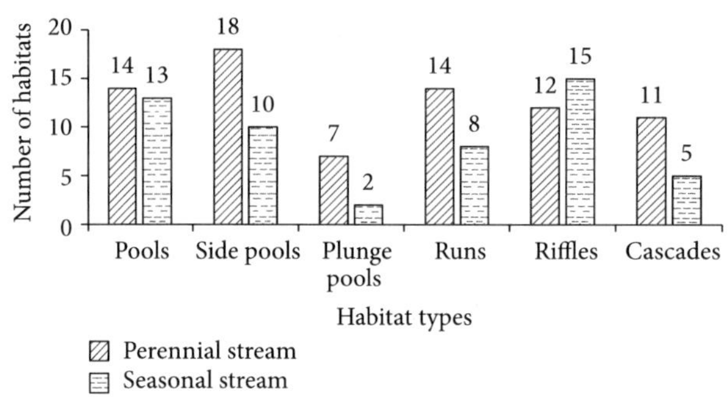

FIGURE 2: Habitats in the streams.

TABLE 4: Physicochemical properties of the water in the creek.

Parameters	Premonsoon	Monsoon	Postmonsoon
Salinity (‰)	28	30	34
pH	7.5	7.8	8.1
Temperature (°C)	26	30	33
Dissolved oxygen ($mg{\cdot}L^{-1}$)	3.89	4.35	3.91

TABLE 5: Physicochemical properties of the water in the sink.

Parameters	Premonsoon	Monsoon	Postmonsoon
Salinity (‰)	29	31	35
pH	7.5	7.7	8.2
Temperature (°C)	27	31	33
Dissolved oxygen ($mg{\cdot}L^{-1}$)	3.62	4.23	3.89

the perennial and 2 plunge pools in the seasonal stream. Totally, 14 runs were found in the perennial stream and 8 runs in the seasonal stream. The perennial stream had 11 cascades and the seasonal stream had 5 cascades. An artificial pond located in the middle of the seasonal stream had a remarkably high diversity of fishes. The two streams were almost similar in terms of habitat parameters but highly varied in their ichthyofaunal assemblages. The creek is located near the bridge to Carbyn's Cove beach and it mixes with the sea. The sink has a vegetation of grasses growing on a deep muddy bottom, containing brackish water which mixes with the sea during high tide.

3.3. Diversity of Fishes in Their Habitats. Species distribution is the manner in which group of species are spread out. The distribution of species over space and time is very useful in understanding the species biodiversity. This study has recorded a total of 1701 individuals constituting 8 orders, 30 families, 42 genera, and 54 species. Among these, 395 individuals represented by 21 species were found in perennial stream, 291 individuals with 11 species were found in seasonal stream, 863 individuals by 48 species were noted in Carbyn's creek, and 152 individuals constituting 14 species were recorded in the sink (Table 6).

Totally, 54 species were identified in all the habitats, with *Liza vaigiensis* (139), *Ambassis interruptus* (124), *Oreochromis mossambicus* (94), *Liza parsia* (78), and *O. niloticus* (76) representing maximum individuals. Among these, 8 species were found in all the habitats, *A. interruptus*, *O. mossambicus*,

Table 6: Species composition of fishes in the study area.

Taxonomy	Species	Perennial stream	Seasonal stream	Creek	Sink
Order: Anguilliformes Family: Anguillidae	*Anguilla* sp.	+	+	–	–
Family: Muraenidae	*Echidna rhodochilus*	+	–	+	–
Order: Clupeiformes Family: Clupeidae	*Sardinella gibbosa*	–	–	+	–
Family: Engraulidae	*Stolephorus commersonnii*	–	–	+	–
	S. indicus	–	–	+	–
	Thryssa baelama	–	–	+	–
Order: Aulopiformes Family: Synodontidae	*Saurida undosquamis*	–	–	+	–
Order: Cyprinodontiformes Family: Hemiramphidae	*Hemiramphus far*	–	–	+	+
Order: Scorpaeniformes Family: Scorpaenidae	*Tetraroge barbata*	–	–	+	–
Order: Perciformes Family: Ambassidae	*Ambassis interruptus*	+	+	+	+
Family: Terapontidae	*Pelates quadrilineatus*	–	–	+	–
	Terapon jarbua	+	–	+	+
Family: Sillaginidae	*Sillago sihama*	–	–	+	+
Family: Cyprinidae	*Esomus* sp.	+	–	–	–
Family: Cichlidae	*Oreochromis mossambicus*	+	+	+	+
	O. niloticus	+	+	+	+
Family: Carangidae	*Carangoides oblongus*	–	–	+	–
	Caranx ignobilis	–	–	+	–
	C. sexfaciatus	–	–	+	–
Family: Leiognathidae	*Leiognathus equulus*	–	–	+	–
	L. fasciatus	–	–	+	–
	L. splendens	+	–	+	+
Family: Lutjanidae	*Lutjanus argentimaculatus*	+	+	+	+
	L. lutjanus	–	–	+	
	L. russellii	–	–	+	–
Family: Gerridae	*Gerres filamentosus*	+	+	+	+
	G. oyena	+	–	+	–
Family: Haemulidae	*Pomadasys kaakan*	–	–	+	–
	P. maculatus	–	–	+	–
Family: Mullidae	*Parupeneus indicus*	–	–	+	–
	Upeneus vittatus	–	–	+	–
Family: Mugilidae	*Liza parsia*	+	–	+	+
	L. vaigiensis	+	+	+	+
Family: Gobiidae	*Glossogobius giuris*	+	+	+	+
	Periophthalmus argentilineatus	–	–	+	–
	Awaous ocellaris	+	+	–	–
	Periophthalmus kalolo	–	–	+	–
	Yongeichthys criniger	–	–	+	–
Family: Eleotridae	*Butis butis*	+	–	+	–
	Eleotris sp.	+	–	–	–
	Ophieleotris sp.	+	+	–	–
	Ophiocara sp.	+	+	–	–

TABLE 6: Continued.

Taxonomy	Species	Perennial stream	Seasonal stream	Creek	Sink
Family: Chaetodontidae	*Chaetodon decussatus*	–	–	+	–
Family: Scatophagidae	*Scatophagus argus*	+	–	+	+
Family: Channidae	*Channa* sp.	+	–	–	–
Family: Siganidae	*Siganus javus*	–	–	+	–
	S. vermiculatus	–	–	+	–
Family: Acanthuridae	*Acanthurus triostegus*	–	–	+	–
	A. xanthopterus	–	–	+	–
Family: Apogonidae	*Apogon hyalosoma*	+	–	+	+
Order: Pleuronectiformes Family: Bothidae	*Psettodes erumei*	–	–	+	–
Family: Soleidae	*Pardachirus pavoninus*	–	–	+	–
Order: Tetraodontiformes Family: Tetraodontidae	*Arothron immaculatus*	–	–	+	–
	Chelonodon patoca	–	–	+	–

– indicates absence; + indicates presence.

TABLE 7: Diversity indices of the ichthyofaunal diversity.

Indices	Perennial stream	Seasonal stream	Creek	Sink
Total number of species	21	11	48	14
Total number of individuals	395	291	863	152
Simpson Index (D_s)	0.93	0.90	0.97	0.90
Shannon-Weiner Index (H')	2.84	2.34	3.59	2.45
Evenness (E)	0.78	0.94	0.77	0.83
Margalef's Diversity Index (D)	3.51	1.76	6.80	2.59

O. niloticus, Lutjanus argentimaculatus, Gerres filamentosus, Liza parsia, L. vaigiensis, and *Glossogobius giuris*. Seven fishes were exclusively fresh water inhabitants, such as *Anguilla* sp., *Awaous ocellaris, Channa* sp., *Echidna rhodochilus, Eleotris* sp., *Ophieleotris* sp., and *Ophiocara* sp.

Major species found in the seasonal stream were *Awaous ocellaris, Oreochromis niloticus,* and *Ophieleotris* sp. The major inhabitants of the Perennial stream were *Esomus* sp., *Channa* sp., and *Ophiocara* sp. *Liza vaigiensis, L. parsia, Oreochromis mossambicus, O. niloticus, Ambassis interruptus, Apogon hyalosoma, Gerres filamentosus, Lutjanus argentimaculatus, Leiognathus splendens, Scatophagus argus, Terapon jarbua,* and *Caranx* sp. were found in both streams.

The total number of species varied among the habitats. The creek had the maximum number of species (48) and the minimum (11) was found at the seasonal stream. Overall individuals were more in creek (863) and less in sink (152). The dominance index (D_s) in creek was the highest (0.97) and lowest values (0.90) were found in seasonal stream and sink. The lowest species diversity was recorded ($H' = 2.34$) in seasonal stream and the highest (3.59) was in creek. The evenness (E) was low (0.77) in creek and high (0.94) in seasonal stream. The species richness (D) was high (6.80) in creek and low (1.76) in seasonal stream (Table 7).

4. Discussion

The fresh water fish fauna of this region is somewhat impoverished due to low habitat diversity and its long isolation from the mainland. There are possibilities that several undiscovered species living in the streams of these islands, particularly in tribal reserves and areas, are inaccessible. Vijay and Priya have observed that many endemic species of Andaman Islands and Nicobar Islands are threatened due to habitat loss and introduction of nonnative species like *Oreochromis mossambicus* [13]. In the present study, apart from *O. mossambicus, O. niloticus* was also recorded. These are hardy species adapted to a wide range of environmental conditions and have marked resistance to environmental perturbations and can disrupt the native species through competition for food and space. The fresh water moray eel *Echidna rhodochilus* is the new report from the Andamans as well as a first-time report from Indian waters.

Vijay and Priya collected 33 species from 77 perennial streams and 1 perennial river in Andaman Islands [13]. Among them, 17 were fresh water species. In the present study, 54 species were identified, among which 7 were exclusively fresh water. This variable percentage of fresh water species reflects a differential ecology of the habitat, which has more seawater influx. The species composition reflects the pattern of fish distribution across varying gradients of salinity. The perennial and seasonal streams are having many similarities and differences in their habitat ecology and faunal diversity. The similarities are in habitat and the number of riffles and pools and differences are the major and minor bends in both streams. The fresh water fauna were obviously found to be very less compared to the creek and the sink due to their confined nature. The perennial stream had continuous water flow even after the rainy season around the year. The major fish species living in this stream are *Esomus* sp. and *Ophiocara* sp. The nonperennial stream had water flow only during rainy season and most of the area dried up during the summer months. The major species in this stream are *Awaous ocellaris*, *Oreochromis niloticus*, and *Ophieleotris* sp.

Both the creek and the sink are brackish water habitats, characterized by a wide fluctuation in salinity due to the influence of the tide and domestic waste water influx. The presence of mangroves along the entire creek has made it highly productive. Due to these factors, the creek had high diversity of fauna with more than 40 species. The creek acts as a breeding and nursery ground for the juveniles of fish and crustaceans. It is a widely accepted fact that mangroves act as nurseries for fishes in both tropical and temperate regions [23].

The diversity indices (H', D_s, and D) showed low values at the sink and the seasonal stream when compared to creek and perennial stream. The evenness index (E) was very less in these areas when compared to other indices. The ichthyofaunal diversity in the perennial stream (2.84) was found to be higher than that of the seasonal stream (2.34). The seasonality can be a deterring factor for encouraging migrations of new populations and sustaining the available individuals, which prevents high diversity. However, when we consider the abundance and evenness of the available species, the seasonal stream was found to be higher (0.94). This could not be correlated with any observed feature. Also, the perennial stream had greater number of pools than the seasonal stream (Figure 2). More pools implies a higher diversity of fish communities [24]. This can be explained as a result of the habitat complexity and/or habitat volume, which is already documented for other temperate and tropical streams [25].

The diversity in the sink was high with regard to the size of the area, but it was less compared to that of the creek. *Liza vaigiensis*, *Ambassis interruptus*, *Oreochromis mossambicus*, *L. parsia*, and *O. niloticus* differed in their distributional patterns on all the sites, whereas other species did not differ much in their distributional patterns.

5. Conclusions

The present study has indicated a wide lacuna in the knowledge of fresh water species and their habitats. It is imperative to formulate conservation strategies in order to protect the native fish species and their habitats. Otherwise it is likely that several fresh water species from Andaman Islands and Nicobar Islands may go extinct before they are even discovered by science.

Conflict of Interests

The authors declare that there is no conflict of interests regarding the publication of this paper.

Acknowledgments

The authors are grateful to Pondicherry University for providing the requisite facilities for the study. They are very thankful to all the fishermen; without their help this work would not have been achieved. The authors are indebted to Mr. A. Saravanan, Department of Disaster Management, Pondicherry University, for preparing the study area map. The corresponding author is very grateful to Mr. B. Ram Kumar, Mr. Sushant, Mr. Snehasis, Mr. Bhaskar, Mr. Chandrakant, and Mr. S. Sai Elangovan for their help.

References

[1] *Committee Constituted to Holistically Address the Issue of Poaching in the Andaman and Nicobar Islands*, Ministry of Environment and Forests, Government of India, New Delhi, India, 2011.

[2] T. I. Ramakrishna, C. R. Sreeraj, C. Raghunathan, R. Raghuraman, P. T. Rajan, and J. S. Yogesh Kumar, *An Account of Additions to the Ichthyofauna of Andaman and Nicobar Islands*, vol. 326 of *Records of the Zoological Survey of India Occasional Papers*, 2010.

[3] Anon, *Report on the Rainfall of Port Blair*, Department of Statistics and Economics. Andaman and Nicobar Administration, 2012.

[4] P. Pande, A. Kothari, and S. Singh, *Directory of National Parks and Sanctuaries in Andaman and Nicobar Islands*, IIPA, New Delhi, India, 1991.

[5] F. Day, "On the fishes of the Andaman Islands," in *Proceedings of the Zoological Society of London*, 1870, pp. 677–705.

[6] S. L. Hora, "Doryichthys insularis," in *The Fresh Water Fish from the Andaman Islands*, N. Annandale and S. L. Hora, Eds., vol. 27 of *Records of the Indian Museum*, pp. 33–41, 1925.

[7] D. D. Mukerji, "Notes on some rare and interesting fishes from the Andaman Islands, with description of two new fresh water gobies," *Records of the Indian Museum*, vol. 37, no. 3, pp. 250–279, 1935.

[8] T. K. Sen, "Further light on fresh water fish fauna of Andaman Islands," *Seafood Export Journal*, vol. 7, no. 2, pp. 31–33, 1975.

[9] A. W. C. T. Herre, "On a collection of Littoral and fresh water fishes from Andaman Islands," *Records of the Indian Museum*, vol. 41, pp. 327–372, 1939.

[10] A. W. C. T. Herre, "A list of the fishes known from the Andaman Islands," *Memoirs of Indian Museum*, vol. 13, pp. 331–403, 1941.

[11] F. P. Koumans, "On the collection of gobioid fishes from Andamans," *Records of the Indian Museum*, vol. 42, pp. 15–18, 1940.

[12] P. K. Talwar, "Fishes of the Andaman and Nicobar Islands: a synoptic analysis," *Journal of Andaman Science Association*, vol. 6, no. 2, pp. 71–102, 1990.

[13] P. Vijay and D. Priya, "A survey of fresh water fishes of the Andaman Islands," *Journal of the Bombay Natural History Society*, vol. 106, no. 1, pp. 264–267, 2009.

[14] R. A. Zampella, J. F. Bunnell, K. J. Laidig, and N. A. Procopio, "Using multiple indicators to evaluate the ecological integrity of a coastal plain stream system," *Ecological Indicators*, vol. 6, no. 4, pp. 644–663, 2006.

[15] P. L. Angermeier and I. J. Schlosser, "Species-area relationships for stream fishes," *Ecology*, vol. 70, no. 5, pp. 1450–1462, 1989.

[16] J. A. Baker and S. A. Foster, *Estimating Density and Abundance of Endemic Fishes in Hawaiian Streams*, Division of Aquatic Resources, Department of Land and Natural Resources, Honolulu, Hawaii, USA, 1992.

[17] APHA, *Standard Methods*, American Public Health Association, Washington, DC, USA, 19th edition, 1995.

[18] W. Fischer and G. Bianchi, Eds., *FAO Species Identification Sheets for Fishery Purposes. Western Indian Ocean (Fishing Area 51)*, vol. 1–6, Danish International Development Agency (DANIDA), Food and Agriculture Organization of the United Nations, Rome, Italy, 1984.

[19] P. K. Talwar and A. G. Jhingran, *Inland Fishes of India and Adjacent Countries (Vols 1-2)*, Oxford & IBH Publishing, New Delhi, India, 1991.

[20] D. V. Rao, K. Devi, and P. T. Rajan, "An account of icthyofauna of Andaman and Nicobar Islands, Bay of Bengal," Records of ZSI, India. Occasional Paper 178, 2000.

[21] R. Froese and D. Pauly, "FishBase," World Wide Web electronic publication, 2014, http://www.fishbase.org/search.php.

[22] O. Hammer, D. A. T. Harper, and P. D. Ryan, "Past: paleontological statistics software package for education data analysis," *Palaeontologia Electronica*, vol. 4, no. 1, p. 9, 2001.

[23] H. M. Austin, "A survey of the ichthyofauna of the mangroves of western Puerto Rico December 1967–August 1968," *Caribbean Journal of Science*, vol. 11, no. 1-2, pp. 27–39, 1971.

[24] F. P. Gelwick, "Longitudinal and temporal comparisons of riffle and pool fish assemblages in a Northeastern Oklahoma Ozark stream," *Copeia*, vol. 1990, no. 4, pp. 1072–1082, 1990.

[25] O. T. Gorman and J. R. Karr, "Habitat structure and stream fish communities," *Ecology*, vol. 59, no. 3, pp. 507–515, 1978.

15

Quality and Conservation of Riparian Forest in a Mountain Subtropical Basin of Argentina

Romina Daiana Fernández,[1] Sergio Javier Ceballos,[1] Ana Lucía González Achem,[2] Margarita del Valle Hidalgo,[3] and Hugo Rafael Fernández[2]

[1] *Instituto de Ecología Regional, UNT-CONICET, CC 34, Yerba Buena, 4107 Tucumán, Argentina*
[2] *Instituto de Biodiversidad Neotropical, UNT-CONICET, CC 34, Yerba Buena, 4107 Tucumán, Argentina*
[3] *Facultad de Ciencias Naturales e IML, Miguel Lillo 205, 4000 Tucumán, Argentina*

Correspondence should be addressed to Ana Lucía González Achem; anagonzalezachem@gmail.com

Academic Editor: Ram Chander Sihag

The aims of this work were to describe the conservation status of riparian forests located in a mountain subtropical basin of Tucumán province, Argentina, and assess how the quality of riparian forests is related with altitude, plant species richness, proportion of exotic species, and Total Suspended Solids (TSS) in adjacent rivers. Composition and species richness of riparian forests were studied at 16 sites located along an altitudinal gradient and TSS was determined from water samples collected in each site. In order to evaluate conservation status of riparian forests, we calculated an index of Quality of Yungas Riparian Forests (QBRy). We recorded 90 plant species at all sites, from which 77% were native. QBRy index was mainly associated with altitude and varied from riparian forests with good preservation or slightly disturbed to those with extreme degradation. At lower altitude, forests were more disturbed, more invaded by exotic plant species, and closer to urban and cropped areas. QBRy was not correlated with species richness or TSS. Like other riparian forests of Argentina, plant species invasion increased their degradation; therefore, future studies should focus on native riparian forests conservation and on the management of invasive plant species, which affect their quality.

1. Introduction

Riparian ecosystems constitute a complex association of organisms and their environment, adjacent to a river [1]. They are also called buffer zones, since one of their main functions is to dampen the impact of human activities on the water body [2]. The riparian ecosystem supports a zone of vegetation that is structural and floristically different from adjacent habitats that they are integrated with [3]. Various human activities have led to the deterioration and/or change at the structure of riparian vegetation. The removal of riparian forests can change the energy dynamics of rivers, from heterotrophic systems (where production is based on the entry of leaves and other materials from land) to autotrophic systems where production is based on algae [4]. These changes directly affect the composition of aquatic biota that feed on organic matter entering from the outside [5].

An issue of special interest is the removal of vegetation cover from the riparian zone, which causes an increase in the load of Total Suspended Solids (TSS) in rivers; the latter directly affects the ecological discharge, which reduces their ability to support life [6]. Poff et al. [7] define ecological discharge as the quality, quantity, and flow regime needed to sustain aquatic ecosystems and the benefits and ecosystem services on which human society depends. Increasing TSS reduces light availability for producers of the aquatic system and consequently modifies the whole food chain [8]. Strand and Merritt [9] observed that land use change in areas surrounding water bodies resulted in decreased riparian vegetation and a consequent increase in nutrients in the water due to the drag of cultivated soil nutrients. The advancement of agricultural activities has thus resulted in the degradation of riparian vegetation and water quality [10].

In Argentina, changes in the use of land surrounding water bodies and pollution have had a deep impact on the functionality of riparian ecosystems [11–17]. These changes have led to the introduction of exotic plant species [18, 19] that may cause the extinction of local plant species [20]. In particular, the mountain rivers of northwestern Argentina, which include the streams of Salí River Basin, our study area, suffer severe environmental degradation, mainly due to their hydrological richness benefits to livestock and agricultural activities. The importance of riparian buffers in the integrity of the aquatic and terrestrial ecosystems is critical, and then determining their ecological and environmental quality is becoming increasingly important [21].

QBR index ("quality of the riparian forest"), proposed by Munné et al. [22], is currently the most widely used measure for the analysis of river banks; it allows describing the quality of vegetation in situ. QBR index is useful for assessing the quality of the riparian forest and the degree of alteration of the waterfront in four separate blocks; three of them are based on riparian vegetation characteristics (coverage, structure, and nature), while the fourth referred to changes in the river channel (for details see [22]). Several authors adapted QBR index according to the study area [23] for riparian forests of Spain; [24] for the Argentinean Patagonia; [25] for Chile; [17, 26] for riparian forests of the Argentinean Yungas. All of them highlight the increasing degradation of riparian areas. In this work, we use the QBRy index by Sirombra and Mesa [26] that amends three of the four items of the original index to take into account the characteristics and changes in the rivers of Yungas (for details see [27]).

The aims of this paper are (1) to describe the conservation status of the riverbanks of rivers and streams in a subtropical mountain basin of Tucumán province, Argentina; (2) to analyze how the quality of riparian vegetation and the species composition are related with altitude, species richness, and proportion of exotic species of the region; and (3) to evaluate the influence of the quality of riparian vegetation in the water content of Total Suspended Solids.

2. Materials and Methods

2.1. Study Area. Salí River Basin is the main hydrological basin of the province of Tucumán, Argentina. It is an endorheic system, covering about 45% of the province, and flows into the Mar Chiquita Lagoon in the province of Córdoba. The climate of the region is subtropical with marked seasonality. The average range of annual precipitation goes from 900 to 1500 mm, concentrated in summer (80%) [28], and the average annual temperature is 14°C.

The study area (Figure 1) includes riparian forests belonging to the Yungas and Dry Chaco phytogeographic provinces [29]. Yungas is the southern-most extension of Neotropical Andean forests [30] and extends along discontinuous mountain ranges at NW Argentina. Yungas ecoregion is a forest which has an area of 700 km extent and 50 km width and has a strong altitudinal gradient that results in floristic stratification [31]. Therefore, Yungas presents three main altitudinal belts, identified as "selva pedemontana" (400–700 m a.s.l.), "selva montana" (700–1500 m a.s.l.), and "bosque montano" (1500–2700 m a.s.l.) [32]. This work was conducted in riparian forests of "selva pedemontana" and "selva montana." The predominant canopy vegetation is dominated by *Blepharocalyx salicifolius* (Myrtaceae), *Ocotea porphyria* (Lauraceae), and *Pisonia ambigua* (Nyctaginaceae); and understory tree composition (5–12 m) is dominated by *Eugenia uniflora* (Myrtaceae), *Piper tucumanum* (Piperaceae), and *Allophylus edulis* (Sapindaceae) [33]. The main economic activities that affect the Yungas are agriculture and livestock [17, 34]. Dry Chaco is one of the largest forest biomes of Argentina. This habitat is characterized by thicket and thorny vegetation and it is the most unexplored and least documented region in the country [35]. The predominant vegetation is deciduous dry forests, but there are also palms, savannas, and halophytic steppes [29]. In Chaco, there are three main subregions distinguished by climatic conditions: "Semiarid Chaco," "Chaco Serrano," and "Arid Chaco" [36]. This work was carried on riparian forests of "Chaco Serrano" and "Semiarid Chaco." Serrano Chaco is dominated by *Schinopsis haenkeana* and *Lithraea molleoides* [29, 36]. Semiarid Chaco is a xerophytic semideciduous forest, which before human intervention had a canopy dominated by *Schinopsis quebracho-colorado* and *Aspidosperma quebracho-blanco* [37]. For over a century, the region has been under sustained degradation and loss of natural heritage; the unplanned extractive use of natural resources is the main cause. In addition to the livestock and logging activities, in the last years, it was developed as an unplanned expansion of agriculture on a large scale [38].

2.2. Characterization of the Riverbank. We worked with a total of 16 sites whose encodings, coordinates, and main characteristics are shown in Table 1. In each sampling point, we established transects of 100 m length by 5 meters' width, on both banks of the watercourse. At each transect, we recorded plant species richness and we calculated QBRy index [15, 26]. The quality riparian forest index (QBR) was designed to study the ecological quality of riparian forests from Spain [22]. The index consists of 4 sections and each of them is assigned a score between 0 and 25. The sum of the 4 sections represents thus the final score and ranks between 0 and 100. Within each section, elements which enhance the quality of riparian forests accumulate points, while those which lead to reductions in the quality of forests are given negative scores. Three of four sections of this index were modified by Sirombra [27] for riparian forests of Yungas rivers. The new index was called "QBRy" (Quality of Yungas Riparian Forests Index) and takes into account the percentage of native and exotic species of trees and shrubs for vegetation cover estimation. These 4 sections are vegetation cover, vegetation structure, vegetation quality, and degree of naturalness of the river channel. The obtained QBRy values allowed us to identify five ranges of quality of the banks: very good (>90); good, slightly disturbed (75–90); intermediate, start of alteration (50–70); bad, strong alteration (30–50); and bad, extreme degradation (<25).

The vegetation sampling was conducted in 2012 and repeated in 2013 to improve estimates of QBRy index and to corroborate the species identification. We used the Catalogue of Vascular Plants of the Southern Cone [38] for identification

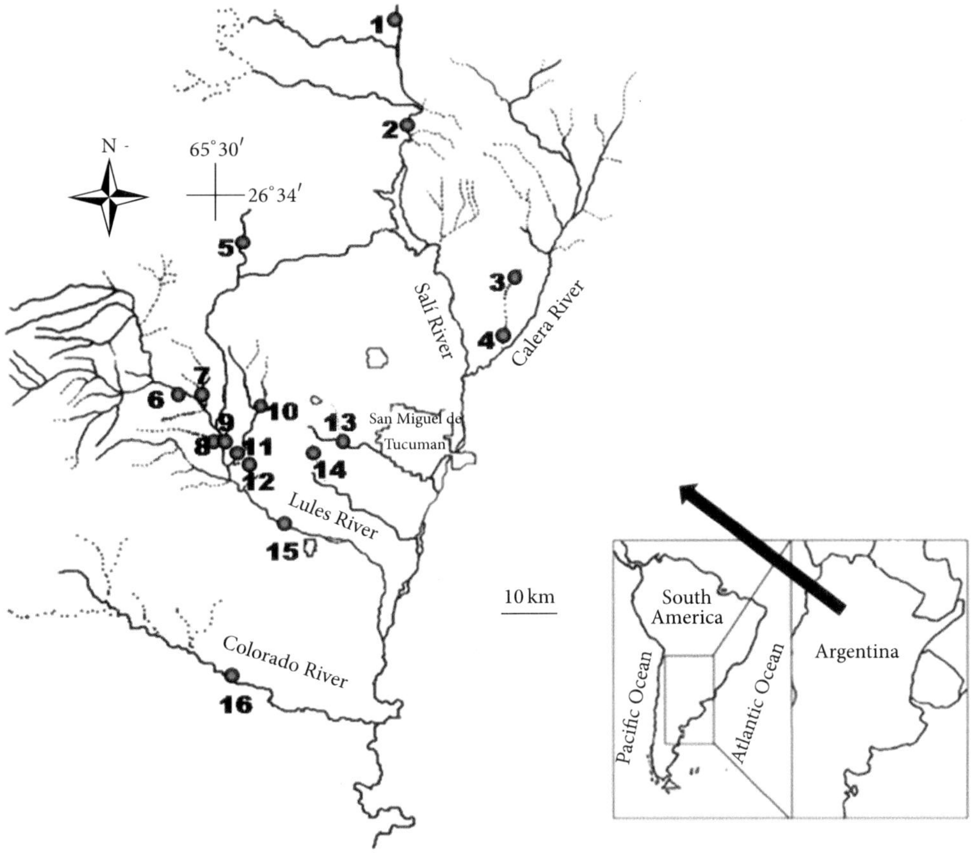

(1) Sa1 (9) Lu3
(2) Sa2 (10) SJ1
(3) LP1 (11) SJ2
(4) LP2 (12) LN1
(5) Ra1 (13) Mu1
(6) Gr1 (14) SP1
(7) Lu1 (15) Lu4
(8) Lu2 (16) Co1

FIGURE 1: Study area. Each numerated spot corresponds to a sampling site. Sites codes are on the left corner. Sites names are in Table 1.

of plant species and Tropicos [39] for updating scientific names.

We collected and preserved surface water samples for analysis of TSS in the laboratory, following IRAM protocols [40]. These data were taken during the years 2012 (October) and 2013 (March, May, and August), to represent dry and wet season. The water for TSS determination was analyzed following Volumetric Method: we put 80 mL of each water sample, previously agitated, into a porcelain capsule and heated it to 105°C for 24 hours. Then, we cooled the capsule in a vacuum container to prevent sample wet. To determine the weight of Total Suspended Solids (TSS) in the water sample, we weighed the empty capsules, then filled them with water sample, and subsequently made a comparison [41, 42]. For sites Ra1, LN1, Mu1, and SP1, we used data from Carrasco [43] corresponding to the years 2011 and 2012.

2.3. Data Analysis. All the sampled sites were ranked using the value of QBRy index, considered as an indicator of the conservation status of riparian forests. We evaluated whether riparian forests with less invasive species, higher species richness, and higher altitudes had a more suitable conservation status. For this purpose, we performed Pearson correlations between QBRy and plant species richness, percentage of exotic species, and altitude.

Changes in species composition were analyzed with a NMDS (Nonmetrical Multidimensional Scaling), based on a Bray Curtis similarity matrix of the sites ($n = 16$) calculated from the presence-absence of plant species. We also included a secondary matrix of altitude. Autopilot mode ("slow and thorough") was used to decide the appropriate number of dimensions [44]. The choice of two dimensions was based on a final stress (an index of the dissimilarities between the distances in the ordination plot and the distances in the matrix of Bray Curtis) equal to 20.4 and different from random (Monte Carlo test = 250 runs, $p = 0.04$) [44]. The NMDS considers that closer sites in the ordination plot have a similar plant species composition.

Table 1: Main characteristics of the sampled sites.

Name of the sites	Codification	Geographical coordinates	Altitude (m a.s.l.)	QBRy	Taxa richness	% native species	% exotic species
Salí River 1	Sa1	26°13′56.27″S; 65°15′10.43″W	765	55	26	90	10
Salí River 2	Sa2	26°28′33.29″S; 65°12′43.70″W	650	35	14	100	0
La Perdiz Stream 1	LP1	26°42′44.00″S; 65°8′39.40″W	529	5	15	21	79
La Perdiz Stream 2	LP2	26°43′8.00″S; 65°9′20.40″W	496	25	9	45	55
Raco River 1	Ra1	26°40′40.00″S; 65°25′39.00W	1168	60	23	91	9
Grande River 1	Gr1	26°46′21.98″S; 65°27′57.59″W	918	30	5	75	25
Lules River 1	Lu1	26°46′23.16″S; 65°27′50.04″W	909	30	8	75	25
Lules River 2	Lu2	26°47′28.20″S; 65°27′8.04″W	848	80	10	100	0
Lules River 3	Lu3	26°47′13.20″S; 65°27′10.20″W	860	80	10	100	0
San Javier River 1	SJ1	26°42′28.78″S; 65°22′44.45″W	1221	60	18	81	19
San Javier River 2	SJ2	26°44′57.48″S; 65°23′13.92″W	995	40	14	65	35
Los Noques River 1	LN1	26°47′10.00″S; 65°27′57.59″W	914	50	15	80	20
Muerto River 1	Mu1	26°48′42.00″S; 65°19′34.00″W	496	5	16	44	54
San Pablo Stream 1	SP1	26°51′50.00″S; 65°20′37.00″W	581	20	15	53	47
Lules River 4	Lu4	26°54′20.08″S; 65°21′38.39″W	453	55	23	79	21
Colorado River 1	Co1	26°59′3.98″S; 65°22′10.95″W	392	5	10	73	27

To assess whether the conservation status of the riparian forests influences the water quality of the adjacent rivers, we derived Pearson correlations between QBRy and TSS. In this case, the analysis considered TSS values for both wet and dry seasons, so it was possible to evaluate the period of the year in which riparian vegetation provides better buffer protection to the rivers.

The data used for all the correlations were transformed to a logarithmic scale to standardize. Given that the tables for correlations presented "zero entries," we added "1" to each value while performing transformation logarithms. Analyses were performed using the statistical software R [45].

3. Results

We identified 90 plant species, 77% of which were native (Table 2). More than half of the species were trees (60%); adding the bushes, the percentage rises to 83%, representing together the two life forms that structure the plant communities in these riparian forests. The species present in most of the sites were *Morus alba* (11 sites), *Acacia aroma* (10 sites), *Baccharis salicifolia* (9 sites), and *Phenax laevigatus* (8 sites). The range of values of the QBRy index varied between 5 and 80 (Table 1). The lower values of the index indicated a bad status and extreme degradation (<25). A noteworthy feature of these riparian forests is that they are located in areas of plains and highly disturbed by the change in land use (urbanized areas and crops).

Intermediate values of QBRy corresponded to recreational or touristic riverbanks (e.g., Raco River and Los Noques River). The highest values (75–90) indicating a good and slightly disturbed riparian forest belonged to two sites of Lules River located at remote areas (far away from human impact but cattle feces were detected in these sites, which is also a disturbance to the habitat).

The QBRy index correlated significantly with altitude ($r = 0.65$, df = 14, and $p = 0.006$) and the percentage of exotic plants ($r = -0.62$, df = 14, and $p = 0.01$). The correlations between QBRy and species richness ($r = 0.14$, df = 14, and $p = 0.59$), TSS in the wet season ($r = -0.09$, df = 14, and $p = 0.71$), and TSS in the dry season ($r = -0.34$, df = 12, and $p = 0.23$) were nonsignificant. The concentration of TSS was

TABLE 2: List of identified taxa, indicating family, habit, and origin (native = N or exotic = E).

Taxa	Family	Habit	Origin
Acacia aroma Gillies ex Hook. & Arn.	Fabaceae	Tree	N
Acacia praecox Griseb.	Fabaceae	Tree	N
Acacia visco Lorentz ex Griseb.	Fabaceae	Tree	N
Aechmea distichantha Lem.	Bromeliaceae	Epiphyte	N
Allophylus edulis (A. St.-Hil., A. Juss. & Cambess.) Hieron. ex Niederl.	Sapindaceae	Tree	N
Anadenanthera colubrina (Vell.) Brenan	Fabaceae	Tree	N
Arundo donax L.	Poaceae	Rattan	E
Aspidosperma quebracho-blanco Schltdl.	Apocynaceae	Tree	N
Baccharis salicifolia (Ruiz & Pav.) Pers.	Asteraceae	Shrub	N
Baccharis tucumanensis Hook. & Arn.	Asteraceae	Shrub	N
Blepharocalyx salicifolius (Kunth) O. Berg	Myrtaceae	Tree	N
Boehmeria caudata Sw.	Urticaceae	Shrub	N
Broussonetia papyrifera (L.) Vent.	Moraceae	Tree	E
Cedrela angustifolia DC.	Meliaceae	Tree	N
Ceiba chodatii (Hassl.) Ravenna	Bombacaceae	Tree	N
Celtis iguanaea (Jacq.) Sarg.	Celtidaceae	Tree	N
Cercidium praecox (Ruiz & Pav. ex Hook.) Harms	Fabaceae	Tree	N
Cestrum strigillatum Ruiz & Pav.	Solanaceae	Shrub	N
Citrus limon (L.) Osbeck	Rutaceae	Tree	E
Citrus sinensis (L.) Osbeck	Rutaceae	Tree	E
Clematis montevidensis Spreng.	Ranunculaceae	Climber	N
Condalia buxifolia Reissek	Rhamnaceae	Tree	N
Cupania vernalis Cambess.	Sapindaceae	Tree	N
Dolichandra unguis-cati (L.) L.G. Lohmann	Bignoniaceae	Climber	N
Enterolobium contortisiliquum (Vell.) Morong	Fabaceae	Tree	N
Eriobotrya japonica (Thunb.) lindl.	Rosaceae	Tree	E
Eucalyptus sp.	Myrtaceae	Tree	E
Eugenia uniflora L.	Myrtaceae	Tree	N
Geoffroea decorticans (Gillies ex Hook. & Arn.) Burkart	Fabaceae	Tree	N
Gleditsia triacanthos L.	Fabaceae	Tree	E
Guadua angustifolia Kunth	Poaceae	Rattan	E
Handroanthus impetiginosus (Mart. ex DC.) Mattos	Bignoniaceae	Tree	N
Ipomoea alba L.	Convolvulaceae	Climber	N
Jacaranda mimosifolia D. Don	Bignoniaceae	Tree	N
Juglans australis Griseb.	Juglandaceae	Tree	N
Juglans regia L.	Juglandaceae	Tree	E
Lantana camara L.	Verbenaceae	Shrub	N
Ligaria cuneifolia (Ruiz & Pav.) Tiegh.	Loranthaceae	Epiphyte	N
Ligustrum lucidum W.T. Aiton	Oleaceae	Tree	E
Lycium sp.	Solanaceae	Shrub	N
Manihot sp.	Euphorbiaceae	Shrub	E
Maytenus vitis-idaea Griseb.	Celastraceae	Tree	N
Mikania cordifolia (L. f.) Willd.	Asteraceae	Climber	N
Mikania periplocifolia Hook. & Arn.	Asteraceae	Climber	N
Morrenia odorata (Hook. & Arn.) Lindl.	Apocynaceae	Climber	N
Morus alba L.	Moraceae	Tree	E
Myrcianthes mato (Griseb.) McVaugh	Myrtaceae	Tree	N
Myrsine coriacea (Sw.) R. Br.	Myrsinaceae	Tree	N
Myrsine laetevirens (Mez) Arechav.	Myrsinaceae	Tree	N

TABLE 2: Continued.

Taxa	Family	Habit	Origin
Nicotiana glauca Graham	Solanaceae	Shrub	N
Ocotea porphyria (Griseb.) van der Werff	Lauraceae	Tree	N
Opuntia ficus-indica (L.) Mill.	Cactaceae	Shrub	E
Opuntia quimilo K. Schum.	Cactaceae	Tree	N
Opuntia sulphurea Gillies ex Salm-Dyck	Cactaceae	Shrub	N
Phenax laevigatus Wedd.	Urticaceae	Shrub	N
Phytolacca bogotensis Kunth	Phytolaccaceae	Shrub	N
Pinus sp.	Pinaceae	Tree	E
Piper hieronymi C. DC	Piperaceae	Shrub	N
Platanus occidentalis L.	Platanaceae	Tree	E
Populus deltoides W. Bartram ex Marshall	Salicaceae	Tree	E
Prosopis sp.	Fabaceae	Tree	N
Proustia cuneifolia D. Don	Asteraceae	Shrub	N
Prunus persica (L.) Batsch	Rosaceae	Tree	E
Psidium guajava L.	Myrtaceae	Tree	E
Quechualia fulta (Griseb.) H. Rob.	Asteraceae	Climber	N
Ricinus communis L.	Euphorbiaceae	Shrub	E
Rubus imperialis Cham. & Schltdl.	Rosaceae	Shrub	N
Ruta chalepensis L.	Rutaceae	Shrub	E
Salix humboldtiana Willd.	Salicaceae	Tree	N
Schinopsis lorentzii (Griseb.) Engl.	Anacardiaceae	Tree	N
Schinus bumelioides I.M. Johnst.	Anacardiaceae	Tree	N
Schinus fasciculatus (Griseb.) I.M. Johnst.	Anacardiaceae	Tree	N
Scutia buxifolia Reissek	Rhamnaceae	Tree	N
Serjania meridionalis Cambess.	Sapindaceae	Climber	N
Solanum paniculatum L.	Solanaceae	Shrub	N
Solanum riparium Pers.	Solanaceae	Tree	N
Suaeda divaricata Moq.	Chenopodiaceae	Shrub	N
Syagrus romanzoffiana (Cham.) Glassman	Arecaceae	Palm	E
Tecoma stans (L.) Juss. ex Kunth	Bignoniaceae	Tree	N
Tessaria absinthioides (Hook. & Arn.) DC.	Asteraceae	Shrub	N
Tillandsia sp.	Bromeliaceae	Epiphyte	N
Tipuana tipu (Benth.) Kuntze	Fabaceae	Tree	N
Trema micrantha (L.) Blume	Celtidaceae	Tree	N
Tripodanthus acutifolius (Ruiz & Pav.) Tiegh.	Loranthaceae	Epiphyte	N
Urera baccifera (L.) Gaudich.	Urticaceae	Tree	N
Urera caracasana (Jacq.) Gaudich. ex Griseb.	Urticaceae	Tree	N
Vassobia breviflora (Sendtn.) Hunz.	Solanaceae	Tree	N
Verbesina suncho (Griseb.) S.F. Blake	Asteraceae	Shrub	N
Zanthoxylum coco Gillies ex Hook. f. & Arn.	Rutaceae	Tree	N
Ziziphus mistol Griseb.	Rhamnaceae	Tree	N

considerably lower at dry station for all the sites, and some of them did not have water at all (Table 3).

The NMDS (Figure 2) did not separate the sites based on identity of the forests (Chaco and Yungas sites were dispersed and mixed in the graph). The sites were ordered mainly based on the dominance of exotic species, locating forests that have less exotic species in the negative side of axis 1 and sites with more invasive species in the positive side of axis 1. The percentage of native and exotic species was related to the altitude of the sites (see the size of the symbols).

4. Discussion

The surveyed plants species in the province of Tucumán were mostly native and included different life forms (i.e., trees, shrubs, palms, rattans, lianas, and epiphytes). In order to achieve a more complete assessment of their conservation status, we strongly believe that a quality analysis of riparian forests should include all the varieties of life forms. The number of species presented here is greater than that observed in other studies in the same province [16, 17, 26], probably

TABLE 3: Average concentration of TSS at humid and dry seasons for each sampled site. The standard deviation for humid season is 0.05 and for dry season it is not informed because of the existence of many sites without water (ww) and so there is no data to inform.

Site codification	Average concentration of TSS (mg/L) in wet season	Average concentration of TSS (mg/L) in dry season
Sa1	46.25	6.25
Sa2	75.00	3.75
LP1	3.75	7.50
LP2	47.50	23.75
Ra1	6.00	ww
Gr1	16.25	7.50
Lu1	36.25	ww
Lu2	58.00	20.00
Lu3	26.75	1.25
SJ1	2.50	3.75
SJ2	6.25	23.75
LN1	2.00	ww
Mu1	4.00	ww
SP1	6.00	6.00
Lu4	41.25	3.75
Co1	106.25	ww

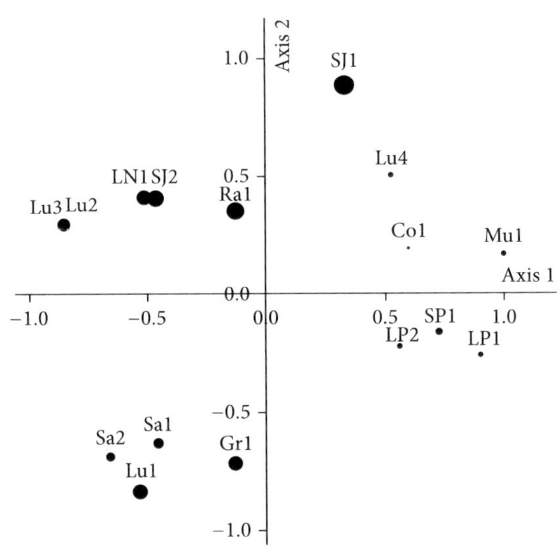

FIGURE 2: NMDS showing the sites in relation to species composition ordering. Increasing size symbols indicate increasing altitude.

because our study area was wider (i.e., species-area effect [46, 47]), comprising both Yungas and Chaco forests, with different species composition. Similarly to most riparian forests in the world, the structure was dominated by woody plants, such as trees and bushes [19]. Exotic species were also present and they were dominant in several sites. The most common exotic species were *Ligustrum lucidum*, *Morus alba*, *Arundo donax*, and *Gleditsia triacanthos*. The main problem we find considering this point is that they can represent a big barrier for the establishment of native vegetation [48]. Previous studies in the same study area showed that these exotic plants have the ability to invade and displace native species, thus reducing their diversity [49, 50]. For example, *G. triacanthos*, one of the most important invasive plant species of Tucumán, caused major physiognomic and structural changes in some areas of Lules River Basin and Sierra de San Javier [16]. These "new forests," dominated by invasive species, acquired xeric characteristics which differ widely from the original riparian forests characteristic from Yungas ecoregion [16]; thus, it is likely that these changes derive in ecological consequences in the area [51]. In other ecoregions of Argentina (e.g., estepa patagonica, Monte, Espinal), invasive species are also changing the structure of riparian communities, with consequences for ecosystem functioning [52]. For example, in Pampean streams of the province of Buenos Aires, *G. triacanthos* modified the structure from grasslands to forests, causing changes in the richness of invertebrates and the composition of aquatic species, probably due to reduced light conditions [53]. On the other hand, exotic leaves may have different physical and chemical characteristics (leaf toughness, leaf area, content of cellulose, hemicelluloses, lignin, C, and N) from native ones, which might negatively influence the activity of detritivores invertebrates [54].

The variation of QBRy values was strongly associated with the altitude of the sampled area, since riparian forests located at lowlands were highly disturbed. Although there are some exceptions [55], it is well known that degradation is higher at lower altitudes, where anthropogenic impact is usually more intense and land is more accessible for different uses [26]. In the study area, riparian forests of lowland rivers are close to urban zones and extensive agriculture areas (e.g., fruit crops and sugarcane) and had a low degree of conservation. In these areas, the anthropogenic impact continues to increase and these forests are becoming narrow strips, for which urgent conservation measures are required in these areas, especially in Yungas piedmont [56]. Other uses, such as recreational activities, did not significantly affect the conservation status of riparian forests, since they usually take place at specific sectors (i.e., not over all river margins) and exhibited intermediate values of QBRy. However, it should be taken into account that recreational activities might also imply alterations in the form of degradation or removal of vegetation cover [57].

The greatest anthropic impact at lower altitude sites and the differences in species composition due to the altitudinal gradient [15] may explain changes in the richness and composition of aquatic and marsh vegetation at the surveyed watercourses. At mountain watercourses, mainly in zones of quiet waters, the vegetation was that typical of flooded and waterlogged habitats [58], with species like *Bacopa monnieri* (Plantaginaceae), *Myriophyllum aquaticum* (Haloragaceae), *Hydrocotyle bonariensis* (Apiaceae), and *Pluchea sagittalis* (Asteraceae). At areas of lower altitude, we found formations of *Heimia salicifolia* (Lythraceae) and *Ludwigia peploides* (Onagraceae) in sediment banks and a less diversity of aquatic and marsh plant communities. This could be due to the fact that these communities are very sensitive to the destabilizing effect of external disturbances [59]. Moreover, runoff during wet season significantly influences the stability and resilience of plant communities at lower altitude areas.

There was no relationship between QBRy and water TSS content in wet and dry seasons. Perhaps other factors are more appropriate than riparian forest quality as determinants of TSS in water. For example, an earlier study made in this area showed that in 2012 (>1.200 mm of precipitations during humid period) the higher concentrations of TSS were directly related to the discharge of rivers and streams [60].

5. Conclusions

Riparian forests surveyed in Tucumán province exhibit a variable conservation status, but most of them present some degree of degradation, mainly due to human impact. Despite the wide variability observed with the QBRy index, some patterns derived from altitude can be identified. The altitudinal gradient determines that riparian forests located in lower areas present a low conservation status, since they coexist with human activities in adjacent areas, with great impact on their conservation. In turn, the altitudinal gradient also determines changes in plant species composition among forests, with a higher percentage of exotic species at lower altitudes. The absence of a relationship between riparian forests quality and TSS in water in the study area suggests that flow is a more relevant factor influencing TSS concentrations. Like other riparian forests of Argentina, degradation is increasing dramatically in the study area. Therefore, habitat conservation and management, particularly at lower altitudes, appear as a priority for future studies.

Competing Interests

The authors Romina Daiana Fernández, Sergio Javier Ceballos, Ana Lucía González Achem, Margarita del Valle Hidalgo, and Hugo Rafael Fernández declare that there are no competing interests regarding the publication of this paper.

Acknowledgments

The authors thank CONICET for supporting their work, as well as Project of CIUNT "Desarrollo de Indices de Calidad de Aguas (ICA) para la Provincia de Tucumán," Code 26/G422. They are grateful to Rafael Escobar, Gabriela Araujo, Guillermo Hankel, Luciana Ruiz, Guillermo Terán, Ana Martínez, Alejandra Cocimano, Ariana Sanchez, Álvaro Galbán, César Moreno, and Laura Rolandi for their selfless help with field and laboratory work. They are especially grateful to Sofia Nanni for the English revision.

References

[1] G. P. Malanson, *Riparian Landscapes*, Cambridge University Press, Cambridge, UK, 1993.

[2] S. Wenger, *A Review of the Scientific Literature on Riparian Buffer Width, Extent and Vegetation*, Office of Public Service and Outreach Institute of Ecology University of Georgia, 1999.

[3] R. L. Vannote, G. W. Minshall, K. W. Cummins, J. R. Sedell, and C. E. Cushing, "The river continuum concept," *Canadian Journal of Fisheries and Aquatic Sciences*, vol. 37, no. 1, pp. 130–137, 1980.

[4] J. D. Allan, *Stream Ecology: Structure and Function of Running Waters*, Chapman & Hall, London, UK, 1995.

[5] M. A. S. Graça, "The role of invertebrates on leaf litter decomposition in streams—a review," *International Review of Hydrobiology*, vol. 86, no. 4-5, pp. 383–393, 2001.

[6] J. E. Clapcott, R. G. Young, J. S. Harding, C. D. Matthaei, J. M. Quinn, and R. G. Death, *Sediment Assessment Methods: Protocols and Guidelines for Assessing the Effects of Deposited Fine Sediment on in-Stream Values*, Cawthron Institute, Nelson, New Zealand, 2011.

[7] N. L. Poff, B. D. Richter, A. H. Arthington et al., "The ecological limits of hydrologic alteration (ELOHA): a new framework for developing regional environmental flow standards," *Freshwater Biology*, vol. 55, no. 1, pp. 147–170, 2010.

[8] G. S. Bilotta and R. E. Brazier, "Understanding the influence of suspended solids on water quality and aquatic biota," *Water Research*, vol. 42, no. 12, pp. 2849–2861, 2008.

[9] M. Strand and R. W. Merritt, "Impacts of livestock grazing activities on stream insect communities and the riverine environment," *American Entomologist*, vol. 45, no. 1, pp. 13–29, 1999.

[10] M. J. Kennard, B. J. Pusey, A. H. Arthington, B. D. Harch, and S. J. Mackay, "Development and application of a predictive model of freshwater fish assemblage composition to evaluate river health in eastern Australia," *Hydrobiología*, vol. 572, no. 1, pp. 33–57, 2006.

[11] L. Pizzolón, M. L. Miserendino, and L. Arias, *Inventario de Perturbaciones Antrópicas Sobre los Ambientes Acuáticos de Chubut*, Resúmenes de las Jornadas Patagónicas de Medio Ambiente, Esquel, Argentina, 1996.

[12] M. L. Miserendino, "Effects of landscape and desertification on the macroinvertebrate assemblages of rivers in Andean Patagonia," *Archiv für Hydrobiologie*, vol. 159, no. 2, pp. 185–209, 2004.

[13] P. F. Temporetti, "Efecto a largo plazo de los incendios forestales en la calidad del agua de dos arroyos en la subregión Andino-Patagónica, Argentina," *Ecología Austral*, vol. 16, no. 2, pp. 157–166, 2006.

[14] M. L. Miserendino, C. Brand, and C. Y. Di Prinzio, "Assessing urban impacts on water quality, benthic communities and fish in streams of the Andes mountains, Patagonia (Argentina)," *Water, Air, and Soil Pollution*, vol. 194, no. 1–4, pp. 91–110, 2008.

[15] M. G. Sirombra, "Q.B.R.y una propuesta ecológica para evaluar la calidad del bosque ribereño en ríos de yungas del Noroeste Argentino," *Serie Monográfica y Didáctica*, vol. 48, p. 248, 2009.

[16] M. G. Sirombra and L. M. Mesa, "Composición florística y distribución de los bosques ribereños subtropicales andinos del Río Lules, Tucumán, Argentina," *Revista de Biología Tropical*, vol. 58, no. 1, pp. 499–510, 2010.

[17] P. A. Quiroga, H. R. Fernández, M. G. Sirombra, and E. Domínguez, "Riparian forests and cattle management problems in Andean subtropical streams: in the search of quality water sustainability," *Lilloa*, vol. 48, no. 1, pp. 36–52, 2011.

[18] W. G. Hood and R. J. Naiman, "Vulnerability of riparian zones to invasion by exotic vascular plants," *Plant Ecology*, vol. 148, no. 1, pp. 105–114, 2000.

[19] D. M. Richardson, P. M. Holmes, K. J. Esler et al., "Riparian vegetation: degradation, alien plant invasions, and restoration

prospects," *Diversity and Distributions*, vol. 13, no. 1, pp. 126–139, 2007.

[20] I. M. Parker and S. H. Reichard, "Critical issues in invasion biology for conservation science," in *Conservation Biology for the Coming Decade*, P. L. Fiedlerand and P. M. Kareiva, Eds., pp. 283–305, Chapman and Hall, 1997.

[21] M. I. Posada-Posada and M. D. P. Arroyave Maya, "Análisis de la calidad del retiro ribereño para el diseño de estrategias de restauración ecológica en el río La Miel, Caldas, Colombia," *Revista EIA*, vol. 12, pp. 117–128, 2015.

[22] A. Munné, C. Solá, and N. Prat, "QBR: Un índice rápido para la evaluación de la calidad de los ecosistemas de ribera," *Tecnología del Agua*, vol. 175, pp. 20–37, 1998.

[23] M. L. Suárez, M. R. Vidal-Abarca, M. M. Sánchez Montoya et al., "Las riberas de los ríos mediterráneos y su calidad: el uso del índice QBR," *Limnetica*, vol. 21, no. 3-4, pp. 135–148, 2004.

[24] A. Kutschker, C. Brand, and M. L. Miserendino, "Evaluación de la calidad de los bosques de ribera en ríos del NO del Chubut sometidos a distintos usos de la tierra," *Ecología Austral*, vol. 19, pp. 19–34, 2009.

[25] A. Palma, R. Figueroa, and V. H. Ruiz, "Evaluación de ribera y hábitat fluvial a través de los índices QBR e IHF," *Gayana*, vol. 73, no. 1, pp. 57–63, 2009.

[26] M. G. Sirombra and L. M. Mesa, "A method for assessing the ecological quality of riparian forests in subtropical Andean streams: QBRy index," *Ecological Indicators*, vol. 20, pp. 324–331, 2012.

[27] M. G. Sirombra, *Diagnóstico y propuestas de gestión ambiental sobre la base del estudio de calidad de ecosistemas rivereños en un sector del pedemonte oriental de la Sierra de San Javier, Provincia de Tucumán, Argentina [Ph.D. thesis]*, Universidad Nacional de Tucumán, Facultad de Ciencias Naturales e Instituto Miguel Lillo, 2012.

[28] H. Hunzinger, "La precipitación horizontal: Su importancia para el bosque y a nivel de cuencas en la Sierra San Javier, Tucumán, Argentina," in *Investigación, Conservación y Desarrollo en Selvas Subtropicales de Montaña*, A. Brown and H. R. Grau, Eds., pp. 53–58, LIEY, Tucumán, Argentina, 1995.

[29] A. L. Cabrera, "Fitogeografía de la República Argentina," *Boletín de la Sociedad Argentina de Botánica*, vol. 14, pp. 1–42, 1971.

[30] A. L. Cabrera and A. Willink, "Regiones biogeográficas en América Latina," in *Biogeografía de América Latina*, E. Chesneau, Ed., Serie de Biología, Monografía no. 13, chapter 8, pp. 29–107, Secretaría de la Organización de los Estados Americanos, Programa de Desarrollo Científico y Tecnológico, 1980.

[31] A. D. Brown, H. R. Grau, L. R. Malizia, and A. Grau, "Argentina," in *Bosques Nublados del Neotrópico*, M. Kappelle and A. D. Brown, Eds., pp. 623–658, INBIO, Santo Domingo de Heredia, Costa Rica, 2001.

[32] A. L. Cabrera, *Regiones Fitogeográficas Argentinas*, vol. 1, Editorial Acme, 1976.

[33] A. Malizia and H. R. Grau, "Liana-host tree associations in a subtropical montane forest of north-western Argentina," *Journal of Tropical Ecology*, vol. 22, no. 3, pp. 331–339, 2006.

[34] A. Grau and A. D. Brown, "Development threats to biodiversity and opportunities for conservation in the mountain ranges of the Upper Bermejo River Basin, NW Argentina and SW Bolivia," *Ambio*, vol. 29, no. 7, pp. 445–450, 2000.

[35] A. Malizia, P. I. Campanello, M. Villagra, and S. Ceballos, "Geographical, taxonomical and ecological aspects of lianas in subtropical forests of Argentina," in *Biodiversity of Lianas, Sustainable Development and Biodiversity*, N. Parthasarathy, Ed., pp. 17–42, 2015.

[36] S. A. Torrella and J. Adámoli, "Situación ambiental de la ecorregión del Chaco Seco," in *La Situación Ambiental Argentina*, A. Brown, U. Martinez Ortiz, M. Acerbi, and J. Corcuera, Eds., pp. 75–82, Fundación Vida Silvestre, Buenos Aires, Argentina, 2006.

[37] F. O. Zuloaga, O. Morrone, and M. J. Belgrano, "Catalogue of the vascular plants of the southern cone (Argentina, Southern Brazil, Chile, Paraguay and Uruguay)," in *Dicotyledoneae: Acanthaceae-Fabaceae (Abarema-Schizolobium)*, vol. 2, p. 985, Missouri Botanical Garden Press, 2008.

[38] The Nature Conservancy (TNC), Fundación Vida Silvestre Argentina (FVSA), Fundación para el Desarrollo Sustentable del Chaco (DeSdel Chaco) and Wildife Conservation Society Bolivia (WCS), Evaluación Ecorregional del Gran Chaco Americano/Gran Chaco Americano Ecorregional Assessment, Fundación Vida Silvestre, Buenos Aires, Argentina, 2005.

[39] TROPICOS, Missouri Botanical Garden, 2015, http://www.tropicos.org.

[40] Instituto Argentino de Normalización y Certificación IRAM 29012-2: Técnicas de muestreo. IRAM 29012-3: Calidad ambiental. Calidad del agua. Muestreo. Parte 3: Guía para la preservación y manipulación de las muestras. IRAM 29012-6: Calidad ambiental. Calidad del agua. Muestreo: Parte 6: Directivas para el muestreo en ríos y cursos de agua. IRAM 29012-14: Calidad ambiental. Calidad del agua. Muestreo: Parte 14: Directivas sobre aseguramiento de la calidad del muestreo y manipulación de agua, 2012, http://www.iram.org.ar/.

[41] J. Rodier, *Análisis de las Aguas. Aguas Naturales. Aguas Residuales. Agua de Mar. Química, Fisicoquímica, Bacteriología, Biología*, Ediciones Omega, Barcelona, Spain, 1989.

[42] APHA, *Standard Methods*, American Public Health Association, Washington, DC, USA, 17th edition, 1998.

[43] J. Carrasco, *Comparación de un modelo paramétrico de producción de sedimentos con valores experimentales, en cuencas de la provincia de Tucumán Argentina [Ph.D. thesis]*, Thesis of Licenciature degree in Ciencias Biológicas de la Facultad de Ciencias Naturales e Instituto Miguel Lillo, Universidad Nacional de Tucumán, Argentina, 2013.

[44] B. McCunne and J. B. Grace, *Analysis of Ecological Communities*, MjM Software, Gleneden Beach, Ore, USA, 2002.

[45] R Core Team, *R: A Language and Environment for Statistical Computing*, R Foundation for Statistical Computing, Vienna, Austria, 2015, http://www.R-project.org/.

[46] R. H. MacArthus and E. O. Wilson, *The Theory of Island Biogeography*, Princeton University Press, 1967.

[47] M. L. Rosenzweig, *Species Diversity in Space and Time*, Cambridge University Press, Cambridge, UK, 1995.

[48] F. D. Shields, A. J. Bowie, and C. M. Cooper, "Control of streambank erosion due to bed degradation with vegetation and structure," *Journal of the American Water Resources Association*, vol. 31, no. 3, pp. 475–489, 1995.

[49] H. R. Grau and R. Aragón, "Árboles invasores de la Sierra de San Javier," in *Árboles Exóticos de las Yungas Argentinas, LIEY-UNT*, H. R. Grau and R. Aragón, Eds., pp. 5–20, 2000.

[50] R. Aragón and M. Groom, "Invasion by *Ligustrum lucidum* (Oleaceae) in NW Argentina: early stage characteristics in different habitat types," *Revista de Biologia Tropical*, vol. 51, no. 1, pp. 59–70, 2003.

[51] P. A. Quiroga and J. P. Juliá, "Estudio de una cuenca de río subtropical de montaña: pautas para su gestión ecosistémica," in *La Cuenca del Río Lules: Una Aproximación Multidisciplinaria a su Complejidad*, H. R. Fernandez and H. Barber, Eds., pp. 159–174, Editorial Edunt, 2011.

[52] P. Gantes, A. V. Marano, and L. Rigacci, "Changes in the decomposition process associated with the invasion of *Gleditsia triacanthos* (honey locust) in pampean streams (Buenos Aires, Argentina)," *Journal of Freshwater Ecology*, vol. 26, no. 4, pp. 481–494, 2011.

[53] A. Giorgi, C. Vilches, M. C. R. Castro et al., "Honey locust (*Gleditsia triacanthos* L. (Fabaceae)) invasion effect on temperature, light and metabolism of a pampean stream," *Acta Biologica Colombiana*, vol. 19, no. 1, pp. 99–105, 2014.

[54] I. W. Ashton, L. A. Hyatt, K. M. Howe, J. Gurevitch, and M. T. Lerdau, "Invasive species accelerate decomposition and litter nitrogen loss in a mixed deciduous forest," *Ecological Applications*, vol. 15, no. 4, pp. 1263–1272, 2005.

[55] E. Valero, J. Picos, and X. Álvarez, "Characterization of riparian forest quality of the Umia River for a proposed restoration," *Ecological Engineering*, vol. 67, pp. 216–222, 2014.

[56] D. E. Prado, "Selva pedemontana: contexto regional y lista florística de un ecosistema en peligro," in *Investigación, Conservación y Desarrollo en Selvas Subtropicales de Montaña*, A. D. Brown and H. R. Grau, Eds., pp. 19–52, L.I.E.Y., UNT, 1995.

[57] C. M. Gualdoni, C. A. Duarte, and E. A. Medeot, "Ecological status of two mountain streams of the south of Córdoba, Argentina," *Ecologia Austral*, vol. 21, no. 2, pp. 149–162, 2011.

[58] S. Rivas-Martínez, "Datos ecológicos sobre la vegetación acuática continental," *Anales del Instituto Botánico Cavanilles*, vol. 32, no. 1, pp. 199–205, 1975.

[59] C. Duarte, C. Montes, S. Agustí, P. Martino, M. Bernués, and J. Kallf, "Biomasa de macrófitos acuáticos en la marisma del parque nacional de Doñana (SW España): importancia y factores ambientales que controlan su distribución," *Limnetica*, vol. 6, pp. 1–12, 1990.

[60] M. C. Reynaga and D. A. Dos Santos, "Rasgos biológicos de macroinvertebrados de ríos subtropicales: patrones de variación a lo largo de gradientes ambientales espacio-temporales," *Ecología Austral*, vol. 22, no. 2, pp. 112–120, 2012.

Avian Community Structure as a Function of Season, Habitat Type, and Disturbance, in Mole National Park, Northern Region (Ghana)

Collins Ayine Nsor[1], **Emmanuel Acquah**[1], **Grace Mensah**[1], **Vincent Kusi-Kyei**[1] **and Samuel Boadi**[2]

[1] *Department of Ecotourism and Forest Recreation, Kwame Nkrumah University of Science and Technology, Kumasi, Ghana*
[2] *Department of Social Forestry, Kwame Nkrumah University of Science and Technology, Kumasi, Ghana*

Correspondence should be addressed to Collins Ayine Nsor; ayineus@yahoo.com

Academic Editor: Ram Chander Sihag

The study assessed factors that influenced bird-habitat preference, diversity, and spatial distribution in Mole National Park. Birds were identified using point count sampling method, while ordination techniques were performed to determine the influence of environmental factors on bird-habitat preference. A total of 4951 individuals belonging to 131 species were identified across the four habitat types in the wet (n = 3033) and dry (n = 1648) seasons. Despite the high abundance and richness of birds in the woodland and shrubland habitats, grassland habitat was the most diverse, due to the high spatial evenness distribution of the birds. Bushfire, patchiness, and animal trampling were the key environmental determinants in bird assemblages and habitat preferences and accounted for 62.02% and 81.82% variations in the two seasons. Rarer birds like White-Backed Vulture (NT) and White-Headed Vulture (CR), Bateleur Eagle (NT), and Woolly-Necked Stork (VU), with high conservation concern, suggest intensification of conservation effort, while the dominance of Sahel Bush-sparrow was probably due to their broad range habitat preferences and ability to adapt to environmental conditions across these habitats. These species could be used as indicators of habitat quality, if future scenarios restrict their movement or distribution in narrow range habitats.

1. Introduction

The study of bird-habitat selection and use has long been established as a functional field of study in bird ecology [1–4]. Early habitat-selection theory was characterized by correlative models of habitat characteristics and species abundance [5, 6], which subsequently evolved into models that involved density dependence (i.e., the "ideal-free distribution" and "ideal-despotic distribution" models) [7, 8]. More recently, habitat-selection studies have shown that many factors, such as landscape structure [9–11], nest predation [12], competition [11], intraspecific attraction [13], food availability, variable climate, diseases, and human activities [14], can influence exactly how "ideal" and "free" animals are while moving through a landscape and selecting habitats. Habitat-selection studies have assumed a new urgency, partially as a result of the importance of incorporating both habitat and demographic information into conservation planning [15], given recent report about global bird population decline [16], largely linked to habitat transformation [17, 18]. But Pulliam [19] and Caughley [15] argue that there is no guarantee that the presence of individuals in a given habitat is positively related to habitat quality. This is so because, in the absence of behavioral or life-history information, there is no way to know if detected differences have any bearing on choices of individuals [12, 20]. Pulliam and Danielson [10] and Martin [20] therefore conclude that habitat preferences are rather assumed to be adaptive without demonstration of increased fitness in preferred habitats.

Notwithstanding these views about bird-habitat relationship, studies have shown particular birds to be associated with particular habitats [21, 22] and therefore they respond quickly to changing habitats [23–25]. Birds have been used as surrogates for assessing the impact of habitat changes [26] and considered as good predictors of habitat quality as they relate to changes that are associated with their habitats [27–29]. Other habitat quality indicators, such as hemeroby, have widely been used in plant ecology to indicate the level of disturbance of the optimal habitat for a species [30]. Hemeroby is scaled on a range of ten- or five-point scores, higher scores of hemeroby meaning higher level of disturbance [30].

Vegetation structure (a typical floristic habitat type) is important in structuring avian communities [31, 32] and, thus, their relative abundance is often associated with vegetation community [33]. Vegetative structure is frequently assumed to be the primary proximate factor determining where and how species use resources. For example, Sage Grouse (*Centrocercus urophasianus*) is strongly associated with sagebrush (*Artemesia tridentata*), Chukar (*Alectoris chukar*) is associated with cheatgrass (*Bromus tectorum*), and Chestnut-backed Chickadees (*Parus rufescens*) appears to follow the distribution of Douglas-fir (*Pseudotsuga menziesii*) in California [34]. In the Sudan-Guinea Savannah biome of Ghana, where continued habitat loss in protected and unprotected areas is thought to threaten bird survival [25], maintenance of vegetation structure is crucial in sustaining bird population. Skowno and Bond [35] observed a strong influence of vegetation structure (i.e., mosaic of trees, shrubs, and open grassland in mesic African savannas) on bird-habitat use compared to floristic composition. However, MacNally [36] suggested that floristic habitat relationships of birds are often influenced by the scale of observation. MacNally [36] concluded that it is important to consider different observational scales, to determine the relative significance of structural and floristic attributes of the habitat.

Although tourists have expressed satisfaction at the wildlife attraction (*i.e.*, large mammals) and the environment [37], bird watching was among the least attractive, possibly due to scanty information about their preferred habitats, borne out habitat transformation. Given their important role in enhancing biodiversity and as bioindicators to ecosystem change [16, 27], birds are among wildlife attractions to MNP and other game reserves, across African landscapes. Thus, understanding bird-floristic habitat preference and the factors that influence their spatial distribution is key in developing right conservation measures that will sustain bird watching, as part of ecotourism drive. Though previous studies have established the presence of diverse avifauna in MNP [38] and are found to be of biogeographical importance, the study did not investigate the habitat preferences and (dis)similarity of bird species assemblage in the various floristic habitat types and factors influencing their habitat preference and spatial distribution. The possible mechanisms underlying the bird species distribution pattern and the uncovering of the possible mechanisms underlying these patterns are critical due to the intensity of land use by humans, which continues to alter the availability, quality, and distribution of habitats supporting native bird species [39]. This study focused on understanding bird community structural assemblages as a function of habitat preference and the predictive factors that influenced their spatial distribution in the MNP in different seasons. Thus, we hypothesized that predictive factors will not alter bird: (i) abundance and richness among habitats, (ii) compositional differences among habitats, (iii) seasonality will not influence changes in abundance and richness, and finally, (iv) processes that structure assemblages in different habitats will not vary among the habitats, along seasonal transition.

2. Materials and Methods

2.1. Study Area. The Mole National Park (MNP), where this study was conducted, is the largest wildlife reserve in Ghana and home to a large proportion of Sudan woodland in the north of Ghana. It is located in the Northern Region of Ghana and covers an area of 4,577 km^2. It is one of the major tourism destinations in the Northern Region, as well as the largest protected area in Ghana. The area lies within geographic coordinates of 9° 11′ and 10° 10′ N and between 1° 22′ and 2° 13′ W and an elevation of 120 to 490 meters above sea level (Figure 1). Average annual temperature is 28°C which varies from 26°C in December to 31°C in March. Average annual rainfall is about 1100 mm, decreasing to 1000 mm in the north of the park [40]. The dominant vegetation type in Mole National Park is open savanna woodland, interspersed with grasses [41]. There are at least 344 different species of birds in Mole National Park and among the key protected sites in Ghana, conserving 37 endemic Sudan-Guinea Savanna biome bird species [40].

2.2. Sampling Procedure. A survey was undertaken within the study area to map out designated vegetation types and the transect points to use for bird census. We adopted point count sampling design employed to census birds, following Bibby et al. [42], Sutherland [43], and Hutto et al. [44] approach in the dry (January-May, 2017) and wet (June-October, 2017) seasons. We counted birds within a 25 m radius surrounding the observer in four plots, covering an area 60 m × 50 m (3000 m^2) each in the four vegetation types (i.e., grassland, woodland, riparian, and shrubland). These vegetation types were selected because they represented the general vegetation composition in Mole National Park. The distance between each plot to the other was 100 m and randomly demarcated. Birds that were detected outside the 25 m radius, but still found within the demarcated boundaries of the plot, were counted and considered as being within the fixed-radius [42, 44]. We chose a 25 m radius because we tended to compare bird abundances, evenness distribution, and richness among the different vegetation types. Point count of birds was done twice in the day, the early hours of the morning (from 06:30-09:30 a.m. and in the afternoon (15:30-17:30 p.m.), when the temperature is low and most of the birds were out feeding, perching, or nesting. Flying over birds were not counted, because we assumed that their habitat preference did not strongly relate to the vegetation types found in Mole National Park. We avoided performing bird count during windy days, since visualization of birds might be difficult. Bird identification was done visually using Birds of Ghana keys developed by Borrow and Demey [45] and by their calls,

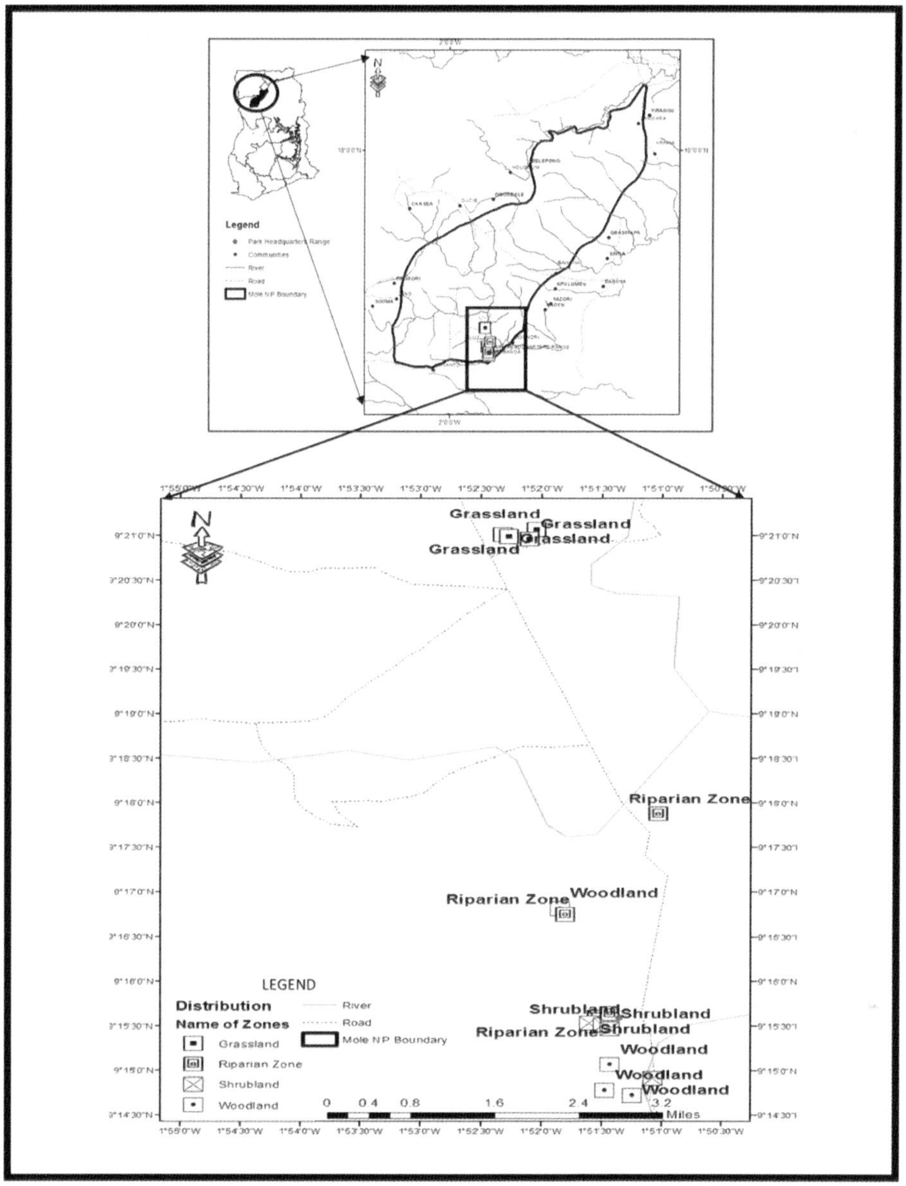

FIGURE 1: Map of Ghana, showing the study area at Mole National Park, in Northern Region.

using the bird sounding technique. The data was collected in eight weeks within the period of November to February 2016-2017. GPS coordinates of the various sampling points were taken. Environmental factors, namely, bushfire, animal trampling, erosion, and patchiness on each sample plot, were assessed to determine how these factors gave rise to variations in bird abundance, evenness, and richness, across the four vegetation types. These environmental factors were assessed on the basis of severity and scope of their threats on birds, following guidelines provided by [46]. A score ranging from 1 to 4 (1 being lesser impact and 4 highest impact) was used to assess scope and severity of every threat. "Scope" more precisely is referred to the percentage of the study area affected by a specific threat within the last 5 years (where 100%) corresponding to the total site area [46], whereas severity is the intensity of a disturbance or a threat in the study area. The scores were assigned as follows: 4: the threat is found throughout (50%) the site area; 3: the threat is spread in 15-50% of the study area; 2: the threat is scattered (5-15%); and 1: the threat is much localized (<5%). Assessment of the area disturbed was carried out within 60 m by 50 m in each habitat type. This is because all the disturbances were observed within this perimeter following a preliminary survey. We further classified the level of disturbances, following the hierarchal threat guidelines provided by Salafsky et al. [47]. They are as follows: threats are considered comprehensive when they contain all possible items at least at higher levels of the hierarchy), consistent (ensuring that entries at a given level of the classification are of the same type, expandable (enabling new items to be added to the classification if they are discovered), and exclusive (allowing any given item to only be placed in one cell within the hierarchy) [46].

3. Data Analysis

3.1. Species Abundance Distribution. Bird abundance as a measure of diversity was quantified using rank abundance model [48]. In each habitat, we listed the number of bird species, say S_1, represented by one individual, and the number of species, say S_K, represented by K individuals, where K denotes the abundance of the most abundant species and $S_1 + \ldots + S_K = S$ [49]. Accordingly, the sequence of relative frequencies $f_r = S_r/S$ $(r = 1 \ldots K)$ constitutes a frequency distribution for the number of individuals per species which is usually referred to as the *species abundance curve* [49].

We then fitted the geometric model (GS) in the species data (raw abundance) using the regression model approach [49], to determine how the communities are assembled in each habitat. This model approach was used in order to test against the null hypothesis (*Ho*) that bird abundance and evenness distribution did not differ in each of the four habitats. All the species in each of the four sample plots per habitat (four habitats) were ranked from the most to the least abundant on the rank abundant curve [49]. Each species rank is plotted on the x-axis and the abundance plotted on the y-axis.

With the geometric series, if a log scale is used for abundance, the species exactly fall along a straight line, according to the model equation $\log A = b_o + b_1 R$, where A is the species abundance, R is the respective rank, and b_0 and b_1 are optimized fitting parameters [50]. Analysis of covariance (ANCOVA) was applied to test for the significant difference of the slope of the SADs for the four habitats, while Pearson's Chi-square test (χ^2) was applied to determine whether an observed distribution along the goodness of fit statistically differed in the GS model. The SAD model is mostly used to measure the impact of disturbance on community structure [51], while the geometric series (a proposed SAD model) represents species distribution with lower evenness and provide a good fit to simple communities characterized by the high dominance of a few species [48]. The shape of the frequency distribution gives an insight into the bird diversity of the communities under study.

Individual-based rarefaction technique [52] was used to compare bird richness across the four habitats (*rarefaction curves*). Rarefaction curves are created by randomly resampling the pool of N samples multiple times and then plotting the average number of species found in each sample $(1, 2 \ldots N)$ [53]. Thus, rarefaction generates the expected number of species in a small collection of n individuals (or n samples) drawn at random from the large pool of N samples. The rarefaction curve, f_n, is defined as

$$f_n = E[X_n] = K - \binom{N}{n}^{-1} \sum_{i=1}^{k} \binom{N - Ni}{n} \quad (1)$$

[52], where X_n is the number of groups still present in the subsample of "n" less than K whenever at least one group is missing from this subsample, N is total number of items, K is total number of groups, and Ni is total number of items in group i $(i = 1, \ldots k)$ [53, 54]. Therefore, the linear model for the GS was fitted for each rarefied run in order to build the 95% confidence limits for the slopes of woodlands.

Rarefaction methods, both sample-based and individual-based, allow for meaningful standardization and comparison of datasets [51] and have been used on bird counts [e.g., [55]].

Finally, Renyi diversity ordering approach [56] was applied to quantify and compare current bird diversity status among the habitats, following initial analysis that resulted in various diversity indices. Thus, Renyi diversity ordering has the ability to harmonize the different techniques and indices developed for biodiversity analysis (e.g., Berger-Parker, Shannon-Wiener, and Simpson's_1-D diversity indices) that makes it complex to select the right tool for comparing biodiversity measurements [48]. Renyi [56] extended the concept of Shannon's entropy [57], by defining the entropy of order α $(\alpha \geq 0, \alpha \neq 1)$ of a probability distribution $(p_1, p_2 \ldots p_s)$. Diversity profile values (H-alpha) were calculated from the frequencies of each component species (proportional abundances pi = abundance of species i/ total abundance) and a scale parameter (α) ranging from zero to infinity as

$$(H_\alpha) = \frac{(\log \sum_{i-1}^{s} pi^\alpha)}{(1 - \alpha)} \quad (2)$$

[58]. All analyses of bird abundance, richness, and diversity ordering were performed using PAST ver. 3.06 software package [59], which contains robust algorithm as indicated in Krebs [60].

A priori test for normality of data distribution was carried out, using Shapiro-Wilk test. This was to determine the appropriate statistical test to subject data (i.e., parametric or nonparametric data analysis approach) [61]. Kruskal-Wallis H test (a rank-based nonparametric test approach) was then used to determine if overall bird abundance significantly differed among the four communities. Mann–Whitney U test pairwise comparison was applied to test for differences in bird richness among the four communities. A paired t-test was used to test for significant differences in bird abundance, richness, and evenness distribution between the wet and dry seasons, using GraphPad Prism 5 statistical software for Windows. Canonical correspondence analysis (CCA) [62] was performed to determine the drivers of habitat quality and its influence on bird preference, using ECOM.exe ver. 1.4 [63]. CCA is a direct method of ordination with the resulting product being the variability of the environmental data, as well as the variability of species data [61].

4. Results

4.1. Species Abundance and Structural Distribution. Shapiro-Wilk test showed that data were not normally distributed ($p = 0.83$). A total of 4951 individual birds belonging to 131 species were identified across the four habitat types in the wet ($n = 3033$) and dry ($n = 1648$) seasons (Table 1). Majority of birds were detected in the woodland habitat (*dry season* = 557, *wet season* = 981), while grassland habitat was the

TABLE 1: Mean number of birds counted at each sampling point count (SPC), across the four habitat types, in the wet (n = 3033) and dry (n = 1648) seasons.

	No. of individuals	SPC 1	SPC 2	SPC 3	SPC 4
			Dry season		
Woodland	557	7.3±1.2	6.8±1.02	6.7±1.1	5.8±1.1
Grassland	262	3.0±0.85	2.8±0.21	3.6±0.33	3.1±0.04
Shrubland	536	6.0±1.08	6.3±1.2	7.1±0.4	5.9±1.05
Riparian zone	293	4.1±1.06	3.5±1.3	3.2±0.9	3.1±1.9
			Wet season		
Woodland	981	7.4±1.4	8.5 ±1.9	8.1±1.6	8.6±1.3
Grassland	447	3.4±2.1	4.7±1.5	2.8±1.7	3.5±1.3
Shrubland	927	8.1±2.2	7.4±1.47	8.2±1.1	6.9±1.04
Riparian zone	678	6.1±0.4	4.9±1.8	6.3±1.6	5.1±0.07

TABLE 2: Results of the geometric series model for the abundance rank distribution of birds of Mole National Park, calculated for Woodland, Grassland, Shrubland, and Riparian habitats. Levels of significant difference (P-values) were shown with superscript alphabet ([a]).

	Dry season			
Sample	Intercept ±S.E.	Slope ±S.E.	R^2	Prob.
Woodland	4.169±1.10	-0.158±0.09	-0.19	0.11
Grassland	6.760±0.86	-0.122±0.09	-0.18	0.20
Shrubland	8.319±0.93	-0.292±0.08	-0.36	0.0007[a]
Riparian	3.551±0.99	-0.009±0.08	-0.01	0.91
Slope of *SAD*: $F_{3,80}$,= 6.53, p (regr): 0.0005				
	Wet season			
Woodland	1.824±0.82	0.231±0.07	0.28	0.001[a]
Grassland	7.501±0.87	0.043±0.11	0.28	0.001[a]
Shrubland	5.816±1.03	0.22±0.08	0.22	0.01[a]
Riparian	3.856±0.97	0.21±0.08	0.23	0.01[a]
Slope of *SAD* = $F_{3,117}$ = 7.32, p(regr) = 0.0001,				

least recorded (*dry season* = 262, *wet season* = 447). Mean bird abundance varied significantly (Hc = 21.72, p<0.0007, *Kruskal-Wallis H test*) in the dry season and ranged between 7.3 ± SE 1.2 and 5.8 ±1.1 in the woodland community and 3.6 ± SE 0.3 and 2.8 ± 0.21 in the grassland community. The wet season also showed significant difference in bird abundance among the four communities (Hc = 27.74, p<0.0004, *Kruskal-Wallis H test*) and ranged between 8.6 ± SE 1.3 and 7.4 ± 1.4 in woodland habitat and 4.7 ± SE 1.5 and 2.8 ± 1.7 in grassland habitat (Table 1). Species abundance distribution (SAD) fitted well in the geometric series distribution (GS) model and showed a general significant difference in the dry ($F_{3,80}$= 6.534, p (regr) = 0.0005, *ANCOVA interactions x species rank*) and wet ($F_{3,117}$ = 7.32, p(regr) = 0.0001, *ANCOVA interactions x species rank*) (Figures 2 and 3, Table 2). However, from individual habitats, we observed significant variations in bird abundance along the slopes of the SAD curve in the shrubland habitat (slope [k] = -0.292±0.08, R^2 = -0.36, $\chi^2 P$ = 0.0007), while grassland (k = -0.122±0.09, R^2 = -0.18 $\chi^2 P$ = 0.20) and riparian habitats (k = -0.009±0.08, R^2 = -0.01, $\chi^2 P$ = 0.91) did not show any significant variations in abundance in the dry season (Table 2, Figure 2). Although, in the wet season, bird abundance generally varied across the habitats, woodland (k = 0.231±0.07, R^2 = 0.28, $\chi^2 P$ = 0.001) and grassland (k = 0.231±0.07, R^2 = 0.28, $\chi^2 P$ = 0.001) habitats revealed the highest variations in bird abundance (Table 2, Figure 3). Comparison of SADs for the four sites allows for the distinguishing of a particular habitat quality, in relation to its influence on bird abundance. Thus, the shape of the rank abundance curve generally revealed differences in bird dominance and evenness from individual habitats, which indicate their relative success at competing for food and space.

Bird community assemblages across the four habitats revealed that species were ranked from the most abundant to the least abundant (Figures 2 and 3). Thus, fewer than six species (Sahel Bush-sparrow, *Gymnoris dentata;* Longed-tail Glossy Starling, *Lamprotornis caudatus;* Northern Black-flycatcher, *Melaenornis edolioides;* European pied flycatcher, *Ficedula hypoleuca;* Senegal Eremomela, *Eremomela pusilla;* and Vinaceous, dove *Streptopelia vinacea*) were the most abundant and detected in all four habitats during the wet season. These species constituted 4.95% of the total sampled (n = 121). In the dry season, the habitat preferences of these species were rather limited, as they did not occur in all habitats. Sahel Bush-sparrow, *Gymnoris dentata* (a typical resident bird), was

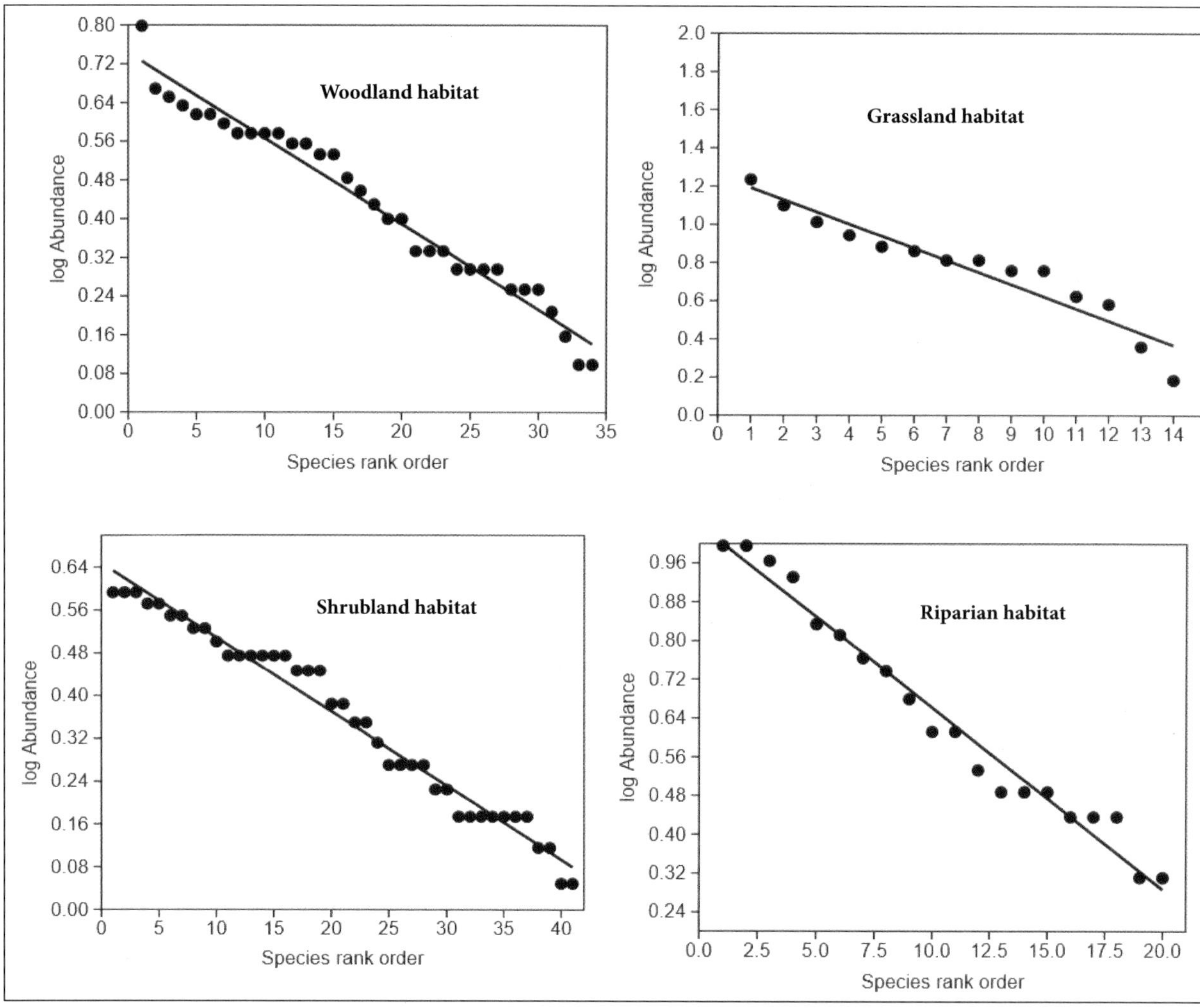

FIGURE 2: Geometric model for bird rank abundance distribution across the four habitats in Mole National Park, in the dry season. Abundance is based on cumulative cover values per species per test site. Notice that SADs are ordered in a decreasing magnitude and plotted against the corresponding rank in this order.

the most abundant during the dry (n = 69) and wet (n = 86) seasons, with a least concern (LC) conservation status. Senegal Eremomela, *Eremomela pusilla* (n = 79, a typical insectivorous bird), was the second most abundant species across the four habitats but was only detected in the wet season. The occurrence of this species during the wet season was expected, as they mostly feed on insects. Rarer species like the Northern Ground-hornbill, *Bucorvus abyssinicus*, was the least ranked on the SAD curve in the grassland (n = 4, dry season), while Bateleur, *Terathopius ecaudatus* (n =1), and Wooly-necked stock, *Ciconia episcopus* (n =1), with a conservation status of near threatened (NT) and vulnerable (VU), respectively, were the least ranked in the woodland and shrubland habitats, in the wet season (Figure 3, Appendix 1). These species including a few rarer species with conservation concern (e.g., White-backed Vulture, *Gyps africanus* (NT), and White-headed Vulture, *Trigonoceps occipitalis* (CR), were confined to less disturbed sections in the woodland habitat.

Helmeted Guinea fowl, *Numida meleagris* (n = 45), was the single most dominant species in the dry season, which reflects on its ability to withstand harsh environmental conditions (e.g., bushfire) that are common in this time of the year. Oriole warbler, *Hypergerus atriceps* (n = 4) and Yellow-fronted Tinkerbird, *Pogoniulus chrysoconus* (n = 4), were the least ranked species in the riparian zone and only encountered in the wet season. They were mostly confined to areas of where insects and fruit-bearing trees were abundant.

4.2. Seasonal Trends in Bird Richness across the Four Habitats. Interpolating SADs in the four habitats, with sample-based rarefaction, we observed a significant increase in bird richness from dry to wet seasons (*t-test* = -9.59, $p<0.002$) (Figure 4). We found bird richness in the dry season to be the highest in the shrubland habitat (n = 40), while in the wet season, it was the woodland habitat (n =73) (Figure 4). Grassland habitat consistently recorded lowest species richness in both dry (n = 14) and wet (n = 35) seasons. Comparing bird richness among habitats in the dry season revealed shrubland habitat as the richest compared to those from grassland (*t-test* = 2.75, $p<0.006$) and riparian

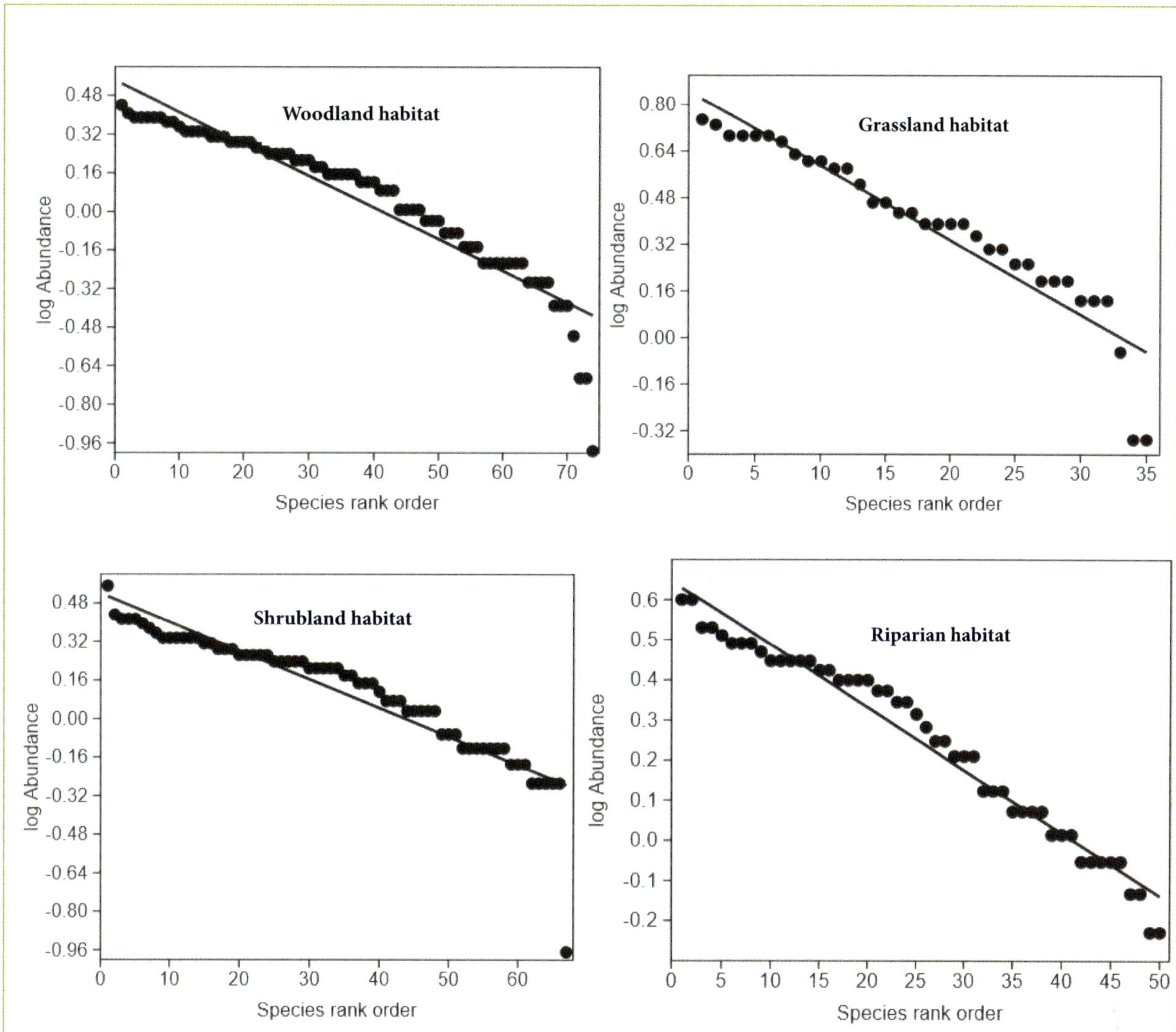

FIGURE 3: Geometric model for bird rank abundance distribution across the four habitats in Mole National Park, in the wet season. Abundance is based on cumulative cover values per species per test site. Notice that SADs are ordered in a decreasing magnitude and plotted against the corresponding rank in this order.

habitats (*t-test* = 2.57, $p<0.011$) but showed no variation with woodland preferred species (*t-test* = -0.19, p = 0.84). In the wet season, we equally observed similar significant variations in bird richness between the woodland and grassland (*t-test* = 4.47, $p<0.00012$) and riparian habitats (*t-test* = 2.37, $p<0.018$). However, bird richness from woodland and shrubland habitats showed no significant difference (*t-test* = 0.078, p= 0.94).

4.3. Bird Diversity Profiles across the Four Habitats. The observed differences in bird community structure (i.e., abundance, evenness, and richness) are shown in the Renyi diversity ordering (Figure 5). Generally, the Renyi diversity curves showed a clear tight bend towards lower values of the diversity index and higher alpha scale values. Although bird diversity did not differ significantly in the dry (*Hc* = 0.087, *p*=0.83, *Kruskal-Wallis test*) and wet (*Hc* = 0.17, *p*=0.98, *Kruskal-Wallis test*) seasons, the profile diversity of species from the grassland habitat appeared marginally high in the two seasons. This observation was linked to the shallower nature of the SAD curves (Figures 2 and 3) and clearly reflected in the Renyi ordering (Figure 5), although species abundance and richness (individual-based rarefaction) were the least in this habitat. Grassland species showed a high-profile alpha (α) values 0.04 and 3.96, which corresponded to diversity indices of 1.99 and 1.51 (dry season) and 1.99 and 1.39 (wet season), respectively. These diversity indices are indicative of sites in the habitats that were richer in species and spatially even distribution.

Woodland and shrubland appeared highly similar in diversity profile and were the lowest, especially in the wet season where the two habitats could barely be distinguished from lower to higher α scale values. The α values of these two habitats ranged between 0.04 and 3.96, with corresponding diversity indices of 1.980–1.32 (Woodland) and 1.981-1.33

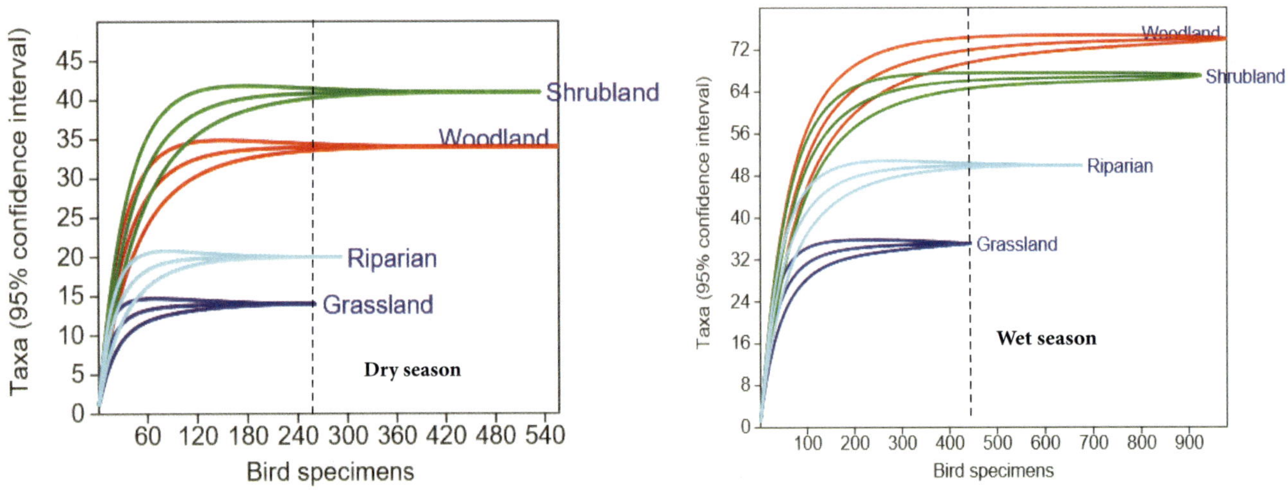

FIGURE 4: Standardized comparison of species richness for two individual-based rarefaction curves. The data represent summary counts of birds that were recorded from the four habitats in Mole National Park. The green, red turquois and blue lines are the rarefaction curves, calculated from (1) [53], with a 95% confidence interval. The dotted vertical lines illustrate a species richness comparison standardized to 262 (dry season) and 447 (wet season) individuals, which was the observed bird abundance in the smallest (grassland) of the four habitats bird data sets.

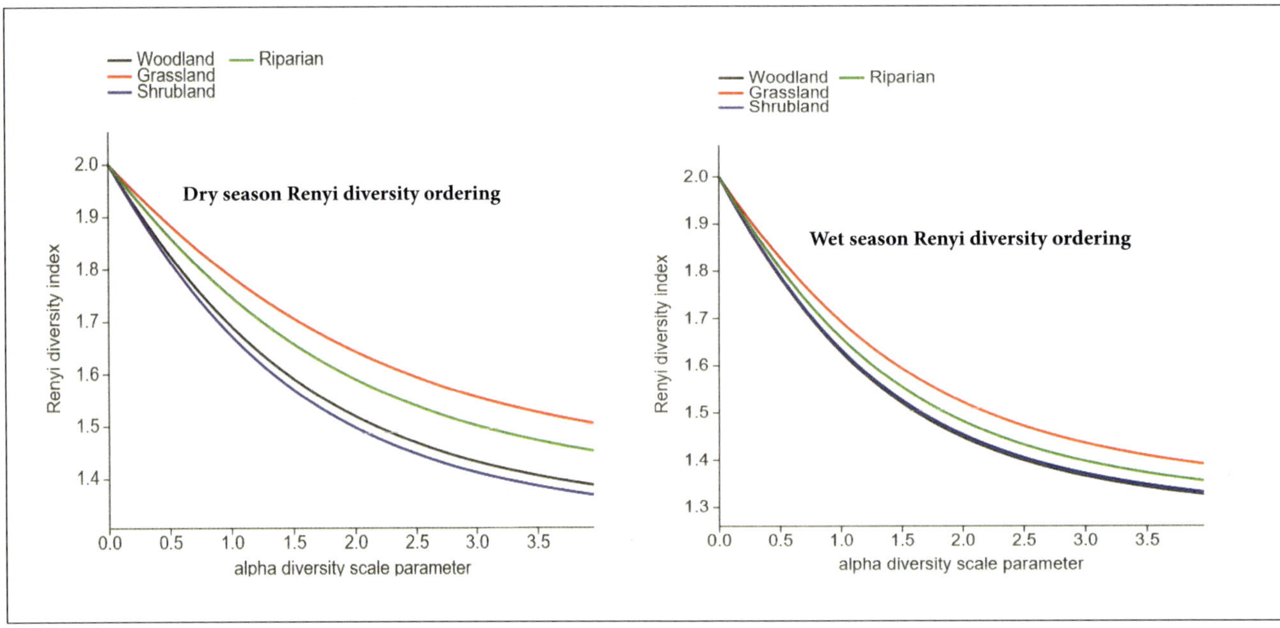

FIGURE 5: Renyi diversity ordering that compares bird evenness and richness for the dry and wet seasons, across the four habitats. Notice that the shape of the curve of each habitat is an indication of their spatial evenness of species and ordered from the lowest to highest diversity.

(shrubland) (Figure 5). These diversity values revealed their close similarity in richness and the proportion of the most abundant species. Overall, we found that the relationship between α values and diversity indices, especially at high infinities ($\alpha > \infty$), across the four habitats had low proportion of diversity indices and dominant species.

4.4. Environmental Predictors of Bird-Habitat Preference. The CCA ordination diagrams present summarized information on bird structural assemblages, mediated by environmental factors across the four habitats. For the dry season, CCA ordination revealed a strong species-environment correlation on axes I and II, where bushfire (r = -0.836, p<0.0001) and erosion (r = 0.638, p<0.0001) affected the abundance and richness of grassland and riparian preferred species, respectively (Figure 6, Table 3). However, in the shrubland and woodland habitats, majority of species like Palm-Nut Vulture, *Gypohierax angolensis*; White-Backed Vulture, *Gyps africanus;* and Brown-Capped Babbler, *Pellorneum fusco-capillus*, tended to be attracted to postincidences of bushfire, characterized by left-over carcass. Rarer species like White-backed Vulture, *Gyps africanus* (CR), were impacted by this

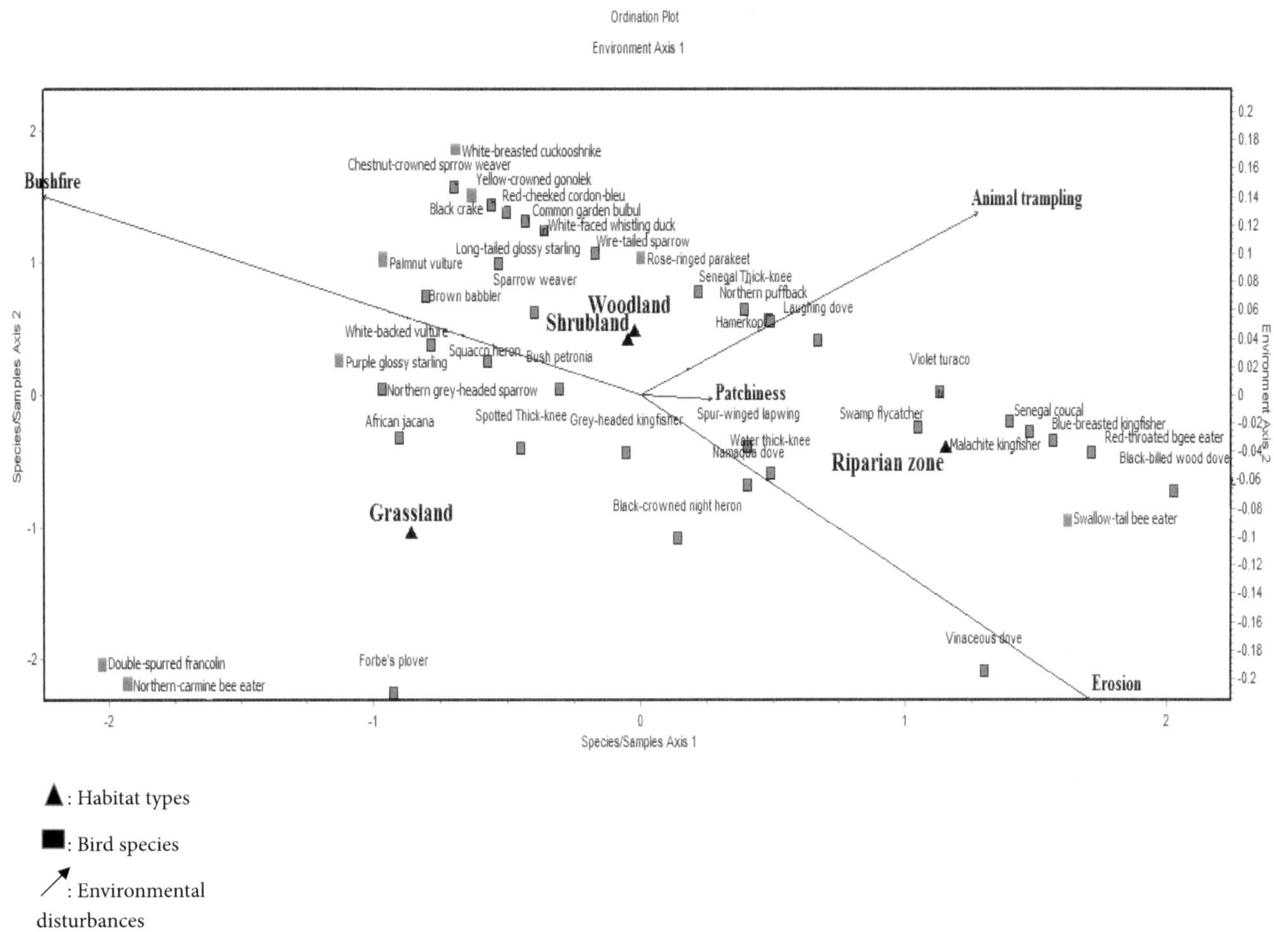

Figure 6: Canonical correspondence (CCA) ordination diagram, showing the relationship between environmental disturbances and bird species across the four habitat types in the dry season. The arrows represent each of the environmental variables plotted pointing in the direction of maximum change of explanatory variables across the four habitats.

disturbance in terms of limited distribution. Species like Spur-winged lapwing, *Vanellus spinosus,* and Namaqua dove, *Oena capensis,* responded to patchiness and animal trampling (r = 0.89, p<0.0001, axis I), in portions along the riparian zone (Figure 6, Tables 3 and 4). Overall, axes I and II jointly accounted for 81.82% variance in bird diversity across the four habitats (Figure 6, Table 3).

In the wet season, bird structural assemblages generally showed an increase in bird richness and spatial distribution across the four habitats (Figure 7), in spite of the strong correlation between species-environmental factors on axis I (i.e., patchiness, r_p = 0.85) and II (i.e., animal trampling, r_p = 0.88) (Figure 7, Table 5). These two axes (58.56% and axis 3.46%) jointly explained 62.02% of variability of bird assemblages caused by the environmental factors (Table 5). While species richness from grassland habitat were notably lowest and affected by erosion condition (following previous animal trampling activities, r = -0.951, p<0.0001) (Figure 7, Table 6), species from the remaining three habitats rather showed more tolerant to patchiness and animal trampling conditions. Example included Bruce's pigeon (*Treron waalia*), African grey Hornbill (*Tockus nasutus*) in the woodland, Red-Billed Hornbill (*Tockus erythrorhynchus*), and Yellow-Crowned Gonolek (*Laniarius barbarus*), and Bush petronia, Sahel Bush-sparrow (*Gymnoris dentata).*

Table 3: Summary of canonical correspondence analysis (CCA) of birds showing the levels of correlation between axes and environmental gradients, variance of species, and species-environmental relationships in the dry season.

	Axis I	Axis II
Canonical Eigenvalue	0.912	0.418
Cumulative % variance	81.22	0.6
% variance explained	81.82	11.3
Pearson corr. coefficient species/env. score	0.976	0.990
Kendall rank corr. of species/env. score	0.666	0.666
Correlations		
(1) Patchiness	0.101	-0.003
(2) Erosion	0.638	-0.215
(3) Animal trampling	0.474	0.128
(4) Bushfire	-0.836	0.140

TABLE 4: Intercorrelations (Pearson) between the environmental variables (soil erosion, animal trampling, tree felling, bushfire, and patchiness) in bird dataset, during the dry season. A correlation above/below ±0.61 is significant at $p = 0.0013{*}{*}{*}$; ±0.45 at $p = 0.01{*}{*}$ and ±0.33 at $p = 0.05{*}$.

	Patchiness	Erosion	Animal trampling	Bushfire
Patchiness	1	-0.588	0.891	-0.523
Erosion	-0.588	1	-0.296	-0.370
Animal trampling	0.891	-0.296	1	-0.756
Bushfire	-0.523	-0.370	-0.756	1

TABLE 5: Summary of canonical correspondence analysis (CCA) of birds showing the levels of correlation between axes and environmental gradients, variance of species, and species-environmental relationships (wet season).

	Axis 1	**Axis 2**
Canonical Eigenvalue	0.273	0.016
Cumulative % variance	58.56	62.02
% variance explained (62.02%)	58.56	3.46
Pearson corr. coefficient species/env. score	0.854	0.880
Kendall rank corr. of species/env. score	0.333	0.666
Correlations		
(1) Patchiness	0.669	0.235
(2) Erosion	0.490	0.705
(3) Animal trampling	0.375	0.789

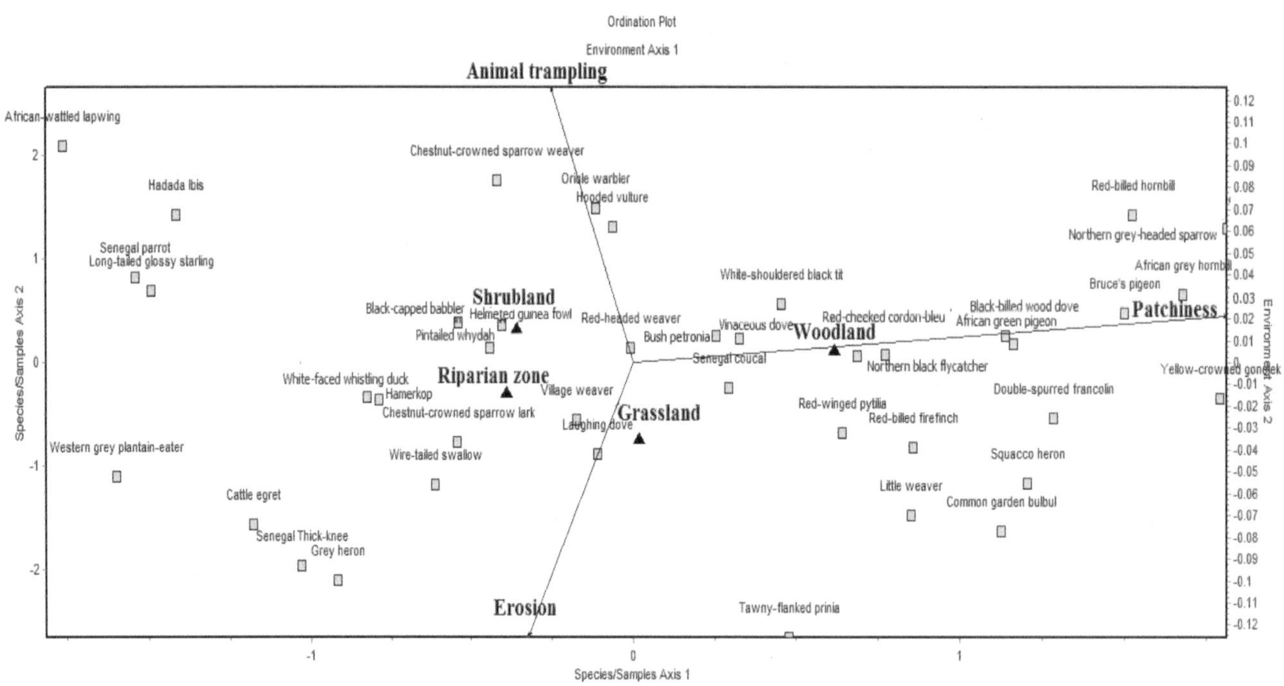

▲: Habitat types

■: Bird species

↗: Environmental disturbances

FIGURE 7: Canonical correspondence 773 (CCA) ordination diagram, showing the relationship between environmental disturbances and bird species across the four habitat types in the wet season. The arrows represent each of the environmental variables plotted pointing in the direction of maximum change of explanatory variables across the four habitats.

TABLE 6: Intercorrelations (Pearson) between the environmental variables (soil erosion, animal trampling, tree felling, bushfire, and patchiness) in bird dataset, during the wet season. A correlation above/below ±0.61 is significant at $p = 0.001***$; ±0.45 at $p = 0.01**$ and ±0.33 at $p = 0.05*$.

	Patchiness	Erosion	Animal trampling
Patchiness	1	-0.336	0.028
Erosion	-0.336	1	-0.951
Animal trampling	0.028	-0.951	1

The change in the longitudinal profile of species assemblages from the four habitats (Figures 6 and 7) reflects similar trend observed in previous analysis in SAD curves and individual rarefaction outcomes (Figures 2, 3, and 4).

5. Discussion

Several studies have shown that vegetation habitat type plays a critical role in structuring bird communities [31, 32, 64, 65]. This was confirmed in our findings where woodland and shrubland habitats supported highest bird abundance and richness, probably due to vegetation cover for nesting/sheltering, availability of food, and less predation accessibility. Despite the high abundance and richness of birds' in the woodland and shrublands, overall impact of bushfire occurrence affected bird diversity therein. This was probably due to their uneven distribution, especially during the dry season, where food resources and quality nesting sites were limited. This scenario tended to increase bird competing for these resources at sections of the habitats where they are found, leading to their uneven spatial distribution. Species evenness distribution is considered as one of the key measures of diversity in ecosystems and has been used by many scientists [e.g., [66, 67]] to quantify biodiversity.

Observed variations in bird assemblages suggest their attraction to habitat-specific, which is influenced by either evolution or environmental modifications in Mole National Park. Birds, like White-shouldered Black tit, *Melaniparus guineensis*; Senegal batis, *Batis senegalensis;* and Yellow-fronted Canary, *Crithagra mozambicus*, that were found nesting on selected thorny woody species like *Ziziphus mauritiana* probably did so to seek protection from predation. The dominance of Sahel Bush-sparrow, *Gymnoris dentata,* and Senegal eremomela, *Eremomela pusilla*, in the dry and wet seasons, was probably because of their broad range habitat preferences and their ability to adapt to environmental conditions inherent across these habitats. These species could be used as indicators of habitat quality, if future scenarios restrict their movement or distribution in narrow range habitats. Sekercioglu [68] reported that birds encountered in various habitats types indicate their tolerance to a wide range of ecological conditions. More so, their population size usually depends on the availability of food, variable climate, and the interactions among the other factors of the environment like human interactions, nest predation, and diseases [14].

Bird abundance decline in the dry season, especially among grassland associates, compared with the remaining three habitats, was probably due to frequency and intensity of bushfires in this habitat, which tended to affect food availability, reduced vegetation cover meant to provide shelter and support for bird nest, and increased in predation accessibility (due to less ground cover). The relatively less impact of bushfire on birds in woodland and shrubland habitats, including rarer species like Wooly-necked stock, *Ciconia episcopus,* and White-backed Vulture, *Gyps africanus* (NT), with high conservation concern, was probably due to less amount of fuel load, early vegetation sprout, following burning (which is used as a management tool to either reduce fuel load or encourage early vegetation growth), and left-over carcass. The survival of these species, following postbushfire occurrence, suggests the critical role that woodland and shrublands habitats play in their functional life stages. These near threatened or vulnerable conservation status of these birds, as a result of anthropogenic disturbances, may have gradually developed adaptive strategies using their proximate cues (i.e., genetic and sensory mechanisms), to survive in their transformed habitats.

Studies in Northern Region of Ghana revealed how bushfire affected bird population in off- reserves, particularly in the dry season [69]. Sallabanks, Walters, and Collazo [70] also reported bird postmortality following bushfires. Despite the adverse effects of bushfire and patchiness on birds observed in this study, other studies rather detected an increase in the abundance of some species like upland sandpipers (*Bartramia longicauda)* and yellow billed kite, *Milvus aegyptius*, following burning and grazing activities [70]. These varied responses of birds to bushfires may be due to their tolerance or resilience level at different burning frequency and intensity. Other environmental factors, namely, predation and food scarcity [71], farming practice and urban development [72–74], habitat fragmentation [72, 75], cutting of mangrove vegetation [76], and deforestation [24], have been reported to affect bird populations in other ecosystems. Recent reports have shown that birds have gone extinct at an exceptionally high rate, estimated to be 1,000 to 10,000 times the natural background rate [16]. The International Union for the Conservation of Nature (IUCN) "*Red List*" have also shown documented evidence that rate of extinction is getting worse among species from small islands across the globe. Thus, the least birds recorded in the grassland habitat partly explained that grassland associate birds are more prone to bushfire outbreak, because of the relative quantity of fuel load and ease of burn, than the woodland, shrubland riparian vegetation.

Notwithstanding the least number of birds per species in grassland, Renyi diversity profile, rather showed this habitat to be the most diverse in both seasons. Ricotta and Avena [66] and Dazzo et al. [67] reported that shallower declining slopes (an indicator of high diversity communities) are more evenly distributed that steep slopes. Thus, the shape of the Renyi diversity profiles is influenced by their evenness [77].

Prediction of future bird-habitat specific preference may be possible because demographic effect of habitat heterogeneity will be modified by habitat-selection behaviors, since the subset of habitats that are actually used will determine the population's demographics [78]. Therefore, bird-floristic habitat preference in Mole National Park could help predict the spatiotemporal to long-term distribution and abundance of birds.

6. Conclusions

Generally, bird species in Mole National Park were more associated with woodland and shrubland habitats and this is reflected in the outcomes of the SAD and individual rarefaction analyses. However, variations in bird structural assemblages suggest their attraction to habitat-specific and mediated by environmental disturbances like bushfire, patchiness, and animal trampling. We also observed seasonality influence on bird abundance and richness dynamics and linked the high abundance in the wet season to availability of food and vegetation cover against predation. Rarer species with high conservation concern (e.g., White-headed Vulture, *Trigonoceps occipitalis* (CR); Wooly-necked stock, *Ciconia episcopus* (VU); and White-backed Vulture, *Gyps africanus* (NT)), in the woodland and shrubland habitats, suggest the need to intensify conservation of woodland habitats, through control of bushfires, reduction of fuel load, and limit elephant destruction of vegetation. Limitation of elephant destruction of these two habitats can be done by constructing more watering holes in the Park, to sparsely distribute the elephant population throughout the reserve. The resultant effect will lead to reducing pressure on the few watering holes found in the woodland and shrubland habitats. These interventions could contribute to increasing bird abundance, richness, and their spatial even distribution.

Conflicts of Interest

The authors declare that there are no conflicts of interest whatsoever, regarding the publication of this research article.

Acknowledgments

The authors are grateful to the park rangers and tour guides at Mole National Park for their assistance during field survey. The study was self-funded by the authors.

References

[1] S. C. Kendeigh, "Community Selection by Birds on the Helderberg Plateau of New York," *The Auk*, vol. 62, no. 3, pp. 418–436, 1945.

[2] G. Svärdson, "Competition and habitat selection in birds," *Oikos*, vol. 1, no. 2, pp. 157–174, 1949.

[3] O. Hilden, "Habitat selection in birds: A review," *Annales Zoologici Fennici*, vol. 2, pp. 53–75, 1965.

[4] W. M. Block and L. A. Brennan, "The habitat concept in ornithology: Theory and applications," *Current Ornithology*, vol. 11, pp. 35–91, 1993.

[5] J. Verner and M. L. Morrison, *Wildlife 2000: Modeling Habitat Relationships of Terrestrial Vertebrates*, C. J. Ralph, Ed., University of Wisconsin Press, Madison, Wis, USA, 1986.

[6] M. L. Rosenzweig, "Habitat selection and population interactions: the search for mechanism," *The American Naturalist*, vol. 137, pp. S5–S28, 1991.

[7] S. D. Fretwell and H. L. Lucas Jr., "On territorial behavior and other factors influencing habitat distribution in birds - I. Theoretical development," *Acta Biotheoretica*, vol. 19, no. 1, pp. 16–36, 1969.

[8] S. D. Fretwell, *Populations in Seasonal Environments*, Princeton University Press, Princeton, NJ, USA, 1972.

[9] J. R. Karr and K. E. Freemark, "Habitat selection and environmental gradients: dynamics in the 'stable' tropics," *Ecology*, vol. 64, no. 6, pp. 1481–1494, 1983.

[10] H. R. Pulliam and B. J. Danielson, "Sources, sinks, and habitat selection: a landscape perspective on population dynamics," *The American Naturalist*, vol. 137, pp. S50–S66, 1991.

[11] L. J. Petit and D. R. Petit, "Factors governing habitat selection by prothonotary warblers: field tests of the fretwell-lucas models 1," *Ecological Monographs*, vol. 66, no. 3, pp. 367–387, 1996.

[12] T. E. Martin, "Nest predation and nest sites," *Bioscience*, vol. 43, no. 8, pp. 523–532, 1993.

[13] H. Pöysä, J. Elmberg, K. Sjöberg, and P. Nummi, "Habitat selection, rules in breeding mallards (Anas platyrhynchos): A test of two competing hypotheses," *Oecologia*, vol. 114, no. 2, pp. 283–287, 1998.

[14] J. Davis, E. Rocky, T. Hall et al., Diet differentiation and habitat selection of birds in forested and clear-cut areas, 2000.

[15] G. Caughley, "Directions in conservation biology," *Journal of Animal Ecology*, vol. 63, no. 2, pp. 215–244, 1994.

[16] Birdlife International, *Threatened Birds of the World*, Lynx Editions/Birdlife International, Barcelona, Spain, 2011.

[17] J. R. Herkert, "The effects of habitat fragmentation on midwestern grassland bird communities," *Ecological Applications*, vol. 4, no. 3, pp. 461–471, 1994.

[18] D. J. Twedt and C. R. Loesch, "Forest area and distribution in the Mississippi alluvial valley: implications for breeding bird conservation," *Journal of Biogeography*, vol. 26, no. 6, pp. 1215–1224, 1999.

[19] H. R. Pulliam, "Sources, sinks and population regulation," *The American Naturalist*, vol. 132, no. 5, pp. 652–661, 1988.

[20] T. E. Martin, "Are microhabitat preferences of coexisting species under selection and adaptive?" *Ecology*, vol. 79, no. 2, pp. 656–670, 1998.

[21] J. D. Brawn, S. K. Robinson, and F. R. Thompson III, "The role of disturbance in the ecology and conservation of birds," *Annual Review of Ecology and Systematics*, vol. 32, pp. 251–276, 2001.

[22] C. L. Seymour and R. E. Simmons, "Can severely fragmented patches of riparian vegetation still be important for arid-land bird diversity?" *Journal of Arid Environments*, vol. 72, no. 12, pp. 2275–2281, 2008.

[23] G. M. Siriwardena, S. R. Baillie, S. T. Buckland, R. M. Fewster, J. H. Marchant, and J. D. Wilson, "Trends in the abundance of farmland birds: A quantitative comparison of smoothed Common Birds Census indices," *Journal of Applied Ecology*, vol. 35, no. 1, pp. 24–43, 1998.

[24] J. R. Krebs, J. D. Wilson, R. B. Bradbury, and G. M. Siriwardena, "The second silent spring?" *Nature*, vol. 400, no. 6745, pp. 611-612, 1999.

[25] W. R. L. Cresswell, J. M. Wilson, J. Vickery, P. Jones, and S. Holt, "Changes in densities of Sahelian bird species in response to recent habitat degradation," *Ostrich*, vol. 78, no. 2, pp. 247–253, 2007.

[26] J. A. Veraart, R. S. de Groot, G. Perelló, N. J. Riddiford, and R. Roijackers, "Selection of (bio) indicators to assess effects of freshwater use in wetlands: a case study of Albufera de Mallorca, Spain," *Regional Environmental Change*, vol. 4, no. 2-3, pp. 107–117, 2004.

[27] T. R. Shankar Raman, G. S. Rawat, and A. J. T. Johnsingh, "Recovery of tropical rainforest avifauna in relation to vegetation succession following shifting cultivation in Mizoram, north-east India," *Journal of Applied Ecology*, vol. 35, no. 2, pp. 214–231, 1998.

[28] N. Chettri, E. Sharma, and D. C. Deb, "Bird community structure along a trekking corridor of Sikkim Himalaya: A conservation perspective," *Biological Conservation*, vol. 102, no. 1, pp. 1–16, 2001.

[29] T. R. Shankar Raman, "Effect of slash-and-burn shifting cultivation on rainforest birds in Mizoram, Northeast India," *Conservation Biology*, vol. 15, no. 3, pp. 685–698, 2001.

[30] G. Fanelli and C. Battisti, "Range of species occupancy, disturbance and generalism: Applying hemeroby metrics to common breeding birds from a regional Atlas," *Vie et Milieu - Life and Environment*, vol. 65, no. 4, pp. 243–250, 2015.

[31] A. P. Gabbe, S. K. Robinson, and J. D. Brawn, "Tree-species preferences of foraging insectivorous birds: Implications for floodplain forest restoration," *Conservation Biology*, vol. 16, no. 2, pp. 462–470, 2002.

[32] S. L. Earnst and A. L. Holmes, "Bird-habitat relationships in interior Columbia Basin shrubsteppe," *Condor*, vol. 114, no. 1, pp. 15–29, 2012.

[33] S. M. Caziani and E. Derlindati, "Abundance and habitat of High Andes Flamingos in Northwestern Argentina," *Waterbirds*, vol. 23, pp. 121–133, 2000.

[34] L. A. Brennan and M. L. Morrison, "Long-term trends of chickadee populations in Western North America," *The Condor*, vol. 93, no. 1, pp. 130–137, 1991.

[35] A. L. Skowno and W. J. Bond, "Bird community composition in an actively managed savanna reserve, importance of vegetation structure and vegetation composition," *Biodiversity and Conservation*, vol. 12, no. 11, pp. 2279–2294, 2003.

[36] R. C. MacNally, "An analysis of density responses of forest and woodland birds to composite physiognomic variables," *Australian Journal of Ecology*, vol. 15, no. 3, pp. 267–275, 1990.

[37] E. Acquah, P. Dearden, and R. Rollins, "Nature-based tourism in Mole National Park, Ghana," *African Geographical Review*, vol. 35, no. 1, pp. 53–69, 2016.

[38] F. Dowsett-Lemaire and R. J. Dowsett, *Ornithological Surveys in Mole National Park*, 2004.

[39] T. L. Hamer, C. H. Flather, and B. R. Noon, "Factors associated with grassland bird species richness: The relative roles of grassland area, landscape structure, and prey," *Landscape Ecology*, vol. 21, no. 4, pp. 569–583, 2006.

[40] Mole National Park (M.N.P.), *Management Plan*, 2011.

[41] K. Schmitt and M. Adu-Nsiah, "Forest resource management project, GWD/IUCN," The Vegetation of Mole National Park, 1993.

[42] C. Bibby, N. D. Burguess, and D. A. Hill, *Bird Census Techniques*, Academic Press, London, UK, 1992.

[43] W. J. Sutherland, *Ecological Census Techniques: A Handbook*, Cambridge University Press, 2006.

[44] R. L. Hutto, S. M. Pletschet, and P. Hendricks, "A fixed-radius point count method for nonbreeding and breeding season use.," *The Auk*, vol. 103, no. 3, pp. 593–602, 1986.

[45] N. Borrow and R. Demey, *Field Guide to the Birds of Ghana (Helm Field Guide)*, A & C Black Publishers Ltd., London, UK, 2010.

[46] C. Battisti, L. Luiselli, and C. Teofili, "Quantifying threats in a Mediterranean wetland: are there any changes in their evaluation during a training course?" *Biodiversity and Conservation*, vol. 18, no. 11, pp. 3053–3060, 2009.

[47] N. Salafsky, D. Salzer, A. J. Stattersfield et al., "A standard lexicon for biodiversity conservation: unified classifications of threats and actions," *Conservation Biology*, vol. 22, no. 4, pp. 897–911, 2008.

[48] A. E. Magurran, "Measuring biological diversity," *Journal of the Torrey Botanical Society*, vol. 131, no. 3, pp. 277-278, 2004.

[49] S. Fattorini, "A simple method to fit geometric series and broken stick models in community ecology and island biogeography," *Acta Oecologica*, vol. 28, no. 3, pp. 199–205, 2005.

[50] S. Fattorini, F. Rigal, P. Cardoso, and P. A. V. Borges, "Using species abundance distribution models and diversity indices for biogeographical analyses," *Acta Oecologica*, vol. 70, pp. 21–28, 2016.

[51] J. S. Gray and F. B. Mirza, "A possible method for the detection of pollution-induced disturbance on marine benthic communities," *Marine Pollution Bulletin*, vol. 10, no. 5, pp. 142–146, 1979.

[52] N. J. Gotelli and R. K. Colwell, "Estimating species richness," *Biological Diversity: Frontiers in Measurement and Assessment*, vol. 12, pp. 39–54, 2011.

[53] N. J. Gotelli and R. K. Colwell, "Quantifying biodiversity: procedures and pitfalls in the measurement and comparison of species richness," *Ecology Letters*, vol. 4, no. 4, pp. 379–391, 2001.

[54] A. F. Siegel, "Rarefaction curves," in *Encyclopedia of Statistical Sciences*, K. Samuel, C. B. Read, N. Balakrishnan, and B. Vidakovic, Eds., 2006.

[55] C. S. Robbins, J. R. Sauer, R. S. Greenberg, and S. Droege, "Population declines in North American birds that migrate to the neotropics," *Proceedings of the National Acadamy of Sciences of the United States of America*, vol. 86, no. 19, pp. 7658–7662, 1989.

[56] A. Rényi, "On measures of entropy and information," in *Proceedings of the Fourth Berkeley Symposium on Mathematics, Statistics and Probability*, vol. 1, pp. 1286–1291, University of California Press, Berkeley, Calif, USA, 1961.

[57] C. E. Shannon, "A Mathematical Theory of Communication," *Bell System Technical Journal*, vol. 27, no. 4, pp. 623–656, 1948.

[58] B. Tóthmérész, "Comparison of different methods for diversity ordering," *Journal of Vegetation Science*, vol. 6, no. 2, pp. 283–290, 1995.

[59] O. Hammer, D. A. T. Harper, and P. Ryan, "PAST: paleontological statistics software package for education and data analysis," *Palaeontologia Electronica*, vol. 4, no. 1, p. 9, 2001.

[60] J. R. Krebs, J. D. Wilson, R. B. Bradbury, and G. M. Siriwardena, "The second silent spring?" *Nature*, vol. 400, no. 6745, pp. 611-612, 1989.

[61] M. Kent and P. Coker, "Vegetation description and analysis," in *A Practical Approach*, John Wiley and Sons Ltd, 1992.

[62] C. J. F. Ter Braak, "Canonical correspondence analysis: a new

eigenvector technique for multivariate direct gradient analysis," *Ecology*, vol. 67, no. 5, pp. 1167–1179, 1986.

[63] P. A. Henderson and R. M. Seaby, *Environmental Community Analysis 1.3*, Pisces Conservation Ltd, Hampshire, UK, 2000.

[64] R. A. H. Draycott, A. N. Hoodless, and R. B. Sage, "Effects of pheasant management on vegetation and birds in lowland woodlands," *Journal of Applied Ecology*, vol. 45, no. 1, pp. 334–341, 2008.

[65] K. C. Tanalgo, J. A. F. Pineda, M. E. Agravante, and Z. M. Amerol, "Bird diversity and structure in different land-use types in lowland south-central Mindanao, Philippines," *Tropical Life Sciences Research*, vol. 26, no. 2, pp. 85–103, 2015.

[66] C. Ricotta and G. C. Avena, "On the information-theoretical meaning of Hill's parametric evenness," *Acta Biotheoretica*, vol. 50, no. 1, pp. 63–71, 2002.

[67] F. B. Dazzo, K. J. Klemmer, R. Chandler, and Y. G. Yanni, "In situ ecophysiology of microbial biofilm communities analyzed by CMEIAS computer-assisted microscopy at single-cell resolution," *Diversity*, vol. 5, no. 3, pp. 426–460, 2013.

[68] C. H. Sekercioglu, "Foreword," in *Handbook of the Birds of the World: Old World Flycatchers to Old World Warblers*, J. del Hoyo, A. Elliott, and D. Christie, Eds., vol. 11, p. 48, Lynx Edicions, Barcelona, Spain, 2006.

[69] C. A. Nsor and E. A. Obodai, "Environmental determinants influencing seasonal variations of bird diversity and abundance in Wetlands, Northern Region (Ghana)," *International Journal of Zoology*, vol. 2014, Article ID 548401, 10 pages, 2014.

[70] R. Sallabanks, J. R. Walters, and J. A. Collazo, "Breeding bird abundance in bottomland hardwood forests: habitat, edge, and patch size effects," *The Condor*, vol. 102, no. 4, pp. 748–758, 2000.

[71] A. F. L. A. Powell, "Effects of prescribed burns and bison (Bos bison) grazing on breeding bird abundances in tallgrass prairie," *The Auk*, vol. 123, no. 1, pp. 183–197, 2006.

[72] M. Allen, *Wetlands and Fire. Water and Rivers Commission. Water Notes for Wetlands Managers. WNS2*, Natural Heritage Trust, Australia, 2000.

[73] W. J. Mitsch and J. G. Gooselink, *Wetlands*, John Wiley & Sons, New York, NY, USA, 3rd edition, 2000.

[74] L. A. Brennan and W. P. Kuvlesky Jr., "North American grassland birds: An unfolding conservation crisis?" *The Journal of Wildlife Management*, vol. 69, no. 1, pp. 1–13, 2005.

[75] L. I. Jacoboski, R. K. Paulsen, and S. M. Hartz, "Bird-grassland associations in protected and non-protected areas in southern Brazil," *Perspectives in Ecology and Conservation*, vol. 15, no. 2, pp. 109–114, 2017.

[76] T. J. O'Connell, L. E. Jackson, and R. P. Brooks, "Bird guilds as indicators of ecological condition in the central Appalachians," *Ecological Applications*, vol. 10, no. 6, pp. 1706–1721, 2000.

[77] D. Attuquayefio and F. Gbogbo, "Prospects of conserving wetlands along the Mukwe lagoon at Nungua in the greater Accra region of Ghana," *West African Journal of Applied Ecology*, vol. 2, no. 1, 2009.

[78] W. B. Kristan III, "The role of habitat selection behavior in population dynamics: Source-sink systems and ecological traps," *Oikos*, vol. 103, no. 3, pp. 457–468, 2003.

17

Comparing the Performance of Protected and Unprotected Areas in Conserving Freshwater Fish Abundance and Biodiversity in Lake Tanganyika, Tanzania

Emmanuel Andrew Sweke,[1,2] Julius Michael Assam,[1] Abdillahi Ismail Chande,[1,3] Athanasio Stephano Mbonde,[1] Magnus Mosha,[4] and Abel Mtui[4,5]

[1] *Tanzania Fisheries Research Institute, P.O. Box 90, Kigoma, Tanzania*
[2] *Graduate School of Fisheries Sciences, Hokkaido University, 3-1-1 Minato-cho, Hakodate 041–8611, Japan*
[3] *College of Science, Informatics and Built Environment, Bagamoyo University, P.O. Box 75254, Dar es Salaam, Tanzania*
[4] *Mahale Mountains National Park, P.O. Box 1374, Kigoma, Tanzania*
[5] *Tanzania National Parks, P.O. Box 3134, Arusha, Tanzania*

Correspondence should be addressed to Emmanuel Andrew Sweke; esweke@yahoo.com

Academic Editor: Michel Couderchet

Marine protected areas have been shown to conserve aquatic resources including fish, but few studies have been conducted of protected areas in freshwater environments. This is particularly true of Lake Tanganyika, Tanzania. To better conserve the lake's biodiversity, an understanding of the role played by protected areas in conserving fish abundance and diversity is needed. Sampling of fish and environmental parameters was performed within the Mahale Mountains National Park (MMNP) and nearby unprotected areas at depths between 5 m and 10 m. Twelve replicates of fish sampling were performed at each site using gillnets set perpendicularly to the shore. Mann-Whitney tests were performed, and the total amount of species turnover was calculated. A total of 518 individual fish from 57 species were recorded in the survey. The fish weight abundance was fivefold greater in the MMNP than in the unprotected areas. Fish abundance and diversity were higher in the MMNP than in the unprotected areas and decreased with distance from it. Our findings confirmed the importance of the protected area in conserving fish resources in Lake Tanganyika. The study provides baseline information for management of the resources and guiding future studies in the lake and other related ecosystems. Management approaches that foster awareness and engage with communities surrounding the MMNP are recommended for successful conservation of the resources in the region.

1. Introduction

The complexity of the Lake Tanganyika (Figure 1) ecosystem makes it unique in the world [1], and it notably contributes to global biodiversity [2, 3]. About 58% of the animal species that inhabit the lake are endemic [4], and fish of the family Cichlidae and Molluscs are very diverse; that is, both of these groups have a high number of species with a substantial proportion of endemic species [5] and show considerable genetic variability within species. However, the lake's biodiversity is vulnerable to anthropogenic threats including fishing [5–10].

Conservation researchers have advocated controlling human activities by establishing protected areas as one of the methods of conserving biodiversity in an ecosystem [9, 11, 12]. This mitigation approach ensures that some essentially unmodified sites exist for buffering against uncertainty such as overfishing [13, 14]. It is well documented that protected areas nurse and harbor many more species compared with unprotected areas [11, 12, 15]. Additionally, studies of freshwater ecosystems have revealed that fish in protected areas are larger than those in unprotected areas [16–18]. Unfortunately, many protected areas are designed to conserve terrestrial

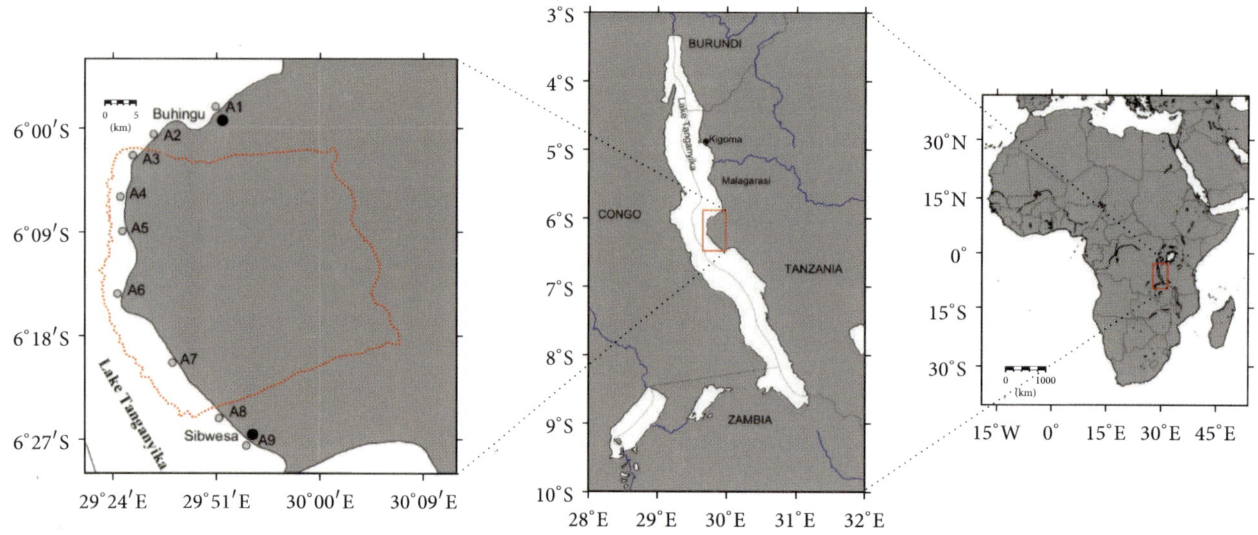

Figure 1: Map of Lake Tanganyika showing study sites within protected area (MMNP: A3–A7) and in unprotected areas (Buhingu: A1-A2 and Sibwesa: A8-A9) during 2008 gillnet survey. The large red polygon in the first frame shows sketch boundaries of MMNP (terrestrial and water strip in grey color and white color, resp.). The blue lines in the middle frame represent rivers.

habitats than aquatic biodiversity, consequently limiting the assessment of spatial and temporal fish abundance and biodiversity in both protected and unprotected areas [19–22].

The Mahale Mountains National Park (MMNP) was established in 1985 to conserve terrestrial biodiversity, with a particular aim of protecting chimpanzees. The MMNP has a total area of 1,613 km^2, of which 96 km^2 is a 1.6 km wide aquatic strip extending along the shore of Lake Tanganyika [23]. All forms of exploitation including fishing are strictly prohibited in the MMNP. Trained park rangers conduct patrols to prevent poaching. However, fishing activities in surrounding waters remain unregulated. The ecological performance of protected areas has been assessed by comparing it with that of nearby unprotected areas [24, 25]. However, studies on the MMNP's biological conservation efficacy and loss of fish abundance and diversity in adjacent areas are scarce.

The purpose of this study was to assess and compare the differences in fisheries resources between protected and unprotected areas. To achieve this, we compared fish abundance and diversity between the MMNP and nearby unprotected areas. Such studies are vitally important to ascertain the spatial and temporal extent of human influence in variations in abundance and diversity in the ecosystem. This baseline information also could contribute valuable knowledge to the global debate concerning the relevance of freshwater protected areas in conserving fisheries resources.

2. Methods

2.1. Study Area Description. The study was conducted in Lake Tanganyika, Tanzania, during May and June 2008. Sampling sites were located within and outside the MMNP (Figure 1). The MMNP is located at the southern edge of the Kigoma region, with an elevation ranging from 2,000 to 2,400 m [26].

2.2. Data Collection. For comparison purposes, the sampling sites were selected based on area status, that is, protected and unprotected, Table 1. Eight sampling sites (four within the protected area (MMNP) and four in unprotected areas) at least 5 km apart were randomly selected. The four outermost sites (i.e., two northernmost and two southernmost) were located in unprotected areas (Buhingu and Sibwesa villages, resp.).

2.3. Environmental Parameters. Four samples for dissolved oxygen (DO) and temperature were taken at each site at 5 m and 10 m water depths using a multiparameter analyzer WTW 340i [27]. The mean values of environmental parameters were statistically analyzed using a one-way ANOVA.

2.4. Sampling of Fish. The sampling of fish was performed at depths between 5 m and 10 m using gillnet set perpendicularly to the shore. Twelve replicates of gillnets, each with a width of 1.37 m and 45 m total length (before mounting), were joined end to end to form one panel of 360 m. The mesh sizes ranged from 1″ to 5″ (inches) (i.e., 25.4 mm to 127.0 mm), at an interval of 0.5″ except for the large mesh sizes (i.e., 1″, 1.5″, 2″, 2.5″, 3″, 3.5″, 4″, and 5″). The first 4 mesh size nets (1″ to 2.5″) were used in pairs. Each net had a length of 30 m after mounting at a hanging ratio of 0.66. Nets of different mesh sizes were used to increase the variability of sizes of fish and minimize sampling errors. The nets were set in the evening and hauled in the following morning. After hauling, the samples were sorted and identified to species level at the shore according to Konings [28]. Total catch, species, and wet weights of fish were recorded at each site. After the experiments, some of the fish samples were used for human consumption and representatives of some species were preserved for reference.

TABLE 1: Environmental parameters mean (±SD) of study sites in protected and unprotected areas in Lake Tanganyika, Tanzania, from gillnet survey during 2008. Samples from site A4 were omitted in the analysis.

Site ID	Site location	Status	DO ($mg{\cdot}L^{-1}$)	Temperature (°C)
A1	Buhingu	Unprotected	8.45 ± 0.21	26.95 ± 0.07
A2	Buhingu	Unprotected	7.30 ± 0.14	26.95 ± 0.07
A3	MMNP	Protected	8.00 ± 0.14	27.05 ± 0.07
A4	MMNP	Protected	7.70 ± 0.42	27.50 ± 0.00
A5	MMNP	Protected	7.35 ± 0.35	27.05 ± 0.07
A6	MMNP	Protected	7.25 ± 0.21	27.20 ± 0.00
A7	MMNP	Protected	8.15 ± 0.07	26.65 ± 0.07
A8	Sibwesa	Unprotected	7.25 ± 0.21	26.00 ± 0.00
A9	Sibwesa	Unprotected	7.50 ± 0.28	26.30 ± 0.00

2.5. Estimation of Abundance of Fish Species. To select appropriate statistical tests, fish mean abundance data and their residuals were tested for normality. The data were not normally distributed and variance groups were heterogeneous. This was followed by a log transformation, but the data still did not assume normal distribution patterns. Owing to this, we performed nonparametric tests after Mann-Whitney (MW) tests to determine pairs of areas with significant differences in fish abundance. The latter test was also performed to determine if habitat types affected fish abundance in the unprotected areas. Additionally, to avoid bias in the analysis, data from the rocky habitat in the MMNP was omitted; that is, data from 8 sites was used: four from within the MMNP and the 4 from the adjacent unprotected areas.

2.6. Calculation of Diversity of Fish Species. Fish biodiversity in both protected and unprotected areas was calculated as absolute species turnover, that is, the total amount of species turnover among the subunits in the dataset [29]:

$$\beta_A = (S_1 - c) + (S_2 - c) + \cdots + (S_n - c), \qquad (1)$$

where β_A is the absolute species turnover (diversity) for area A, S_1 is the total number of species recorded in the first site, S_2 is the total number of species recorded in the second site, S_n is the total number of species recorded in the nth site, and c is the number of species common to both sites.

Statistical tests for the environmental parameters and fish abundance were performed at significance level, $P = 0.05$, using STATISTICA (version 8, StatSoft Inc., 2010).

3. Results

3.1. Environmental Parameters. There were no significant differences in dissolved oxygen ($F_{1,16} = 0.73$, $P = 0.41$) and temperature ($F_{1,16} = 2.60 \times 10^{-3}$, $P = 0.96$) between the protected and unprotected areas (Table 1).

3.2. Abundance (Number) of Individual Fish. The mean number of individuals of fish in sites within the MMNP was higher than in the unprotected areas (Mann-Whitney, $P < 0.05$). In the survey, 518 individual fish from 57 species were recorded: 40 species were recorded in the MMNP and 39 were recorded in the unprotected areas (Table 2). The species belonged to nine families (number of species in parentheses): Cichlidae (41), Mastacembelidae (3), Mochokidae (3), Claroteidae (3), Latidae (2), Mormyridae (2), Clupeidae (1), Cyprinidae (1), and Poeciliidae (1). *Haplotaxodon microlepis* and *Bathybagrus graueri* were the most abundant species recorded in both the protected and unprotected areas (Table 2). The former ranked the first in abundance in both the MMNP and unprotected areas, and the latter was the sixth and fourth in the MMNP and unprotected areas, respectively. Five of the 10 most abundant species in the MMNP area were not recorded in the unprotected areas. On the other hand, only two species (*Synodontis* spp. and *Bathybates graueri*) which were among the most abundant ones in the unprotected areas were not recorded in the MMNP area. The two unprotected areas (Buhingu and Sibwesa) showed significant difference in the mean fish abundance (Mann-Whitney, $P < 0.05$). However, mean fish abundance was not significantly different between habitat types, particularly rocky and sandy types in the unprotected areas (Mann-Whitney, $P = 0.16$).

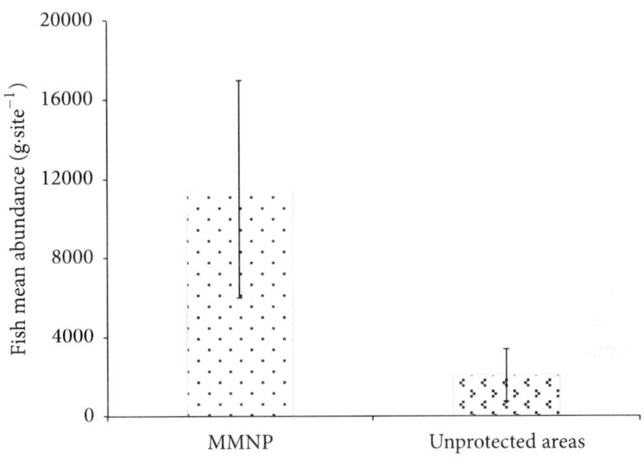

FIGURE 2: Mean fish abundance (wet weight ± SD) within protected area (MMNP) and unprotected areas (Buhingu and Sibwesa) in Lake Tanganyika, Tanzania, during 2008 gillnet survey.

3.3. Weight Abundance. Figure 2 shows the mean wet weight of the fish samples in both the MMNP and the unprotected

Table 2: Mean abundance number of individuals (±SE) of fish species sampled in sites within the protected area (MMNP) and unprotected areas (Buhingu and Sibwesa) in Lake Tanganyika, Tanzania, during 2008 gillnet survey.

Fish species	Protected area	Unprotected areas
Haplotaxodon microlepis	9.00 ± 4.60	5.75 ± 4.42
Limnotilapia dardennii	7.75 ± 2.50	0.75 ± 0.48
Barbus sp.	7.75 ± 2.17	0
Lates angustifrons	6.50 ± 1.66	0
Bathybates hornii	6.00 ± 2.48	0.25 ± 0.25
Limnothrissa miodon	5.50 ± 4.52	0
Bathybagrus graueri	5.00 ± 1.78	3.00 ± 2.68
Lamprichthys tanganicanus	4.25 ± 1.93	0
Petrochromis moshi	3.50 ± 3.50	1.25 ± 1.25
Boulengerochromis microlepis	3.50 ± 0.87	0.25 ± 0.25
Lepidiolamprologus elongatus	3.25 ± 2.63	0
Plecodus paradoxus	3.00 ± 0.82	1.00 ± 0.71
Cyathopharynx foai	2.75 ± 1.80	0.50 ± 0.29
Lamprologus callipterus	2.00 ± 0.82	0.25 ± 0.25
Grammatotria lemairii	1.50 ± 1.19	0.25 ± 0.25
Petrochromis orthognathus	1.50 ± 0.96	0
Lepidiolamprologus cunningtoni	1.50 ± 0.87	0
Ophthalmotilapia nasuta	1.25 ± 1.25	1.75 ± 1.44
Synodontis petricola	1.25 ± 0.95	0.25 ± 0.25
Tropheus moorii	1.00 ± 1.00	0
Cyphotilapia frontosa	0.75 ± 0.75	0.50 ± 0.50
Lates mariae	0.75 ± 0.75	0.25 ± 0.25
Aulonocranus dewindti	0.75 ± 0.48	2.00 ± 2.00
Plecodus straeleni	0.75 ± 0.48	1.25 ± 0.95
Neolamprologus mustax	0.75 ± 0.48	0.50 ± 0.50
Tylochromis polylepis	0.75 ± 0.48	0.50 ± 0.50
Neolamprologus fasciatus	0.75 ± 0.48	0.50 ± 0.50
Perissodus microlepis	0.75 ± 0.48	0
Lepidiolamprologus profundicola	0.75 ± 0.48	0
Xenotilapia melanogenys	0.50 ± 0.50	0
Lobochilotes labiatus	0.50 ± 0.29	0
Xenotilapia sima	0.25 ± 0.25	1.25 ± 1.25
Lamprologus lemairii	0.25 ± 0.25	0.75 ± 0.75
Xenotilapia spiloptera	0.25 ± 0.25	0.25 ± 0.25
Petrochromis famula	0.25 ± 0.25	0
Bathybates ferox	0.25 ± 0.25	0
Mastacembelus cunningtoni	0.25 ± 0.25	0
Auchenoglanis occidentalis	0.25 ± 0.25	0
Tropheus annectens	0.25 ± 0.25	0
Simochromis diagramma	0.25 ± 0.25	0
Synodontis sp.	0	5.25 ± 4.61
Bathybates graueri	0	3.25 ± 3.25
Phyllonemus typus	0	2.00 ± 2.00
Ectodus descampsii	0	1.50 ± 1.50
Petrochromis fasciolatus	0	1.25 ± 1.25
Mastacembelus sp.	0	1.00 ± 1.00
Neolamprologus toae	0	1.00 ± 0.41
Telmatochromis brichardi	0	0.75 ± 0.75
Benthochromis tricoti	0	0.75 ± 0.75
Mormyrus sp.	0	0.25 ± 0.25
Mastacembelus moorii	0	0.25 ± 0.25

TABLE 2: Continued.

Fish species	Protected area	Unprotected areas
Synodontis multipunctatus	0	0.25 ± 0.25
Hippopotamyrus sp.	0	0.25 ± 0.25
Altolamprologus compressiceps	0	0.25 ± 0.25
Petrochromis ephippium	0	0.25 ± 0.25
Lepidiolamprologus attenuatus	0	0.25 ± 0.25
Neolamprologus pleuromaculatus	0	0.25 ± 0.25
Total number of individual fish	351	167
Average number of individual fish per site	87.75 ± 46.16	41.75 ± 37.47

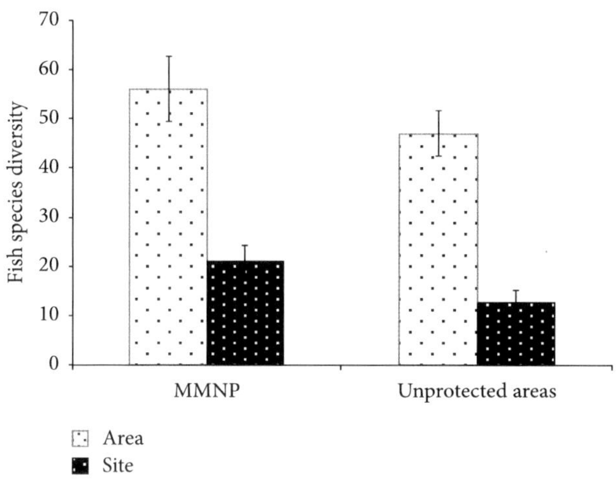

FIGURE 3: Mean fish species diversity (±SD) calculated as absolute species turnover in protected area (MMNP) and unprotected areas (Buhingu and Sibwesa) in Lake Tanganyika, Tanzania, during 2008 gillnet survey.

areas. There was a significant difference (Mann-Whitney, $P < 0.05$) in weight abundance between the MMNP and the unprotected areas. The abundance in the MMNP was fivefold greater than in the unprotected area. The highest weight (19,300 g·site^{-1}) was recorded in the MMNP, whereas the lowest (844 g·site^{-1}) was recorded in unprotected areas (Buhingu).

3.4. Fish Species Diversity. The mean species diversity turnover was higher in the MMNP than in the unprotected areas (Figure 3). The number of species within the MMNP's sites was almost double that of the unprotected areas. The sites with the highest and lowest species diversity were recorded in the MMNP ($n = 30$) and unprotected areas ($n = 8$), respectively. The fish diversity decreased with distance from the MMNP where 23 and 16 species were recorded in Buhingu and Sibwesa, respectively.

4. Discussion

The main findings of this study are that there are higher fish species diversity and abundance of fish in the MMNP compared with the unprotected areas. We discuss these findings in the subsequent text, to illustrate the performance differences between the MMNP and the unprotected areas in conserving fish abundance and biodiversity in the lake.

4.1. Environmental Parameters. Insignificant differences in the environmental parameters between protected and unprotected areas ruled out their effects on fish species diversity and abundance in the areas. Upwelling driven by strong southeast winds that prevail during the sampling period (dry season, i.e., May–September) [30, 31] might have attributed to the similarity in the parameters.

4.2. Abundance (Number) of Individual Fish. The dominance of different fish species within the MMNP suggests that the area is richer in fish species than the adjacent unprotected areas. Fish species in the families Clupeidae, Cyprinidae, and Poeciliidae were recorded only in MMNP and were among the most abundant species. Conversely, sites in the unprotected areas were characterized by few species that notably contributed to the abundance in these areas. This was exemplified by the two most dominant species *Haplotaxodon microlepis* and *Aulonocranus dewindti* that (out of 17 species in the area) constituted about 50% of fish counts in Sibwesa.

4.3. Weight Abundance. The current findings affirm the assumed abundance of socially and economically important fish species such as *Lates angustifrons*, *Boulengerochromis microlepis*, and *Limnotilapia dardennii* in the protected area. There was higher mean abundance within the protected area than in adjacent unprotected areas (Figure 2), which suggests that the MMNP harbors larger individual fish compared to the unprotected areas. In other words, individuals of exploited species were relatively larger in protected areas than in open-access areas. Fish size is used as an indicator of fishing pressure; fish are likely to be smaller in areas that are more heavily fished [12, 15, 32]. These are some positive indications of the effectiveness of the MMNP in conserving fish abundance and diversity in the lake.

4.4. Fish Species Diversity. The differences in fish diversity between the MMNP and unprotected areas (Figure 3) were probably due to differences in management of the resources. Whereas all forms of exploitation including fishing are strictly prohibited in the MMNP, fishing activities in surrounding waters are high [8, 10]. Poor enforcement of existing fisheries regulations by the local and regional authorities may also be attributed to low fish diversity in the unprotected areas. For instance, the use of prohibited beach seine nets was confirmed at Sibwesa during the survey. This could be one of the reasons why the sites in this area recorded the lowest fish species diversity. Identification of any signal of change in the ecosystem that will lead to taking immediate management and conservation measures is especially important [33].

5. Conclusions

Comparisons of fish abundance, especially of fish weight, and diversity showed that the MMNP plays an important role in conserving fisheries resources in the region. Extending the park or establishing more other protected areas could certainly enhance biodiversity in the region. However, this should be carefully considered because it might redirect fishing pressure to unprotected areas and cause more management challenges [34]. We believe that resources conservation approaches such as protected areas in the region cannot be successful through stringent laws and regulations but by creating awareness of the importance of protected areas among stakeholders. We advocate for socioecological studies, particularly on how communities surrounding the protected area should be engaged in enhancing sustainable fisheries resources management and conservation in the region.

Limitations of this study include inadequate sampling in rocky habitats in the MMNP, particularly in river mouths because they harbor dangerous animals such as Nile monitors, crocodiles, and hippopotamuses. Owing to this, we could not compare the influence of habitat types on fish abundance and diversity in the areas. Additionally, we could not find information on fish abundance and biodiversity before and after the addition of the water strip to the protected area. This precludes the comparison in a spatial and temporal basis of the variables in the areas. A similar paradox has been reported in other studies in freshwater protected areas [19–22]. Importantly, therefore, our work provides baseline information for the management of fisheries resources in Lake Tanganyika and guides future studies in the area.

Competing Interests

The authors declare that there are no competing interests regarding the publication of this paper.

Acknowledgments

This study was funded by the Tanzania National Parks (TANAPA), European Union (EU), and Frankfurt Zoological Society (FZS). The authors thank the Mahale Mountains National Park staff for logistics and support during sampling. They are grateful to Robert Wakafumbe, George Kazumbe, Chande Rashid, and Jumanne Nuhu for their tireless efforts during data collection. They also thank Christopher Mulanda Aura, John Richard Bower, Adam Smith, Ismael Aaron Kimirei, and Samwel Mchele Limbu for their useful comments on the drafts of this paper.

References

[1] G. Hanek, E. J. Coenen, and P. Kotilainen, "Aerial frame survey of lake Tanganyika fisheries," FAO/ FINNIDA Research for the Management of the Fisheries on Lake Tanganyika, 1993.

[2] L. C. Beadle, *The Inland Waters of Tropical Africa: An Introduction to Tropical Limnology*, Longman, London, UK, 1974.

[3] A. S. Cohen, L. Kaufman, and R. Ogutu-Ohwayo, "Anthropogenic threats, impacts and conservation strategies in the African Great Lakes-a review," in *Paleoclimatology East African Lakes*, T. C. Johnson and E. O. Odada, Eds., The Limnology, Climatology and Paleoclimatology of the East African Lakes Gordon and Breach, pp. 575–624, Amsterdam, The Netherlands, 1996.

[4] G. W. Coulter, "Lake Tanganyika," in *Speciation in Ancient Lakes*, K. Martens, B. Goddeeris, and G. Coulter, Eds., Advances in Limnology, pp. 13–18, Schweizerbart Science, 1994.

[5] H. Mölsä, J. E. Reynolds, E. J. Coenen, and O. V. Lindqvist, "Fisheries research towards resource management on Lake Tanganyika," *Hydrobiologia*, vol. 407, pp. 1–24, 1999.

[6] E. J. Coenen, "Catch per unit effort (CPUE) study for different areas and fishing gears of Lake Tanganyika," FAO/FINNIDA Research for the Management of the Fisheries of Lake Tanganyika, 1995.

[7] G. Patterson and J. Makin, *The State of Biodiversity in Lake Tanganyika: A Literature Review*, Natural Resources Institute, Chatham, UK, 1998.

[8] I. A. Kimirei, Y. D. Mgaya, and A. I. Chande, "Changes in species composition and abundance of commercially important pelagic fish species in Kigoma area, Lake Tanganyika, Tanzania," *Aquatic Ecosystem Health and Management*, vol. 11, no. 1, pp. 29–35, 2008.

[9] C. Sturmbauer, "The Great Lakes in East Africa: biological conservation considerations for species flocks," *Hydrobiologia*, vol. 615, no. 1, pp. 95–101, 2008.

[10] M. Van der Knaap, "Comparative analysis of fisheries restoration and public participation in Lake Victoria and Lake Tanganyika," *Aquatic Ecosystem Health and Management*, vol. 16, no. 3, pp. 279–287, 2013.

[11] D. Pauly, V. Christensen, S. Guénette et al., "Towards sustainability in world fisheries," *Nature*, vol. 418, no. 6898, pp. 689–695, 2002.

[12] F. Vandeperre, R. M. Higgins, J. Sánchez-Meca et al., "Effects of no-take area size and age of marine protected areas on fisheries yields: a meta-analytical approach," *Fish and Fisheries*, vol. 12, no. 4, pp. 412–426, 2011.

[13] G. W. Allison, S. D. Gaines, J. Lubchenco, and H. P. Possingham, "Ensuring persistence of marine reserves: catastrophes require adopting an insurance factor," *Ecological Applications*, vol. 13, no. 1, pp. S8–S24, 2003.

[14] F. R. Gell and C. M. Roberts, "Benefits beyond boundaries: the fishery effects of marine reserves," *Trends in Ecology and Evolution*, vol. 18, no. 9, pp. 448–455, 2003.

[15] R. Sahyoun, S. Bussotti, A. Di Franco, A. Navone, P. Panzalis, and P. Guidetti, "Protection effects on Mediterranean fish assemblages associated with different rocky habitats," *Journal of the Marine Biological Association of the United Kingdom*, vol. 93, no. 2, pp. 425–435, 2013.

[16] R. A. Sanyanga, C. Machena, and N. Kautsky, "Abundance and distribution of inshore fish in fished and protected areas in Lake Kariba, Zimbabwe," *Hydrobiologia*, vol. 306, no. 1, pp. 67–78, 1995.

[17] L. Mancini, P. Formichetti, A. Anselmo, L. Tancioni, S. Marchini, and A. Sorace, "Biological quality of running waters in protected areas: the influence of size and land use," *Biodiversity and Conservation*, vol. 14, no. 2, pp. 351–364, 2005.

[18] J. Cucherousset, J.-M. Paillisson, A. Carpentier et al., "Freshwater protected areas: an effective measure to reconcile conservation and exploitation of the threatened European eels (*Anguilla anguilla*)?" *Ecology of Freshwater Fish*, vol. 16, no. 4, pp. 528–538, 2007.

[19] R. L. Pressey, G. L. Whish, T. W. Barrett, and M. E. Watts, "Effectiveness of protected areas in north-eastern New South Wales: recent trends in six measures," *Biological Conservation*, vol. 106, no. 1, pp. 57–69, 2002.

[20] R. Abell, S. Blanch, C. Revenga, and M. Thieme, "Conservation of aquatic ecosystems," in *Encyclopedia of Inland Waters*, G. E. Likens, Ed., pp. 249–258, Elsevier, Oxford, UK, 2009.

[21] D. Rodríguez-Olarte, D. C. Taphorn, and J. Lobón-Cerviá, "Do protected areas conserve neotropical freshwater fishes? A case study of a biogeographic province in Venezuela," *Animal Biodiversity and Conservation*, vol. 34, no. 2, pp. 273–285, 2011.

[22] J. M. Scott, F. W. Davis, R. G. McGhie, R. G. Wright, C. Groves, and J. Estes, "Nature reserves: do they capture the full range of America's biological diversity?" *Ecological Applications*, vol. 11, no. 4, pp. 999–1007, 2001.

[23] T. Nishida, *The Chimpanzees of the Mahale Mountains: Sexual and Life History Strategies*, University of Tokyo Press, Tokyo, Japan, 1980.

[24] J.-F. Mas, "Assessing protected area effectiveness using surrounding (buffer) areas environmentally similar to the target area," *Environmental Monitoring and Assessment*, vol. 105, no. 1–3, pp. 69–80, 2005.

[25] B. C. Chessman, "Do protected areas benefit freshwater species? A broad-scale assessment for fish in Australia's Murray-Darling Basin," *Journal of Applied Ecology*, vol. 50, no. 4, pp. 969–976, 2013.

[26] JICA, *Mahale: Study for the Proposed Mahale Mountains National Park*, Japan International Cooperation Agency, 1980.

[27] R. G. Wetzel and G. E. Likens, *Limnological Analyses*, Springer, New York, NY, USA, 2000.

[28] A. Konings, *Tanganyika Cichlids in Their Native Habitat*, Cichlid Press, Lauenau, Germany, 1998.

[29] J. S. Albert and E. R. Roberto, *Historical Biogeography of Neotropical Freshwater Fishes*, University of California Press, Berkeley, Calif, USA, 2011.

[30] P.-D. Plisnier, D. Chitamwebwa, L. Mwape, K. Tshibangu, V. Langenberg, and E. Coenen, "Limnological annual cycle inferred from physical–chemical fluctuations at three stations of Lake Tanganyika," *Hydrobiologia*, vol. 407, pp. 45–58, 1999.

[31] J. R. Corman, P. B. McIntyre, B. Kuboja et al., "Upwelling couples chemical and biological dynamics across the littoral and pelagic zones of Lake Tanganyika, East Africa," *Limnology and Oceanography*, vol. 55, no. 1, pp. 214–224, 2010.

[32] S. E. Lester, B. S. Halpern, K. Grorud-Colvert et al., "Biological effects within no-take marine reserves: a global synthesis," *Marine Ecology Progress Series*, vol. 384, pp. 33–46, 2009.

[33] S. A. Black, "System behaviour charts inform an understanding of biodiversity recovery," *International Journal of Ecology*, vol. 2015, Article ID 787925, 6 pages, 2015.

[34] J. K. Abbott and A. C. Haynie, "What are we protecting? Fisher behavior and the unintended consequences of spatial closures as a fishery management tool," *Ecological Applications*, vol. 22, no. 3, pp. 762–777, 2012.

18

Feasibility of Community Management of Miombo Woodlands for Carbon Project in Southern Highlands of Tanzania

Z. J. Lupala,[1] L. P. Lusambo,[2] and Y. M. Ngaga[2]

[1] *Ministry of Natural Resources and Tourism, Forestry Training Institute, Arusha, Tanzania*
[2] *Forest Economics, Sokoine University of Agriculture, Morogoro, Tanzania*

Correspondence should be addressed to Z. J. Lupala; zachlupala@yahoo.com

Academic Editor: Daniel I. Rubenstein

In response to the pressing global challenges of climate change, community based management of miombo woodlands in Tanzania is promoted for carbon credit project development. However, evidence on its feasibility is scanty and questionable. This study examined the economic feasibility of carbon credit project development in community based forest management (CBFM) using four similar miombo woodlands from Southern highlands. The analysis was based on 144 sample plots from managed woodlands and 100 plots from business as usual (BAU). Allometric equation was applied to convert biomass to carbon per hectare. Improved carbon stock was determined and its economic value ascertained based on global voluntary carbon markets. Project feasibility analysis was performed using discounted cash flow, internal rate of return, and benefits/costs methods. Annual opportunity cost and variable costs were subtracted from total revenue to obtain annual net profit. The annual rate of return on investment was calculated by dividing profits by total costs. It was revealed that carbon stock improved significantly in CBFM compared to BAU ($P < 5$ %). The improvement had positive net present value and benefit-cost ratio of 1.83. Moreover, sensitivity analysis showed that if any unexpected situation occurs, the project will still be of worthiness. The findings are useful to enrich the debate on carbon credit development under community based management of miombo woodlands in Tanzania.

1. Introduction

Miombo woodland is among the most important ecosystems for socioeconomic and environmental roles [1, 2]. It is home to megafauna assemblage and among the world's high-biodiversity hotspots [3, 4]. The role of miombo woodland in global carbon cycle is increasingly recognized [2, 5–8]. Between 0.5 and 0.9 tons of carbon ha^{-1} $year^{-1}$ are being sequestered in miombo woodland, with maximum sequestration in young miombo [5, 8, 9]. Moreover, livelihoods of significant number of people depend on miombo woodland through provision of many products and services [10]. Unfortunately, its capacity to provide these products and services is decreasing rapidly each year due to deforestation and forest degradation [11].

The increasing rate of deforestation and forest degradation is a major concern for many countries in the world [11]. For example in Tanzania, the extent of deforestation and forest degradation is devastating, about 372,816 ha/year [12]. Miombo woodland suffers the most if compared to other tropical forest biomes [11, 13, 14]. This raised necessity to promote sustainable forest management through community based forest management [10, 15]. Community based forest management (CBFM) was developed to combat deforestation and promote sustainable forest management in developing countries [16–18]. The underlying principle is that communities can be best positioned to manage and protect forest resources provided it is their interest to do so [17, 19] (Larson, 2004).

Through CBFM, local communities establish clear rules for sustainable resource use and respond to deforestation signals more rapidly [2, 20]. This is important for the successful management of African miombo woodlands. According to Dewees et al. [10], sustainable management of miombo woodlands is important for three reasons: (i) they sequester enormous amounts of carbon; (ii) they support livelihoods to millions of people and provide a renewable source of energy, that is, fuel wood and charcoal; and (iii) their successful management would contribute to poverty alleviation by supporting and strengthening local livelihood strategies. In times of stress these woodlands serve as an insurance against famine by offering a source of wild foods and fruits and other useful products [10, 21].

By virtue of involving local forest users under CBFM, there is widely reported improvement of forest condition, social cohesion, and rural incomes in Tanzania and elsewhere (e.g., see [22–25] (Antinori and Rausser, 2008)). This has attracted the United Nation program on Reduced Emission from Deforestation and Forest Degradation and the role of conservation, sustainable management, and enhancement of carbon stock (REDD+) to adopt CBFM as an important toolkit [26]. Indeed, this is an important framework for sustainable forest management and climate change mitigation in many developing countries including Tanzania [27].

REDD+ is expected to bring about economic development through transfer of payments, reduce poverty, and boost capacity of developing countries to mitigate climate change [1, 28]. If successful, carbon mitigation projects can produce cobenefits by conserving biodiversity, ecosystem services, improving forest governance, and the welfare of forest dependent communities [26, 29]. This potential, however, has not been well taped in miombo woodlands of Tanzania and its evidence base was found to be weak and questionable (Behera, 2009; Bowler et al., 2012).

Furthermore, there are increasing doubts over the viability and effectiveness of CBFM for carbon project development [30–32]. The main concern is that benefits from carbon project development may not be sufficient to cover various costs imposed on local communities [30, 33]. Rigorous literature review has shown that there is a need for feasibility analysis on carbon project development possibilities in miombo woodlands of Tanzania [2]. This will ensure better-informed policy direction, strategy, and practice which guarantee climate change mitigation and livelihood options [2, 26]. The need for feasibility analysis of CBFM for carbon project development in order to support informed judgment and decision making about its merit is important (e.g., [30–32]).

Whether the project shall contribute to the future improvement in social and economic welfare of the community and promotes sustainable forest management needs to be addressed. "This is a key motivation for economic feasibility analysis of carbon project development in community based managed miombo woodlands and their subsequent implications for sustainable forest management in Tanzania." According to Angelsen et al., [28] the economic viability of this kind of projects may depend on meeting important criteria such as effectiveness, efficiency, equity, and cobenefits (3E+).

Effectiveness requires that net carbon emission reductions should be positive and significant. Likewise the stored carbon needs to be "additional" if compared to business as usual scenario [28]. Efficiency implies that emission reductions are less costly relative to other options. This implies that carbon benefits should be greater than the costs borne by different stakeholders to make this instrument incentive-compatible [35]. Equity implies issues of social justice (fairness and rights) associated with implementation of the REDD+ projects at local levels [36]. Together with biodiversity conservation implications, all these are referred to as cobenefits of REDD+ projects [28]. The understanding is useful for project developers initiating carbon project and also informs climate change mitigation strategies such as those triggering development of economic incentives for promotion of sustainable forest management.

TABLE 1: Land use categories and criteria used in selection of study sites.

S/N	Category	Criteria for selection
1	CBFM	(i) Management plan and by-laws for forest management activities (ii) Experience from expert opinion that the forest is under CBFM (iii) Patrols and management meetings explaining management activities (iv) Presence of Village Environmental Committee (VEC) taking care of the forest (v) Tree vegetation comprise miombo woodlands
2	*BAU	(i) No records of forest management activities (ii) Located adjacent to CBFM and tree composed of miombo woodlands (iii) Open access for local use and extraction of resources is unregulated (iv) Experience from expert that nothing have been done to regulate forest extraction

*Business as usual is a projections of future emission levels in the absence of any policy changes or REDD activities [34] while with CBFM the laws are enforced by community and sustainable use controlled.

2. Materials and Methods

This study examined the economic feasibility of carbon project development based on the extent of carbon stock in CBFM of miombo woodland relative to a business as usual scenario in Southern Highlands of Tanzania. A counterfactual design of identifying matching treatment (CBFM) and control groups (BAU) based on similarities in observed characteristics was used [31]. This was based on assumption that treatment and control are not related to other factors that may have determined outcome (Ferraro, 2009). Community based managed miombo woodland (CBFM) and its adjacent business as usual (BAU) land use categories were selected using specific criteria and land use characteristics. In order to ensure a set of statistically matching comparison, however, the criteria such as similar topographical and forest characteristics, socioeconomic forest use, infrastructure, and market access were established (Table 1).

In addition, geographic proximity and historical and ecological information were used to separate the effects of management from those of possible confounding factors. With this in mind, four CBFM managed miombo woodlands and three adjacent reference scenarios in Mufindi and Iringa rural districts of southern highlands of Tanzania were studied (Table 2 and Figure 1).

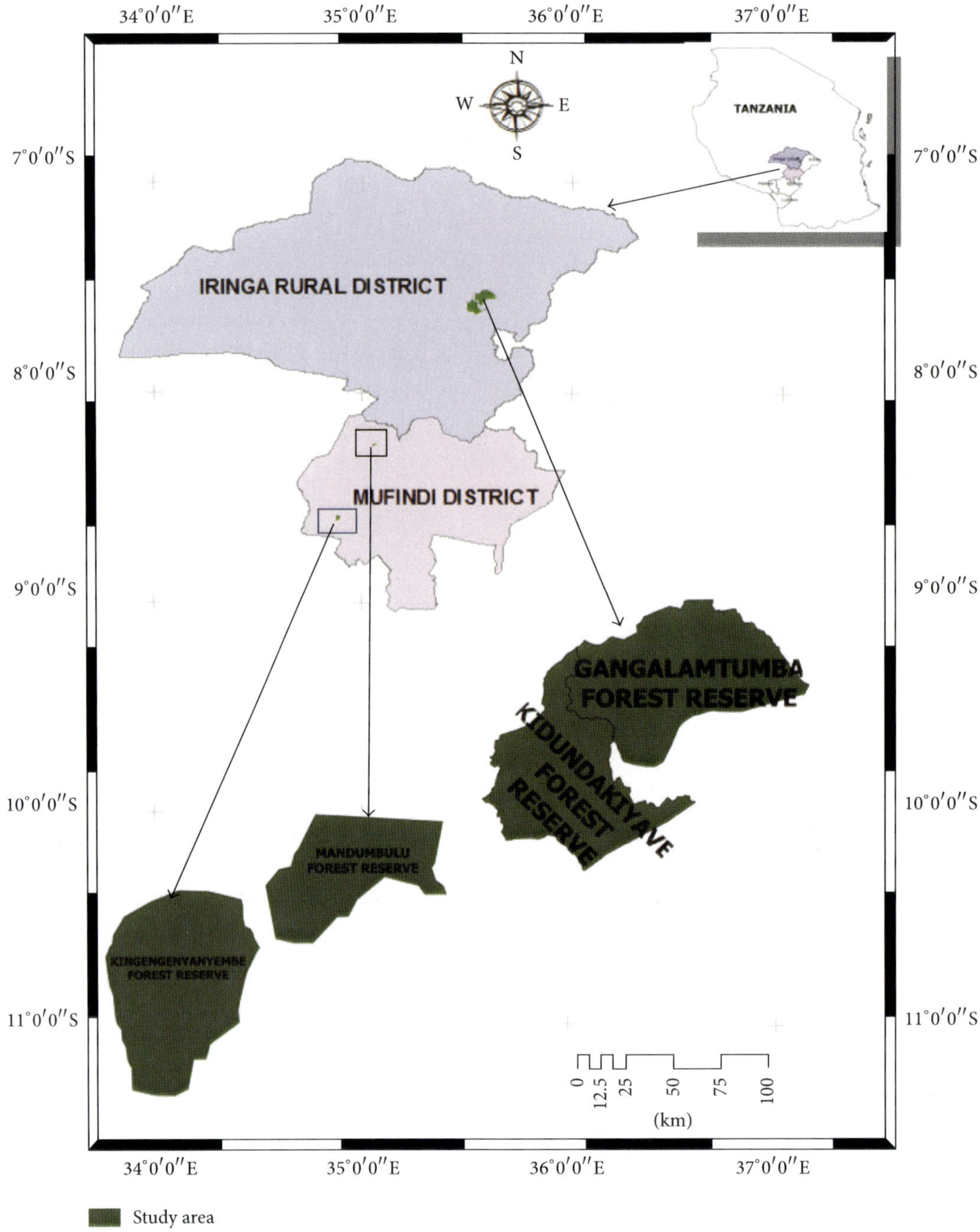

FIGURE 1: Geographical location of study districts and respective studied forest reserves of Southern Highlands of Tanzania.

2.1. Data Collection and Analysis. The procedure for forest inventory was based on MacDicken [37], MNRT [38], and from good practice guidance for land use, land use change, and forestry [39]. Required number of sample plots was determined through $n = (\mathrm{CV}^2 * t^2)/E^2 = ((\mathrm{SD}/X)^2 * t^2)/E^2$, where CV is coefficient of variation; t the value from the students' distribution table at $n - 1$ degrees of freedom, SD standard deviation, X mean of basal area, and E precision, and in this case 10% error was adopted (Zahabu, 2008). The first plot was located at 100 m from the forest edge while subsequent plots were located systematically along transect lines at intervals of 200 m. This helped to cover gradients, species composition, and structural characteristics. Concentric circular plot with a maximum radius of 15 m was used for the aim of increasing the accuracy of the measurement and sampling intensity of various tree sizes and to save time [12]. In each sample plot (0.07 ha), plot number, slope, aspect, vegetation type, and coordinates were recorded to help in characterization of the woodlands. In total, 244 sample plots were examined, including 100 plots in business as usual as reference scenario. In each plot, *Radius 2 m* is measured and recorded diameter of all trees with DBH ≥ 1 cm,

TABLE 2: Location of miombo woodlands, management practices, and its climatic characteristics.

District	Name of miombo under CBFM	Name of compared (BAU)	Location	Altitude (m)	Dominant soil type	Mean temp (°C)/year	Mean rain (mm)/year
Iringa rural district	Gangalamtumba	Makota	7°35′S. 35°35′E.	850–1,300	Sandy clay loam soils	19.8	617
	Kidundakiyave	Makota	7°590′S. 35°79′E.	850–1,300	Clay alluvial soils	21.5	600
Mufindi district	Mandumburu	Ngombe	8°43′S. 35°64′E.	1180–1500	Sandy loam soils	24.06°C	630
	Kingegenyanyembe	Kingegenyanyembe	8°.00′S. 35°15′E.	1600–1800	Sand clay soils	22.56°C	584

Radius 5 m is all trees with DBH ≥ 5 cm, *Radius 10 m* is all trees with DBH ≥ 10 cm, and *Radius 15 m* is all trees with DBH ≥ 20 cm.

Estimation of amount of wood extraction was done at stump level. The number of felled trees was measured from stumps at 30 cm above ground. Three sample trees in each plot from DBH classes were randomly selected and measured for basal diameter (BD) and (DBH) in cm and height (H) of the metre. Measured basal diameter was determined as new stump if harvested within year 2014 (inventory year) and old if harvested in year 2013 and below based on stump conditions observed. The distinction of the age of stumps was established by the color and freshness of exposed wood, the size of sprouts/coppices, and the presence of fire scorch on exposed wood.

In each case, trees and stumps measured were identified using vernacular names with the help of knowledgeable local elders who are well acquainted with ethnobotany and aspect of wood utilization and management in the respective village. The criteria used for identification of the harvested species were coppice growth, wood and bark characteristics, and the symmetry of the stump. Moreover, focus group discussion and key informant interview were used to obtain perception and interest related to carbon project development. A feedback workshop for committee members was organized in Iringa town to discuss on how carbon project development venture might be and how they will be organized for the case of carbon project.

2.2. Estimation of Carbon Stock and Sequestration. The selection of appropriate allometric equations for computation of both aboveground biomass (AGB) and belowground biomass (BGB) was done through pretested locally available allometric equations. The general allometric equation for AGB $= 0.1027D^{2.4798}$ and BGB $= 0.2113D^{1.9838}$ computation by Mugasha et al. [40] was used, where biomass is total tree biomass (kg) and **D** is tree dbh (cm). This equation includes trees greater than 1 cm diameter at breast height (dbh) and it has the advantage of requiring only dbh as an independent variable. Among others, the equation included miombo woodlands from Iringa region and also had R^2 of 95% making it most reliable. The use of local allometric equations for areas with similar geographical and vegetation type is recommended in the literature (Brown, 2003) [39, 40]. Biomass was then converted to carbon using biomass carbon ratio of 0.48 and then multiplied by 3.67 to get equivalent tCO_2/ha [12]. In addition, the total tree volume was calculated from the allometric equation developed by Mauya et al. [41]. The equation was $\mathbf{V} = 0.00016D^{2.463}$, where **V** is tree volume (m^3) and **D** is tree dbh (cm) ($R^2 = 0.87$). Diameter at breast height of stump was obtained by regression of height to basal diameter of sample trees developed (Ht = 4.234 + 0.254 (bd), $R^2 = 0.78$, $P < 0.0001$).

2.3. Economic Feasibility Analysis for Carbon Project Development. This study used Benefit-Cost Analysis (BCA) as economic analysis techniques to help with the decision whether the project is worth undertaking. Economic feasibility depends on (a) existing carbon stock; (b) baseline used here referred to as business as usual to determine deforestation rate; (c) price of carbon stock avoided; (d) costs of implementation to reduced emission. The market value of carbon stock and sequestration was calculated assuming a ton of CO_2 removal worth 5 USD [42]. Based on this, carbon credit was evaluated using net present value (NPV), internal rate of return, and benefit/cost ratio [43]. The cost of carbon sequestration in CBFM is the cost of improved forest management, including the cost of establishing CBFM and maintaining the system, plus the opportunity costs associated with the lost benefits of alternative systems (business as usual or open access). The costs are represented on a per-hectare basis, as are the carbon sequestration potentials.

Thus on an annual basis, the annual gross carbon income = Emission avoided annually ∗ $/ton of carbon while annual net income from emission avoided = Gross carbon income − (opportunity + transaction + management costs). Once the economic values of benefits and costs have been derived, the results were presented in three traditional forms: NPV, B/C ratio, and internal rate of return (IRR) [44]. Sensitivity analysis for the key parameters which might have greater influence on the project development was calculated to check the effect of unexpected situation to the feasibility results.

3. Results and Discussion

3.1. The Observed General Condition of Studied Miombo Woodlands. The mean stem density was highest (4118 ± 2811 stems ha^{-1}) in Kingegenyanyembe and lowest (1982 ± 1563

TABLE 3: Stocking parameters in miombo woodlands based on management practices.

Forest name	District	Management	N (ha^{-1})	G (m^2/ha^{-1})	V (m^{3}ha^{-1})
Mandumburu	Mufindi	CBFM (n = 52)	2716 ± 2169	9.01 ± 2.81	1.63 ± 1.17^a
Ngombe	Mufindi	BAU (n = 40)	3216 ± 2254	10.71 ± 4.84	1.86 ± 1.30^a
Kingegenyanyembe	Mufindi	CBFM (n = 30)	4118 ± 2811	10.59 ± 5.37	1.76 ± 1.11^a
Kingegenyanyembe	Mufindi	BAU (n = 30)	2522 ± 1636	5.14 ± 3.09	0.66 ± 0.49^b
Gangalamtumba	Iringa rural	CBFM (n = 30)	1982 ± 1563	11.56 ± 3.69	4.51 ± 2.42^a
Kidundakiyave	Iringa rural	CBFM (n = 32)	2655 ± 2207	9.07 ± 2.88	1.85 ± 1.48^a
Makota	Iringa rural	BAU (n = 30)	2237 ± 2119	8.45 ± 3.98	1.23 ± 0.71^b

Note. n = sample size, N is the number of stems ha^{-1}, G is the basal area (m^{2}ha^{-1}), and V is the volume (m^{3}ha^{-1}); numbers after ± are 95% confidence limits (products of standard errors of the mean and t-value at 95% confidence level) and means with the same letter are not significantly different using Bonferroni t-test (α = 0.05).

TABLE 4: Effects of woodland management on biomass and carbon stock in studied sites.

Name of miombo woodland	Management	AGB (t/ha^{-1})	CO_2eq (t/ha^{-1})	BGB (t/ha)	CO_2eq (t/ha^{-1})	Total CO_2 (t/ha)
Mandumburu	CBFM	1.10 ± 0.80	2.01 ± 1.46	0.54 ± 0.30	0.99 ± 0.54	3.00 ± 2.00^a
Ngombe	BAU	1.25 ± 0.88	2.30 ± 1.62	0.62 ± 0.36	1.13 ± 0.66	3.43 ± 2.28^a
Kingegenyanyembe	CBFM	1.21 ± 0.74	2.22 ± 1.36	0.61 ± 0.28	1.11 ± 0.51	3.33 ± 1.87^a
Kingegenyanyembe	BAU	0.33 ± 0.45	0.60 ± 0.82	0.19 ± 0.22	0.35 ± 0.40	1.95 ± 1.22^b
Gangalamtumba	CBFM	3.06 ± 1.65	5.61 ± 3.02	1.29 ± 0.61	2.36 ± 1.11	7.97 ± 4.13^a
Makota	BAU	0.83 ± 0.48	1.52 ± 0.89	0.43 ± 0.23	0.78 ± 0.42	2.30 ± 1.31^b
Kidundakiyavye	CBFM	1.25 ± 1.01	2.29 ± 1.85	0.61 ± 0.37	1.10 ± 0.68	3.39 ± 2.53^a
Makota	BAU	0.83 ± 0.48	1.52 ± 0.89	0.43 ± 0.23	0.78 ± 0.42	2.30 ± 1.31^b

Note. Numbers after ± are 95% confidence limits (products of standard errors of the mean and t-value at 95% confidence level) and means with the same letter are not significantly different using Bonferroni t-test (α = 0.05).

stems ha^{-1}) in Gangalamtumba, both CBFM managed miombo woodlands. Likewise, in business as usual (BAU) counterpart, the highest was (3216 ± 2254 stems ha^{-1}) in Ngombe and lowest (2237 ± 2119 stems ha^{-1}) in Makota. This observation indicates miombo woodlands under both scenarios are densely populated and their difference is statistically not significant (P > 5%). However, basal area comparison between management practices was of significant difference (P < 5%). The average basal area ranged from 11.56 ± 3.69 in CBFM to 5.14 ± 3.09 in BAU (Table 3). The observations suggest that CBFM managed woodlands are composed of larger diameter trees compared to BAU practices. Stem density observed can be related to previous deforestation and forest degradation of miombo woodlands. According to Chidumayo, [8] stem density and basal area correlate with disturbance in miombo woodlands. Moreover, there is domination of larger trees in CBFM with average diameter of 12.26 ± 2.92 compared to 10.32 ± 2.54 from BAU and their overall volume estimated differed significantly (P < 5%).

Further analysis on vegetation structure portrayed reverse J-shaped curve in both CBFM and BAU. This suggests existence of more trees in the smaller diameter growing to lager diameter classes. It also indicates continuous recruitment in miombo woodlands and therefore possibility for sustainability. Other studies conducted in Tanzanian miombo woodlands demonstrated similar vegetation structure [23, 40, 45]. This provides insight into how the recovery of miombo woodlands can be used in the emerging C markets as a way of maximizing rural people's benefits from forest management.

3.2. Effects of Community Based Management on Biomass for Carbon Stock. In order to identify forest management strategies that maximize carbon storage, below- and above-ground carbon stock was first determined prior to feasibility analysis. The overall highest average carbon stock observed was in Gangalamtumba CBFM (7.97 ± 4.13 t/ha) and lowest in Kingegenyanyembe BAU (1.95 ± 1.22 t/ha). The sites comparison revealed statistically significant differences between CFM and BAU (Table 4). The exception was observed in Mandumburu CBFM and its counterpart Ngombe BAU with no significant deference, and this was not expected. This could be because leakage effect of CBFM rules impacted larger village landscape which improved biomass for carbon stock in both businesses as usual and CBFM practice.

Further analysis using stump measurement and extrapolation from basal diameter to diameter at breast height suggest incidence of extraction revealed more emission in BAU (6.88±0.62 t/ha) as compared to CBFM (3.95±0.41 t/ha) (Table 5). However, this result also suggests that forest extraction activities are occurring in land use management with different intensity. In addition, the result is supported by research that suggests that strengthened management rules and tenure in community forest help reduce forest product

Table 5: Extent of woodland extraction and its corresponding emission in CBFM as compared to reference scenario ($n = 244$).

Management practice	Density (N/ha)	Basal area (m^2/ha)	Volume removal (m^3/ha)	Biomass removal (t/ha)	Carbon stock emitted (t/ha)
In CBFM	167 ± 24	5.31 ± 2.78	9.63 ± 2.35	7.89 ± 0.3	3.95 ± 0.41
In BAU	431 ± 49	8.29 ± 3.32	15.41 ± 5.09	13.77 ± 0.8	6.88 ± 0.62

Note. N is the number of stems ha^{-1}, G is the basal area (m^2ha^{-1}), and V is the volume (m^3ha^{-1}); numbers after ± are 95% confidence limits (products of standard errors of the mean and t-value at 95% confidence level).

Table 6: Forest area and its potential gross annual income from avoided CO_2e emissions.

Study site	Forest area (ha)	Change in carbon stock (t/ha/yr)	CO_2e (t/ha/yr)	Potential annual income ($/ha/yr)	Annual value for whole forest area ($)
Kingegenyanyembe	459.6	1.38	5.07	25.35	11,650.86
Gangalamtumba	6,065	5.67	20.81	104.05	631,063.25
Kidundakiyave	4904	1.09	4.00	20.00	98,080
Mandumburu	450	−0.43	−1.58	−7.89	−3,551

extraction and improve forest condition [22]. Interestingly, recovery trajectory of miombo woodlands after extraction is rapidly from disturbances caused by human activities [2, 46]. Elsewhere, Syampungani (2009) have reported higher capacity to resprout/regenerate in harvested stands compared to mature, more or less untouched stands suggesting that miombo woodlands are dynamic ecosystems reacting well to appropriate levels of exploitation.

Miombo woodland tree species have extensive vertical and horizontal root systems which facilitate regeneration after cutting [8]. There is availability of stump coppices, root suckers, or suppressed saplings in the herb layer at the time of clearing [8, 47]. This enables the woodlands to recover rapidly, however depending on the intensity of extraction and subsequent land use [2].

3.3. Estimated Potential Gross Revenue from Carbon Project Development. The determinants of feasibility of carbon project development include the amount of carbon stock observed, historical business as usual, carbon price, and associated costs [31, 48]. Carbon price negotiations are still going on [26, 49]. The voluntary carbon market is currently one of the main frameworks within which carbon stock can be sold from CBFM [23, 48]. In this market, the minimum price per ton of carbon is about 5 USD during the last quarter of 2015 (Bayon et al., 2007) [23]. However, offsets that deliver complementary benefits such as biodiversity conservation and poverty reduction are likely to command a premium (Bayon et al., 2007) [26]. Using this minimum price (5 USD/tCO_2e) from voluntary carbon market, different study sites portrayed different potential in carbon credit sales. The Gangalamtumba miombo woodland revealed higher revenue (631,063.25 $), followed by Kidundakiyave (98,080 $) and Kingegenyanyembe (11,650.86 $), and Mandumburu had negative results (−3,551 $) (Table 6).

These results show potential gross annual income from avoided carbon stock emission with clear differences between study sites in terms of carbon stock change, area, and expected income. The differences in carbon stock change could be contributed to management effectiveness and the condition of the baseline used. However, negative result from Mandumburu could be related to baseline used and the issue of effectiveness in forest management.

3.3.1. Estimation of the Annual Net Benefits from Avoided Carbon Stock Emissions. The annual net benefits are calculated as the difference between the annual gross revenue and total annual costs. Taking into account important costs linked to management, opportunity, and transaction cost elements, net benefit that can be anticipated from carbon credit project development in miombo woodlands of Tanzania was determined. This analysis revealed higher transaction and opportunity costs in Gangalamtumba and Kidundakiyave (Table 7). Management cost is based on local control and already developed CBFM institutions as established in Tanzania (Zahabu, 2008). In order to be profitable the gross annual revenue from sales of carbon should be high enough to cover all the costs related CBFM and carbon credit project development.

However, estimates of these cost elements may change over time and these changes must be taken into account [49]. For example, one of the main challenges related to opportunity cost is to know if what is paid is close to the alternative [48]. Further studies are still necessary to be able to generate local and national cost rates that can be useful in making cost estimates for carbon project development process. In this paper it was plausible to apply local cost estimation likely to be considered in carbon project development in Tanzania.

3.4. Carbon Project Evaluation for Economic Feasibility Decision. Before elaborating the feasibility results, the assumption which was used for this analysis considers only incremental costs and incremental benefit and therefore the net benefits and cost are the difference between business as usual situation and the project situation. Carbon project period is 20 years and discount rate is 10% per year based on central bank of Tanzania [50]. Then project evaluation was presented in three traditional forms of net present value (NPV), internal

TABLE 7: Estimated costs and net benefit of improved carbon stock in voluntary market.

Study site	Management cost ($/yr)	Opportunity cost ($/yr)	Transaction cost ($/yr)	Total annual cost ($)	Value of avoided carbon stock ($/yr)	Net benefit from improved carbon stock ($/yr)
Kingegenyanyembe	1,580	2,389.92	919.2	4,889.12	11,650.86	6,761.74
Gangalamtumba	1,580	31,538	12,130	45,248	631,063.25	585,815.25
Kidundakiyave	1,580	25,500.8	9,808	36,888.8	98,080	61,191.20
Mandumburu	1,580	2,340	900	4,820	−3,551	−8,371.00
Overall						*645,397.19*

TABLE 8: Feasibility test based on estimated cost and benefits of improved carbon stock in CBFM.

Operation	Year	Cost ($/ha)	Benefit ($/ha)	Discount factor	PV (cost) ($/ha)	PV (benefit) ($/ha)	NPV ($/ha)
	1	*7.43*	*13.58*	0.91	7	12	6
	5	*37.15*	*67.9*	0.62	23	42	19
	10	*74.3*	*135.8*	0.39	29	52	24
	15	*111.45*	*203.7*	0.24	27	49	22
	20	*148.6*	*271.6*	0.15	22	40	18
Interest rate (/y)	*0.10*			Total	107	196	89
					Benefit-cost ratio	1.83	

rate of return (IRR), and benefit-cost ration (B/C ratio) (Table 8).

Net present value (NPV) = 89 ($/ha) and benefit-cost ration equals 1.83 which is more than 1. Based on these criteria it is worth developing carbon project under CBFM of Tanzanian miombo woodlands. Moreover there is possibility of additional revenue from CBFM management apart from carbon credit sale. One study of CBFM found total revenues gaining to each village just about US$ 540 and US$ 720 per village per year [51]. The average revenue from forest product is about US$ 630 per year; this additional revenue emanated from sustainable collection of miombo woodland products and provision of other services. Together with carbon credit the amount could be very significant return from CBFM to cover the costs imposed to the local communities and therefore contribute to the climate change mitigation and sustainable development. Therefore, this study proves that carbon project development will contribute to the future improvement in social and economic welfare of the community and thus incentivize CBFM in Tanzania.

3.5. Performing Sensitivity Analysis. The key parameters with greater influence on the results of this study include management, opportunity, and transaction cost elements as well as carbon stock from improved management. Therefore to calculate the effect of probability of change in key variable, this study performed simulations on the following scenario.

Scenario 1. Because of the long time period required to improve forest biomass and develop carbon project for climate change mitigation, the costs and benefits can be influenced by the discount rate used [44] (Fisher et al., 2011). Discounting presumes that meeting the need of current generation is more important than meeting the need of future generation. Essentially, discounting involves imposing a major cost to future for the sake of a relatively small gain today. In order to assess the sensitivity of proposed project, however, the use of higher discount rates above the one used in economic analysis of this study is recommended [44]. Increasing discount rate from 10% to 15%, the results revealed net present value (NPV) = 55 USD/ha and benefit-cost ration (B/C) = 1.83. Therefore, sensitivity analysis in both scenarios shows net present values are positive, the IRR are also acceptable, and B/C ratio is more than 1. Based on these results, if any unexpected situations occur, the project will still be of worthiness.

4. Conclusions and Recommendations

CBFM improved biomass for carbon storage which is financially attractive. The study observed potential benefits received from carbon offset outweigh associated costs. In miombo woodlands under CBFM carbon sequestration project will obviously be an additional income source apart from other sustainably extracted nontimber forest products. This study observed positive cash flow which can be harnessed to improve livelihoods, social services, and infrastructure development. Despite the challenges of determining additionality, permanence, and reduced leakage, carbon project development is a viable option combining conservation and development. However factors such as institutional and policy context were not captured in this study and therefore need to be examined.

In combination with these factors and observed results carbon project can be more effective and feasible. Furthermore carbon market should be properly regulated to be privileged to community based intervention. For example, the question of whether the net benefits accrued should be reflected in a lower price to the emitter or increased benefits to the CBFM is open to debate. Moreover, research is important to optimize carbon credit in CBFM for suggesting innovative ways of costs reduction and maximize income. The empirical knowledge about how local community actually depends on miombo woodlands for their livelihood and how this translates to reliable carbon emission estimates is also important research gap. Despite these, miombo woodlands management that generates carbon revenue especially through CBFM may provide a viable and advantageous socioeconomic development alternative.

Conflicts of Interest

The authors declare that there are no conflicts of interest regarding the publication of this paper.

Acknowledgments

The authors acknowledge the financial support and cooperation received from Climate Change Impact, Mitigation and Adaptation (CCIAM), a Royal Norwegian Sponsored Program at Sokoine University of Agriculture, Morogoro, Tanzania, and the Ministry of Natural Resources and Tourism (MNRT), Forestry and Beekeeping Division (FBD), Tanzania. The authors acknowledge the district forest departments in Mufindi and Iringa rural district including local communities involved in management of miombo woodlands for proving them with data and information needed.

References

[1] T. R. Baker, J. P. G. Jones, O. R. R. Thompson et al., "How can ecologists help realise the potential of payments for carbon in tropical forest countries?" *Journal of Applied Ecology*, vol. 47, no. 6, pp. 1159–1165, 2010.

[2] Z. J. Lupala, L. P. Lusambo, and Y. M. Ngaga, "Management, growth, and carbon storage in miombo woodlands of Tanzania," *International Journal of Forestry Research*, vol. 2014, Article ID 629317, 11 pages, 2014.

[3] R. Vinya, *Stem hydraulic architecture and xylem vulnerability to cavitation for miombo woodlands canopy tree species [Ph.D. dissertation]*, University of Oxford, 2010.

[4] J. P. Croxall, S. H. M. Butchart, B. Lascelles et al., "Seabird conservation status, threats and priority actions: a global assessment," *Bird Conservation International*, vol. 22, no. 1, pp. 1–34, 2012.

[5] M. Williams, C. M. Ryan, R. M. Rees, E. Sambane, J. Fernando, and J. Grace, "Carbon sequestration and biodiversity of regrowing miombo woodlands in Mozambique," *Forest Ecology and Management*, vol. 254, no. 2, pp. 145–155, 2008.

[6] P. K. T. Munishi, S. Mringi, D. D. Shirima, and S. K. Linda, "The role of miombo woodlands of Southern Highlands of Tanzania as carbon sinks," *Journal of Ecology and the Natural Environment*, vol. 2, no. 12, pp. 261–269, 2010.

[7] I. M. Bond, B. Chambwera, M. Jones, M. Chunduma, and I. Nhantumbo, *REDD+ in Dryland Forests: Issues and Prospects for Pro-Poor REDD in the Miombo Woodlands of Southern Africa Natural Resources*, Natural Resource Issues no. 21, International Institute for Environment and Development, London, UK, 2010.

[8] E. N. Chidumayo, "Forest degradation and recovery in a miombo woodland landscape in Zambia: 22 years of observations on permanent sample plots," *Forest Ecology and Management*, vol. 291, pp. 154–161, 2013.

[9] L. Miles, K. Kabalimu, B. Bahane et al., *Carbon, Biodiversity and Ecosystem Services: Exploring Benefits. Tanzania*, United Nations Environment Programme World Conservation Monitoring Centre, Cambridge, UK, 2009.

[10] P. A. Dewees, B. M. Campbell, Y. Katerere et al., "Managing the miombo woodlands of southern Africa: policies, incentives and options for the rural poor," *Journal of Natural Resources Policy Research*, vol. 2, no. 1, pp. 57–73, 2010.

[11] FAO, *Global Forest Resources Assessment 2010: Country Report*, United Republic of Tanzania, Food and Agriculture Organization of the United Nations, Rome, Italy, 2010.

[12] URT, *National Forest Resources Monitoring and Assessment (NAFORMA) of Tanzania Mainland: Main Report*, The Ministry of Natural Resources and Tourism, United Republic of Tanzania, 2015, http://www.tfs.go.tz/resources/view/naforma-report-2015.

[13] W. F. Mwase, Å. Bjørnstad, J. M. Bokosi, M. B. Kwapata, and B. Stedje, "The role of land tenure in conservation of tree and shrub species diversity in miombo woodlands of southern Malawi," *New Forests*, vol. 33, no. 3, pp. 297–307, 2007.

[14] F. L. Godoy, K. Tabor, N. D. Burgess, B. P. Mbilinyi, J. J. Kashaigili, and M. K. Steininger, "Deforestation and CO_2 emissions in coastal Tanzania from 1990 to 2007," *Environmental Conservation*, vol. 39, no. 1, pp. 62–71, 2012.

[15] URT, *National Forest Policy*, Ministry of Natural Resources and Tourism, Dar es Salaam, Tanzania, 1998.

[16] W. D. Sunderlin, "Poverty alleviation through community forestry in Cambodia, Laos, and Vietnam: An assessment of the potential," *Forest Policy and Economics*, vol. 8, no. 4, pp. 386–396, 2006.

[17] A. Maryudi, R. R. Devkota, C. Schusser et al., "Back to basics: Considerations in evaluating the outcomes of community forestry," *Forest Policy and Economics*, vol. 14, no. 1, pp. 1–5, 2012.

[18] C. Schusser, "Who determines biodiversity? An analysis of actors' power and interests in community forestry in Namibia," *Forest Policy and Economics*, vol. 36, pp. 42–51, 2013.

[19] K. K. Shrestha and P. McManus, "The embeddedness of collective action in Nepalese community forestry," *Small-Scale Forestry*, vol. 6, no. 3, pp. 273–290, 2007.

[20] R. K. Pokharel, "Factors influencing the management regime of Nepal's community forestry," *Forest Policy and Economics*, vol. 17, pp. 13–17, 2012.

[21] J. Paavola, "Livelihoods, vulnerability and adaptation to climate change in Morogoro, Tanzania," *Environmental Science & Policy*, vol. 11, no. 7, pp. 642–654, 2008.

[22] L. Mbwambo, T. Eid, R. E. Malimbwi, E. Zahabu, G. C. Kajembe, and E. Luoga, "Impact of decentralised forest management on forest resource conditions in Tanzania," *Forests, Trees and Livelihoods*, vol. 21, no. 2, pp. 97–113, 2012.

[23] Z. J. Lupala, L. P. Lusambo, Y. M. Ngaga, and A. A. Makatta, "The land use and cover change in miombo woodlands under community based forest management and its implication to climate change mitigation: a case of southern highlands of Tanzania," *International Journal of Forestry Research*, vol. 2015, 11 pages, 2015.

[24] S. Charnley and M. R. Poe, "Community forestry in theory and practice: Where are we now?" *Annual Review of Anthropology*, vol. 36, pp. 301–336, 2007.

[25] B. B. K. Chhetri, F. H. Johnsen, M. Konoshima, and A. Yoshimoto, "Community forestry in the hills of Nepal: Determinants of user participation in forest management," *Forest Policy and Economics*, vol. 30, pp. 6–13, 2013.

[26] P. Newton, J. A. Oldekop, G. Brodnig, B. K. Karna, and A. Agrawal, "Carbon, biodiversity, and livelihoods in forest commons: synergies, trade-offs, and implications for REDD+," *Environmental Research Letters*, vol. 11, no. 4, Article ID 044017, 2016.

[27] UN-REDD Programme, *Background Analysis of REDD Regulatory Frameworks*, Forest Degradation in Developing Countries, Geneva, Switzerland, 2009.

[28] A. Angelsen, M. Brockhaus, M. Kanninen, E. Sills, W. D. Sunderlin, and S. Wertz-Kanounnikoff, Eds., *Realising REDD+: National Strategy and Policy Options*, CIFOR, Bogor, Indonesia, 2009.

[29] J. W. den Besten, B. Arts, and P. Verkooijen, "The evolution of REDD+: an analysis of discursive-institutional dynamics," *Environmental Science & Policy*, vol. 35, pp. 40–48, 2014.

[30] T. Treue, Y. M. Ngaga, H. Meilby et al., "Does participatory forest management promote sustainable forest utilisation in Tanzania?" *International Forestry Review*, vol. 16, no. 1, pp. 23–38, 2014.

[31] B. P. Sharma, M. Nepal, B. S. Karky, S. K. Pattanayak, and P. Shyamsundar, "Baseline considerations in designing REDD+ pilot projects: evidences from nepal," *SANDEE Working Paper*, 2015.

[32] E. Sungusia and J. F. Lund, "Against all policies: landscape level forest restoration in Tanzania," *World Development Perspectives*, vol. 3, pp. 35–37, 2016.

[33] K. John, D. S. A. Silayo, and A. Vatn, "The cost of managing forest carbon under REDD+ initiatives: a case of kolo hills forests in kondoa district, dodoma, tanzania," *International Journal of Forestry Research*, 12 pages, 2014.

[34] A. Angelsen, *How do we set the reference levels for REDD payments? In Moving ahead with REDD: issues, options and implications*, Center for International Forestry Research (CIFOR), Bogor, Indonesia, 2008.

[35] C. Luttrell, L. Loft, M. F. Gebara et al., "Who should benefit from REDD+? Rationales and realities," *Ecology and Society*, vol. 18, no. 4, article 52, 2013.

[36] L. Peskett and K. Todd, *Putting REDD+ Safeguards and Safeguard Information Systems into Practice. Policy Brief*, vol. 3, UN-REDD, Geneva, Switzerland, 2013.

[37] K. MacDicken, "A guide to monitoring carbon storage in forestry and agroforestry projects," Tech. Rep. 1611 N, Winrock International Institute for Agricultural, 1997.

[38] MNRT, *National Forestry Resources Monitoring and Assessment (NAFORMA) of Tanzania, Field Manual, Biophysical Survey*, Ministry of Natural Resources & Tourism, Dar Es Salaam, Tanzania, 2010.

[39] IPCC, *Good Practice Guidance for Land-Use Changes and Forestry*, Institute of Environmental Strategies, Kanagawa, Japan, 2003.

[40] W. A. Mugasha, T. Eid, O. M. Bollandsås et al., "Allometric models for prediction of above- and belowground biomass of trees in the miombo woodlands of Tanzania," *Forest Ecology and Management*, vol. 310, pp. 87–101, 2013.

[41] E. W. Mauya, W. A. Mugasha, E. Zahabu, O. M. Bollandsås, and T. Eid, "Models for estimation of tree volume in the miombo woodlands of Tanzania," *Southern Forests*, vol. 76, no. 4, pp. 209–219, 2014.

[42] Peters-Stanley, "Back to the future: state of the Voluntary carbon markets," Forest Trends, Ecosystem Marketplace Report, 2012.

[43] M.-G. Piketty, I. Drigo, P. Sablayrolles, E. A. De Aquino, D. Pena, and P. Sist, "Annual cash income from community forest management in the Brazilian Amazon: Challenges for the future," *Forests*, vol. 6, no. 11, pp. 4228–4244, 2015.

[44] R. Perman, Y. Ma, M. Common, D. Maddison, and J. McGilvray, *Natural Resource and Environmental Economics*, Pearson, Edinburgh, UK, 4th edition, 2011.

[45] E. E. Mwakalukwa, H. Meilby, and T. Treue, "Floristic composition, structure, and species associations of dry miombo woodland in Tanzania," *ISRN Biodiversity*, vol. 2014, Article ID 153278, 15 pages, 2014.

[46] F. K. Kalaba, C. H. Quinn, A. J. Dougill, and R. Vinya, "Floristic composition, species diversity and carbon storage in charcoal and agriculture fallows and management implications in Miombo woodlands of Zambia," *Forest Ecology and Management*, vol. 304, pp. 99–109, 2013.

[47] P. G. H. Frost, "The ecology of miombo woodlands," in *The Miombo in Transition: Woodlands and Welfare in Africa*, pp. 11–57, Center for International Forest Research (CIFOR), Bongor, Indonesia, 1996.

[48] J. Pelletier, N. Gélinas, and M. Skutsch, "The place of community forest management in the REDD+ landscape," *Forests*, vol. 7, no. 8, article 170, pp. 1–24, 2016.

[49] D. J. Sonwa, J. H. Nlom, and S. G. Neba, "Valuation of forest carbon stocks to estimate the potential for result-based payment under REDD+ in Cameroon," *International Forestry Review*, vol. 18, supplement 1, pp. 119–129, 2016.

[50] URT, *Tanzania's National REDD-Redness Programme*, Division of Environment, Vice-President's Office, Dar es Salaam, Tanzania, 2009.

[51] J. C. Ribot, "Forestry and democratic decentralization in sub-Saharan Africa. A rough review," in *Governing Africas Forests in a Globalized Word*, L. A. German, A. Karsenty, and A. M. Tiani, Eds., pp. 29–55, Center for International Forestry Research, Earth Scan, London, UK, 2010.

19

Logging Activity Adversely Impacts Primate Diversity and Density in the Kwabre Rainforest of Ghana

Emmanuel Danquah and Elvis Hackman Tetteh

Faculty of Renewable Natural Resources, College of Agriculture and Natural Resources, Kwame Nkrumah University of Science and Technology, Kumasi, Ghana

Correspondence should be addressed to Emmanuel Danquah; emmanueldanquah@gmail.com

Academic Editor: Ram Chander Sihag

Knowledge on the impacts of logging activity on inhabitant primate species in Kwabre Rainforest, Ghana, is vital for the development of a comprehensive conservation and management plan. With this background, primate density and diversity were recorded along line transects in logged and unlogged areas (strata) to assess the impact of logging activity on these parameters. Six distinct primate species were confirmed including Roloway monkey (*Cercopithecus roloway*, listed as endangered in the IUCN List of Threatened Species), white-naped mangabey (*Cercocebus lunulatus*, vulnerable), and Geoffroy's black-and-white colobus (*Colobus vellerosus*, vulnerable). There was a significant difference (Mann-Whitney U test: $U = 36.0$, $p < 0.01$) in primate encounter rates between the logged and unlogged strata with higher species diversity in unlogged stratum ($H = 2.91$) compared to the logged stratum ($H = 1.44$). Regression analysis indicated a significant effect ($r^2 = 0.945$, $p < 0.01$) of logging on primate encounter rates. Our results suggest that logging activity can alter composition of primate communities. One option to forestall further forest degradation and its adverse effects on primates would be to grant the Kwabre Rainforest protected area status under Ghanaian law and manage it under an integrated conservation plan that includes neighbouring Ankasa Conservation Area in Ghana and Tanoé Forest in Cote d'Ivoire.

1. Introduction

The ability of many wildlife species to thrive in densely human populated areas depends largely on their level of tolerance for anthropogenic disturbances like deforestation and habitat fragmentation [1, 2]. Primates are believed to be suitable biological indicators for assessing the effects of logging on wildlife habitats due to their strong connection with forest cover and vegetative complexity [3–5]. Apart from their responses to changes in vegetation structure which makes them good indicators, primates were selected over other species because their habitat requirements are better known and can be sampled with relatively simple methods like transects [6, 7].

In Ghana, the Kwabre Rainforest is a unique community-owned forest and it is reported to host several primate species of international conservation importance, including the Roloway monkey (*Cercopithecus roloway*), white-naped mangabey (*Cercocebus lunulatus*), and Geoffroy's black-and-white colobus (*Colobus vellerosus*) [8, 9]. These species are listed on the International Union for the Conservation of Nature's (IUCN) Red List of Threatened Species [10] and Appendix I of the Convention on International Trade in Endangered Species (CITES) [11, 12]. Unfortunately, the Kwabre Rainforest is exposed to an increasing risk of deforestation from illegal logging and farming activity. These activities, when unchecked, may adversely affect several species, particularly canopy dwelling primates. Existing studies of the impact of anthropogenic forest disturbance on primate populations have provided valuable insights into (i) effects of selective logging on primate density [13–18]; (ii) the role of habitat disturbances in diet selection and primate abundance [14, 19]; and (iii) behavioural and physiological responses of primates to habitat alteration [20, 21]. In Ghana, no one has yet undertaken a quantitative assessment of the impacts of logging on primate abundance. This information is however vital to understand the sensitivities and habitat requirements of the species in order to develop an informed management plan.

The aim of this study was to examine the impact of logging activity on primate species in the Kwabre Rainforest in western Ghana. To meet this aim, censuses were conducted to (1) compare primate density and species composition between the logged and unlogged strata and (2) evaluate the effect of logging intensity on primate encounter rates. The null hypothesis predicts no significant difference in primate density and species composition between logged and unlogged areas. Conclusions drawn will contribute to managing primate conservation projects in this region and help to dictate future logging programmes in other rainforests. This study provides three important advantages over the earlier studies. One, it focuses on different groups of primate species, rather than just one species [22–24]. Two, it incorporates in its design a control area (an unlogged stratum) that was completely without any logging activity. This factor allowed a fairly objective assessment of the relative effects of logging activity on different primate species. Lastly, the study has both local and regional relevance, because the study area is a potentially key corridor between two designated national parks in Ghana and neighbouring Cote d'Ivoire, and this region of southwestern Ghana could be one of the most important areas for wildlife protection in West Africa.

2. Materials and Methods

2.1. Study Area. The Kwabre Rainforest, Ghana (50 km^2; lat. 5°06′–5°16′N; long. 2°40′–2°47′W; Figure 1), is a community-owned swamp forest in the Western Region of Ghana and forms part of the Upper Guinean Forest Block. It is located close to the Ghana-Cote d'Ivoire international border, opposite to the Tanoé Forest. The area forms part of a network of forest reserves in southwestern Ghana which constitute Ghana's portion of the proposed transboundary elephant migration corridor between Ghana and Ivory Coast [25]. In addition, the Kwabre Rainforest makes up a crucial corridor for wildlife between the Ankasa Conservation Area in Ghana and the Tanoé community-managed forest in Cote d'Ivoire [9] (Figure 2). The area also forms part of the range of several flagship endangered primates, including the crowned mangabey, Roloway monkey, and Miss Waldron's red colobus, now considered extinct by primatologists [9, 26].

Selective logging of timber species by chainsaw operators and bushmeat hunting perpetrated by people coming from outside the local communities is destroying the original pristine habitat surrounding the Kwabre Rainforest area. Most of the logged areas are easily accessible from the main Takoradi-Elubo highway (Figure 1). The area is also exposed to an increasing risk of deforestation from farming for cash crops including rubber, palm oil, and cocoa, logging and mining activity, expanding fringe communities, and human encroachment on forest resources. The recent discovery of crude oil off the coast of Western Region with its attendant high local immigration as a result of the influx of job seekers has increased the human population density in the area.

2.2. Data Collection. The study area was classified into logged and unlogged strata (Figure 1) based on an initial reconnaissance to map out the presence or absence of logging activity. Primate surveys were conducted in the two strata using line transects. A total of six 500 m line transects (3 km) were established in each stratum. Transects were monitored consecutively over a survey period of two weeks in February 2014. The survey period comprised alternating each field day with a nonfield day in order to stagger disturbance of survey team on primate behaviour. Thus, a total of seven field days were realized over the survey period, resulting in a total of 21 km of transects walked in each stratum for the entire study area.

The starting point of each transect was determined systematically on a gridded map of the study area. The average distance between any two line transects was 3 km. Line transect surveys began early in the morning around 6:30 a.m. and lasted on average for 5-6 hours. Primate calls were used to access the presence and type of primate species in the vicinity of the transects. However, only direct sightings of primates were recorded. All sightings where estimation of group size was difficult to assess were disregarded. When primates were encountered, the number of individuals (group size), species, time, GPS (Garmin GPSmap 62) location, and perpendicular distance to primate(s) were recorded. The distance along transects was measured with a tape measure. The perpendicular ground distance from the transect center line to an estimated central point directly beneath where the primates were initially sighted was also measured using a tape measure. Binoculars were used when necessary to confirm species identity. An index of logging activity per km was generated by recording the number of logged trees (tree stumps) observed along each transect.

2.3. Data Analysis. Primate density was calculated using DISTANCE Version 4.1 software package [27]. The best model selection was based on the values of Akaike's Information Criterion (AIC) and recommendations by [28]. Encounter rates were calculated in the form of Kilometric Indices of Abundance (KIA), where the number of recorded primates and indicators of logging incidence are presented according to the distance walked during the survey. The Mann-Whitney U test was used to test for any significant difference in primate encounter rates between the logged and unlogged strata. Primate species richness and diversity for each stratum was calculated using EstimateSWin800 Version 8.0.0 software package [29]. Bootstrap Richness Estimator was used for species richness and Shannon-Weiner index for species diversity.

The effect of logging intensity on primate encounter rates was tested using regression analysis in Canoco version 4.5 [30]. Statistical analyses using the program Spatial Analysis in Macroecology [31] were conducted to correct for normality and spatial autocorrelation.

3. Results

3.1. Conservation Status of Primates at Kwabre Forest

Six Species of Diurnal Primates. Roloway monkey (*Cercopithecus roloway*), white-naped mangabey (*Cercocebus lunulatus*), Geoffroy's black-and-white colobus (*Colobus vellerosus*), olive colobus (*Procolobus verus*), lesser spot-nosed monkey (*Cercopithecus petaurista*), and Lowe's monkey (*Cercopithecus lowei*) were recorded in the survey (Table 1).

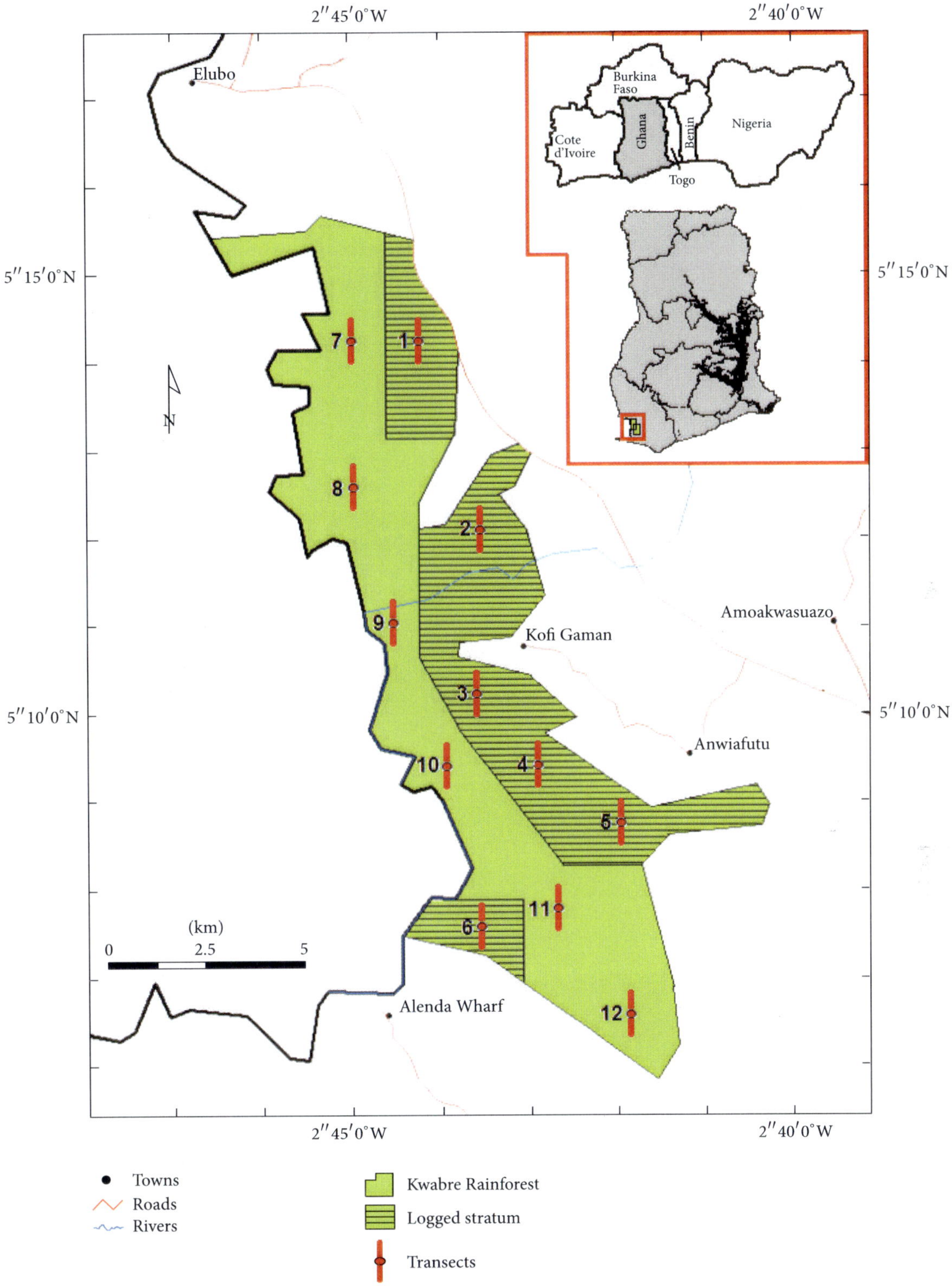

FIGURE 1: Map of Kwabre Rainforest showing the logged stratum.

Roloway monkey is listed as an endangered species on the IUCN Threatened Species List, white-naped mangabey and Geoffroy's black-and-white colobus are listed as vulnerable species, and olive colobus is listed as near-threatened species [10]. The spot-nosed monkey and Lowe's monkey are considered to be of least concern and are of local conservation importance (Schedule 1) under Ghanaian law (L.I. 1284-Wildlife Conservation (Amendment) Regulations, 1983).

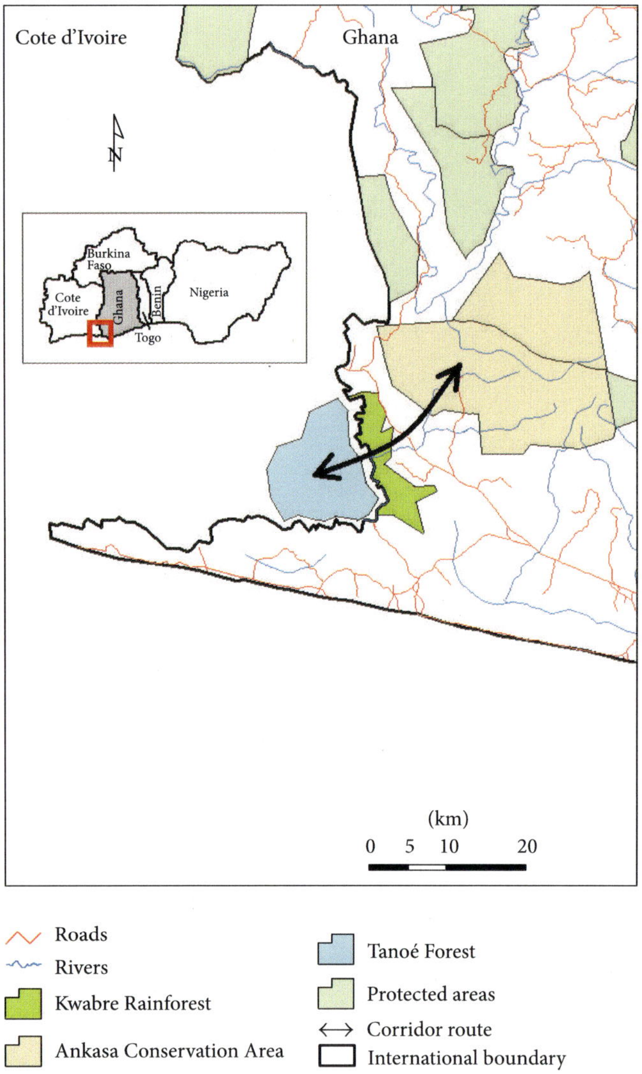

FIGURE 2: Map of the proposed transboundary elephant migration corridor between Ghana and Ivory Coast.

TABLE 1: IUCN status of primate species found in the Kwabre Rainforest.

Common name	Scientific name	IUCN status
Roloway monkey	*Cercopithecus roloway*	EN
White-naped mangabey	*Cercocebus lunulatus*	VU
Black-and-white colobus	*Colobus vellerosus*	VU
Olive colobus	*Procolobus verus*	NT
Spot-nosed monkey	*Cercopithecus petaurista*	LC
Lowe's monkey	*Cercopithecus lowei*	LC

EN: endangered; VU: vulnerable; NT: near threatened; LC: least concern.

3.2. Primate Diversity. Primate species composition varied across the logged and unlogged strata but was not significantly different (Mann-Whitney U test: $U = 36.0$, $p > 0.05$). The unlogged stratum was found to be more diverse than logged stratum when various richness and diversity indices were applied to the data (Table 2). Comparably, these indices were very low in the logged stratum.

TABLE 2: Primate species diversity index generated by EstimateSWin800.

Stratum	Shannon-Weiner mean	Standard deviation
Logged	1.44	0.22
Unlogged	2.91	0.16

TABLE 3: Pooled primate density parameters estimated for the logged and unlogged strata.

Density parameters	Logged stratum	Unlogged stratum
Density (km^{-2})	349.93	883.68
Percent coefficient of variation [CV (%)]	22.39	10.64
Upper confidence limit (CL)	205.29	685.17
Lower confidence limit (CL)	596.47	1139.70

None of the species were unique to any of the defined logging categories, but the generally higher primate species richness in the unlogged stratum in comparison to logged stratum may reflect primates' general adaptation to mature forest systems with taller trees and denser canopy covers.

3.3. Primate Density. Primate density was significantly lower (less than 50%) in the logged stratum compared to the unlogged stratum (Table 3). The half normal + cosine model without truncation gave the best fit to the data. Note that DISTANCE produces estimates with asymmetrical confidence limits.

There was also a significant difference in the individual primate encounter rates (Mann-Whitney U test: $U = 36.0$, $p < 0.01$) across the two categories of strata, that is, logged and unlogged (Table 4). The percentage of individual primates encountered was considerably greater in the unlogged (73%) than in logged (27%) stratum; however, primate group sizes for the same species were similar for both strata (Table 4). All species apart from the black-and-white colobus showed greater densities in the unlogged stratum.

3.4. Relationship between Primate Density and Logging Intensity. The number of primates encountered was inversely related to logging intensity ($r^2 = 0.945$, $p < 0.01$). The highest encounter rates of primates were observed in slightly logged or unlogged forests (Figure 3).

4. Discussion

The results suggest that logging could negatively influence primate communities through the modification of primate density and species composition. Therefore, the null hypothesis of no significant difference in primate density and species composition between logged and unlogged areas is rejected. Such inference must, however, be made cautiously as logging

TABLE 4: Encounter rates of primates and logging intensity in the Kwabre Rainforest.

Stratum/transects	Lowe's monkey	Spot-nosed monkey	White-naped mangabey	Roloway monkey	Olive colobus	Black-and-white colobus	Total primates	Primates per km	Logging intensity per km
Logged									
1	0 (0)	0 (0)	0 (0)	0 (0)	0 (0)	0 (0)	0	0	10
2	1 (11)	1 (8)	0 (0)	0 (0)	0 (0)	2 (6, 5)	30	9	4
3	0 (0)	1 (6)	0 (0)	0 (0)	1 (3)	3 (6, 5, 4)	24	7	3
4	1 (8)	0 (0)	1 (5)	0 (0)	0 (0)	1 (7)	20	6	7
5	2 (7, 9)	0 (0)	0 (0)	0 (0)	1 (4)	1 (6)	26	7	6
6	1 (6)	0 (0)	0 (0)	1 (6)	1 (3)	2 (6, 7)	28	6	5
Mean	8.2	7.0	5.0	6.0	3.3	5.7	25.6	7.0	5.8
Total	*41*	*14*	*5*	*6*	*10*	*52*	128	35	*35*
Unlogged									
7	2 (12, 9)	1 (10)	1 (5)	0 (0)	2 (4, 3)	1 (7)	50	14	0
8	3 (10, 9, 11)	2 (9, 11)	1 (6)	1 (6)	1 (4)	0 (0)	66	19	0
9	1 (7)	2 (7, 9)	1 (3)	0 (0)	1 (5)	1 (5)	36	10	0
10	2 (9, 8)	1 (8)	1 (4)	2 (4, 3)	2 (3, 3)	2 (6, 5)	53	15	0
11	4 (11, 8, 12, 6)	3 (8, 6, 9)	0 (0)	1 (5)	1 (4)	1 (7)	76	22	0
12	2 (6, 8)	3 (11, 9, 7)	2 (5, 3)	1 (4)	1 (3)	1 (5)	61	17	0
Mean	9.0	8.6	4.3	4.4	3.6	5.8	57.0	16	0
Total	*126*	*104*	*26*	*22*	*29*	*35*	*342*	*97*	*0*

Numbers outside brackets indicate the number of primate groups and numbers in brackets indicate group sizes and logged trees per km.

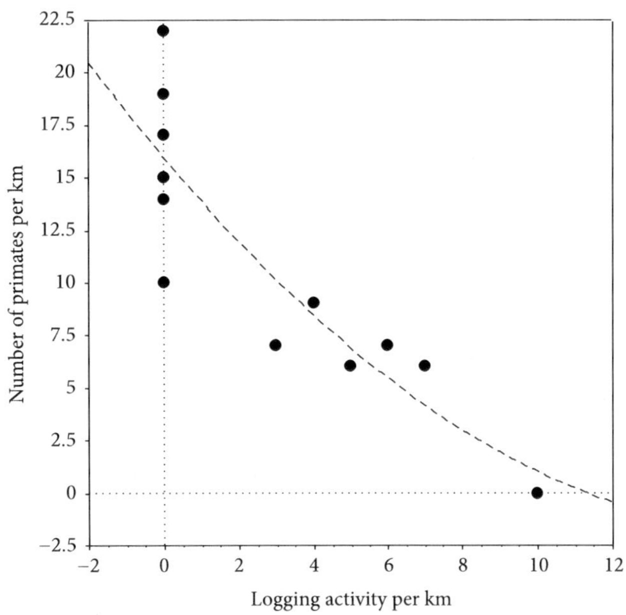

$Y = 15.974 - 2.137 * X + .064 * X^2; R^2 = .769$

FIGURE 3: Relationship between encounter rates of primates and logging intensity.

may not be the only factor influencing primate species composition and density. Although both strata possess similar topography and vegetation type, logged areas had recent logging disturbances that resulted in reduced canopy cover and denser understory cover. A factor that may have also influenced our findings could be the proximity of some fringe communities to the forest edge. During the study, it was observed that logged areas were closer to fringe communities and farms, making such areas more susceptible to human disturbances such as poaching [17, 32]. Some of these differences in human and ecological variables may have influenced our findings. One reason that may have accounted for higher numbers of black-and-white colobuses in the logged stratum is that they are folivores [14] and as such may show preference for degraded areas where the open canopy permits growth of diverse understory. Apart from changes in density and diversity, logging might have also affected the behaviour of primates [13, 20, 33]. For instance, primates exhibited extreme shyness including restricting calls in the logged stratum compared to the unlogged stratum when teams encountered primate groups. The apparent differences in the incidence of alarm calls between groups of monkeys encountered between the two strata could affect our census findings to a small extent. However, previous studies on primate surveys [14, 24, 34] have shown that many replications are needed at each transect before the group density estimates can be considered reliable. In the current study, each transect was sampled more than once and this could increase the reliability of our results.

Our study draws attention to matters of both national and international conservation concern. The evidence of unauthorized human activities recorded during this study indicates clearly that the community-based management system practiced at Kwabre Rainforest may not be effective in preventing poaching and other illegal human activities in the forest. If current threats continue unabated, then rapid population decline of primates is expected in the near future.

Given its considerably intact forest cover, Ghana is in a position to implement important forest conservation initiatives in the subregion [35, 36]. Within Ghana, the general vicinity of the Kwabre Rainforest is very significant from a regional viewpoint because it constitutes part of a network of forest reserves in southwestern Ghana which constitutes Ghana's portion of the proposed crucial corridor for wildlife between the Ankasa Conservation Area in Ghana and the Tanoé Forest in Cote d'Ivoire [9].

In this regional setting, the future management of the Kwabre corridor could be very important. At present, farms and villages are flourishing along the Kwabre corridor, rapidly increasing local human pressure. Concerns about the future of Kwabre corridor are particularly important because interest to protect wildlife in the area has waned and Kwabre Rainforest could face a sudden increase in hunting, logging, and slash-and-burn farming. Instead of acting as a community forest buffer zone and possible faunal corridor between two key international protected areas (Ankasa Conservation Area in Ghana and Tanoé Forest in neighbouring Cote d'Ivoire, Figure 2), the larger Kwabre corridor area could instead become a major population sink for wildlife [13, 37, 38]. This would reinforce adverse edge effects and genetic isolation of the protected areas, increasing the probability of local extinctions of species of conservation interest [39, 40].

5. Conclusions and Recommendations

Kwabre Rainforest is an important site where careful wildlife management can support major biodiversity conservation in Ghana; hence, more effort should be channeled into preventing further habitat degradation of the larger Kwabre Rainforest corridor area. One possibility would be to grant the Kwabre Rainforest protected area status under Ghanaian law and to manage it under the umbrella of an integrated conservation plan that includes its two adjacent international protected areas. Such a plan can be achieved by designating the entire area as a biosphere reserve, which would permit planned development outside reserves while affording high protection for protected areas [41]. With effective long-term management, this region of southwestern Ghana could remain one of the most important areas for wildlife in West Africa.

Competing Interests

The authors declare that there are no competing interests regarding the publication of this paper.

Acknowledgments

Many thanks are due to the traditional head and people of Kwabre.

References

[1] B. Czech, P. R. Krausman, and P. K. Devers, "Economic associations among causes of species endangerment in the United States," *BioScience*, vol. 50, no. 7, pp. 593–601, 2000.

[2] D. S. Wilcove, C. H. Mclellan, and A. P. Dobson, "Habitat fragmentation in the temperate zone," in *Conservation Biology: The Science of Scarcity and Species Richness*, M. E. Soule, Ed., pp. 237–256, Sinauer Associates, Sunderland, Mass, USA, 1998.

[3] A. G. Chiarello, "Density and population size of mammals in remnants of Brazilian Atlantic forest," *Conservation Biology*, vol. 14, no. 6, pp. 1649–1657, 2000.

[4] R. K. Laidlaw, "Effects of habitat disturbance and protected areas on mammals of Peninsular Malaysia," *Conservation Biology*, vol. 14, no. 6, pp. 1639–1648, 2000.

[5] M. A. Lopes and S. F. Ferrari, "Effects of human colonization on the abundance and diversity of mammals in eastern Brazilian Amazonia," *Conservation Biology*, vol. 14, no. 6, pp. 1658–1665, 2000.

[6] E. Carrillo, G. Wong, and N. A. D. Cuarón, "Monitoring mammal populations in Costa Rican protected areas under different hunting restrictions," *Conservation Biology*, vol. 14, no. 6, pp. 1580–1591, 2000.

[7] M. Conroy and D. Nichols, "Designing a study to assess mammalian diversity," in *Measuring and Monitoring Biological Diversity. Standard Methods for Mammals*, D. Wilson, R. Cole, J. Nichols, R. Rudran, and M. Foster, Eds., pp. 41–49, Smithsonian Institution Press, London, UK, 1996.

[8] WAPCA, *Annual Report. West African Primate Conservation Action*, WAPCA, Accra, Ghana, 2012.

[9] D. Osei, R. H. Horwich, and J. M. Pittman, "First sightings of the Roloway Monkey (*Cercopithecus diana* roloway) in Ghana in ten years and the status of other endangered primates in Southwestern Ghana," *African Primates*, vol. 10, pp. 25–40, 2015.

[10] The IUCN Red List of Threatened Species, Version 2015.2, 2015, http://www.iucnredlist.org/.

[11] W. S. McGraw and J. F. Oates, "Roloway monkey Cercopithecus diana roloway (Schreber, 1774)," in *Primates in Peril: The World's Most Endangered Primates 2012–2014*, C. Schwitzer, R. A. Mittermeier, A. B. Rylands et al., Eds., pp. 14–16, IUCN SSC Primate Specialist Group (PSG), International Primatological Society (IPS), Conservation International (CI), and Bristol Zoological Society, Arlington, Va, USA, 2014.

[12] R. A. Mittermeier, J. Ratsimbazafy, A. B. Rylands et al., "Primates in peril: the world's 25 most endangered primates, 2006–2008," *Primate Conservation*, vol. 22, pp. 1–40, 2007.

[13] D. Swift, "The consequences of logging on primate density and diversity in Kibale National Park, Uganda," *Bioscience Horizons*, vol. 5, pp. 1–6, 2012.

[14] C. A. Chapman, L. J. Chapman, L. Naughton-Treves, M. J. Lawes, and L. R. Mcdowell, "Predicting folivorous primate abundance: validation of a nutritional model," *American Journal of Primatology*, vol. 62, no. 2, pp. 55–69, 2004.

[15] J. P. Skorupa, *The effects of selective timber harvesting on rain-forest primates in Kibale Forest, Uganda [Ph.D. thesis]*, University of California, Davis, Davis, Calif, USA, 1988.

[16] T. T. Struhsaker, *Ecology of an African Rainforest: Logging in Kibale and the Conflict between Conservation and Exploitation*, University Press of Florida, Gainesville, Fla, USA, 1997.

[17] P. Mayor, P. Pérez-Peña, M. Bowler, P. E. Puertas, M. Kirkland, and R. Bodmer, "Effects of selective logging on large mammal populations in a remote indigenous territory in the northern peruvian amazon," *Ecology and Society*, vol. 20, no. 4, pp. 36–44, 2015.

[18] A. J. Plumptre and V. Reynolds, "The effect of selective logging on the primate populations in the Budongo Forest Reserve, Uganda," *Journal of Applied Ecology*, vol. 31, no. 4, pp. 631–641, 1994.

[19] C. Mammides, M. Cords, and M. K. Peters, "Effects of habitat disturbance and food supply on population densities of three primate species in the Kakamega Forest, Kenya," *African Journal of Ecology*, vol. 47, no. 1, pp. 87–96, 2009.

[20] C. Schwitzer, L. Glatt, K. A.-I. Nekaris, and J. U. Ganzhorn, "Responses of animals to habitat alteration: an overview focussing on primates," *Endangered Species Research*, vol. 14, no. 1, pp. 31–38, 2011.

[21] R. Rimbach, A. Link, M. Heistermann, C. Gómez-Posada, N. Galvis, and E. W. Heymann, "Effects of logging, hunting, and forest fragment size on physiological stress levels of two sympatric ateline primates in Colombia," *Conservation Physiology*, vol. 1, no. 1, Article ID cot031, pp. 1–11, 2013.

[22] R. F. W. Barnes, K. L. Barnes, M. P. T. Alers, and A. Blom, "Man determines the distribution of elephants in the rain forests of northeastern Gabon," *African Journal of Ecology*, vol. 29, no. 1, pp. 54–63, 1991.

[23] R. F. W. Barnes, K. Beardsley, F. Michelmore, K. L. Barnes, M. P. T. Alers, and A. Blom, "Estimating forest elephant numbers with dung counts and a geographic information system," *Journal of Wildlife Management*, vol. 61, no. 4, pp. 1384–1393, 1997.

[24] S. A. Lahm, R. F. W. Barnes, K. Beardsley, and P. Cervinka, "A method for censusing the greater white-nosed monkey in northeastern Gabon using the population density gradient in relation to roads," *Journal of Tropical Ecology*, vol. 14, no. 5, pp. 629–643, 1998.

[25] L. Sebogo and R. F. W. Barnes, "Action plan for the management of trans-frontier elephant conservation corridors in West Africa," Tech. Rep., IUCN/AfESG, Ouagadougou, Burkina Faso, 2003.

[26] J. F. Oates, M. Abedi-Lartey, W. S. McGraw, T. T. Struhsaker, and G. H. Whitesides, "Extinction of a West African red colobus monkey," *Conservation Biology*, vol. 14, no. 5, pp. 1526–1532, 2000.

[27] L. Thomas, J. L. Laake, S. Strindberg et al., *Distance 4.0, Release 1. Research Unit for Wildlife Population Assessment*, University of St Andrews, St Andrews, Scotland, 2002.

[28] S. T. Buckland, D. R. Anderson, K. P. Burnham, J. L. Laake, D. L. Borchers, and L. Thomas, *Introduction to Distance Sampling: Estimating Abundance of Biological Populations*, Oxford University Press, Oxford, UK, 2001.

[29] R. K. Colewell, *EstimateS: Statistical Estimation of Species Richness and Shared Species from Samples*, Version 8, 2006, http://purl.oclc.org/estimates.

[30] C. J. F. Ter Braak and P. Imilauer, *Canoco Reference Manual and CanoDraw for Windows User's Guide: Software for Canonical Community Ordination (Version 4.5)*, Microcomputer Power, Ithaca, NY, USA, 2002.

[31] T. F. L. V. B. Rangel, J. A. F. Diniz-Filho, and L. M. Bini, "Towards an integrated computational tool for spatial analysis in macroecology and biogeography," *Global Ecology and Biogeography*, vol. 15, no. 4, pp. 321–327, 2006.

[32] W. F. Laurance, B. M. Croes, L. Tchignoumba et al., "Impacts of roads and hunting on central African rainforest mammals," *Conservation Biology*, vol. 20, no. 4, pp. 1251–1261, 2006.

[33] C. A. Chapman, S. R. Balcomb, T. R. Gillespie, J. P. Skorupa, and T. T. Struhsaker, "Long-term effects of logging on African primate communities: a 28-year comparison from Kibale National Park, Uganda," *Conservation Biology*, vol. 14, no. 1, pp. 207–217, 2000.

[34] S. A. Lahm and J. P. Tezi, "Assessment of the communities of medium-sized and large arboreal and terrestrial mammals in the Rabi-Toucan region of the Ngove-Ndogo Hunting Domain and southwestern Loango National Park," *Bulletin of the Biological Society of Washington*, vol. 12, pp. 383–416, 2006.

[35] W. F. Laurance, A. Alonso, M. Lee, and P. Campbell, "Challenges for forest conservation in Gabon, Central Africa," *Futures*, vol. 38, no. 4, pp. 454–470, 2006.

[36] A. Kamdem-Toham, A. W. Adeleke, N. D. Burgess et al., "Forest conservation in the Congo Basin," *Science*, vol. 299, no. 5605, article 346, 2003.

[37] A. J. Novaro, K. H. Redford, and R. E. Bodmer, "Effect of hunting in source-sink systems in the Neotropics," *Conservation Biology*, vol. 14, no. 3, pp. 713–721, 2000.

[38] C. A. Peres, "Synergistic effects of subsistence hunting and habitat fragmentation on amazonian forest vertebrates," *Conservation Biology*, vol. 15, no. 6, pp. 1490–1505, 2001.

[39] R. Woodroffe and J. R. Ginsberg, "Edge effects and the extinction of populations inside protected areas," *Science*, vol. 280, no. 5372, pp. 2126–2128, 1998.

[40] W. F. Laurance, T. E. Lovejoy, H. L. Vasconcelos et al., "Ecosystem decay of Amazonian forest fragments: a 22-year investigation," *Conservation Biology*, vol. 16, no. 3, pp. 605–618, 2002.

[41] A. Kaus, "Environmental perceptions and social relations in the Mapimí Biosphere Reserve," *Conservation Biology*, vol. 7, no. 2, pp. 398–406, 1993.

Morphological Variations in *Tamarindus indica* LINN. Fruits and Seed Traits in the Different Agroecological Zones of Uganda

Jaspher Okello,[1] **John Bosco Lamoris Okullo**,[2] **Gerald Eilu**,[2] **Philip Nyeko**,[2] **and Joseph Obua**[2]

[1]*National Agricultural Research Organisation, P.O. Box 295, Entebbe, Uganda*
[2]*School of Forestry, Environmental and Geographical Sciences, Makerere University, P.O. Box 7062, Kampala, Uganda*

Correspondence should be addressed to Jaspher Okello; jaspherokello2016@gmail.com

Academic Editor: Daniel I. Rubenstein

An investigation was carried out on variations in the morphological traits of *Tamarindus indica* LINN. fruits (length, breadth, mass, and pulp mass) and seeds (number and mass) from the different agroecological zones and land use types of Uganda. Fruits were collected from the two land use types in the three agroecological zones and measured for various morphological traits. The study sites were located between 593 and 1,096 meters above sea level. ANOVA was used to test the differences in morphological traits of fruits and seeds between agroecological zones and land use types. The morphological traits relationship was determined using Pearson Correlation Coefficient (R). There were significant ($P \leq 0.05$) variations among the *T. indica* fruits and seed morphological traits within the agroecological zones and land use types. Lake Victoria Crescent agroecological zone recorded higher fruits morphological traits values. Wild and on-farm land use types were superior in fruits and seed traits, respectively. Many seeds per fruit (17) were recorded in Uganda (Eastern agroecological zone) than those recorded elsewhere. Significant uphill positive linear correlations between all morphological traits ($P \leq 0.05$) were observed, with the strongest relationships being between seed mass and seed number ($R=0.79$), fruit mass and fruit length ($R=0.75$), pulp mass and fruit mass ($R=0.73$), and seed mass and fruit mass ($R=0.73$). However, fruit breadth presented slightly weaker positive linear correlations with all other morphological traits. This is the first quantitative evaluation of *T. indica* morphological traits variation in Uganda. Dispersion/variation and correlation relationships suggest that all the studied morphological traits can be used for selection of plus trees for tree breeding improvement such as yield per tree. The observed variations are probably attributed to influence of agroecological zones' factors, environmental factors, climate, land use types, and farming systems, a reflection of *T. indica* adaptation to different conditions showing high genetic and phenotypic differences to be exploited.

1. Introduction

Tamarindus indica LINN. (Linnaeus 1753) is a fruit tree species that belongs to the Order FABALES, family FABACEAE syn. LEGUMINACEAE, subfamily CAESALPINIOIDEAE, and Genus TAMARINDUS. The genus Tamarindus is monotypic, containing a single species, *Tamarindus indica* [1]. Commonly, *Tamarindus indica* is known by the following local names: English names (*Madeira mahogany, Indian date,* and *tamarind*); trade name (*tamarind*); Ateso, Eastern Uganda (*Epeduru*); Luo, Northern Uganda (*cwao*); Luganda, Lake Victoria Crescent zone (*mukoge*), as well as Lugbara and Madi dialects of West Nile (*Iti*) in Uganda. The *T. indica* tree is widespread throughout the tropics and subtropics. It originated from tropical Africa [2] and native to many tropical African countries including Uganda, where it naturally dominates six out of nine agroecological zones. The fruit tree is naturally widespread on-farms and in the wild habitats in the tropics [1] as well as in Uganda [3] and grows well over a wide range of soil and climatic conditions, ranging from low altitude woodland savanna of about 590 masl (lowest point in Uganda at Albert Nile) to high altitude Mt. Elgon ranges of above 2,000 masl [4].

The ecological range of *T. indica* is very wide, corresponding to the polymorphous character of the species. According to Okello et al. [3–5], the species is frequently associated with shrubs, deciduous thickets, termite mounds, agricultural landscapes, abandoned homesteads (settlements), and public places. While El-Siddig et al. [1] reported that the *T. indica* fruits from different regions of Sudan are morphologically different, Gebauer et al. [6] reported that in some regions of Sudan, *T. indica* fruits are sweeter than others. The variations in size, shape, flavor, and colour of fruits and seeds have also been reported in many other studies elsewhere [7–9]. The multipurpose long-lived fruit tree that supports the livelihoods of many poor rural communities is useful for meeting cultural, social, economic, ecological, and many other significant needs [4]. The species provides households' food security, timber, nutritional needs, and treatment and prevention of diseases (meningitis, fever, cough, intestinal disorders) and combine well with cereal and perennial crops on-farms. In on-farms, *T. indica* provides shade for human, animals, and crops; ornamental and avenue tree; windbreak or shelterbelt; soil fertility improvement (mulch and N fixation), animal fodder in zero grazing systems, and energy needs (firewood and charcoal).

The *T. indica* can serve as good natural and cheap sources of macro and trace minerals, proximate and physiochemical composition. Its seeds contain more minerals (especially Zn, Fe, Mg, P, Na, and Ca), crude oil, carbohydrates, calorific values, acid, and peroxide values than the pulp. While the pulp contains more proximate properties (moisture content, total ash, and crude protein), ascorbic acid, and K than the seeds [3–5]. However, Uganda has not yet developed the *T. indica* agroprocessing industries that would provide the ready-made products for locally available markets and exports. The native tree legume species has been neglected and indiscriminately grows without much attention in many places like on-farm, wild, roadsides, riverine, market places, and mountainous areas. The study of morphological traits of natural populations is often considered to be useful step in the study of the genetic variability [10]. Morphological characteristics information of *T. indica* is also useful in determining the domestication, protection, consumption, commercialization, and development of local industries. Additionally, morphological variability also signifies the adaptation of the species to the environment and it may be genetically or environmentally determined [8].

Until recently when *T. indica* species was prioritized for conservation and product development to support livelihoods in sub-Saharan Africa and other African countries, *T. indica* utilisations have depended largely upon wild populations, with no conservation strategies [11]. The wild habitats are threatened by anthropogenic activities such as charcoal burning for household incomes, food and firewood gathering, effects of wild fires, construction materials, animal fodder and increasing conversion to farmlands that have led to serious species genetic erosion. Much as several studies were carried out on Uganda's *T. indica* species [3–5, 12–14], there is still inadequate scientific information on fruits and seed morphological traits' relationships between agroecological zones and land use types. When made available, such information can guide the selection of *T. indica* tree species for various uses and breeding improvement. This study therefore aimed at evaluating the variations of *T. indica* fruits and seeds' morphological traits between the different agroecological zones and land use types in Uganda. It was, thus, hypothesised that there were no significant differences in the morphological traits of *T. indica* fruits and seeds across the agroecological zones of Uganda.

2. Materials and Methods

2.1. Description of Agroecological Zones of Uganda. Uganda, officially known as the Republic of Uganda, is located in the East African region on the edge of the Equator. It occupies total area of 241,038 km^2, averages about 1,100 meters above sea level, and lies between latitude 1°30′N and 32°00′E. Uganda's geographical features comprise plateaus, highlands, mountains, rolling hills, flat lands, rivers, lakes, and wetlands. The climate is tropical and generally rainy with defined two dry seasons. The predominant rocks were formed between 3,000 and 6,000 million years ago (pre-Cambrian era). However, in some western and eastern parts of the country, there are major developments of younger rocks, which are either sediments or volcanic in origin ranging from 135 million years ago (cretaceous period) to the present day [15]. The type and character of the soils of Uganda are defined by a number of parameters such as the nature of the parent rock, age of the form, and climate, especially the amount of moisture.

The country is divided into nine different agroecological zones, based on agroclimatic factors (rainfall totals and distribution) and soils (productivity and fertility). Topography, temperatures, moisture, and vegetation cover are the secondary factors considered to be uniform in each agroecological zone but differs between zones [16]. The wet and dry seasons are fairly marked with dry season between late November and March. The present study on *T. indica* was conducted in the three different agroecological zones of Uganda where *T. indica* trees are grown both on-farms and in the wild. Each of the selected agroecological zone was represented by a district (Figure 1). In addition, the sampled agroecological regions represented by Moyo, Nakasongola, and Soroti districts are more than 200 kilometres apart, to avoid having the same offspring and having the effects of agroecological zones' factors.

2.2. Description of the Study Areas. The study was carried out in three agroecological zones of Uganda, namely, West Nile, Lake Victoria Crescent, and Eastern out of the nine agroecological zones. Each selected agroecological zone was represented by the district of Moyo, Nakasongola, and Soroti, respectively. Attributes of each study area (district) within the agroecological zone are summarised in Table 1.

2.3. Sampling Design. The sampling sites from each of the agroecological zones were chosen from over 200 km apart that took into consideration ecological factors of the area, influence of climate or environmental factors and farming systems [4]. The agroecological zones are also separated by major physical features such as water bodies. Each zone was stratified into two major land use types: on-farm (current crop fields) and wild lands or fallow fields.

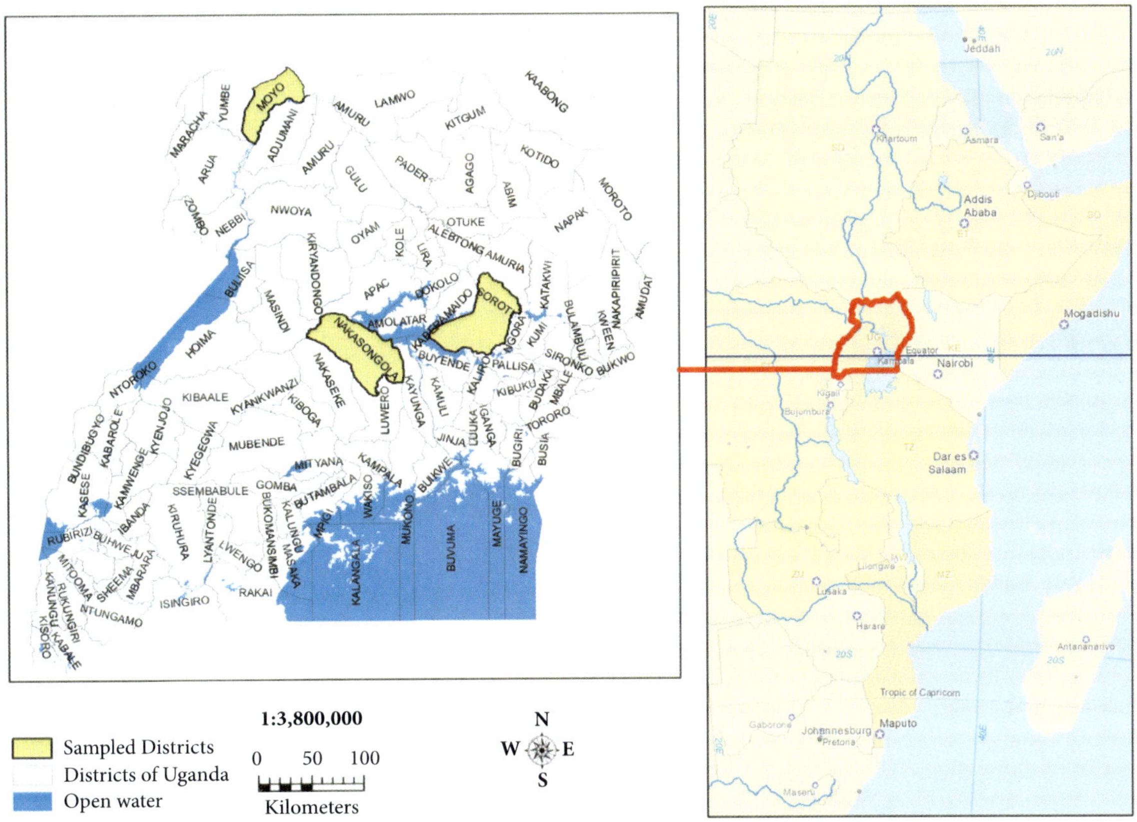

Figure 1: Map of Uganda showing sample sites in different agroecological zones.

The on-farm lands are current farm lands for agricultural crops (annual or perennial) while the wild or fallow fields have not been cultivated for five or more years prior to this study. Three agroecological zones were selected based on the large naturally occurring population of *T. indica*. Each agroecological zone was represented by a district—an administrative unit comprising of many subcounties. In each district, four subcounties were randomly selected, thus making four subcounties per district (zone). In each selected subcounty, five *T. indica* trees were selected from each of the two land use types, thus a total of twenty mature trees were selected from each land use type in each zone (district). Selected trees were separated from each other by more than 200 meters apart to minimize the risk of sampling closely related individuals (siblings). Selection criteria included ease of access, well developed crowns, good tree health (absence of obvious signs of pests, diseases, and fire disturbances), and presence of good-looking fruits [3].

2.4. Data Collection. The *T. indica* fruits were collected during the 2007 fruiting season from December 2007 to March 2008. The canopy of each sample tree was divided into three levels (top, middle, and bottom). The trees were climbed using ladders to collect ripe fruits which were selected by gently squeezing to check if ready. Ripe fruits were considered to be those with scurfy brown, woody, fragile shell with brown pulp and blackish-brown, and hard shiny seeds [3]. Eight fruits were randomly collected from each canopy level (two from each of the cardinal compass directions). In total, 24 fruits were harvested from each tree. A total of 480 fruits were, therefore, harvested from 20 trees per land use type, giving 960 fruits per agroecological zone (represented by district) and 2,880 fruits for the two land types in each agroecological zone. The fruits collected from each sampled tree, land use, and agroecological zone were bagged and labelled to avoid loss of tree identity and assessed for fruit physical characterization. Each sampled tree was labelled and its location recorded using a Global Position System (GPS) unit to determine its altitude. Fruits collected from each tree were pooled, kept in white polythene bags, and labelled according to tree number and canopy levels, for example, T_1C_1, T_1C_2, and T_1C_3 (representing tree number and tree canopy number). The fruits were allowed to dry at room temperature for five days before storing them in polythene bags. The bags were labelled with a permanent marker for ease of identification. The fruits were then taken to Makerere University's College of Agricultural and Environmental Sciences laboratory where they were stored in cool and dry place for further processing.

2.5. Quantitative Trait Measurements. All fruits collected were washed with distilled water to remove foreign particles and other inert materials and allowed to dry in the sun for about one hour. After this, the following descriptive

TABLE 1: Summary of study areas.

Attributes	Details of Studied District Representing Agro-ecological Zones		
Agroecological zone	West Nile	Lake Victoria Crescent	Eastern
District	Moyo	Nakasongola	Soroti
Location	03°39N 31°43′E	01°18′N 32°30′E	01°43′N 33°36′E
Geographical location	North Western Uganda	Central Uganda	Eastern Uganda
Land area (km^2)	2,059	3,512	2,663
Altitude (masl)	900 - 1,500	1,000 - 1,400	1,036 - 1,127
History	District formed in 1956	District created in 1997	District formed in 1962
Major geological formations	Gneiss, alluvial deposits, & schist, quartzite & marble	Buganda-Toro System geological formation	Granites, mignalites, gneiss, schists and quartzites
Major soil types	Vertisols, Leptosols, alluvial deposits & Ferralsols	Bululi and Lwampanga soil catena	Serere and Amuria catena, Metu complex and Usuk series
Economic activities	Subsistence agriculture	Livestock and subsistence agriculture	Subsistence agriculture
Rainfall (mm)	1,500 - 1,700	600 - 1,000	1,000 - 1,500
Rainfall patterns	March-June; August-November	March-July; August-November	March - June; August – November
Dry season	Late November to early March	Late November to early March	December to February
Min. and max. temp.	23.7°C and 30°C	18°C and 35°C	18 and 31.3°C
Climate	Modified equatorial	Tropical type	Tropical savanna
Historical sites	Limestone burning hole at Arapi, Metu tourist attraction cave, Mt. Otze, Kotilo Sacred Forest, Emin Pasha's underground Fort at Dufile, Era Forest, Moki and Metu springs, Albert Nile	White rhinoceroses' sanctuary, bird watching, Nakasongola hills	Soroti rock, East African Civil Aviation Academy, Lake Bisina, Namulekwa island, Mulondo and Mugarama landing sites, Bugondo and Ogera hills, Omeleku rocks, Agu swamp

Source: [3–5, 17–20].

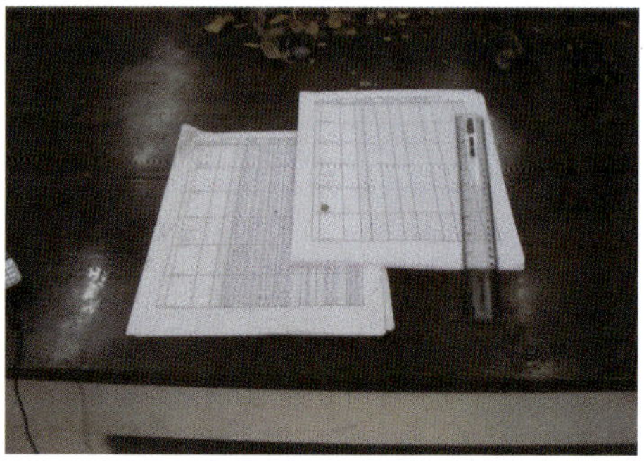

FIGURE 2: Morphological triplicate measurements.

FIGURE 3: Assorted clean pulp for analysis.

parameters were determined in triplicate and the average was determined: fruit length (cm), fruit breadth (cm), individual fruit total mass (g), total seeds mass (g), total pulp mass (g), and number of seeds per fruit (Figure 2). Length of fruit was determined as the distance from the panicle to the end of fruit on the outer surface (measured with the help of a thread and meter ruler). Measurements for curved fruits were made on the outer side of the curve. Fruit breadth was determined by using a vernier calliper graduated in 0.1 cm. Pulp mass was obtained by weighing the pulp per fruit using an electric weighing scale to 100% accuracy and graduated to 2 decimal places (0.00 g) after removing the seeds. Depulping of fruits was done using a pair of scissors. Total seed mass and number of filled (sound) seeds per fruit were determined in accordance with procedures of ISTA [20]. To determine seed mass, seeds were manually removed from the fruits (depulped), counted to determine the numbers per fruit, later weighed in triplicate, and recorded. Decomposing fruits, pulp, and seeds damaged by insects and other objects were excluded from the samples after measurement to avoid contamination of whole samples (Figure 3).

TABLE 2: *T. indica* fruits and seeds morphological/quantitative traits.

Morphology Traits	Unit	Measures	Three Agro-ecological Zones			Land Use Types	
			Lake Victoria C.	Eastern	West Nile	Wild	On-farm
Fruit length	(cm)	Min – Max	4.30 - 29.80	4.20 - 22.50	4.30 - 24.10		
		Mean±SD	11.73±3.3[a]	10.67±2.9[b]	12.48±2.9[c]	12.00±0.51[a]	11.40±0.45[b]
Fruit breadth	(cm)	Min – Max	1.10 - 3.00	1.20 - 2.60	1.30 - 2.70		
		Mean±SD	1.98±0.3[a]	1.76±0.3[b]	1.87±2.0[c]	1.90±0.21[a]	1.80±0.05[b]
Fruit mass	(g)	Min – Max	1.40 - 56.80	2.40 - 45.60	2.00 - 32.50		
		Mean±SD	14.30±7.4[a]	13.40±6.1[b]	13.17±5.3[b]	13.80±0.49[a]	13.60±0.86[a]
Pulp mass	(g)	Min – Max	0.19 - 22.00	0.94 - 19.30	0.27 - 15.80		
		Mean±SD	4.58±2.8[a]	5.25±2.8[b]	5.24±2.4[b]	5.20±0.36[a]	4.90±0.15[b]
Total seed No.	#	Min – Max	1 - 15	1 – 17	1 – 14		
		Mean±SD	5.59±2.8[a]	5.84±2.7[b]	6.53±2.8[c]	5.90±0.59[a]	6.10±0.67[a]
Total seeds mass per fruit	(g)	Min – Max	0.18 - 17.15	0.19 - 11.93	0.20 - 12.20		
		Mean±SD	4.22±2.5[a]	3.36±1.9[b]	3.88±2.1[c]	3.80±0.15[a]	3.90±0.27[a]

Same superscript letters in each row are not statistically significant according to LSD test at P≤0.05.

2.6. Data Analysis. Each of the fruit and seed samples was aggregated by land use type and agroecological zone. Analyses were carried out in triplicate to investigate relationships between agroecological zones and land use types. Two-Way Analysis of Variance (ANOVA) (with agroecological zone and land use types as fixed factors) was used to analyse the data using IBM statistic version 16.0. The Least Significant Difference (LSD) in a Post Hoc Test was used to separate means at 5% using the General Linear Model (GLM) for Post Hoc Multiple Comparisons of Observed Means. Pearson Correlation Coefficient (R) was calculated to determine the relationship among the morphological characteristics in *T. indica* fruits and seeds. Additionally, the morphological traits dispersion/variability was calculated to describe the variability of the study population.

3. Results

3.1. Quantitative Variations of T. indica Fruits and Seed Morphological Traits. There were significant (P≤0.05) differences between agroecological zones and all morphological traits (Table 2). There were also significant differences in the *T. indica* morphological characteristics between land use types. The results indicate that West Nile agroecological zone had the longest mean fruits of 12.48±2.9 but the maximum value (22.50 cm) was witnessed in the sample from Lake Victoria Crescent zone and higher mean seed numbers (6.53±2.8) with the maximum number (17) being recorded in eastern zone. Lake Victoria Crescent zone had wider mean fruits (1.98±0.3), heavier mean fruits (14.30±7.4), and heavier mean seeds (4.22±2.5), while Eastern zone had heavier mean pulp (5.25±2.8). In addition, all the fruit trait values were higher in the wild land use type compared to all seed traits that recorded higher values in on-farm land use type (Table 2).

3.2. Morphological Traits, Agroecological Zones, and Land Use Types. The observed differences in values of *T. indica* fruits and seeds morphological characteristics between the land use types and agroecological zones showed that most of *T. indica* samples from the Lake Victoria Crescent agroecological zone and wild land use type had higher morphological values compared to other zones and on-farm land use type. Only seed number consistently showed higher average values in West Nile than in Eastern and Lake Victoria Crescent agroecological zones, respectively (Figures 4 and 5).

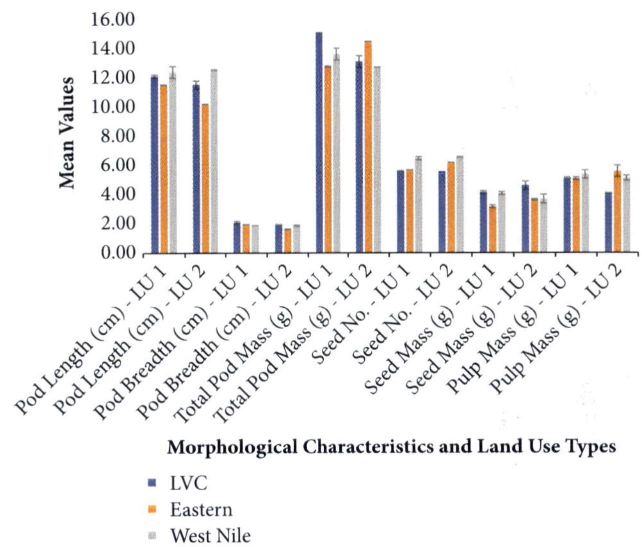

FIGURE 4: Variations of *T. indica* morphological characteristics by land use types. **Key:** LU 1, land use 1 (wild); LU 2, land use 2 (on-farm).

3.3. Correlations Between T. indica Fruits and Seeds Morphological Traits. Table 3 shows the correlations (P≤0.05) between the *T. indica* fruits and seeds morphological traits. Strong uphill positive linear relationships were observed between seed mass and seed number (R=0.78), total fruit mass and fruit length (R=0.73), pulp mass and total fruit mass (R=0.72), and seed mass and total fruit mass (R=0.72).

Table 3: Pearson coefficient (R) of correlation between *T. indica* morphological traits.

Morphotypes		Fruit Length	Fruit Breadth	Fruit Mass	Seed Number	Seed Mass
Fruit Breadth	Pearson Correlation	**0.480****				
	Sig. (2-tailed)	0				
	N	2880				
Fruit Mass	Pearson Correlation	**0.749****	**0.491****			
	Sig. (2-tailed)	0	0			
	N	2880	2880			
Seed Number	Pearson Correlation	**0.580****	**0.222****	**0.626****		
	Sig. (2-tailed)	0	0	0		
	N	2880	2880	2880		
Seed Mass	Pearson Correlation	**0.527****	**0.338****	**0.725****	**0.786****	
	Sig. (2-tailed)	0	0	0	0	
	N	2880	2880	2880	2880	
Pulp Mass	Pearson Correlation	**0.556****	**0.382****	**0.731****	**0.558****	**0.564****
	Sig. (2-tailed)	0	0	0	0	0
	N	2880	2880	2880	2880	2880

** Correlation is significant at $P \leq 0.05$ level (2-tailed).

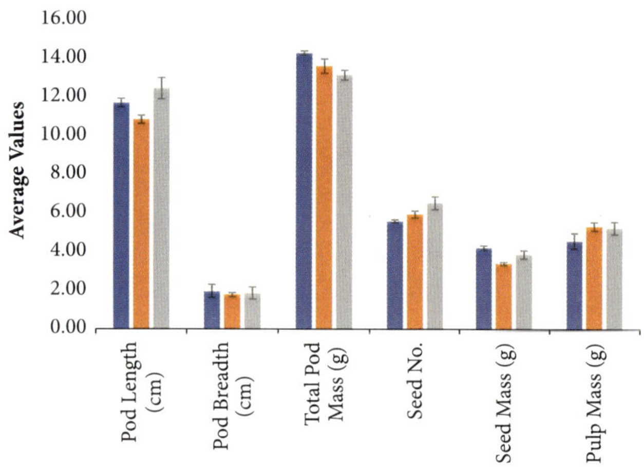

Figure 5: Variations of *T. indica* morphological characteristics by agroecological zones.

There were strong uphill positive relationships observed for seed number and fruit mass (R=0.63), seed number and fruit length (R=0.58), seed mass and fruit length (R=0.53), pulp mass and fruit length (R=0.56), pulp mass and seed mass (R=0.56), fruit mass and fruit breadth (R=0.49), fruit breadth and fruit length (R=0.48), pulp mass and fruit breadth (R=0.38), and seed mass and fruit breadth (R=0.34). A weak uphill positive linear relationship was observed for seed number and fruit breadth (R=0.22).

4. Discussion

4.1. Quantitative Variations in T. indica Fruits and Seed Traits

4.1.1. Fruits and Seed Traits Quantitative Variations. Statistical analyses indicate that most of the significant morphological variation occurs among the agroecological zones and land use types, although morphological variation is more significant in the agroecological zones than in the land use types. Many factors may have contributed to the divergence; for example, the reproductivc and vegetative traits usually manifest different patterns of variations. Variations in *T. indica* fruit morphological traits have been reported by other authors [21–25]. Studies on fruits and seeds morphological traits in many tree species natural populations have also been reported by Okwu et al. [7]; Fatima et al. [8]; Mingfei et al. [26]; Dadegnon et al. [27]; Mkwezalamba et al. [28]; and Abasse et al. [29]. In accordance with the findings of Mingfei et al. [26], it is certainly the case that many factors contribute to the relative amounts of variation observed among populations of the same species, particularly when populations have been separated for long period of time. The observed variations suggest that different factors affect morphological traits, at least in the different agroecological zones and land use types such as environmental and climatic factors.

Fruit length varied at 4.20-29.80 cm, fruit breadth at 1.10-3.00 cm, fruit mass at 1.40-56.00 g, pulp mass at 0.19-22.00 g, total seed number at 1-17, and seed mass at 0.18-17.15 g. It is evident from this study that the dispersion/variation observed in fruit traits—mass (22.13), length (10.11), and pulp mass (9.13)—was higher than the dispersion/variation observed in seed traits—number (6.44) and mass (6.40)—respectively. These differences reflect that fruit traits, especially fruit mass, integrate all other fruit and seed traits. This suggest that variations exist in *T. indica* fruit traits especially fruit mass, and these variations also cut across into the seed number—the longer the fruits, the more the fruit pulp and the likelihood of the fruit producing many seeds, or the shorter the fruits, the less the amount of fruit pulp and the more the likelihood of the fruit producing few seeds. Similar observations were documented [7] for *C. albidum*. The tree variation observed in all the traits studied is typical of that found in an outbreeding species and the trend appears to be consistent with fruits of a particular outbreeding species.

The observed variation can also be further emphasized by the strong uphill positive linear correlations observed between fruit mass and fruit length (R=0.75), pulp mass and fruit mass (R=0.73), and seed mass and seed number (R=0.79). The large variation in the fruit and seed traits suggests that there are considerable opportunities for phenotypic selection of the species for quantitative and qualitative improvement in productivity. These variations observed could be attributed to the different genetic, effects of environment and climatic factors as well as growing conditions of the tree species because all the agroecological zones were significantly different. The diversity among these attributes could potentially form the basis for future breeding programmes, thereby contributing significantly towards hunger alleviation, nutritional benefits, and income generation, in particular among the rural poor people of Uganda.

4.1.2. Fruit Length. Longer fruits (4.30-29.80 cm) were documented from Lake Victoria Crescent zone. These results show that fruit length values are within those obtained by Orwa et al. [30] but higher than values obtained for other species such as those by Mkwezalamba et al. [28], Orsanic et al. [31], and Jarret and Berke [32] for *Sclerocarya birrea* (A. Rich.), *Sorbus torminalis* L. Crantz, and *Capsicum chinense* Jacq., respectively. Orsanic et al. [31] found out that the higher the altitude of the tree population, the longer the fruits, which confirms the findings of this study that documented that Lake Victoria Crescent (1050-1093 masl) and Eastern (1030-1096 masl) zones are located at higher altitudes than West Nile (593-997 masl) zone. Abasse et al. [29] also reported that phenotypic variation in fruits and seeds may reflect natural selections related to rainfall gradient influence on adaptability of different fruit tree populations. The variation of *T. indica* fruit length shows that superior phenotype could be ideal candidates for multiplication in the event that bigger and longer fruits are preferred. Mayavel et al. [21] found out that it might be inferred that fruit length, breadth, weight, pulp weight, and other traits could be considered as most important yield contributing traits in red tamarind.

Many research works conducted with fruits and seeds of tree species from different botanical families, such as cucumber tree (*Averrhoa bilimbi* L.) of Oxalidaceae family [33], *Campomanesia adamantium* of Myrtaceae family [34], and majestic heaven lotus (*Gustavia augusta* L.) from Lecythidaceae family [35], indicate that longer and wider fruits usually present higher values of fruit mass and seeds. These authors' results are in total agreement with the results of our study which showed that longer, wider, and heavier fruits have more seeds. However, it is not clear whether the species with long and large *T. indica* fruits may be more productive than those with narrow and short fruits. The observed variations could probably be attributed to environmental and climatic factors, agroecological zones factors, and farming systems, which partly agrees with the work of Abasse et al. [29]. A positive and quite strong correlation between the fruit length and all studied traits obtained means that fruit length is important and determinant for all the traits studied in tree breeding programme. In cases where *T. indica* provenances with the longer and wider fruits can be selected, it would mean that more preferred products can be obtained in large quantities and *T. indica* trees that produce these fruits can be recommended for tree per yield improvement.

4.1.3. Fruit Breadth. The *T. indica* wider fruits (1.10-3.00 cm) were documented from Lake Victoria Crescent zone. Other authors such as Mkwezalamba et al. [28] and Orsanic et al. [31] documented values for *Sclerocarya birrea* (A. Rich.) Hochst. and *Sorbus torminalis* L. Crantz, respectively, which are within the results of our study. Meanwhile, higher values were documented for other species [29, 32] for *Balanites aegyptiaca* and *Capsicum chinense* Jacq., respectively. Fruit length and breadth are important morphological characteristics for plants characterization because biometric studies of fruits and seeds and their phenotypical correlations can allow the quantitative evaluation of a relevance character in relation to another [36]. The generated significant variations in fruit production between provenances is an indication that superior provenances of *T. indica* can be selected for domestication, breeding improvement, and commercialization purposes. This study found that fruit breadth is a least important trait for consideration to any tree breeding improvement. The findings concur with Orsanic et al. [31] who documented that when determining fruit shape, width seems to be a less important property than length.

4.1.4. Fruit Mass. *T. indica* samples from Lake Victoria Crescent zone recorded heavier fruits (1.40-56.80 g) while lighter fruits (2.00-32.50 g) were recorded from West Nile zone. Fruit size, thickness, shiny color, and shape are some important qualitative characteristics for consumer acceptability on the market. Our study documented lower values than those reported by Bilcke et al. [22]. Higher values were documented for other species [28, 29, 31, 32] for *Sclerocarya birrea* (A. Rich.), *Balanites aegyptiaca*, *Sorbus torminalis* L. Crantz, and *Capsicum chinense* Jacq., respectively. According to İmrak et al. [37], fruit and seed size are determined by the genetic characteristics of each cultivar and also influenced by plant management. Since fruit size is used for grading produce, large fruits would normally attract premium prices due to their greater capability to store photosynthates having a lower rate of respiration [38]. During fruit growth, fruit size usually increases, and such increase is determined by the cultivar and by cell elongation during anthesis and after anthesis [38].

The significant differences in fruit mass as found in this study are probably a result of genetic and environmental factors. Positive and strong correlations between the fruit mass and fruit length (R=0.75), pulp mass (R=0.73) and seed mass (R=0.73), and fruit breadth (R=0.49) and seed number (R=0.73) mean that larger and longer fruits should be collected as they give many and heavy seeds and heavier pulp. Our study findings show that *T. indica* fruit mass is the most important parameter for tree breeding improvement. Similar results were documented by Mayavel et al. [21] who found out that fruit mass is among the 15 characters taken into consideration for yield per tree improvement in red tamarind in Tamil Nadu.

4.1.5. Pulp Mass. Heavy pulp (0.10-22.00 g) and lighter pulp (0.27-15.80 g) were recorded in Lake Victoria Crescent and West Nile agroecological zones, respectively. It is clear from this study that *T. indica* produces pulp of different mass depending on the agroecological zone and land use types. This could probably be caused by agroecological zone, land use, genetic, phenotypic, environmental, and climatic factors. According to Mkwezalamba et al. [28], variations in fruit pulp weight usually depend on provenances. Mingfeli et al. [26] also reported that key environmental factors and genetic divergence can also coaffect the phenotypic variation in vegetative and reproductive traits which is an observed variation in *T. indica*. In general, both vegetative and reproductive traits may adapt to these responses differently with respect to local environment and climate.

The strong correlations documented between pulp mass and fruit mass imply that the bigger the fruit, the more the pulp mass of that given fruit. Since pulp is the single most important *T. indica* product which is variously used [3, 4], the *T. indica* industry in Uganda should first target the agroecological zone (such as the Lake Victoria Crescent) which produces bigger fruits for research and development. Such programme should focus on both on-farm and wild *T. indica* populations which have shown significant differences between these land use types. This variability can be important in selecting the best pulp mass for further research and multiplications instead of relying on natural regeneration that will continue producing similar offspring. This is so because such correlations are the basis for conservation of the local species both on-farms and in the wild habitats.

4.1.6. Total Seed Number. The total number of *T. indica* seeds per fruit documented in this study ranges from 1 to 17. Seeds are the foundation for any tree multiplication programme. The seeds are important to plants because they ensure that the gene pool continues to the next generation. Seed numbers can be emphasized when many seeds are preferred for breeding programme as well as nutritional (proximate, mineral, and physicochemical) properties. The highest number of seeds per fruit (17) recorded in this study is more than those reported in other studies carried out elsewhere, such as [1, 2]. The high number of seeds indicates how important the samples are for nutritional product development as well as developing improved *T. indica* cultivars for specific environments (and possibly agroecological zones) in Uganda.

The high number of seeds revealed in this study could be related to agroclimatic factors, soil conditions, farming systems, and geological formations, which differ across the various agroecological zones in Uganda. Soils in Eastern agroecological zone are made of mainly Serere and Amuria catena, Metu complex, and Usuk series [17]. Many are of ferralitic type (sandy sediments and sandy loams) which are well drained, friable, and exhibiting low cation exchange capacity and high anion absorptive capacity. This could have been responsible for the seed numbers observed in this study. Also, Raddad [39] documented that clay and sand types usually have effects on the seed number and seed weight of *Acacia senegal*. In addition, the study documented a positive and strong correlation between the seed number and seed mass (R=0.79), fruit mass (R=0.63), fruit length (R=0.58), and pulp mass (R=0.56), which means many seeds in a long fruit pod produce heavier seed mass and heavy fruits and pulp, which is more preferred by the local communities in the study areas and will go a long way in tree development.

4.1.7. Total Seed Mass. The amount of total seed mass per fruit is higher (0.18-17.15 g) in Lake Victoria Crescent zone than in West Nile and Eastern agroecological zones, respectively. In this study, the total seed mass per fruit is higher than that reported by Abraham et al. [23] while Bilcke et al. [22] reported higher values than values reported in our study. However higher values were reported by Mkwezalamba et al. [28] and Orsanic et al. [31] for *Sclerocarya birrea* (A. Rich.) Hochst and *Sorbus torminalis* L. Crantz, respectively. The findings of this study corroborate with results obtained by Aline et al. [40] who documented higher average values of length, width, thickness, and seed fresh mass among *Prunus persica* cultivars. Mkwezalamba et al. [28] and Kadu et al. [41] also reported that bigger seeds are the best for selection for early seedling growth and development.

It was possible to deduce that the bigger the fruit fresh mass is, the bigger the seed mass will be. Seeds with bigger mass present better germination capacity, standardization, and seedling emergence [42]. Such characteristics are useful to help the selection of new genotypes to be used as rootstocks, which, in addition to the standardization of fruit lots with higher fresh mass, allows obtaining seed lots with higher mass and with maximum germination power, an implication for the reduction of management and improving the homogeneity of seedling lots. This is so because agroecological zone with a greater number of seeds per fruits (fresh and dry weights) tends to have high pungency and seeds and heavier mass could have more food reserves. Although large food reserves could prolong seed viability and seed storage and enhance seed germination and vigorous seedlings growth, more studies are still required to find out whether larger seed (in this study) would produce high quality seedlings, nutritional values, and taste differently compared to lighter seed of *T. indica* in the different zone.

4.2. Agroecological Zones, Land Uses, and T. indica Morphological Variations. In plants, the phenotypic variation is generally a response to differences in climatic conditions, which reflects either adaptive evolution or phenotypic plasticity, or a combination of the two [43]. Many studies that documented phenotypic changes in natural populations attributed them to climate change. In addition, Parmesan [44] and

Chen et al. [45] have examined and compared intraspecific phenotypic variation across the geographic ranges of selected species. Within the same agroecological zone, we reported significant differences in land use types (on-farm and wild) and altitudes where the *T. indica* individuals grow, which could have also been the possible cause of the obtained differences. Maranz and Wiesman [46] reported that important variability in *Vitellaria paradoxa* fruit and seed sizes in relation to climate gradient was found throughout the entire sub-Saharan Africa zone located at the northern part of the equator.

In this study, sampled *T. indica* fruits were from altitudes that ranged from 593 to 1,096 masl (1,050-1,093; 1,030-1,096; and 593-997 masl for Lake Victoria Crescent and Eastern and West Nile zones, respectively). This shows that, even within the same agroecological zone, there are differences in altitude where *T. indica* grows. Similar findings were reported earlier by Mayavel et al. [21] who documented *T. indica* 21 genotypes of red tamarind at Tamil Nadu between elevations 3-773 masl while Orwa et al. [30] found out that *T. indica* trees usually grow at altitude between 0 and 1,500 masl and mean annual rainfall from 350 to 2,700 mm. In addition to altitude, *T. indica* grows at different soils (productivity and fertility) and soil types, topography, vegetation cover, climate (such as rainfall totals and distribution), geological formation, rainfall, and farming systems or practices [3–5].

The observed differences in *T. indica* morphological trait characteristics probably indicate the influence of agroecological zones, environmental factors, climatic factors (such as elevation, topography, effects of geography, surface of the earth, climate change over time, and prevailing wind), and land use types on the *T. indica* trees, which differs in these zones and concurs with [1] that reported the substantial effects of the environmental and land use factors. A study by Fatima et al. [8] indicated that parameters that affect growth, development, and health condition of a tree range from physical factors (climate, soils, water balance, and topography) and biological factors (diseases and pests) to land use types. It was, however, not possible to state which of these factors and others have had a greater effect in producing these variations. Because the flowers of *T. indica* are predominantly open-pollinated and usually nonuniform, with wild land use trees probably having more pollinators than those on-farms, great variations in a wide range of characteristics such as seedlings, tree size and structure, leaf and fruit form, age to bearing, and fruit quality are bound to be exhibited. This could be the reason why there were significantly bigger and heavier fruits in the wild than on-farms.

4.3. Correlation Relationships among T. indica Morphological Traits. The correlation and path analysis studies are important in fruit crops like tamarind, mango, and citrus [21], where its quantity and quality traits are important in improvement programme. Correlation studies between traits of a crop are of interest to the breeders in planning the hybridization programme and evaluating the individual plants in segregating populations [47]. Understanding the existing correlations between fruit and seed characteristics allows relating the behaviour of a variable through analysis of the other, even if it does not give an exact position of the relative importance of direct and indirect effects of the various characters on yield [47]. Together it will provide knowledge on interrelationships and relative contribution of independent characters on dependent variables which enables a tree breeder to apply suitable selection procedures in tree improvement programme. The knowledge on nature and magnitude of interrelationship among yield and its contributing factor is mandatory for the simultaneous improvement of characters and yield [21]. Such correlation studies can be used in indirect selection of plus trees. Selection can be used on most morphological characteristics to select traits of interest which are difficult to assess [21]

All fruits and seeds morphological traits presented significant and uphill positive strong correlations. Similar results were reported by Dresch et al. [34] who observed a positive and moderate correlation between fruit width with seed thickness, and similar correlation types were documented [28] for *Sclerocarya birrea* fruits. Correlations witnessed among all traits reported are important because they show that all these traits can be used indirectly to select plus trees for yield per tree and the development of various nutritious components of the species. On the other hand, the weak or low positive relationships that exist in some characteristics could probably mean that indirect selection may not be possible. Association of characteristics determined by correlation may not provide an exact picture of the relative significance of direct and indirect influence of each of the yield components towards yield [47]. Khan et al. [47] reported that the phenotypic and genotypic variations of the yield components are considerably high in some species such as pointed gourd, which points to the possibility of developing a variety with high yield. It is certainly the case that many factors contribute to the relative amounts of variation observed among populations of the same species.

The numbers of fruits and seeds are important biological properties which influence the ability to generatively propagate particular tree species. A positive and quite strong correlation between the number of seeds and seed mass (R=0.78) was obtained, indicating that the seeds mass per plant would increase with the increase in seed number. Even then, the low number of seeds per fruit can be explained by low density of *T. indica* trees in the studied populations and by small isolated populations with low gene flow between the neighbouring populations as well as by the decrease in gene numbers. Inadequate silvicultural practices of *T. indica* trees especially in the wild population also make the *T. indica* trees grow in stressful ecological conditions (such as frequent fires), which can result in reduced fertilization. *T. indica's* bisexual flowers exhibit protogynous dichogamy, with flowering occurring in dry season where there is both self- and cross-pollination but the latter is much more effective. However, these processes are affected by these anthropogenic activities.

5. Conclusion

There were significant differences in the morphological traits of *T. indica* fruits and seeds across the agroecological zones of Uganda. Lake Victoria Crescent agroecological zone had higher fruit morphological traits values than other zones. *T. indica* traits from wild and on-farm land use types were superior in fruits and seed trait values, respectively. Many seeds per fruit, up to 17, were recorded in Uganda's Eastern agroecological zone than those recorded elsewhere. All fruits and seed morphological traits showed variabilities within the agroecological zones and land use types. Strong significant uphill positive linear correlations between all morphological traits ($P \leq 0.05$) were observed, with the strongest relationships being between seed mass and seed number ($R=0.78$), total fruit mass and fruit length ($R=0.73$), pulp mass and total fruit mass ($R=0.72$), and seed mass and total fruit mass ($R=0.72$). The dispersion/variation and correlated relationships suggest that all the studied morphological traits can be used for selection of plus trees for tree breeding improvement. This study also suggests that it may not be challenging for the tree breeding improvement to simultaneously increase *T. indica* fruits and seed traits in the same breeding population and will not require separate ideotypes for the traits. The observed variations are probably attributed to influence of agroecological zones factors, environmental factors, climate, land use types, and farming systems, a reflection of *T. indica* species adaptation to different conditions showing high genetic and phenotypic differences to be exploited. This study represents the first quantitative evaluation of *T. indica* morphological traits variation and should form the basis to guide the indirect selection of plus tree species for valuable nutritional requirements in both wild and on-farm land uses. The study findings were based on data collected in a single dry season of the trees' lifespan which possibly possesses uncertainty in variations among agroecological zones and land use types whose uses keep changing. In addition, the exact start of dry season period which is slightly different from one zone to another does vary due to climate change. There is therefore a need to carry out more studies to fully understand variations which may occur due to seasonality.

Conflicts of Interest

The authors declare that there are no conflicts of interest in any form regarding the publication of this paper.

Acknowledgments

We are most grateful for having received financial support from Norwegian Agency for Development Cooperation (NORAD) through Makerere University. The authors are also indebted to the field assistants, laboratory staff, friends, and the farmers whose fields were visited during data collection that made the compilation leading to the development of this manuscript possible.

References

[1] K. El-Siddig, H. P. M. Gunasena, B. A. Prasad et al., *Tamarind (Tamarindus indica L.). Fruits for the Future 1*, International Centre for Underutilised Crops, Southampton, UK, 2006.

[2] E. M. Yahia and N. K. E. Salih, *Tamarind (Tamarindus indicaL.)*, Woodhead Publishing Limited, 2011.

[3] J. Okello, J. B. L. Okullo, G. Eilu, P. Nyeko, and J. Obua, "Mineral composition of Tamarindus indica LINN. (tamarind) pulp and seeds from different agro-ecological zones of Uganda," *Journal of Food Science and Nutrition*, vol. 5, no. 5, pp. 959–966, 2017.

[4] J. Okello, J. B. L. Okullo, G. Eilu, P. Nyeko, and J. Obua, "Physicochemical Composition of Tamarindus indica LINN. (Tamarind) in the Agro-Ecological Zones of Uganda," *Journal of Food Science and Nutrition*, vol. 6, no. 5, pp. 1179–1189, 2018.

[5] J. Okello, J. B. L. Okullo, G. Eilu, P. Nyeko, and J. Obua, "Proximate Composition of Wild and On-Farm Tamarindus indica LINN. Fruits in The Agro-Ecological Zones of Uganda," *Journal of Food Processing and Technology*, 2018.

[6] J. Gebauer, K. El-Siddig, and G. Ebert, "The Potential of Underutilized Fruit Trees in Central Sudan," in *Proceedings of the International Agricultural Research for Development*, Deutscher Tropentag, Witzenhausen, 2002.

[7] C. Okwu, E. G. Oboho, E. K. Asaah, E. S. Osazuwa, S. O. Igberaese, and Z. Tchoundjeu, "Phenotypic Variations in Fruits and Seed Traits of Chrysophyllum albidum in three Agroecological Zones in Nigeria," *Scientific Research Journal (SCIRJ*, vol. 5, no. 10, pp. 39–50, 2017.

[8] F. El Ayadi, F. Msanda, F. Baniaameur, and A. El Mousadik, "Morphological and Shape Fruits Variability of Acacia tortilis ssp. Raddiana (Savi) Brenan in South Morocco," *International Journal of Plant Breeding and Genetics*, vol. 6, no. 4, pp. 151–167, 2012.

[9] C. R. Elevitch and H. I. Manner, *Traditional Tree Initiative - Species Profiles for Pacific Island Agroforestry (http//www.traditionaltree.org)*, Artocarpus heterophyllus, jackfruit, 2006.

[10] N. Kaushik, S. Kumar, K. Kumar, R. S. Beniwal, N. Kaushik, and S. Roy, "Genetic variability and association studies in pod and seed traits of Pongamia pinnata (L.) Pierre in Haryana, India," *Genetic Resources and Crop Evolution*, vol. 54, no. 8, pp. 1827–1832, 2007.

[11] FAO, *Food and Agriculture Organization of The United Nations*, Forest Genetic Resources, Rome, Italy, 2004, http://www.fao.org/3/a-y5901e.pdf.

[12] J. Lakor, G. Elepu, M. J. Buyinza, and P. Nyeko, "Analysis of Tamarindus (Tamarindus indica L.) Value Chain in Uganda: Identification of Opportunities and Constraints to Its Commercialization and Domestication," *Journal of Agriculture and Environmental Sciences*, vol. 5, no. 1, pp. 101–112, 2016.

[13] M. Masette, A. Candia, and A. Grace Ocheng, "The commercial viability of Tamarind (Tamarindus indica L) fruit based products for improved incomes among farmers in Northern and Eastern Uganda," *African Journal of Food Science and Technology*, vol. 06, no. 06, pp. 167–176, 2015.

[14] P. Nyadoi, P. J. B. L. Okori, J. Okullo et al., "Tamarinds' (Tamarindus indica L.) niche tree species diversity characterisation reveals conservation needs and strategies," *International Journal of Biodiversity and Conservation*, vol. 1, no. 4, pp. 151–176, 2009.

[15] *NEMA, National Environment Management Authority, State of the Environment Report for Uganda 2006-2007*, Kampala, Uganda, 2007.

[16] *GoU, Government of The Republic of Uganda, Increasing Incomes through Exports: A Plan for Zonal Agricultural Production, Agro-processing and Marketing*, Kampala, Uganda, 2004.

[17] *UDIH, Uganda Districts Information Handbook, Expanded Edition 2007 - 2008*, Fountain Publishers, Kampala, Uganda, 2007.

[18] *NEMA, National Environment Management Authority, District State of Environment Report for Moyo District*, Moyo, Uganda, 2004.

[19] *NEMA, National Environment Management Authority, District State of Environment Report for Nakasongola District*, Uganda, 2004.

[20] ISTA (International Seed Testing Association), "International Rules for Seed Testing: Rules 1996," *Seed Science and Technology*, vol. 24, pp. 1–335, 1996.

[21] A. Mayavel, B. Nagarajan, K. Muthuraj, A. Nicodemus, and R. Prabhu, "Correlation and Path Coefficient Analysis of Selected Red Tamarind (Tamarindus indica var rhodocarpha) Genetic Resources," *International Journal of Current Microbiology and Applied Sciences*, vol. 7, no. 04, pp. 794–802, 2018.

[22] N. Van den Bilcke, K. Alaerts, S. Ghaffaripour, D. J. Simbo, and R. Samson, "Physico-chemical properties of tamarind (Tamarindus indica L.) fruits from Mali: Selection of elite trees for domestication," *Genetic Resources and Crop Evolution*, vol. 61, no. 2, pp. 537–553, 2014.

[23] Z. Abraham, S. K. Malik, G. E. Rao, S. L. Narayanan, and S. Biju, "Collection and characterisation of Malabar tamarind [Garcinia cambogia (Gaertn.) Desr.]," *Genetic Resources and Crop Evolution*, vol. 53, no. 2, pp. 401–406, 2006.

[24] P. Soloviev, T. D. Niang, A. Gaye, and A. Totte, "Variability of fruit physicochemical characters for three harvested woody species in Senegal: Adansonia digitata, Balanites aegyptiaca and Tamarindus," *Fruits*, vol. 59, no. 2, pp. 109–119, 2004.

[25] B. Fandohan, A. E. Assogbadjo, R. Glèlè Kakaï, T. Kyndt, and B. Sinsin, "Quantitative morphological descriptors confirm traditionally classified morphotypes of *Tamarindus indica* L. fruits," *Genetic Resources and Crop Evolution*, vol. 58, no. 2, pp. 299–309, 2011.

[26] M. Ji, J. Deng, B. Yao et al., "Ecogeographical variation of 12 morphological traits within Pinus tabulaeformis: the effects of environmental factors and demographic histories," *Journal of Plant Ecology*, vol. 10, no. 2, pp. 386–396, 2017.

[27] S. Dadegnon, C. Gbemavo, C. Ouinsavi, and N. Sokpon, "Morphological Variation and Ecological Structure of Chrysophyllum albidum G. Don," *International Journal of Plant & Soil Science*, vol. 5, no. 1, pp. 25–39, 2015.

[28] I. Mkwezalamba, C. R. Y. Munthali, and E. Missanjo, "Phenotypic Variation in Fruit Morphology among Provenances of Sclerocarya birrea (A. Rich.) Hochst," *Journal of Forestry Research*, vol. 2015, Article ID 735418, 8 pages, 2015.

[29] T. Abasse, J. C. Weber, B. Katkore, M. Boureima, M. Larwanou, and A. Kalinganire, "Morphological variation in Balanites aegyptiaca fruits and seeds within and among parkland agroforests in eastern Niger," *Agroforestry Systems*, vol. 81, no. 1, pp. 57–66, 2011.

[30] C. Orwa, A. Mutua, R. Kindt, R. Jamnadass, and S. Anthony, *Agroforestree Database: a tree reference and selection guide version 4.0*, 2009, http://www.worldagroforestry.org/sites/treedbs/treedatabases.asp/.

[31] M. Oršanić, D. Drvodelić, T. Jemrić, I. Anić, and S. Mikac, "Variability of morphological and biological characteristics of Wild Service Tree (Sorbus torminalis (L.) Crantz) fruits and seeds from different altitudes," *Periodicum biologorum*, vol. 111, no. 4, pp. 495–504, 2009.

[32] R. L. Jarret and T. Berke, "Variation for fruit morphological characteristics in a Capsicum chinense Jacq. germplasm collection," *HortScience*, vol. 43, no. 6, pp. 1694–1697, 2008.

[33] M. A. Santos, L. F. Braga, R. M. Rondon Neto, and A. M. Sorato, "Aspectos morfológicos e fisiológicos da germinação e morfometria de frutos e sementes de swartzia recurva poeep (fabaceae)," *Ciencia & Natura*, vol. 37, no. 3, pp. 34–54, 2015.

[34] D. M. Dresch, S. d. Scalon, T. E. Masetto, and M. d. Vieira, "Germinação e vigor de sementes de gabiroba em função do tamanho do fruto e semente," *Pesquisa Agropecuária Tropical*, vol. 43, no. 3, pp. 262–271, 2013.

[35] R. M. Silva, R. d. Ribeiro, D. J. Coutinho, S. I. Silva, and M. I. Gallão, "Caracterização de frutos, sementes, plântulas e germinação de Jeniparana," *Revista Ceres*, vol. 61, no. 5, pp. 746–751, 2014.

[36] S. A. Felizardo, A. D. Freitas, N. D. Marques, and D. A. Bezerra, "Características biométricas de frutos e sementes de Oenocarpus bataua Mart. com procedência em Almeirim, Pará," *Revista Verde de Agroecologia e Desenvolvimento Sustentável*, vol. 10, no. 2, p. 09, 2015.

[37] B. İmrak, A. B. Küden, E. Tanriver, and E. Kafkas, "Volatile and some fruit quality characteristics of new promising peach genotypes," *Acta Scientiarum Polonorum, Hortorum Cultus*, vol. 14, no. 2, pp. 3–12, 2015.

[38] J. A. Barrera, M. S. Hernández, L. M. Melgarejo, O. Martínez, and J. P. Fernández-Trujillo, "Physiological behavior and quality traits during fruit growth and ripening of four Amazonic hot pepper accessions," *Journal of the Science of Food and Agriculture*, vol. 88, no. 5, pp. 847–857, 2008.

[39] E. A. Y. Raddad, "Ecophysiological and genetic variation in seedling traits and in first-year field performance of eight Acacia senegal provenances in the Blue Nile, Sudan," *New Forests*, vol. 34, no. 3, pp. 207–222, 2007.

[40] A. das Graças Souza, O. J. Smiderle, V. M. Spinelli, R. O. de Souza, and V. J. Bianchi, "Correlation of biometrical characteristics of fruit and seed with twinning and vigor of Prunus persica rootstocks," *Journal of Seed Science*, vol. 38, no. 4, pp. 322–328, 2016.

[41] C. A. C. Kadu, M. Imbuga, R. Jamnadass, and I. K. Dawson, "Genetic management of indigenous fruit trees in southern Africa: a case study of *Sclerocarya birrea* based on nuclear and chloroplast variation," *South African Journal of Botany*, vol. 72, no. 3, pp. 421–427, 2006.

[42] R. C. Colombo, V. Favetta, L. Y. Yamamoto et al., "Biometric description of fruits and seeds, germination and imbibition pattern of desert rose [Adenium obesum (Forssk.), Roem. & Schult.]," *Journal of Seed Science*, vol. 37, no. 4, pp. 206–213, 2015.

[43] J. Merilä and A. P. Hendry, "Climate change, adaptation, and phenotypic plasticity: The problem and the evidence," *Evolutionary Applications*, vol. 7, no. 1, pp. 1–14, 2014.

[44] C. Parmesan, "Ecological and evolutionary responses to recent climate change," *Annual Review of Ecology, Evolution and Systematics*, vol. 37, pp. 637–669, 2006.

[45] I.-C. Chen, J. K. Hill, R. Ohlemüller, D. B. Roy, and C. D. Thomas, "Rapid range shifts of species associated with high levels of climate warming," *Science*, vol. 333, no. 6045, pp. 1024–1026, 2011.

[46] S. Maranz and Z. Wiesman, "Evidence for indigenous selection and distribution of the shea tree, Vitellaria paradoxa and its potential significance to prevailing parkland savanna tree patterns in sub-Saharan Africa north of the equator," *Journal of Biogeography*, vol. 30, no. 10, pp. 1505–1516, 2003.

[47] A. S. M. M. R. Khan, M. Y. Kabir, and M. M. Alam, "Variability, Correlation Path Analysis of Yield and Yield Components of Pointed Gourd," *Journal of Agriculture and Rural Development*, vol. 7, no. (1&2), pp. 93–98, 2009.

Physicochemical Parameters and Fish Assemblages in the Downstream River of a Tropical Hydroelectric Dam Subjected to Diurnal Changes in Flow

Lee Nyanti,[1] Noor-Iskandar Noor-Azhar,[1] Chen-Lin Soo,[1] Teck-Yee Ling,[1] Siong-Fong Sim,[1] Jongkar Grinang,[2] Tonny Ganyai,[3] and Karen-Suan-Ping Lee[3]

[1] *Faculty of Resource Science and Technology, Universiti Malaysia Sarawak, 94300 Kota Samarahan, Sarawak, Malaysia*
[2] *Institute of Biodiversity and Environmental Conservation, Universiti Malaysia Sarawak, 94300 Kota Samarahan, Sarawak, Malaysia*
[3] *Research and Development Department, Sarawak Energy Berhad, 93050 Kuching, Sarawak, Malaysia*

Correspondence should be addressed to Lee Nyanti; nyantilee@gmail.com

Academic Editor: Michel Couderchet

The downstream river of the tropical Batang Ai Dam is experiencing diurnal flow fluctuation due to power generation operation. Three samplings were conducted to collect the water quality and fish assemblage data and one sampling was conducted to study the hydrological characteristics of the downstream river. The results show that the downstream river is extremely shallow and moves slowly when the power generation is halted and no water is discharged from the powerhouse. Significant correlations between river flow and pH, dissolved oxygen (DO), total suspended solids (TSS), and five-day biochemical oxygen demand (BOD_5) indicated that those parameters were influenced by the water discharge regulation. Fish abundance was low in upstream segment but it gradually increased as distance from the dam increased and was significantly correlated with DO. Fish diversity in the downstream river was influenced by the river depth and chlorophyll *a* concentration. The most abundant fish species, *Cyclocheilichthys apogon*, exhibited a negative allometric growth pattern, while *Hampala macrolepidota* and *Hemibagrus planiceps* exhibited an isometric growth pattern. Fulton's condition factor (K) values for 89.4% of *H. planiceps* were ≤ 1, indicating poor to extremely poor conditions.

1. Introduction

Impoundment of a river often causes dramatic changes in the physicochemical characteristics and biological structure of its downstream river [1–5]. Studies have shown that connectivity disruptions coupled with erratic flow pattern change the morphology and physicochemical characteristics of a river [6–11]. Domination of fish species that is well adapted to the new environment and elimination of sensitive species could alter the composition of fish assemblages and lower the fish diversity in the downstream river [12, 13].

In Malaysia, about 80 dams have been built and hydropower dams are the most common in the state of Sarawak [14]. As more hydroelectric dams are being planned in the state of Sarawak, there has been an increased effort to understand the impacts of hydroelectric dams on the tropical aquatic ecosystem. It is also of scientific importance to investigate and evaluate the hydrological and physicochemical characteristics of a downstream river induced by river impoundment and operation of hydroelectric dam, which could have an impact on the fish community. In addition, knowledge on fish assemblages in the regulated downstream river of a dam coupled with growth pattern and well-being of fishes in the stream is also crucial for ecosystem conservation and management.

The Batang Ai Dam is the oldest dam built for electricity supply in the state of Sarawak, Malaysia. The reservoir has been impounded for more than 30 years. Several studies on water quality of the reservoir have been conducted [15–18]. However, knowledge on its downstream river below the powerhouse is scarce. The influence of the impoundment and the erratic discharge of the reservoir water during power generation on its downstream river remain unknown. Hence, this study aimed to establish the baseline data on hydrological

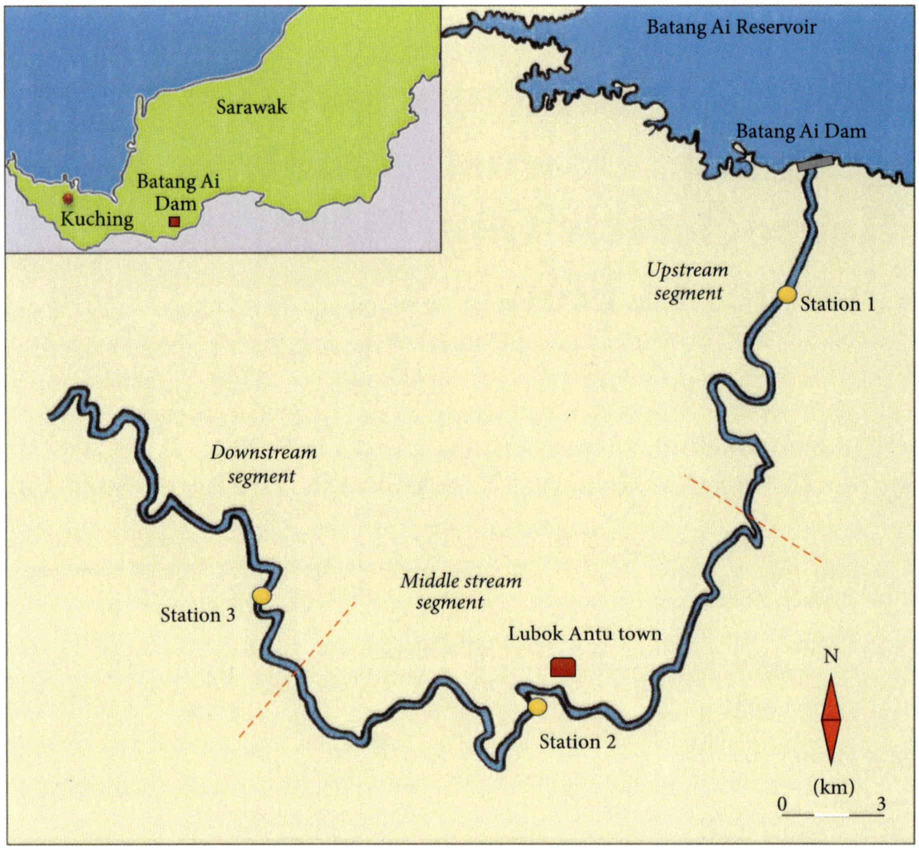

FIGURE 1: Location of the three segments and sampling stations in the study area.

characteristics, physicochemical parameters of water, and fish fauna composition in the downstream river, which was subjected to diurnal discharge of reservoir water during power generation. Additionally, the length-weight relationship and Fulton's condition factor of the three most dominant fish species in the downstream river were determined.

2. Materials and Methods

2.1. Study Area and Sampling Location. The present study was conducted in the downstream river of the Batang Ai Dam located in Lubok Antu, Sarawak, Malaysia, approximately 250 km southeast of Kuching (Figure 1). The Batang Ai Dam was constructed across Batang Ai (Batang means large river) and was commissioned in 1985. It serves as a hydropower dam and the reservoir is also used for aquaculture and tourism activities. The stretch of the river that was studied started at the powerhouse and stretched 30 km downstream from it. This stretch was divided into three segments, namely, upstream segment, middle stream segment, and downstream segment. Each segment covered a distance of about 10 km. The river was subjected to diurnal fluctuation of flow, where the discharge through turbines mainly occurs during daytime. Subsistence agricultural activities and settlements in longhouses are located along the river. In addition, the middle stream segment of the river passes through a small town, Lubok Antu, as well as a road.

2.2. Field Samplings and Laboratory Analysis. Data on fish fauna and water quality were collected in October 2014, January 2015, and February 2017. The hydrological characteristics for each segment were studied in February 2017. The width, depth, and velocity of the river were measured at three different time frames in a day by using a range finder (Bushnell Elite 1500), a depth finder (Speedtech), and a flow meter (Flowatch FL-03), respectively.

pH, temperature, electrical conductivity, dissolved oxygen (DO), and turbidity were measured in triplicate by using a pH meter with a temperature probe (SDL100 Extech Instrument), a conductivity meter (Lutron CD-4303), a DO meter (SDL150 Extech Instrument), and a turbidity meter (TU-2016 Lutron), respectively. Triplicate water samples were taken for the analyses of total suspended solids (TSS), five-day biochemical oxygen demand (BOD_5), and chlorophyll *a* (chl *a*) following the procedure of Jenkins et al. [19]. TSS was calculated as the difference between the initial weight and final weight of the 0.7 μm retention glass microfibre filter (Advantec) after the filtration of one liter of water sample and drying to constant weight at 105°C. BOD_5 was determined as the difference between the initial and final DO concentrations after a five-day long incubation of the sample at 20°C. Chl *a* was determined from samples filtered through a 0.45 μm retention glass microfibre filter (Sartorius) and extracted for 24 h using 90% (v/v) acetone.

Fish samples were collected using two types of gill nets, namely, monofilament gill net and three-layered gill net. The

TABLE 1: Hydrological characteristics in the three segments of Batang Ai below the dam.

Parameter	Upstream			Middle stream			Downstream		
	0840	1500	2000	0740	1020	1806	0940	1222	1530
Mean depth, m	0.55	0.90	0.96	1.22	0.89	2.17	0.79	0.96	0.72
Mean velocity, m/s	0.07	0.50	1.00	0.22	0.37	0.88	0.68	0.80	0.60
Flow, m^3/s	1.90	26.23	53.50	15.25	18.56	86.14	17.29	25.30	13.74

monofilament gill net consisted of four different mesh sizes (2.5 cm, 5.1 cm, 7.2 cm, and 10.2 cm) and each had a length of 45.0 m. The three-layered gill net had mesh sizes of 14.0 cm, 4.0 cm, and 14.0 cm with a length of 65.0 m. For each sampling trip, equal numbers of nets were placed and left for two nights in each segment and the nets were checked twice daily during daytime. The fish specimens collected were fixed in 10% formalin and later transferred into 70% ethanol for preservation. Fish specimens were sorted and enumerated according to sampling station and species identified according to the taxonomic keys [20, 21] and an online global database of fish species [22]. The total length (TL) and body weight (BW) of the three most abundant fish species were measured to the nearest millimeter and gram, respectively.

2.3. Data and Statistical Analysis. The flow, mean depth, and mean velocity of the river were calculated according to Chapra [23]. The water quality parameters collected in the three segments were compared by using one-way ANOVA, where the sampling trip was treated as a random effect in the analysis followed by Tukey's test, where $P \leq 0.05$ indicated significant difference. The fish data collected during the three sampling trips at each segment were pooled to calculate the fish abundance in each segment. Species diversity, evenness, and richness were calculated using the following formulas.

Shannon's diversity index (H) [24] is given as follows:

$$H = \sum \frac{N_i}{N} \log_e \frac{N_i}{N}. \quad (1)$$

Pielou's evenness (J) [25] is given as follows:

$$J = \frac{H}{\log_e S}. \quad (2)$$

Margalef's index of species richness (D) [26] is given as follows:

$$D = \frac{S-1}{\log_e N}, \quad (3)$$

where N is the sample size, N_i is the number of specimens per species, and S is the total number of species.

The length-weight relationship (LWR) and the general well-being of fish expressed by Fulton's condition factor (K) of the three most abundant species were determined. The LWR of fish species was established using the formula $\text{BW} = a\text{TL}^n$ by expressing it in the logarithmic form as $\log \text{BW} = \log a + n \log \text{TL}$ [27]. The statistical significance of the relationship was determined by using linear regression analysis with $P \leq 0.05$, where the constants "a" and "n" were obtained. Fulton's condition factor (K) [28] was calculated using the following formula:

$$K = \frac{\text{BW} \times 100}{\text{TL}^3}. \quad (4)$$

Pearson's correlation analysis was performed to elucidate significant relationship among the biotic and abiotic parameters at $P \leq 0.10$. All the statistical analyses were carried out using the Statistical Package for Social Sciences (SPSS version 24, SPSS Inc., 1995).

3. Results and Discussion

3.1. Hydrological Characteristics of the Downstream River of Batang Ai Dam. The hydrological characteristics of the downstream river of the dam varied substantially in a day due to the regulation of water discharge through the turbines during power generation and no water was released from the spillway. Prior to the discharge in the morning, the river was shallow and moved slowly, particularly in the upstream segment, with a mean depth of 0.55 m, a mean velocity of 0.07 m/s, and an estimated discharge of 1.90 m^3/s (Table 1). The mean velocity and flow gradually increased towards the downstream direction with the highest mean velocity of 0.68 m/s and the highest flow of 17.29 m^3/s observed in the downstream segment. When the water was discharged from the reservoir into the downstream river during power generation, the mean depth, mean velocity, and flow increased substantially in the upstream and middle stream segments. The mean depth, mean velocity, and flow in the upstream segment were found to increase to 0.90 m, 0.50 m/s, and 26.23 m^3/s, respectively, at 1500 hrs. The water discharge continues to affect the mean velocity and flow in the upstream segment, where they increased up to 1.00 m/s and 53.50 m^3/s, respectively, at 2000 hrs. In middle stream segment, the impact of the water discharge was observed at around 1800 hrs, where the mean depth, mean velocity, and flow increased up to 2.17 m, 0.88 m/s, and 86.14 m^3/s, respectively. There was no noticeable variation in the hydrological characteristics of the downstream segment as the discharge had not reached the downstream segment yet.

3.2. Physicochemical Parameters of the Water of the Downstream River of Batang Ai Dam. Water quality at the downstream river was significantly different ($P \leq 0.05$) among segments along the river (Table 2). The lowest pH value of 6.4 was recorded in the upstream segment and significantly increased ($P \leq 0.05$) to 6.8 in the middle stream and downstream segments. The pH value was classified as Class II according to the Water Quality Index (WQI) of Malaysia [29]. The lower pH value in the upstream segment is due to the

TABLE 2: Means and standard deviations of the physicochemical characteristics of the downstream river of Batang Ai.

Parameter	Upstream	Middle stream	Downstream
pH	$6.4^{a} \pm 0.2$	$6.8^{b} \pm 0.2$	$6.8^{b} \pm 0.4$
Temperature	$29.0^{c} \pm 1.4$	$28.0^{b} \pm 1.6$	$26.8^{a} \pm 0.7$
Conductivity	$23.9^{b} \pm 5.3$	$25.4^{c} \pm 5.0$	$21.8^{a} \pm 5.7$
DO	$5.9^{a} \pm 1.7$	$6.2^{b} \pm 1.5$	$6.3^{c} \pm 1.4$
Turbidity	$3.8^{a} \pm 0.8$	$4.6^{a} \pm 1.3$	$15.4^{b} \pm 14.9$
TSS	$4.4^{a} \pm 0.5$	$11.7^{ab} \pm 13.9$	$12.4^{b} \pm 11.6$
BOD_5	$3.5^{b} \pm 0.2$	$2.6^{a} \pm 1.4$	$2.6^{a} \pm 1.2$
Chl *a*	$2.5^{a} \pm 0.9$	$2.1^{a} \pm 0.4$	$2.4^{a} \pm 1.4$

Mean followed by different letters indicated significant difference at $P \leq 0.05$.

intake of lower pH water from hypolimnion of the reservoir during power generation. In Batang Ai Dam, the water intake for power operation is about 14 m below the surface, where the pH is predominantly acidic [15]. The increase in pH value at middle stream and downstream segments is attributable to dilution effect [30]. Water temperature ranged from 26.8 ± 0.7°C to 29.0 ± 1.4°C, where it decreased significantly ($P \leq 0.05$) with distance from the dam. Changes in water temperature of the downstream river of a hydropower plant occurred when warmer or colder water of a reservoir was being released into its downstream river [31, 32]. Bobat [31] stated that the downstream water of dams and hydropower plants was generally warmer than upstream water because of the passing of water from pipelines, penstock, turbine, and cooling system into the downstream river. However, Ling et al. [16] showed that the water temperature of river located at outflow of the Batang Ai reservoir was colder than the reservoir water. Hence, it is hypothesized that warm water in upstream segment in the present study is due to direct solar radiation in the shallow and exposed river. As river depth increased in middle stream and downstream segments, the heat capacity per unit surface area increased and thus a decrease in water temperature was observed [33].

Mean conductivity values at the segments ranged from 21.8 ± 5.7 μS/cm to 25.4 ± 5.0 μS/cm and they were significantly different ($P \leq 0.05$) (Table 2). The highest value of conductivity was observed in the middle stream segment. Similar to pH value, the DO values in the downstream river also increased significantly with distance from the dam ($P \leq 0.05$), that is, from 5.9 ± 1.7 mg/L in upstream segment to 6.3 ± 1.4 mg/L in downstream segment. The increase in DO value further from the powerhouse indicates that the DO content was first impacted by the low DO from the reservoir but was later diluted by river water with higher DO content as well as reaeration due to faster water flow along the river. The mean DO values met the minimum requirement for healthy sensitive aquatic organisms (5 mg/L) at all segments and are classified as Class II. However, low DO values of less than 5 mg/L were observed occasionally in the downstream river during samplings. This is due to the lower value of DO from deeper water column of the reservoir. At the water intake depth of 14 m, DO has been reported to be 3.92 mg/L [15]. Factors responsible for the large variations in DO value among trips were seasonal variations in precipitation, which resulted in changes in river flow rate and the differences in discharge during power generation. This is further supported by the significant correlation ($r = +0.993$; $P \leq 0.10$) between DO and flow observed in the present study.

Turbidity and TSS values were low in the downstream river, ranging from 3.8 ± 0.8 NTU to 15.4 ± 14.9 NTU and from 4.4 ± 0.5 mg/L to 12.4 ± 11.6 mg/L, respectively. Both parameters were classified as Class I and/or Class II according to WQI. The low turbidity and TSS values in the present study were different from the high turbidity and TSS values observed in the downstream river of a new hydroelectric dam [30]. However, they are consistent with low turbidity (2–15 NTU) and TSS values (2–7 mg/L) recorded in the Batang Ai reservoir at 10 and 20 m depths [17], as suspended solids have settled at this 30-year-old reservoir [16]. Among the segments studied, turbidity and TSS values were significantly higher ($P \leq 0.05$) in downstream segment than in upstream segment. TSS was significantly and positively correlated ($r = +0.999$; $P \leq 0.05$) with flow (Table 3), indicating resuspension of sediment under higher flow rate in downstream segment, which was likely the cause of higher turbidity and TSS values in the downstream segment.

BOD_5 showed an opposite trend with turbidity and TSS values, where the mean decreased significantly from 3.5 mg/L in upstream segment to 2.6 mg/L in middle stream and downstream segments. BOD_5 concentration was classified as Class III in upstream segment and Class II in middle stream and downstream segments. The gradual decrease of BOD_5 concentration further downstream indicates that BOD_5 was most probably contributed by the reservoir water as high BOD_5 concentration (6.33 mg/L) had been reported in the outflow of the reservoir [16]. This is further supported by the significant and negative correlation ($r = -0.993$; $P \leq 0.10$) between BOD_5 and flow. Chl *a* concentration was relatively constant along the river with a mean value of 2.3 mg/m^3 as the analysis shows that there was no significant difference ($P > 0.05$) among the segments.

3.3. Fish Assemblage of the Downstream River of the Batang Ai Dam. A total of 436 fish individuals belonging to eight families, 13 genera, and 17 species were captured in the downstream river of Batang Ai (Table 4). The results show that the most dominant family is Cyprinidae (67.9%) followed by Bagridae (11.2%). The three most dominant species were *Cyclocheilichthys apogon* (48.2%) and *Hampala macrolepidota* (11.0%) from the family Cyprinidae and *Hemibagrus planiceps* (10.8%) from the family Bagridae. Shannon's diversity index of 1.9 and the richness index of 2.6 show that fish diversity and species richness in the downstream river were low [34].

TABLE 3: Correlation between abiotic and biotic parameters at downstream river of Batang Ai ($N = 3$).

	pH	Temp.	Cond.	Turb.	DO	TSS	BOD_5	Chl *a*	Flow	Mean depth	Abundance	*H*	*D*
pH	1.000												
Temp.	−0.839	1.000											
Cond.	−0.096	0.623	1.000										
Turb.	0.553	−0.917	−0.883	1.000									
DO	0.971	−0.945	−0.332	0.737	1.000								
TSS	0.997**	−0.879	−0.174	0.617	0.987*	1.000							
BOD_5	−1.000**	0.839	0.096	−0.553	−0.971	−0.997**	1.000						
Chl *a*	−0.693	0.189	−0.651	0.218	−0.500	−0.634	0.693	1.000					
Flow	0.993*	−0.899	−0.217	0.650	0.993*	0.999**	−0.993*	−0.600	1.000				
Mean depth	0.774	−0.304	0.556	−0.100	0.599	0.721	−0.774	−0.993*	0.691	1.000			
Abundance	0.920	−0.985	−0.478	0.835	0.987*	0.948	−0.920	−0.356	0.961	0.464	1.000		
H	0.713	−0.216	0.630	−0.190	0.524	0.656	−0.713	−1.000**	0.622	0.996*	0.382	1.000	
D	0.531	0.017	0.793	−0.413	0.312	0.462	−0.531	−0.979	0.423	0.947	0.156	0.973	1.000

$^{*}P \le 0.10$; $^{**}P \le 0.05$.

TABLE 4: Summary of the fish assemblages at the downstream river of Batang Ai.

Family	Species	Upstream		Middle stream		Downstream		Total	
		Count	%	Count	%	Count	%	Count	%
Bagridae	*Hemibagrus nigriceps*	0	0	2	1	0	0	2	0.5
	Hemibagrus planiceps	15	17	16	10	16	8	47	10.8
Channidae	*Channa striata*	1	1	6	4	1	1	8	1.8
Clariidae	*Clarias batrachus*	3	3	2	1	2	1	7	1.6
Cyprinidae	*Barbonymus schwanenfeldii*	3	3	1	1	2	1	6	1.4
	Cyclocheilichthys apogon	45	52	70	46	95	48	210	48.2
	Cyclocheilichthys armatus	1	1	2	1	0	0	3	0.7
	Hampala macrolepidota	5	6	16	10	27	14	48	11.0
	Osteochilus enneaporos	0	0	0	0	1	1	1	0.2
	Osteochilus schlegelii	0	0	1	1	0	0	1	0.2
	Osteochilus waandersii	0	0	6	4	4	2	10	2.3
	Puntius binotatus	3	3	2	1	5	3	10	2.3
	Rasbora caudimaculata	3	3	4	3	0	0	7	1.6
Eleotridae	*Oxyeleotris marmorata*	5	6	5	3	8	4	18	4.1
Mastacembelidae	*Mastacembelus erythrotaenia*	0	0	5	3	3	2	8	1.8
Osphronemidae	*Osphronemus goramy*	3	3	14	9	23	12	40	9.2
Toxotidae	*Toxotes microlepis*	0	0	1	1	9	5	10	2.3
Total number of species		11	100	16	100	13	100	17	100
Total number of fish caught		87		153		196		436	
Shannon's diversity index, *H* (*e*)		1.7		1.9		1.7		1.9	
Margalef's richness index, *D* (*e*)		2.2		3.0		2.3		2.6	
Pielou's evenness index, *J* (*e*)		0.7		0.7		0.7		0.7	

The highest diversity and species richness were found in the middle stream segment. In the downstream river, an evenness index of 0.7 shows that the fish compositions are considered to be more evenly distributed [25].

The present study shows that fish abundance increased with an increasing distance from the dam. A total of 87, 153, and 194 individuals were caught in the upstream, middle stream, and downstream segments, respectively. Reduced biotic abundance, diversity, and productivity in tail water may be due directly to flow variation or indirectly to factors related to flow variation such as changes in river depth and temperature [35]. In the present study, fish abundance was significantly and positively correlated with DO ($r = +0.987$; $P \le 0.10$). This indicates that lower DO content in the upstream segment plays a significant role in the reduction of fish abundance in that segment. The fish abundance increased when DO improved in the downstream direction. As DO was influenced by the river flow, fish abundance in the study area was indirectly influenced by the flow as well. However, there is no strong relationship found between fish abundance and flow in the present study. Similarly, Growns [12] reported that the abundance of individual fish species was only weakly correlated with the index of flow deviation and river hydrology and thus explained only a small part of variation in fish assemblage structure. The species diversity was found to be significantly and positively correlated with the river mean depth ($r = +0.996$; $P \le 0.10$). The result indicates that fish assemblage in a shallow river in the upstream segment is less diverse than a deeper river in middle stream and downstream segments.

Table 5 summarizes the LWR analysis conducted on the three most dominant species in the downstream river of

TABLE 5: Summary of regression analyses of length-weight relationships of the three most abundant species in the downstream river of Batang Ai.

Species	log BW = log *a* + *n* log TL			Parameter estimates		Standard error	P	95% confidence interval		LWR: $W = aL^n$
	N	R^2	P					Lower bound	Upper bound	
Cyclocheilichthys apogon	208	0.844	**0.001**	log *a*	−1.482	0.092	**0.001**	−1.663	−1.302	$W = 0.033L^{2.678}$
				n	2.678	0.080	**0.001**	2.520	2.836	
Hampala macrolepidota	46	0.875	**0.001**	log *a*	−1.916	0.211	**0.001**	−2.341	−1.490	$W = 0.012L^{3.014}$
				n	3.014	0.172	**0.001**	2.668	3.361	
Hemibagrus planiceps	47	0.901	**0.001**	log *a*	−2.119	0.191	**0.001**	−2.503	−1.734	$W = 0.008L^{3.039}$
				n	3.039	0.150	**0.001**	2.737	3.341	

Significant values ($P \leq 0.05$) are indicated in bold.

TABLE 6: Summary of Fulton's condition factor (K) of the three most abundant species in the downstream river of Batang Ai.

Species	N	Fulton's condition factor (K)				Percentage of individuals					
		Min	Max	Mean	SD	≤ 0.8	$0.8 < x \leq 1$	$1 < x \leq 1.2$	$1.2 < x \leq 1.4$	$1.4 < x \leq 1.6$	>1.6
Cyclocheilichthys apogon	208	0.77	3.09	1.48	0.53	0.5	1.4	23.1	46.2	12.5	16.3
Hampala macrolepidota	46	0.89	4.05	1.35	0.59	0.0	21.7	45.7	8.7	8.7	15.2
Hemibagrus planiceps	47	0.66	1.41	0.86	0.15	46.8	42.6	6.4	2.1	2.1	0.0

the Batang Ai Dam. Significant linear regressions with R^2 close to the value of one show that TL and BW of each species fitted the linear model well. The LWRs for *C. apogon*, *H. macrolepidota*, and *H. planiceps* are expressed as $W = 0.033L^{2.678}$, $W = 0.012L^{3.014}$, and $W = 0.008L^{3.039}$, respectively. The most dominant species, *C. apogon*, exhibited a negative allometric growth (coefficient "n" value < 3), whereas *H. macrolepidota* and *H. planiceps* exhibited an isometric growth as indicated by the "n" values of 3.

Fulton's condition factor (K) value gradually decreased from the most abundant species, *C. apogon* (1.48 ± 0.53), to the third most abundant species, *H. planiceps* (0.86 ± 0.15) (Table 6). Around 46.2% of the K values of *C. apogon* were found in the range of 1.2–1.4, which indicates that nearly half of the fish in this species were in good condition. On the other hand, 45.7% of *H. macrolepidota* scored K values between 1.0 and 1.2, indicating that nearly half of this species caught were in fair condition. However, most of the K values of *H. planiceps* (89.4%) were ≤1, indicating poor to extremely poor conditions [36]. Food availability and environmental condition are most likely the reasons for the well-being of those species. The most abundant fish species, *C. apogon*, thrived better in the downstream river of Batang Ai compared to other species.

4. Conclusions

The present study shows that water discharge regulation of the Batang Ai Dam affected the hydrological characteristics, physicochemical parameters of water, and fish assemblages of its downstream river. The downstream river was extremely shallow and moved slowly but the mean depth, velocity, and flow of the river increased substantially when water was discharged from the dam during power generation. Physicochemical parameters of water, particularly pH, DO, TSS, and BOD_5 concentrations, were impacted by the dam water discharge as indicated by the significant correlation between those parameters and flows. The three most abundant fish species were *C. apogon* (48.2%) and *H. macrolepidota* (11.0%) of family Cyprinidae followed by *H. planiceps* (10.8%) of family Bagridae. Fish abundance was low in upstream segment but gradually increased in the downstream direction, away from the dam. Fish abundance was significantly correlated with DO, whereas the fish diversity was influenced by the river depth. Length-weight relationship analyses show that *C. apogon* exhibited negative allometric growth, whereas *H. macrolepidota* and *H. planiceps* exhibited isometric growth.

Conflicts of Interest

The authors declare that there are no conflicts of interest regarding the publication of this paper.

Acknowledgments

The authors appreciate the financial support provided by Sarawak Energy Berhad through Grant no. GL (F07)/SEB/5A/2013(28) and the facilities provided by Universiti Malaysia Sarawak.

References

[1] P. McCully, "Rivers no more: the environmental effects of dams," in *Silenced Rivers: The Ecology and Politics of Large Dams*, P. McCully, Ed., pp. 29–64, Zed Books, London, UK, 1996.

[2] W. Wildi, "Environmental hazards of dams and reservoirs," in *Near Curriculum in Natural Environmental Science*, vol. 88 of *Geneva: Section des sciences de la Terre et de l'environnement*, pp. 187–197, Geneva, Switzerland, 2010.

[3] Y. Yi, Z. Yang, and S. Zhang, "Ecological influence of dam construction and river-lake connectivity on migration fish habitat in the Yangtze River basin, China," *Procedia Environmental Sciences*, vol. 2, no. 5, pp. 1942–1954, 2010.

[4] Q. Lin, "Influence of dams on river ecosystem and its countermeasures," *Journal of Water Resource and Protection*, vol. 03, no. 01, pp. 60–66, 2011.

[5] M. W. Beck, A. H. Claassen, and P. J. Hundt, "Environmental and livelihood impacts of dams: common lessons across development gradients that challenge sustainability," *International Journal of River Basin Management*, vol. 10, no. 1, pp. 73–92, 2012.

[6] G. L. Wei, Z. F. Yang, B. S. Cui et al., "Impact of dam construction on water quality and water self-purification capacity of the Lancang River, China," *Water Resources Management*, vol. 23, no. 9, pp. 1763–1780, 2009.

[7] Y. Zhang, J. Xia, T. Liang, and Q. Shao, "Impact of water projects on river flow regimes and water quality in Huai River Basin," *Water Resources Management*, vol. 24, no. 5, pp. 889–908, 2010.

[8] M. Wiatkowski, "Influence of słup dam reservoir on flow and quality of water in the Nysa Szalona river," *Polish Journal of Environmental Studies*, vol. 20, no. 2, pp. 469–478, 2011.

[9] H. Guo, Q. Hu, Q. Zhang, and S. Feng, "Effects of the three gorges dam on yangtze river flow and river interaction with poyang lake, China: 2003-2008," *Journal of Hydrology*, vol. 416-417, pp. 19–27, 2012.

[10] Q. Zhao, S. Liu, L. Deng et al., "Landscape change and hydrologic alteration associated with dam construction," *International Journal of Applied Earth Observation and Geoinformation*, vol. 16, no. 1, pp. 17–26, 2012.

[11] L. Feng, X. Sun, and X. Zhu, "Impact of floodgates operation on water environment using one-dimensional modelling system in river network of Wuxi city, China," *Ecological Engineering*, vol. 91, pp. 173–182, 2016.

[12] I. Growns, "The influence of changes to river hydrology on freshwater fish in regulated rivers of the Murray-Darling basin," *Hydrobiologia*, vol. 596, no. 1, pp. 203–211, 2008.

[13] X. Li, S. Dong, Q. Zhao, and S. Liu, "Impacts of manwan dam construction on aquatic habitat and community in middle reach of lancang river," *Procedia Environmental Sciences*, vol. 2, no. 5, pp. 706–712, 2010.

[14] A. M. Latifah and L. Met, "An ecological evaluation approach for dam project development in Malaysia," *Life Science Journal*, vol. 11, no. 7, pp. 225–237, 2014.

[15] T. Y. Ling, D. P. Debbie, N. Lee, I. Norhadi, and J. J. E. Justin, "Water quality at Batang Ai Hydroelectric Reservoir (Sarawak, Malaysia) and implications for aquaculture," vol. 2, pp. 23–30, 2012.

[16] T.-Y. Ling, L. Nyanti, C.-K. Leong, and Y.-M. Wong, "Comparison of water quality at different locations at Batang Ai Reservoir, Sarawak, Malaysia," *World Applied Sciences Journal*, vol. 26, no. 11, pp. 1473–1481, 2013.

[17] T. Y. Ling, T. Z. E. Lee, and L. Nyanti, "Phosphorus in batang ai hydroelectric dam Reservoir, Sarawak, Malaysia," *World Applied Sciences Journal*, vol. 28, no. 10, pp. 1348–1354, 2013.

[18] L. Nyanti, K. M. Hii, A. Sow, I. Norhadi, and T. Y. Ling, "Impacts of aquaculture at different depths and distances from cage culture sites in Batang Ai hydroelectric dam reservoir, Sarawak, Malaysia," *World Applied Sciences Journal*, vol. 19, no. 4, pp. 451–456, 2012.

[19] D. Jenkins, J. J. Connors, and A. E. Greenberg, *Standard Methods for the Examination of Water and Wastewater*, American Public Health Association, Washington DC, USA, 21st edition, 2005.

[20] T. R. Roberts, *The Freshwater Fishes of Western Borneo (Kalimantan Barat, Indonesia)*, California Academy of Science, San Francisco, Calif, USA, 1989.

[21] M. Kottelat, A. J. Whitten, S. Kartikasari, and S. Wirjoatmodjo, *Freshwater Fishes of Western Indonesia and Sulawesi*, Periplus E., Hong Kong, China, 1993.

[22] R. Froese and D. Pauly, "FishBase. World Wide Web electronic publication," http://www.fishbase.org/, 2016.

[23] S. C. Chapra, *Surface Water-Quality Modeling*, McGraw- Hill Inc, New York, NY, USA, 1997.

[24] C. E. Shannon and W. Weaver, *The Mathematical Theory of Communication*, University of Illinois press, Urbana, Illinois, USA, 1963.

[25] E. C. Pielou, "Species-diversity and pattern-diversity in the study of ecological succession," *Journal of Theoretical Biology*, vol. 10, no. 2, pp. 370–383, 1966.

[26] R. D. Margalef, *Perspectives in Ecological Theory*, University of Chicago Press, Chicago, USA, 1968.

[27] E. D. Le Cren, "The length-weight relationship and seasonal cycle in gonad weight and condition in the perch (*Perca fluviatilis*)," *Journal of Animal Ecology*, vol. 20, no. 2, pp. 201–219, 1951.

[28] R. Froese, "Cube law, condition factor and weight-length relationships: history, meta-analysis and recommendations," *Journal of Applied Ichthyology*, vol. 22, no. 4, pp. 241–253, 2006.

[29] Department of Environment, "Malaysia Environmental Quality Report 2014," Tech. Rep., Kuala Lumpur, Malaysia, 2014.

[30] T.-Y. Ling, C.-L. Soo, T. L.-E. Heng, L. Nyanti, S.-F. Sim, and J. Grinang, "Physicochemical characteristics of river water downstream of a large tropical hydroelectric dam," *Journal of Chemistry*, vol. 2016, Article ID 7895234, pp. 1–7, 2016.

[31] A. Bobat, "Chapter 2: thermal pollution caused by hydropower plants," in *Energy Systems and Management, Springer Proceedings in Energy*, A. N. Bilge, Ed., pp. 19–33, Springer International Publishing, Switzerland, 2015.

[32] D. Carpentier, J. Haas, M. Olivares, and A. de la Fuente, "Modeling the multi-seasonal link between the hydrodynamics of a reservoir and its hydropower plant operation," *Water (Switzerland)*, vol. 9, no. 6, article no. 367, pp. 1–18, 2017.

[33] J. C. Carron and H. Rajaram, "Impact of variable reservoir releases on management of downstream water temperatures," *Water Resources Research*, vol. 37, no. 6, pp. 1733–1743, 2001.

[34] A. E. Magurran, *Ecological Diversity and Its Measurement*, Croom Helm, London, UK, 1st edition, 1988.

[35] R. M. Cushman, "Review of ecological effects of rapidly varying flows downstream from hydroelectric facilities," *North American Journal of Fisheries Management*, vol. 5, pp. 330–339, 1985.

[36] C. Barnham and A. Baxter, "Condition factor, K, for salmonid fish," *Fisheries Notes*, pp. 1–3, 1998.

Permissions

All chapters in this book were first published in IJE, by Hindawi Publishing Corporation; hereby published with permission under the Creative Commons Attribution License or equivalent. Every chapter published in this book has been scrutinized by our experts. Their significance has been extensively debated. The topics covered herein carry significant findings which will fuel the growth of the discipline. They may even be implemented as practical applications or may be referred to as a beginning point for another development.

The contributors of this book come from diverse backgrounds, making this book a truly international effort. This book will bring forth new frontiers with its revolutionizing research information and detailed analysis of the nascent developments around the world.

We would like to thank all the contributing authors for lending their expertise to make the book truly unique. They have played a crucial role in the development of this book. Without their invaluable contributions this book wouldn't have been possible. They have made vital efforts to compile up to date information on the varied aspects of this subject to make this book a valuable addition to the collection of many professionals and students.

This book was conceptualized with the vision of imparting up-to-date information and advanced data in this field. To ensure the same, a matchless editorial board was set up. Every individual on the board went through rigorous rounds of assessment to prove their worth. After which they invested a large part of their time researching and compiling the most relevant data for our readers.

The editorial board has been involved in producing this book since its inception. They have spent rigorous hours researching and exploring the diverse topics which have resulted in the successful publishing of this book. They have passed on their knowledge of decades through this book. To expedite this challenging task, the publisher supported the team at every step. A small team of assistant editors was also appointed to further simplify the editing procedure and attain best results for the readers.

Apart from the editorial board, the designing team has also invested a significant amount of their time in understanding the subject and creating the most relevant covers. They scrutinized every image to scout for the most suitable representation of the subject and create an appropriate cover for the book.

The publishing team has been an ardent support to the editorial, designing and production team. Their endless efforts to recruit the best for this project, has resulted in the accomplishment of this book. They are a veteran in the field of academics and their pool of knowledge is as vast as their experience in printing. Their expertise and guidance has proved useful at every step. Their uncompromising quality standards have made this book an exceptional effort. Their encouragement from time to time has been an inspiration for everyone.

The publisher and the editorial board hope that this book will prove to be a valuable piece of knowledge for researchers, students, practitioners and scholars across the globe.

List of Contributors

Ismael Martins Pereira, Jovan Martins Rios, Vagner Santiago do Vale, Ademilson Coneglian and Ednaldo Cândido Rocha
Universidade Estadual de Goiás (UEG), Câmpus Ipameri, Ipameri, GO, Brazil

Vera Lúcia Gomes-Klein
Department de Botânica, Universidade Federal de Goiás (UFG), Avenida Esperança, S/N, Câmpus Samambaia (Campus II),Goiânia, GO, Brazil

Fatmawati Patang, Agoes Soegianto and Sucipto Hariyanto
Department of Biology, Faculty of Sciences and Technology, Universitas Airlangga, Campus C Unair., Jl. Dr. Ir. Soekarno, Surabaya, Indonesia

Geoffrey W. Kinga and Wilfred O. Odadi
Department of Natural Resources, Egerton University, Egerton, Kenya

John Mironga
Department of Geography, Egerton University, Egerton, Kenya

Stephanie Stiegel and JasminMantilla-Contreras
Institute of Biology and Chemistry, University of Hildesheim, 31141 Hildesheim, Germany

Evgeny Abakumov, Ekaterina Maksimova and George Shamilishviliy
Department of Applied Ecology, Saint Petersburg State University, 16th Line 29, Vasilyevsky Island, Saint Petersburg 199178, Russia

Anna Tsibart
Department of Landscape Geochemistry and Soil Georgaphy, Moscow State University, Vorobyovy Gory 1, Moscow 119991, Russia

Collins Ayine Nsor
Department of Ecotourism and Forest Recreation, Kwame Nkrumah University of Science and Technology, Kumasi, Ghana

Edward Adzesiwor Obodai
Department of Fisheries and Aquatic Sciences, University of Cape Coast, University Post Office, Cape Coast, Ghana

Cyrus M. Kavwele and Mwangi J. Kinyanjui
Department of Natural Resources, Karatina University, Karatina, Kenya

Johnstone K. Kimanzi
Department of Wildlife Management, University of Eldoret, Eldoret, Kenya

Sushma Singh
Department of Environmental Sciences, HNB Garhwal University, Srinagar, Garhwal, Uttarakhand 246174, India

Mudasir Youssouf
Centre for Environmental Science and Technology, Central University of Punjab, Bathinda, Punjab 151001, India

Zubair A. Malik
Department of Botany and Microbiology, HNB Garhwal University, Srinagar, Garhwal, Uttarakhand 246174, India

Rainer W. Bussmann
Saving Knowledge, Casilla 13092, La Paz, Bolivia

Galina I. Kobanova, Vadim V. Takhteev, Olga O. Rusanovskaya and Maxim A. Timofeyev
Institute of Biology at Irkutsk State University, Irkutsk State University, Irkutsk 664003, Russia

Anita Bhatnagar, Pooja Devi and M. P. George
Department of Zoology, Kurukshetra University, Kurukshetra 136119, India

M. P. George
Delhi Pollution Control Committee, ISBT Building, Delhi, India

Ramesh P. Sapkota and Peter D. Stahl
Department of Ecosystem Science & Management and Program in Ecology, University of Wyoming, 1000 E University Avenue, Laramie, WY, USA

Ramesh P. Sapkota and Krishna Hengaju
Department of Environmental Science, Amrit Campus, Tribhuvan University, Kathmandu, Nepal

Kedar Rijal
Central Department of Environmental Science, Tribhuvan University, Kathmandu, Nepal

John T. Hunter
School of Environmental and Rural Science, University of New England, Armidale, Australia

Animesh Dey and P. S. Chaudhuri
Earthworm Research Laboratory, Department of Zoology, Tripura University, Suryamaninagar, West Tripura 799022, India

M. Arun Kumar, S. Venu and G. Padmavati
Department of Ocean Studies and Marine Biology, Pondicherry University, Brookshabad Campus, P.O. Bag No. 26, Chakkaragaon (Post), Port Blair 744112, India

Romina Daiana Fernández and Sergio Javier Ceballos
Instituto de Ecología Regional, UNT-CONICET, CC 34, Yerba Buena, 4107 Tucumán, Argentina

Ana Lucía González Achem and Hugo Rafael Fernández
Instituto de Biodiversidad Neotropical, UNT-CONICET, CC 34, Yerba Buena, 4107 Tucumán, Argentina

Margarita del Valle Hidalgo
Facultad de Ciencias Naturales e IML, Miguel Lillo 205, 4000 Tucumán, Argentina

Collins Ayine Nsor, Emmanuel Acquah, Grace Mensah and Vincent Kusi-Kyei
Department of Ecotourism and Forest Recreation, Kwame Nkrumah University of Science and Technology, Kumasi, Ghana

Samuel Boadi
Department of Social Forestry, Kwame Nkrumah University of Science and Technology, Kumasi, Ghana

Emmanuel Andrew Sweke, Julius Michael Assam, Abdillahi Ismail Chande and Athanasio Stephano Mbonde
Tanzania Fisheries Research Institute, Kigoma, Tanzania

Emmanuel Andrew Sweke
Graduate School of Fisheries Sciences, Hokkaido University, 3-1-1 Minato-cho, Hakodate 041-8611, Japan

Abdillahi Ismail Chande
College of Science, Informatics and Built Environment, Bagamoyo University, Dar es Salaam, Tanzania

Magnus Mosha and Abel Mtui
Mahale Mountains National Park, Kigoma, Tanzania

Abel Mtui
Tanzania National Parks, Arusha, Tanzania

Z. J. Lupala
Ministry of Natural Resources and Tourism, Forestry Training Institute, Arusha, Tanzania

L. P. Lusambo and Y. M. Ngaga
Forest Economics, Sokoine University of Agriculture, Morogoro, Tanzania

Emmanuel Danquah and Elvis Hackman Tetteh
Faculty of Renewable Natural Resources, College of Agriculture and Natural Resources, Kwame Nkrumah University of Science and Technology, Kumasi, Ghana

Jaspher Okello
National Agricultural Research Organisation, Entebbe, Uganda

John Bosco Lamoris Okullo, Gerald Eilu, Philip Nyeko and Joseph Obua
School of Forestry, Environmental and Geographical Sciences, Makerere University, Kampala, Uganda

Lee Nyanti, Noor-Iskandar Noor-Azhar, Chen-Lin Soo, Teck-Yee Ling and Siong-Fong Sim
Faculty of Resource Science and Technology, Universiti Malaysia Sarawak, 94300 Kota Samarahan, Sarawak, Malaysia

Jongkar Grinang
Institute of Biodiversity and Environmental Conservation, UniversitiMalaysia Sarawak, 94300 Kota Samarahan, Sarawak, Malaysia

Tonny Ganyai and Karen-Suan-Ping Lee
Research and Development Department, Sarawak Energy Berhad, 93050 Kuching, Sarawak, Malaysia

Index